CATALYSIS OF
ORGANIC REACTIONS

CHEMICAL INDUSTRIES

A Series of Reference Books and Textbooks

Consulting Editor

HEINZ HEINEMANN

1. *Fluid Catalytic Cracking with Zeolite Catalysts*, Paul B. Venuto and E. Thomas Habib, Jr.
2. *Ethylene: Keystone to the Petrochemical Industry*, Ludwig Kniel, Olaf Winter, and Karl Stork
3. *The Chemistry and Technology of Petroleum*, James G. Speight
4. *The Desulfurization of Heavy Oils and Residua*, James G. Speight
5. *Catalysis of Organic Reactions*, edited by William R. Moser
6. *Acetylene-Based Chemicals from Coal and Other Natural Resources*, Robert J. Tedeschi
7. *Chemically Resistant Masonry*, Walter Lee Sheppard, Jr.
8. *Compressors and Expanders: Selection and Application for the Process Industry*, Heinz P. Bloch, Joseph A. Cameron, Frank M. Danowski, Jr., Ralph James, Jr., Judson S. Swearingen, and Marilyn E. Weightman
9. *Metering Pumps: Selection and Application*, James P. Poynton
10. *Hydrocarbons from Methanol*, Clarence D. Chang
11. *Form Flotation: Theory and Applications*, Ann N. Clarke and David J. Wilson
12. *The Chemistry and Technology of Coal*, James G. Speight
13. *Pneumatic and Hydraulic Conveying of Solids*, O. A. Williams
14. *Catalyst Manufacture: Laboratory and Commercial Preparations*, Alvin B. Stiles
15. *Characterization of Heterogeneous Catalysts*, edited by Francis Delannay
16. *BASIC Programs for Chemical Engineering Design*, James H. Weber
17. *Catalyst Poisoning*, L. Louis Hegedus and Robert W. McCabe
18. *Catalysis of Organic Reactions*, edited by John R. Kosak
19. *Adsorption Technology: A Step-by-Step Approach to Process Evaluation and Application*, edited by Frank L. Slejko
20. *Deactivation and Poisoning of Catalysts*, edited by Jacques Oudar and Henry Wise
21. *Catalysis and Surface Science: Developments in Chemicals from Methanol, Hydrotreating of Hydrocarbons, Catalyst Preparation, Monomers and Polymers, Photocatalysis and Photovoltaics*, edited by Heinz Heinemann and Gabor A. Somorjai
22. *Catalysis of Organic Reactions*, edited by Robert L. Augustine

78. *The Desulfurization of Heavy Oils and Residua, Second Edition, Revised and Expanded,* James G. Speight
79. *Reaction Kinetics and Reactor Design: Second Edition, Revised and Expanded,* John B. Butt
80. *Regulatory Chemicals Handbook,* Jennifer M. Spero, Bella Devito, and Louis Theodore
81. *Applied Parameter Estimation for Chemical Engineers,* Peter Englezos and Nicolas Kalogerakis
82. *Catalysis of Organic Reactions,* edited by Michael E. Ford

ADDITIONAL VOLUMES IN PREPARATION

The Chemical Process Industries Infrastructure: Function and Economics, James R. Couper, O. Thomas Beasley, and W. Roy Penney

Elements of Transport Phenomena, Joel Plawsky

CATALYSIS OF ORGANIC REACTIONS

edited by

Michael E. Ford

Air Products and Chemicals
Allentown, Pennsylvania

CRC Press
Taylor & Francis Group
Boca Raton London New York

CRC Press is an imprint of the
Taylor & Francis Group, an **informa** business

CRC Press
Taylor & Francis Group
6000 Broken Sound Parkway NW, Suite 300
Boca Raton, FL 33487-2742

First issued in paperback 2019

ISBN-13: 978-0-367-39800-2

**Visit the Taylor & Francis Web site at
http://www.taylorandfrancis.com**

**and the CRC Press Web site at
http://www.crcpress.com**

To my parents

EHF and GIF

Preface

Catalysis of Organic Reactions is a compendium of 58 chapters that originated from contributed papers and posters presented at the 18th Conference on Catalysis of Organic Reactions in Charleston, South Carolina. The symposium was sponsored by the Organic Reactions Catalysis Society (ORCS). These proceedings document recent, novel developments in the study of catalysis as it pertains to organic synthesis and its application in industrial processes. Over the years, the ORCS Conference has provided a forum for chemists and engineers from chemical and pharmaceutical industries and from academia to present and discuss their work on the application of catalysis in organic synthesis.

Written by over 200 experts from industry and academia, *Catalysis of Organic Reactions* surveys a wide range of homogeneous and heterogeneous catalysis for industrial and pharmaceutical chemicals. Topics addressed include practical aspects of catalytic processes, recent developments in enantioselective hydrogenation, heterogeneously and homogeneously catalyzed hydrogenations and oxidation, environmental catalysis, solid acid-base catalysis, methods for immobilization of homogeneous catalysts, carbonylation, hydroaminomethylation, and a variety of other topics.

The 18th Conference recognizes the generous support of Searle, which, with the sponsorship of Mike Scaros, published the conference preprints. The generosity of the corporate sponsors, Air Products & Chemicals, Dow Chemical, E.I. DuPont, W. R. Grace & Co.-Conn., Johnson Matthey, Eli Lilly & Co., Merck & Co., Monsanto, Parr Instruments, Phillips Petroleum, and Uniroyal Chemical is also gratefully acknowledged.

I wish to express my appreciation to all the contributors for their superb efforts in the preparation and presentation of the papers and posters, and to the Session Chairs for their introduction of the speakers. In addition, I want to personally thank the Executive and Editorial Boards of the Organic Reactions Catalysis Society for their time, encouragement and help in organizing this conference. Additional thanks go to all the referees who gave their time to review and improve the manuscripts.

Michael E. Ford

Acknowledgments

The Organic Reactions Catalysis Society gratefully acknowledges the generous support of the following companies. The support of each company was instrumental in allowing our society to fund the 18th Conference and to assist the graduate students, post-graduates and academicians who participated in the conference.

        Air Products and Chemicals, Inc.
        Dow Chemical Co.
        E. I. DuPont Co.
        W. R. Grace & Co.-Conn.
        Johnson Matthey
        Eli Lilly & Co.
        Engelhard Corporation
        Merck Corp
        Monsanto Co.
        Phillips Petroleum
        Uniroyal Chemical Co.

Our refreshments during the morning and afternoon meeting were provided by the **Parr Instrument Company**. We wish to acknowledge and thank the Parr Instrument Company for this valuable service.

The Society also wishes to thank **Air Products and Chemicals, Inc.** for its generous support by providing the Editor's time, supplies and services needed to organize the meeting. The Society would also like to thank **G. D. Searle** for publishing the preprints for this conference.

Contents

Contents

Thomas A. Johnson and Douglas P. Freyberger

Reactivity and Surface Analysis Studies on the Deactivation of
Raney Nickel During Adiponitrile Hydrogenation   229
Alan M. Allgeier and Michael W. Duch

Catalytic Hydrogenation of Benzonitrile over Raney Nickel:
Influence of Reaction Parameters on Reaction Rates and Selectivities   241
Oliver Gerald Degischer, F. Roessler, and P. Rys

Catalytic Hydrogenation of Lignin Aromatics Using Ru-Arene
Complexes   255
*Terrance Y. H. Wong, Russell Pratt, Carolyn G. Leong, Brian R.
James, and Thomas Q. Hu*

Reaction Pathways in the Catalytic Reductive N-Methylation
of Polyamines   267
Richard P. Underwood and Richard V. C. Carr

Reductive Alkylation of 2-Methylglutaronitrile with Palladium
Catalysts   279
Frank E. Herkes and Jay L. Snyder

The Synthesis of Amines by Catalytic Hydrogenation of Nitro
Compounds   293
*E. Auer, Monika Berweiler, M. Gross, J. Pietsch, Daniel Ostgard, and
Peter Panster*

Hydrogenation of Cinnamaldehyde on Ru-MCM and Ru-beta
Catalysts   301
*V. I. Pârvulescu, V. Pârvulescu, S. Kaliaguine, U. Endruschat,
B. Tesche, and H. Bönnemann*

Novel Preparation of 5α-Dihydroethisterone from
Androst-4-ene-3,17-dione   307
*Mike G. Scaros, Peter K. Yonan, Kalidas Paul, John Schulz, and
Jae C. Park*

Selective Epoxidation of Allylic Alcohols with Amine-Modified
Titania-Silica Aerogels   315
Marco Dusi, Carsten Beck, Tamas Mallat, and Alfons Baiker

Contents

Contents

Contributors

Alan M. Allgeier, E. I. DuPont de Nemours and Co., Inc. Wilmington, Delware

Varinder K. Aggarwal, University of Sheffield, Sheffield, United Kingdom

Michael D. Amiridis, University of South Carolina, Columbia, South Carolina

Stephen Anderson, Seton Hall University, South Orange, New Jersey

E. Angelescu, University of Bucharest, Bucharest, Romania

C. Ansems, Engelhard de Meern, De Meern, The Netherlands

John B. Appleby, Air Products and Chemicals, Inc., Allentown, Pennsylvania

E. Auer, Degussa-Huels AG, Hanau-Wolfgang, Germany

Robert L. Augustine, Seton Hall University, South Orange, New Jersey

Alfons Baiker, ETH-Zentrum, Zurich, Switzerland

M. Banciu, Polytechnic University of Bucharest, Bucharest, Romania

C. F. J. Barnard, Johnson Matthey, Reading, United Kingdom

Mihály Bartók, Hungarian Academy of Sciences and József Attila University, Szeged, Hungary

Matthias Beller, Institut für Organische Katalyseforschung, Rostock, Germany

Carsten Beck, ETH-Zentrum, Zurich, Switzerland

P. H. Berben, Engelhard de Meern, De Meern, The Netherlands

Monika Berweiler, Degussa Huels AG, Hanau-Wolfgang, Germany

I. Beul, Degussa Huels AG, Hanau-Wolfgang, Germany

H. Bönnemann, Max-Plank-Institut für Kohlenforschung, Mulheim/Ruhr, Germany

J. A. M. Brandts, Engelhard de Meern, De Meern, The Netherlands

L. Breytenbach, Sasol Technology, Sasolburg, South Africa

Richard D. Broene, Bowdoin College, Brunswick, Maine

H. Buchold, Lurgi Öl Gas Chemie GmbH, Frankfurt, Germany

M. J. Burk, Chirotech Technology Ltd., Cambridge, United Kingdom

M. Campanati, University of Bologna, Bologna, Italy

Richard V. C. Carr, Air Products and Chemicals, Inc., Allentown, Pennsylvania

W. Eamon Carroll, Air Products and Chemicals Inc., Allentown, Pennsylvania

G. L. Castiglioni, Lonza Intermediates and Additives, Scanzorosciate, Italy

Fabrizio Cavani, University of Bologna, Bologna, Italy

Baoshu Chen, Degussa-Hüls Corporation, Calvert City, Kentucky

Zhiyu Chen, Universitat Autonoma de Barcelona, Barcelona, Spain

Agnès Choplin, Institut de Recherches sur la Catalyse-CNRS, Villeurbanne, France

K. T. Chuang, University of Alberta, Edmonton, Alberta, Canada

James H. Clark, University of York, North Yorkshire, United Kingdom

A. A. Clifford, University of Leeds, Leeds, United Kingdom

S. Coman, University of Bucharest, Bucharest, Romania

M. P. Coogan, The University of Sheffield, Sheffield, United Kingdom

R. D. Culp, Eastman Chemical Co., Longview, Texas

Susan R. Cyganiak, Searle/Monsanto, Skokie, Illinois

P. Eugene Dai, Shell Chemical Co., Houston, Texas

Francis P. Daly, Apyron Technologies, Inc., Atlanta, Georgia

L. De Gioia, Dipartimento di Biotecnologie e Bioscienze, Università degli Studi di Milano-Bicocca, Milan, Italy

Oliver Gerald Degischer, Laboratory of Chemical Engineering and Industrial Chemistry, ETH Zürich, Zurich, Switzerland

Francis M. de Rege, Los Alamos National Laboratory, Los Alamos, New Mexico

T. J. Devon, Eastman Chemical Company, Longview, Texas

Johannes G. de Vries, DSM Research, Geleen, The Netherlands

F. Donati, University of Bologna, Bologna, Italy

J. G. Donkervoort, Engelhard de Meern, De Meern, The Netherlands

S. dos Santos, Institut de Recherches sur la Catalyse-CNRS, Villeurbanne, France

Michele T. Drexler, University of South Carolina, Columbia, South Carolina

Michael W. Duch, E. I. DuPont Co., Wilmington, Delware

Marco Dusi, ETH-Zentrum, Zurich, Switzerland

Alexander J. Dyakonov, A. W. Spears Research Center, Greensboro, North Carolina

K. Ekman, Smoptech, Turku, Finland

U. Endruschat, Max-Plank-Institut für Kohlenforschung, Mulheim/Ruhr, Germany

Misty L. Ernstberger, Degussa-Hüls Corporation, Calvert City, Kentucky

Károly Felföldi, University of Szeged, Szeged, Hungary

R. Fieldhouse, Zeneca Agrochemicals, Grangemouth, United Kingdom

Yvette M. Fobian, Monsanto Co., St. Louis, Missouri

Andreas Freund, Degussa Huels AG, Hanau-Wolfgang, Germany

Douglas P. Freyberger, Air Products and Chemicals, Inc., Allentown, Pennsylvania

J. Froelich, Johnson Matthey, Reading, United Kingdom

C. Fumagalli, Lonza Intermediates and Additives, Scanzorosciate, Italy

Patrycja V. Galka, University of Saskatchewan, Saskatoon, Canada

Yujing Gao, Seton Hall University, South Orange, New Jersey

A. Gerlach, Chirotech Technology Ltd., Cambridge, United Kingdom

D. A. Grider, Lorillard Tobacco Co., Greensboro, North Carolina

K. G. Griffin, Johnson Matthey, Reading, United Kingdom

Mike Grolmes; Centaurus Technology, Inc., Simpsonville, Illinois

M. Gross, Degussa-Huels AG, Hanau-Wolfgang, Germany

Viktor Háda, Technical University of Budapest, Budapest, Hungary

Noel Hallinan, Millenium Petrochemicals, Cincinnati, Ohio

Fei He, Tianjin University, Tianjin, P. R. China

László Hegedûs, Technical University of Budapest, Budapest, Hungary

Frank E. Herkes, E. I. DuPont de Nemours and Co., Wilmington, Delaware

J. Herwig, Celanese GmbH-Werk Ruhrchemie, Oberhausen, Germany

James Hinnenkamp, Millenium Petrochemicals, Cincinnati, Ohio

S. H. Holmes, Institut de Recherches sur la Catalyse-CNRS, Villeurbanne, France

G. B. Howe, Research Triangle Institute, Research Triangle Park, North Carolina

Thomas Q. Hu, Pulp and Paper Research Institute of Canada, Vancouver, Canada

Alexei V. Iretskii, Georgia Institute of Technology, Atlanta, Georgia

Vasilios H. Iskos, Searle/Monsanto, Skokie, Illinois

S. David Jackson, Synetix, Billingham, Cleveland, United Kingdom

Brian Robert James, University of British Columbia, Vancouver, British Columbia, Canada

B. W.-L. Jang, Research Triangle Institute, Research Triangle Park, North Carolina

Thomas A. Johnson, Orefield, Pennsylvania

R. V. H. Jones, Zeneca Agrochemicals, Grangemouth, United Kingdom

P. K. Kahol, Wichita State University, Wichita, Kansas

S. Kaliaguine, Universite Laval, Quebec, Canada

John F. Knifton, Shell Chemical Co., Houston, Texas

Thomas R. Kowar, Pharmacia, Skokie, Illinois

Heinz-Bernhard Kraatz, University of Saskatchewan, Saskatoon, Saskatchewan, Canada

Jürgen Ladebeck, Süd Chemie AG, Bruckmühl, Germany

Wugeng Liang, University of Alberta, Edmonton, Alberta, Canada

S. Ligi, University of Bologna, Bologna, Italy

Carolyn G. Leong, University of British Columbia, Vancouver, Canada

Andy Longinow, Wiss, Janney, Elstner Associates, Inc., Northbrook, Illinois,

Bernhard Lücke, Institut für Angewandte Chemie Berlin, Berlin, Germany

Sanjay V. Malhotra, New Jersey Institute of Technology, Newark, New Jersey

Tamas Mallat, ETH-Zentrum, Zurich, Switzerland

D. Maripane, Sasol Technology, Sasolburg, South Africa

Andreas Martin, Institute for Applied Chemistry Berlin-Adlershof, Berlin, Germany

Tibor Máthé, Technical University of Budapest, Budapest, Hungary

Timothy J. McCarthy, Degussa-Hüls Corporation, Piscataway, New Jersey

B. J. McCormick, Wichita State University, Wichita, Kansas

T. Mebrahtu, Air Products and Chemicals Inc., Allentown, Pennsylvania

Andrea Mereu, University of Sheffield, Sheffield, United Kingdom

Kim Mniszewski, Triodyne Fire and Explosion Engineers, Inc., Oakbrook, Illinois

Konrad Möbus, Degussa Huels AG, Hanau-Wolfgang, Germany

J. Moineau, Laboratoire de Synthèse Asymétrique, CNRS-UCBL, Villeurbanne, France

G. L. Monks, MEL Chemicals, Manchester, United Kingdom

T. Monti, University of Bologna, Bologna, Italy

David A. Morgenstern, Pharmacia Co., St. Louis, Missouri

David K. Morita, Los Alamos National Laboratory, Los Alamos, New Mexico

Rainer Müller, Hoffmann-La Roche Ltd., Basel, Switzerland

Bruce D. Nacker, Searle/Monsanto, Skokie, Illinois

Apostolos A. Nikolopoulos, Research Triangle Institute, Research Triangle Park, North Carolina

D. J. Nightingale, University of York, York, United Kingdom

R. S. Oakes, University of Leeds, Leeds, United Kingdom

David S. Oburn, Monsanto Co., St. Louis, Missouri

D. J. Olsen, Eastman Chemical Co, Kingsport, Tennessee

Rosa M. Ortuno, Universitat Autonoma de Barcelona, Barcelona, Spain

Daniel J. Ostgard, Degussa-Hüls AG, Hanau-Wolfgang, Germany

Kevin C. Ott, Los Alamos National Laboratory, Los Alamos, New Mexico

Peter Panster, Degussa Huels AG, Hanau-Wolfgang, Germany

Jae C. Park, Searle/Monsanto, Skokie, Illinois

V. Pârvulescu, University of Bucharest, Bucharest, Romania

V. I. Pârvulescu, University of Bucharest, Bucharest, Romania

Kalidas Paul, Searle/Monsanto, Skokie, Illinois

R. Peltonen, Smoptech, Turku, Finland

A. Petride, Institute of Organic Chemistry, Bucharest, Romania

József Petró, Technical University of Budapest, Budapest, Hungary

O. Piccolo, Chemi SpA, Cinisello Balsamo MI, Italy

J. Pietsch, Degussa-Huels AG, Hanau-Wolfgang, Germany

K. Pillay, Sasol Technology, Sasolburg, South Africa

G. Pozzi, Laboratoire de Synthèse Asymétrique, CNRS-UCBL, Villeurbanne, France

Russel Pratt, University of British Columbia, Vancouver, Canada

Peter M. Price, University of York, York, United Kingdom

R. Psaro, Dipartimento di Chimica I.M.A., Milan, Italy

F. Quignard, Institut de Recherches sur la Catalyse-CNRS, Villeurbanne, France

R. Ramatsebe, Sasol Technology, Sasolburg, South Africa

Dorai Ramprasad, Air Products and Chemicals Inc., Allentown, Pennsylvania

Nicoletta Ravasio, Consiglio Nazionale Delle Ricerche, Milan, Italy

Christopher Mark Rayner, University of Leeds, Leeds, United Kingdom

Tiberius Regula, Süd Chemie AG, Bruckmühl-Heufeld, Germany

C. Rehren, Degussa Huels AG, Hanau-Wolfgang, Germany

Christopher N. Robb, Searle/Monsanto, Skokie, Illinois

George W. Roberts, North Carolina State University, Raleigh, North Carolina

Stefan Röder, Degussa Huels AG, Hanau-Wolfgang, Germany

Felix Roessler, Hoffmann-La Roche, Kaiseraugst, Switzerland

P. Rys, ETH Zurich, Zurich, Switzerland

Mike G. Scaros, Pharmacia, Skokie, Illinois

John Schulz, Searle/Monsanto, Skokie, Illinois

Sheldon C. Sherman, Georgia Institute of Technology, Atlanta, Georgia

N. Shezad, University of Leeds, Leeds, United Kingdom

Eric H. Shreiber, North Carolina State University, Raleigh, North Carolina

D. Sinou, Laboratoire de Synthèse Asymétrique, CNRS-UCBL, Villeurbanne, France

Gerard V. Smith, Southern Illinois University, Carbondale, Illinois

Jay L. Snyder, E. I. DuPont de Nemours and Co., Wilmington, Delaware

S. H. Sookraj, Sasol Technology, Sasolburg, South Africa

J. J. Spivey, Research Triangle Institute, Research Triangle Park, North Carolina

Rachel A. Stenson, The University of Sheffield, Sheffield, United Kingdom

Massoud S. Stephan, DSM Research, Geleen, The Netherlands

R. Subramanian, Research Triangle Institute, Research Triangle Park, North Carolina

M. Sundell, Smoptech, Turku, Finland

John D. Super, Dixie Chemical Company, Inc., Pasadena, Texas

László Szepesy, Technical University of Budapest, Budapest, Hungary

Kornél Szőri, Department of Organic Chemistry, University of Szeged, Szeged, Hungary

Thomas Tacke, Degussa-Huels, Corp., Calvert City, Kentucky

Setrak K. Tanielyan, Seton Hall University, South Orange, New Jersey

B. Tesche, Max-Plank-Institut für Kohlenforschung, Mulheim/Ruhr, Germany

Anthony C. Testa, Seton Hall University, South Orange, New Jersey

Bèla Török, University of Szeged, Szeged, Hungary

F. Trifirò, University of Bologna, Bologna, Italy

William Tumas, Los Alamos National Laboratory, Los Alamos, New Mexico

Antal Tungler, Budapest University of Technology and Economics, Budapest, Hungary

Richard P. Underwood, Air Products and Chemicals, Inc., Allentown, Pennsylvania

Angelo Vaccari, University of Bologna, Bologna, Italy

A. Valentini, University of Bologna, Bologna, Italy

John W. Venitz, Searle/Monsanto, Skokie, Illinois

Francis J. Waller, Air Products and Chemicals, Inc., Allentown, Pennsylvania

S. R. Watson, University of Hull, Hull, United Kingdom

G. Webb, University of Glasgow, Glasgow, Scotland, United Kingdom

Stephen C. Webb, Air Products and Chemicals, Inc., Allentown, Pennsylvania

P. B. Wells, University of Hull, Hull, United Kingdom

J. F. White, Engelhard Corporation, Beachwood, Ohio

Mark G. White, Georgia Institute of Technology, Atlanta, Georgia

T. Y. H. Wong, University of British Columbia, Vancouver, Canada

Genhui Xu, Tianjin University, Tianjin, P. R. China

H. Yang, Beijing Polytechnic University, Beijing, P. R. China

Hong Yang, Seton Hall University, South Orange, New Jersey

Peter K. Yonan, Searle/Monsanto, Skokie, Illinois

N. C. Young, University of Glasgow, Glasgow, Scotland, United Kingdom

F. Zaccheria, Dipartimento di Chimica I.M.A., Milan, Italy

Kui Zhang, Tianjin University, Tianjin, P. R. China

B. Zimmermann, Technische Universität München, Garching, Germany

The Raney®-Nickel

József Petró

Department of Organic Chemical Technology, Technical University of Budapest, H-1521 Budapest, Hungary

Abstract

A report is given about the research activities carried out on Raney-nickel at the Technical University of Budapest. The topics are as follows: Ni alloys, the nickel hydrogen bond and catalyst preparation/production.

Introduction

First of all, I express my sincere thanks for the Executive Committee of the Organic Reaction Catalysis Society and the Chemical Division of W.R. Grace & Co. for selecting me as recipient of the fifth "Murray Raney Award". I know that this award is a great honour not only for me but for all my colleagues and for our University, as well. In my talk I am going to give you selected examples of our activities in this specific field.

Many excellent publications and patents were published on these topics. We referred to many of them in our papers, but in this special talk an overview will be given based upon our 30 years of research.

This is not a pure scientific presentation because our interest was focused partially on industrial projects as we had close contacts with the Hungarian Pharmaceutical and Fine Chemicals Industry. Our research philosophy was the following: carry out basic research, monitor new results and discoveries, which meet the criteria of technical feasibility and potential industrial interest.

More than 70 years have elapsed since the discovery of skeleton catalysts /1925/ by Murray Raney, but their importance and significance did not decrease. This is also indicated by several hundred publications and patents relevant to the project. However, in this period the essential steps of technology, suitable for industrial production, remained nearly the same.

My presentation is divided into the following topics:
1. the structure of alloys,
2. metal-hydrogen bonding,
3. catalysts, preparation/production.

2 Petró

I. Alloy structures (1-3)

    1. Structure, by optical and X-ray measurements
    2. Other metals instead of aluminium
    3. Promoters

The alloy of skeleton catalysts determines the potential catalytic properties. More than 90 alloys were investigated in our laboratory and in some of them the aluminium was partially or totally replaced by silicon or zinc.

    For simple optical investigation of metal alloy metal microscopy is used. By etching the polished surface with an alkaline or acidic solution, various contours can be developed. Fig. 1. shows the micrograph of a Ni-Al and a Ni- Si alloy, containing Mo and Cr promoters, taken in polarised light after alkaline and acidic etching. The two featured structures clearly show the separation of phases.

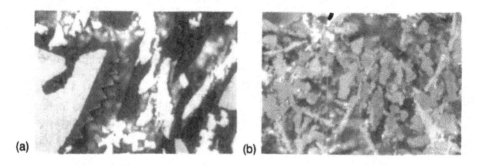

(a)   (b)

Figure 1 Raney-nickel alloys (a) promoted (50% Al, 46% Ni, 2,5% Mo, 1,5% Cr) basic etching (magnification x 2000), (b) promoted Si based (50% Si, 46% Ni, 2,5% Mo, 1,5% Cr). Hydrochloric and nitric acid etching, violet polarised light (magnification x 500).

    X-ray diffraction can give a more precise information on the metal or alloy phases present in the substance. Fig 2. shows the diffractogram of a standard Raney-Ni alloy (50% Al-50% Ni) and of the catalysts prepared from it. Fig. 3. is the diffractogram of a Raney-Ni alloy containing three promoters Mo, Cr, Co and of the catalyst prepared from it by a „solid phase" process (see later). It is obvious that the diffractogram of the alloy is very rich in lines. Only a few phases, mainly those, which are present in larger quantities, can be identified obviously.

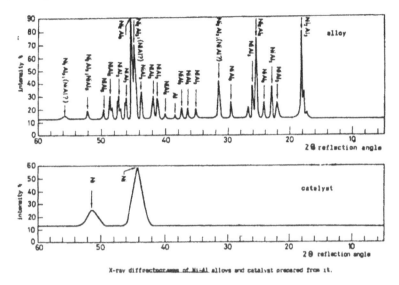

Figure 2 X-ray diffractograms of a Raney-nickel alloy (50% Al, 50% Ni) and the catalyst prepared from it.

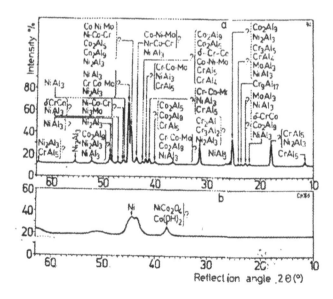

Figure 3 X-ray diffractograms of Co, Cr and Mo promoted Raney-Ni-alloy (a) and catalyst (b) prepared from it.

Fig. 4. shows the pictures of an alloy containing 50 wt% of Al, 50 wt% Ni and of the catalyst. The lighter is a spot on the „electron" picture the higher of the atomic number or/and the concentration of the given element, and vice versa. The γ-phases of light colour are separated from one another by β-and α-phases of Ni, the latter contains only a few % Al. The quantitative distribution of the phase rich in Al (β and α phase, $NiAl_3$) and rich in Ni (γ phase, Ni_2Al_3)) can be estimated on the basis of the microprobe scans showing the Al and Ni distribution respectively.

Using Mo and Cr promoters the number of phases increases, as can be observed on the „electron" picture in Fig. 5. The analysis of the scans shows that promoters have at least two effects. On one hand, they shift the „equilibrium" of the Al-Ni phases to the expense of the γ-phase towards the β-phase, on the other hand, new phases are formed with the promoters. The distribution of Mo is uneven, it is alloyed with Cr, and mainly with aluminium. A Mo-Cr-Ni phase can also be observed, in which Al is present in a small quantity or it is totally absent. The Cr fraction, unalloyed with Mo, is uniformly distributed in the Al and Ni.

The structure of the alloy mentioned above becomes more complicated in the presence of a small amount of silicon (Fig. 6.). Almost the total quantity of Mo is alloyed with Si, a fraction of Cr is also present in this alloy, and this Mo-Si-Cr alloy is to be found in the phase rich in Al, in which the quantity of Ni is minimal. Some of Cr and Si is present in the nickel-aluminium phase.

We were interested in silicon containing alloys because we recognised that small amounts of Si added to the alloys diminish the pyrophority of the catalysts prepared from them. Fig. 7. shows the curves of a microprobe line analysis for a catalyst containing Mo and Cr promoters where only Mo-Cr phase can be detected.

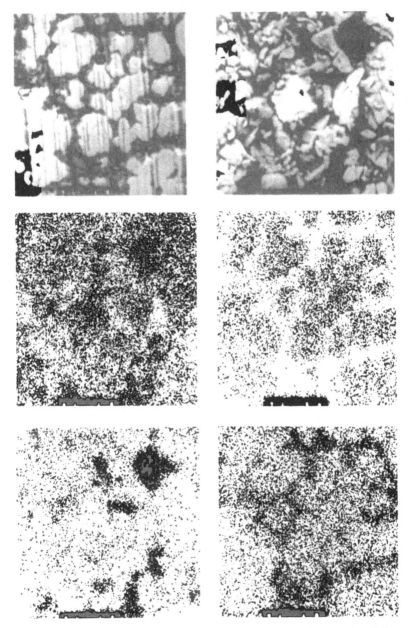

Figure 4 Microprobe images (x 1200) of a Raney-Ni alloy on the right and catalyst on the left. From above downwards: „electron" picture, Al, Ni.

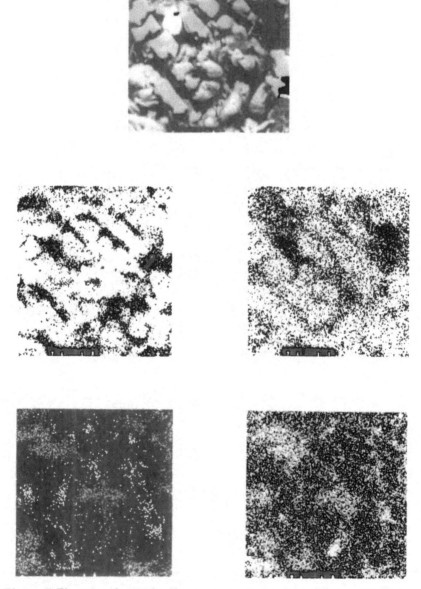

Figure 5 Electron microprobe line analysis of a promoted Raney-Ni catalyst. From above downwards. On the left side: „electron" picture, Ni, Co. On the right: Al, Cr, Mo.

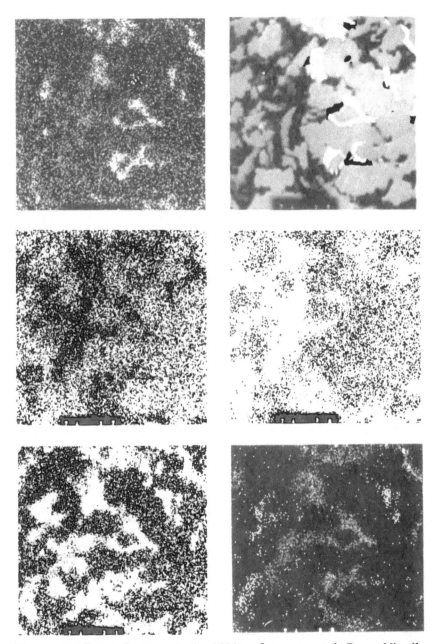

Figure 6 Microprobe images (x 1200) of a promoted Raney-Ni alloy (composition, m%: Al 48, Ni 46, Mo 2,5, Cr 1,5 Si 2). From above downwards: right side: „electron" picture, Al, Mo, on the left: Si, Ni, Cr.

8

Petró

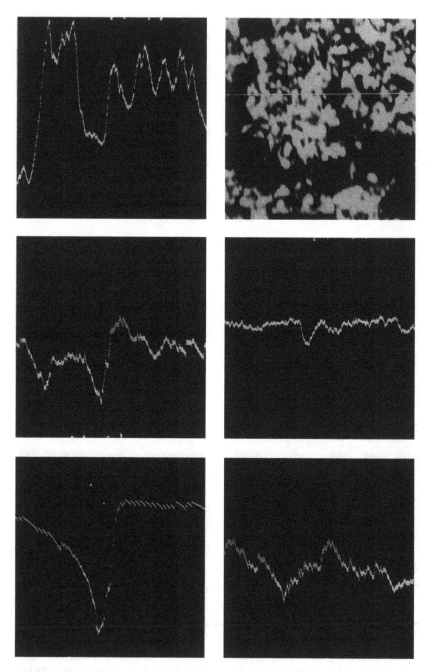

Figure 7 Microprobe line analysis of a Raney-Ni, promoted with Mo and Cr.

II. The metal-hydrogen bond (4-20)

The metal-hydrogen bond was investigated by three different methods:
1. Magnetic method
2. Thermodesorption
3. Electrochemical polarisation

It is accepted that Raney-Ni is the most active among the nickel catalysts prepared by different procedures. It is attributed to the specific metal-hydrogen system, which is formed during preparation, where nascent hydrogen saturates the nickel resulting in sponge-like structure, physically and a relatively sensitive catalyst, chemically.

Hydrogen is sorbed on hydrogenation catalysts in the course of their preparation and use. On the various active sites hydrogen is bound with various energies. These differ from each other according to the energy of the metal - hydrogen bonds. The amount of hydrogen sorbed and the strength of the interaction in the metal-hydrogen system significantly effect the catalytic properties. Opinions differ concerning the nature and character of the relationship between sorbed hydrogen and catalytic properties.

There are authors who think that catalytic activity is due to the metal hydrogen system that should be regarded as the catalyst. Numerous experimental data show that catalytic reactions proceed with different kinds of hydrogen and their role depends on the substrate and the experimental conditions.

The amount and nature of the adsorbed hydrogen can be influenced by the method of catalyst preparation. The conditions of leaching affect strongly the catalytic activity. TPD curves of some catalysts are depicted on Fig. 8. One should notice the differences in bond-strength and quantities of adsorbed hydrogen. In the hydrogenation of the carbonyl group the amount of weakly bound, whereas in that of nitrobenzene the amount of strongly bound hydrogen could be correlated with the catalytic activity. However the forms of hydrogen may transform into one another and can be supplemented partially from the gas phase. Some examples are comprised in literatures 18-20.

10

Petró

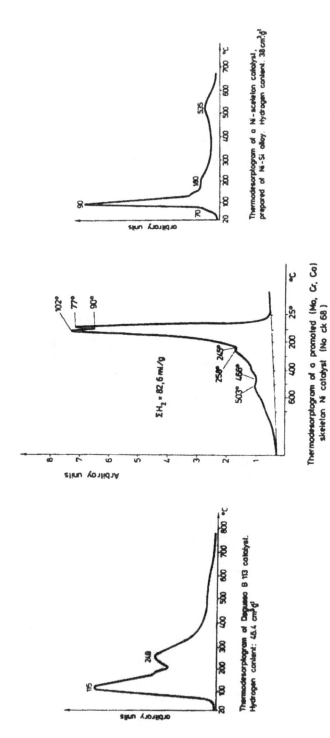

Figure 8 Thermodesorptograms of catalysts prepared from different alloys.

The TPD curves of Raney-nickel shows that at 373 K the loosely bound hydrogen is completely released. Estimated from magnetic and microcalorimetric measurements, in a repeated treatment with hydrogen only one tenth of the released hydrogen is resorbed.

Raney-Ni is ferromagnetic, the magnetization depends on its particle size and hydrogen content. As a result of thermodesorption and magnetic measurements of Raney-Ni the increase of magnetization, calculated in Bohr magnetons caused by the desorption of one hydrogen atom could be determined. The weakly bound hydrogen caused much smaller change in magnetization than the strongly bound.

The dry catalyst adsorbs substantially less hydrogen, upon flushing with argon. At room temperature the catalyst releases its hydrogen content almost completely within 2 h, and its magnetization approaches that of the catalysts heat-treated at higher temperatures. This also proves that a large part of hydrogen detected by thermodesorption is sorbed on the catalyst, and does not arise from a reaction between water, bound in aluminium hydroxides, and the aluminium and nickel. As hydroxides release their water content only above 473 K, the water and Al reaction becomes possible only above this temperature.

On the other hand, the aluminium and nickel may react at higher temperatures with the water present, since the magnetization measured after heat treatment in Ar or N_2 atmosphere is lower than that of samples subjected to more prolonged drying. Our measurements indicate that the Curie temperature of the catalysts treated at increasing temperatures tends towards the Curie temperature of bulk nickel, but does not reach it in either case (lower by more than 80K). It appears therefore that even after heat treatment at 400°C, the catalysts are in a finely dispersed state, as also shown by the diffraction patterns, but the nickel-hydrogen system, which is characteristic for Raney-nickel, was decomposed. The finding that the Curie temperature of all catalysts prepared by alkaline extraction from Ni-Al alloys decreases upon heat treatment between 250 and 300°C is surprising (Fig. 9.).

This decrease of the Curie temperature takes place after heat treatment at 250-300°C in the case of catalysts prepared from the Ni-Al alloy. These temperature ranges coincides with the temperatures of water loss in aluminium hydroxide and hydrogen desorption. From this phenomenon, we conclude that residual hydroxides have a stabilising effect on the structure. As they become dehydrated, the size of nickel particles in the catalyst will decrease and the particles will disintegrate. Some of our results on thermodesoprtion and magnetic measurement are summarised in Table 1. No such phenomenon was observed with the catalyst prepared from the Ni-Si alloy (Fig. 9.).

Table 1 Results for Raney nickel samples prepared by various procedures

Symbol of catalyst	Ferro-magnetic nickel (%)	Specific surface (m²/g)	Activity (cm³ H₂/g cat·min)		Desorbed hydrogen (cm³/g) (°C, temp. of desorption)			Magnetic change upon desorption				Surface calcd. from desorbed hydrogen (m²/g)	
			eugenol	nitro-benzene	I	II	Total	$\frac{\mu_B}{H\ atom}$ I	II	$\frac{\Delta\sigma}{\sigma_\infty}$ (%)	$\Delta\Theta$	H_{2I}	Total
T 1	11.0	43	2.4	1.9	10 (185)	5 (537)	15	0.63		68	93	27	27
T 2	13.2	82	7.2	2.2	12 (157)	28 (360)	40	0.48		77.5	100	25	100
T 3	44.0	90	13.3	4.2	27 (127)	59 (262)	86	0.34	0.44	69	117	40	185
T 4	62.0	127	20.6	5.8	29 (120)	63 (252)	92	0.32	0.41	49	51	41	154
T 5*	34.0	115	23.5	6.8	40 (100)	73 (252)	113	0.16	0.29	70	68	28	121
DEG	74.0	94	26.2	7.2	49 (111)	37 (221)	86	0.19	0.44	31	-	41	113
CH	43.1	51	22.9	8.1	14 (80)	44 (257)	58	0.20	0.49	54.5	78	12	107
Na 60	49.1	54	15.5	5.1	20 (52)	61 (241)	82	0.32	0.39	53	42	28	125
K 60	46.6	54	17.6	4.7	32 (60)	50 (248)	82	0.20	0.35	43	36	28	105
K 180	50.2	79	20.5	4.6	36 (61)	58 (250)	94	0.22	0.48	60	58	35	158

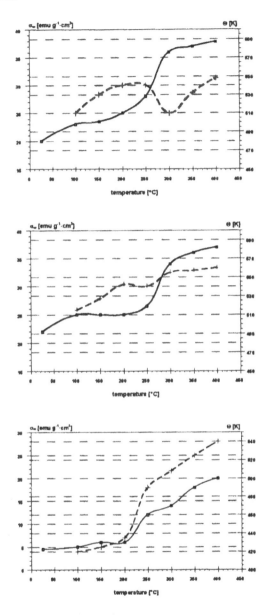

Figure 9 Saturation magnetisation(■) and Curie temperature(+) vs. temperature of heat treatment (a) standard Raney-Ni, (b) Raney-Ni prepared in solid phase, (c) Raney-Ni prepared from Si-Ni alloy.

The main conclusions are the followings: (*i*) Raney nickel catalysts prepared by various procedures contain two kinds of hydrogen. The one desorbes with maximum rate between 50 and 130°C and causes an increase of 0.16 to 0.34 B.M. whereas the other desorbes between 150 and 260°C increases the magnetization of the catalyst with 0.30 to 0.65 B.M. per hydrogen atom. Accordingly, the Ni-H bond strength is characterised not only by the temperature of maximum desorption rate but also by the specific change in magnetization occurring upon desorption, (*ii*) the characteristics of the two kinds of hydrogen cannot be sharply distinguished and vary continuously between certain limits, (*iii*) upon more extensive leaching (at higher temperatures, for a longer time and with a more concentrated alkali solution) catalysts of increasingly coarse structure, larger particle size and higher stability are obtained. The change of particle size was also monitored by measuring Curie temperature: the larger the size, the higher the Curie temperature (the structure approached that of the bulk nickel).

The metal-hydrogen system was investigated electrochemically also. During the electrochemical polarization of Raney-nickel several electrochemical processes can take place parallel, the equilibrium is questionable. Nevertheless, this method can give also some useful information for the practical chemist.

The measurements are fairly simple. The Raney-nickel catalyst saturated with hydrogen has zero potential. If it is polarised till + 200 mV (in water containing NaOH, on a Pt electrode equipped with an electromagnet, to ensure the electric contact of the catalyst with it) the loosely bound hydrogen is oxidized; above that potential the oxidation of irreversibly bound hydrogen and nickel takes place.

From the amount of charge consumed in the oxidation of adsorbed hydrogen its amount could be determined. On addition of a substrate into a reaction mixture containing Raney-nickel catalyst, even in hydrogen atmosphere the electrochemical potential of the catalyst shifts toward positive direction (the substrate „oxidizes" the hydrogen adsorbed on the catalyst). The value of the electrochemical potential depends on the substrate molecule and the reaction conditions (e.g. concentrations, temperature, pressure, stirring rate, etc.). During the hydrogenation one should choose the conditions so that the potential of the catalyst remains in that 0-+200mV interval. In that case the catalyst performs well and its lifetime may become longer.

III. The catalysts (21-36)

Working with a standard Raney-nickel catalyst, there are some possibilities for changing catalytic performance:

– *changing the pH*: some experts claim that small amount of acids loosen whereas bases strengthen the metal-hydrogen bond. The adsorption of both reduces the quantity of adsorbed hydrogen.

– *adsorption of metal ions*: stirring Raney-nickel catalyst with the solution of metal ions, depending upon the ion concentration the activity (and selectivity) is enhanced or diminished. If the standard electrochemical-potentials of those ions (metal-ion reaction) are more negative than that of the reversible hydrogen electrode (e.g. transition metal ions) they will not be reduced to metal on the surface and operate as ions. Fig.10. shows some positive examples.

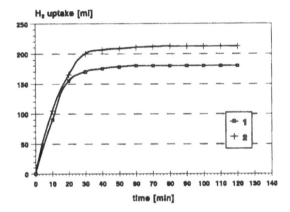

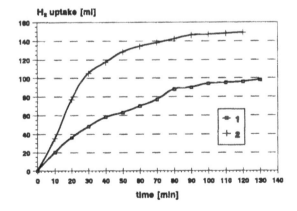

Figure 10 Effect of metal salts on the hydrogenation activity of 1 ml wet Raney-nickel (a) benzaldehyde: 1 without salt, 2 with 16 mg of $KMnO_4$, (b) nitrobenzene: 1 without salt, 2 with 13 mg of $NiCl_2$.

 – *adsorption of organic molecules*: stirring the catalyst with diluted
solution of organic (possibly unsaturated) molecules in inert atmosphere,
depending on their adsorption strength, different Ni-adsorbed hydrogen ratio and
consequently different catalytic properties could be achieved (Fig. 11).

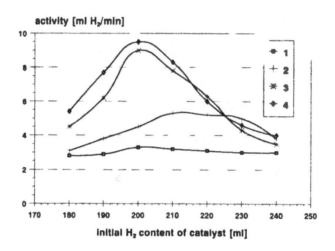

Figure 11 Changes in the activity of Raney-nickel catalyst by decreasing its
hydrogen content with acetone. Substrates are: 1. acetone, 2. benzylcyanide, 3.
nitrobenzene, 4. eugenole.

 – *effect of alloyed promoters*: in the frame of an industrial project we
investigated the effects of metal promoters with alloying (more than 90
variations.). It turned out that the surface area, the total amount of adsorbed
hydrogen, the ratio between strongly and loosely bonded hydrogen were affected
(Fig. 12). As seen, the chromium promoters enhance more drastically the amount
of loosely bound (reversible) hydrogen: the maximum is around 8 m% of
chromium, when practically this is the only form detected. On the other hand
manganese has similar effect on the strongly bound form.

 At the reduction of acetophenone the specific hydrogenation activity was
the highest with chromium promoted catalyst. It adsorbed the maximum amount
of loosely bound hydrogen.

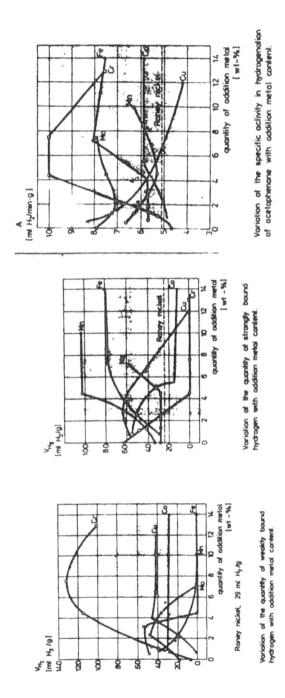

Figure 12 Change of adsorbed hydrogen forms and activity of Raney-nickel on the effect of promoters (a) weakly bound, hydrogen, (b) strongly bound hydrogen, (c) specific activity in hydrogenation of acetophenone.

We developed a skeleton-Ni catalyst whose specific activity in many reactions is higher than that of the catalyst used up till now. The stability of the new catalyst is also high. During this development two factors played an important role: the elaboration of a new process for the preparation of the catalysts and the promotion of nickel.

The essence is that the alloy powder is decomposed in solid phase as follows. The alloy powder is homogenised in a kneading device with a small quantity of NaOH, and under continuous stirring of the powder, water is added at a rate to moisten uniformly the substance without formation of a liquid phase. Decomposition is performed by water, NaOH works as a catalyst, so from aluminium mainly solid aluminium hydroxide is formed at the end of the process. When decomposition is terminated, giving about equal volume of water to the mass, two solid phases are separated, the lower of which contains the catalyst, while aluminium hydroxide in the form of a white precipitate is in the upper phase, and can be easily separated. The operation is finished by a brief treatment in a hot alkaline solution where hydrogen evolution is already insubstantial.

The aluminium hydroxide partly remains in the catalyst. Accordingly, the specific weight of the catalysts is lower than that of the conventional one, so by mixing it can be easily dispersed in all the volume of the reaction mixture and even it is easily settled and filtered. The nickel content of the catalyst is about 80%. Since the whole quantity of the alloy reacts with the NaOH for the same period, the substance is uniform. The catalyst formed has a good thermal stability of nickel crystallite size (Fig. 13).

In a catalyst containing Mo-Cr-Co promoters the changes in composition from the surface into the bulk of the grains has been investigated by SIMS. Results in Fig. 14. show that in the external zone of about 5-15 nm of the grains the concentration of Ni, Cr and Co is somewhat lower, while that of Mo is about the same as in the bulk.

This result can be explained by the peritectic formation of the grains. Technical advantages are as follows: there is no foaming during the leaching process. The volume of the necessary equipment and the time of operation are about half of that needed for conventional catalyst preparation. The specific amount of NaOH is also lower. On introducing the new process, existing equipment can be used, only an additional kneading device is needed.

Decreasing the pyrophority is achieved with adding small amount of nitric acid to the catalyst suspended in water. Then the pH becomes more alkaline because the nitric acid is reduced to ammonia.

The process discussed above offers another way for preparation of catalysts, as well.

Catalysts on oxide support are generally prepared by the joint precipitation of the hydroxides. This is followed by the washing and calcination of the hydroxides and the reduction of the metal oxide. Another possibility is the impregnation of the support, where the insolubilization of the substance in water applied, washing, drying and reduction at elevated temperature are the usual steps.

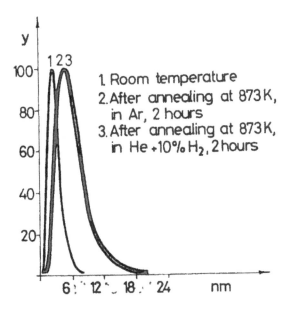

Figure 13 Crystallite size distribution of Mo-Cr-Co promoted Raney-Ni catalyst, prepared in solid phase.

Both ways involve several steps, have a high volume demand and comprise various difficult operations and a reduction of nickel oxide requires relatively high temperature (e.g. 620-670 K) and some nickel-oxide remains unreduced.

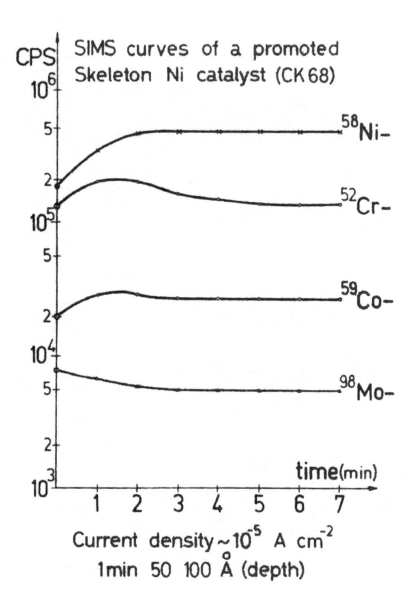

Figure 14 SIMS curves of a Mo-Cr-Co promoted Raney-Ni catalyst prepared in solid phase.

If the alloy is decomposed in solid phase, pure metal on oxide support can also be obtained in a single step. But the catalyst prepared in this way is not Raney-nickel anymore, but more active than the one produced by high temperature reduction, from nickel-oxide.

With the new processes a family of catalysts can be produced from alloys (Fig. 15.).

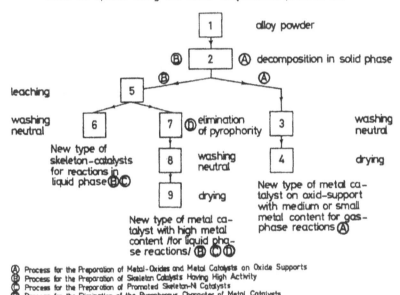

Block-diagram of a new process by wich a catalyst family could be spread using the same fundamental process ⑦.

Ⓐ Process for the Preparation of Metal-Oxides and Metal Catalysts on Oxide Supports
Ⓑ Process for the Preparation of Skeleton Catalysts Having High Activity
Ⓒ Process for the Preparation of Promoted Skeleton-N Catalysts
Ⓓ Process for the Elimination of the Pyrophorous Character of Metal Catalysts

Figure 15 Block-diagram of a process by which Raney-Ni and alumina supported metal catalyst is produced in solid phase.

Non-pyrophoric, alumina supported Raney-nickel

The leaching process of the alloy can be achieved so that high portion of the aluminium becomes part of the catalyst as support, so it does not separate at all and does not dissolve in the diluted NaOH applied. The catalyst formed consist of about 27-30 m% nickel, not pyrophoric even in the presence of volatile organic solvents and can be dried in air (e.g. 60°C).

In the new process the alloy powder is added at once in a 1.25% solution of NaOH, heated to 100°C for about half an hour. Then some 12% aqueous solution of NaOH is added to it and the mixture is cooled down. Some other features that differ from the catalyst produced by standard procedure are in Table 2.

The specific catalytic activity of the catalyst tested in liquid phase hydrogenations was somewhat higher in comparison with the standard one (Table 3) and its stability was good (Table 4).

Table 2

	Alloy [a] (kg)	NaOH (kg)	Reaction water [b]	Reaction time (min)	The wet catalyst	
					Volume (cm^3)	Weight (kg)
Standard procedure	1	2,1	8	270	350	0,5
New process	1	0,195	7	70	2000	2,2

[a] About 50 m% Ni and 50% Al, particle size 45 μ.
[b] Reaction water without water used for washing neutral.

Table 3

No. Sr. serie	Volume under water (cm^3)	Wet. weight (g)	Specific activities for nitrobenzene	acetophenone
149/A	102	110	16	8,3
147	100	112	13	11
147/1	112	110	16	13
147/2	110	110	33	17
147/3	100	119	18	9
147/4	104	117	25	17
147/5	108	116	17	9

Table 4 Specific catalytic activities of fresh and ageing samples

No SR serie	Fresh catalyst		Ageing catalyst		
	Nitro-benzene	Aceto-phenone	Ageing time in month	Nitro-benzene	Aceto-phenone
147/11[a]	27	8,1	6	26,6	10
147/7[a]	28	8,5	17	22,5	8
139[b]	25	22,4	18	32	25

[a] particle of the alloy~ 45 μm
[b] particle of the alloy ~ 35μm

Some advantages at industrial scale production:

Since the catalyst is not pure Ni but an alumina supported one, its specific weight is slightly above 1 and therefore it is easy to stir and get it more homogeneously dispersed in the reaction volume. It is also easy to filter.

The surface is polar owing to the alumina support.

The process is fairly simple and can be applied to the production of all Raney-type metal catalysts.

The time needed for the production of the same amount of catalyst is shorter than the standard one. Thus, the capacity of the equipments is higher.

At scaling up one should not encounter difficulties, because the process is less exothermic and the foaming is more moderate.

There is no need for considerable investment for a Raney-Ni producer (standard apparatus can be used).

Since a smaller amount of chemicals are used, environmental problems are reduced. To wash the catalyst neutral is easier.

According to rough assessments, the production cost is smaller compared to that of a standard catalyst.

The process is already scaled up. The pharmaceutical industry achieved good results on lab scale.

Patent applications were filed in several countries.

The structure of the new catalyst

X-ray diffractogram (Fig. 16.) reveals that nickel exists in three different forms in the catalyst. Two broad peaks appear in the diffraction pattern at $2\Theta=44.5°$ and $51.5°$. The first one has to be attributed to the (111) reflection of metallic nickel.

The crystallite size was found to be 54 Å according to the Scherrer equation and around 40 Å determined by electron microscopy. The peak at $2\Theta = 51,5$ is too low in intensity to obtain meaningful $\beta 1/2$ values but may represent smaller particles.

The broad background peak ranging from about $2\Theta = 35°$ to $55°$ may be indicative of an "X-ray amorphous" nickel phase. The alumina is present as crystalline gibbsit with a few bayerit An electron microscope picture is shown in Fig. 17. The particle size D_{111} of a standard Raney-Ni crystallites amounts to 168 Å.

The sample No. 1 was two weeks, whilst No. 2 was 6 months old. Both show an intensive SAXS scattering and from the data, one can calculate the particle size of the catalyst (in this case nickel on alumina).

The Guinier radius which is characteristic of the particle size is about 4.2 nm in both samples (Fig. 18.). From the data very smooth surface and open chain like structure is deduced. Other data originating from the measurement are summarised in Table 5.

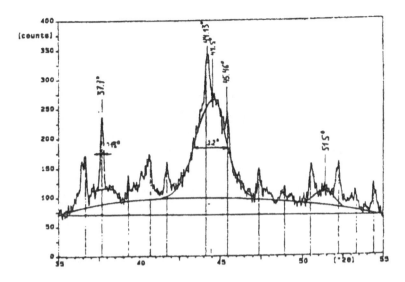

Figure 16 X-ray diffractogram of the new catalyst.

Table 5 Parameters of new catalysts derived from electron microscopy

Parameters	Fresh catalyst	Catalyst stored for 6 months
correlation length, l_c	4.72	4.87
inhomogeneity length, l_i nm	1.87	2.07
average domain thickness,nm in the phase of Ni	22.5	25
average domain thickness, nm in the phase of gibbsite	2.0	2.25
specific surface area, m²/g	53±8	48±7
surface fractal dimension, D_s	2.16±0.2 *no surface*	2.15±0.18 *fractals*
mass fractal dimensions, D_m	1.47±0.05 *the structure*	1.48±0.03 *is "chain-like"*

Figure 17. Electron microscope picture of a non-pyrophoric Raney-Ni sample. Magnification: 10^5

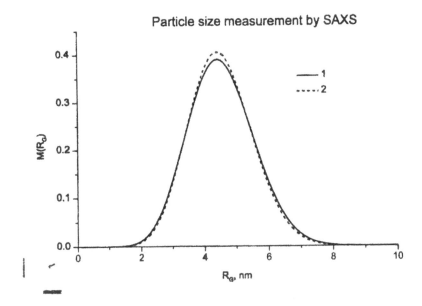

Figure 18 Distribution curves of Guiner radius of samples No1 and No 2.

Table 6 Data derived from adsorption isotherm of nitrogen

Sample	Conventional Raney-nickel	Fresh catalyst	Catalyst stored for 6 months
BET surface (m^2/g)	1.13	17.2	14.2
Total pore volume (cm^3/g)	$6.33 \cdot 10^{-3}$	$2.77 \cdot 10^{-2}$	$2.16 \cdot 10^{-2}$
Average pore radius (nm)	11.2	3.23	3.04
Adsorption energy (kJ/mol)	16.0	13.3	13.9

SEM/EDX measurement reveals that the Ni content stretches from 0-42 m% in the catalyst particles which is attributed to the different alloy phases present in the nickel-aluminium alloy.

Images show nickel-free particles that might be formed from the aluminium (Fig.19.) of nickel poor ($NiAl_3$) (Fig.20.) and nickel rich (Ni_2Al_3) (Fig.21.) phases.

28   Petró

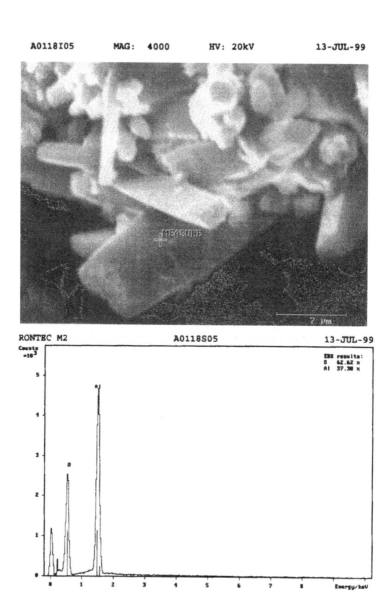

Figure 19. SEM/EDX image of a nickel-free catalyst particle magnification: 4000

Result Report SEM/RÖNTEC-Spectrometer

```
--------------------------------------------------------------
       Specimen: Ni-Al2O3
   Magnification:
    High Voltage:20kV
--------------------------------------------------------------
   Electron Image:  N=4000
```

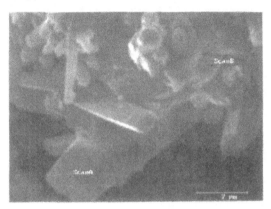

Spectrum: átlag

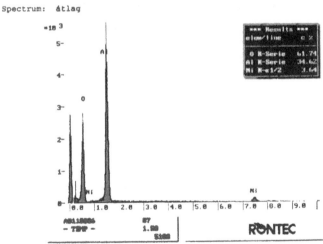

Figure 20. SEM/EDX image of a nickel poor catalyst particle, magnification: 3000

Result Report SEM/RÖNTEC-Spectrometer

```
............................................................
        Specimen: Ni-Al2O3
Magnification:
  High Voltage:20kV
............................................................
Electron Image:  N=3000
```

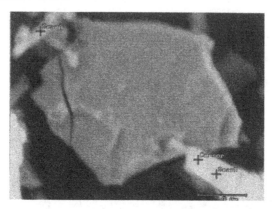

Spectrum: átlag

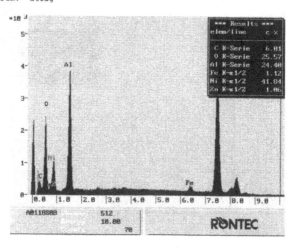

Figure 21. SEM/EDX image of nickel rich particle

Calculated form the adsorption/desorption isotherms, the mesopore distribution of the catalysts particles characterised by a sharp peak giving a radius of about 20 Å (Fig.22.).

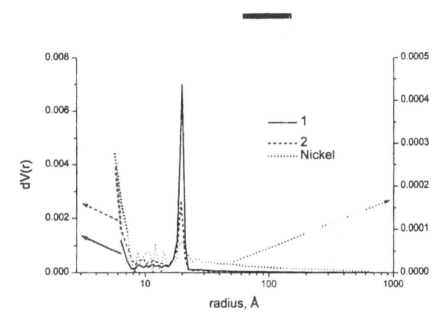

Figure 22 Pore size distribution of samples No1 and 2 and standard Raney type catalyst

Further data derived from the nitrogen adsorption isotherms are listed in Table 6. It is seen that the pore volume of both samples No 1 and 2 is much higher than that of the standard Raney type nickel catalyst.

Ageing effects are monitored by changing in the specfic surface area and the pore size distribution as well. Within half year the BET surface area decreased by about 17% and the height of the distribution peak became smaller as well (Table. 4.).

Acknowledgments

The work was supported continuously by the Hungarian Scientific Committee.

References

1. S. Békássy, J. Petró, E. Kristyák, A. Csanádi, A. Kálmán: Metallographical study of novel non-pyrophoric nickel skeleton catalysts, *Acta Chim Hung.* (Budapest) **88**, 375 (1976).

2. J. Petró: A novel method for the preparation and production of skeleton catalysts. *Preparation of catalysts II., Scientific bases for the preparation of heterogeneous catalysts*, B. Delmon, P. Grange, P. Jacobs, G. Poncelet, eds., Amsterdam, Elsevier, 1979, p. 641.

3. J. Petró, S. Békássy, A. Tungler, J. Heiszmann: Skeletal-Ni Catalysts, *Proceedings of "Europa Cat-1, Montpellier"* Vol. 2, 1993, p. 854.

4. J. Heiszman, J. Petró, A. Tungler, T. Máthé, Z. Csűrös: Thermodesorption and magnetic study of the hydrogen content of Raney-nickel, *Acta Chim Hung.* (Budapest) **86**, 117 (1975).

5. A. Tungler, J. Petró, T Máthé, Z. Csűrös: Magnetic measurements on industrial Ni-catalysts, *Magyar Kémiai Folyóirat* **78**, 434 (1972) (in Hung.)

6. A. Tungler, J. Petró, T. Máthé, J. Heiszman, S. Békássy: Nickel particle size and hydrogen content in skeleton catalysts, *Acta Chim Hung.* (Budapest) **89**, 31 (1976).

7. J. Heiszman, S. Békássy, J. Petró: Thermodesorption study of novel, non-pyrophoric nickel skeleton catalysts, *Acta Chim. Hung.* (Budapest) **86**, 347 1975).

8. S. Békássy, J. Heiszman, J. Petró: Structure and hydrogen sorption properties of skeleton catalysts prepared with various dissolvagle metals. *Heterogeneous Catalysis. Proceedings of the Fourth International Symposium.* Varna, 1979, Part I, p. 111.

9. J. Heiszman, S. Békássy, A. Tungler, J. Petró: Etude de hydrogene sorbé sur la surface des catalyseurs squelettes, *Proceedings of the Second Int. Congress on Hydrogen in Metals*, Paris, Oxford, Pergamon, 1977, Vol. 2., 4A13 1.

10. A. Tungler, J. Petró, T. Máthé, Z. Csűrös, K. Lugosi: Magnetic studies on Ni catalysts, *Acta Chim. Hung.* (Budapest) **79**, 289 (1973).

11. A. Tungler, S. Békássy, J. Petró: Magnetic study of novel, non-pyrophoric Raney nickel catalysts, *Acta Chim. Hung.* (Budapest) **86**, 359 (1975).

12. A. Tungler, S. Békássy, J. Petró, T. Máthé: Hydrogen adsorption on dried Raney-nickel, *Acta Chim. Hung.* (Budapest) **105**, 73 (1980).

13. A. Tungler, J. Petró, T. Máthé: Magnetic studies of the adsoprtion on nickel catalyst, *React. Kinet. Catal. Lett.* **19**, 181 (1982).

14. S. Békássy, Gy. Liptai, J. Petró: Study of normal nonpyrophoric nickel skeleton catalysts by derivatograph, *Thermochimica Acta* **11**, 45 (1975).

15. Z. Csűrös, J. Petró, É. Polyánszky: Electrochemical methods and their application to Raney-nickel, *Periodica Polytechnica* **12**, 251 (1968).

16. J. Heiszman, S. Békássy, J. Petró, G. Bidló, L. Bezur: Study of the effect of heat treatment on Raney nickel by magnetic electrochemical and thermo-desorption methods, *Acta Chim. Hung.* (Budapest) **89**, 151 (1976).
17. A. Tungler, J. Petró, T. Máthé, J. Heiszman, F. Buella, Z. Csűrös: Study of the effect of heat-treatment on Raney-nickel by magnetic, electrochemical and thermodesorption methods, *Acta Chim. Hung.* (Budapest) **89**, 151 (1976).
18. Z. Csűrös, J. Petró: Studies on the behaviour of Raney -nickel catalyst in hydrogenation processes, as a function of its hydrogen content, *Acta Chim. Hung.* Acad. Sci. Hung **29**, 321 (1961).
19. Z. Csűrös, Zs. Dusza, J. Petró: Correlations between the hydrogen content, sorption power and activity of Raney-nickel catalysts, *Acta Chim. Hung.* Acad. Sci. Hung. **30**, 461 (1962).
20. Z. Csűrös, J. Petró: Investigation of the behaviour of various substrates with the bound hydrogen of Raney-Ni, *Acta Chim. Hung.* **22**, 87 (1960).
21. Z. Csűrös, J. Petró, J.Vörös: Effect of alkali additives on the activity of Raney-nickel, *Periodica Polytechnica* (Ch) **1**, 169 (1957).
22. Z. Csűrös, J. Petró: Effect of various alkiles applied for preparations, on the hydrogenation activity of Raney-nickel, *Acta Chim. Hung.* **17**, 289 (1958).
23. Z. Csűrös, J. Petró: Investigations of the changes in the activity and effectiveness of Raney-nickel on the addition of acids, *Acta Chim. Hung.* **19**, 379 (1959).
24. Z. Csűrös, J. Petró J.: Investigation on the action of nickel, copper and manganese salts on the hydrogenation activity and effectiveness of Raney-Ni, *Acta Chim. Hung.* **20**, 129 (1959).
25. Z. Csűrös, J. Petró, J. Heiszman: Investigation of the effect of cobalt salts on the activity and effectiveness of Raney-nickel, *Acta Chim. Hung.* **22**, 73 (1960).
26. Z. Csűrös, J. Petró, S. Holly: Interaction of Raney nickel and substrate in hydrogenation reactions, *Acta Chim. Hung.* **29**, 351 (1961).
27. Z. Csűrös, J. Petró, S. Holly: Effect of additives on hydrogen sorbed by Raney-Ni, *Acta Chim. Hung.* **29**, 419 (1961).
28. Z. Csűrös, J. Petró: Possibilities of increasing the activity of Raney-nickel type catalysts, *Periodica Polytechnica* (Ch) **3**, 123 (1959).
29. Z. Csűrös, J. Petró, J. Heiszman: Investigation of the activity of catalysts prepared from Al-Ni-Co and Al-Ni-Cr alloys, *Acta Chim. Hung.* **17**, 289 (1958).
30. Z. Csűrös, J. Petró, J. Heiszman: Untersuchungen mit aluminothermisch hergestellten Co-, Fe-, Cu- und Mn-hältigen Katalysatoren, *Periodica Polytechnica* (Ch) **10**, 405 (1966).
31. Z. Csűrös, J. Petró, J. Heiszman: Untersuchung von mit Bor promovierten Raney-Nickel Katalysatoren, *Periodica Polytechnica* (Ch) **10**, 421 (1966).

32. Z. Csűrös, J. Petró, J. Nádas: Changes in the hydrogenation activity of Raney-nickel as a function of the temperature and duration of extraction, *Periodica Polytechnical* (Ch) **1**, 153 (1957).
33. Z. Csűrös, Zs. Dusza, J. Petró, L. Erdey, F. Paulik: The alkali used as extractant and of the hydrogen content on the activity of Raney-Ni, *Acta Chim. Hung.* **42**, 131 (1964).
34. Z. Csűrös, J. Petró, V. Kálmán, L. Erdey, F. Paulik: Catalytic properties of Raney-nickel as a function of its preparation, *Periodica Polytechnica* (Ch) **10**, 27 (1966).
35. J. Petró: A process for the preparation of supported non-pyrophoric Raney-catalysts. Patent applications in Hungary, Germany, USA and Japan (1998).
36. J. Petró, A. Bóta, H. Beyer, I. Dódony, E. Kálmán, K. László: A new alumina supported not pyrophoric Raney-type Ni catalyst, *Appl. Catal. A* **190**, 73 (2000).

Selecting Between Batch Slurry and Continuous Fixed Bed Hydrogenation

John D. Super

Dixie Chemical Co., Inc., 10701 Bay Area Blvd., Pasadena, TX 77507

Abstract

Considering primarily technoeconomic aspects, and some qualitative considerations, this paper builds the basis for selecting between batch and continuous hydrogenation units. As a function of annual capacity, capital investment and cost plus profit are developed for both cases. The economic breakeven point is at about 15,000,000 pounds per year of product.

Discussion

In the worldwide chemical industry, there are many batch slurry hydrogenators, often in multi-product facilities; and fewer continuous fixed bed units, usually found in the larger firms employed in single product units, sometimes in multi-product units. To address the decision made between building a batch slurry unit or a continuous fixed bed unit, first a base case or center point will be developed for each. This paper will deal primarily with the technoeconomic and qualitative issues, not the many important technical aspects covered recently by Murthy in an excellent article (Ref 1).

The product basis for both cases is the hydrogenation of an organic with molecular weight of 100, requiring one mole of H2 for the reaction. This is typically 15-30 kcal/gm mole exothermic heat of reaction (Ref 2). For the purposes of this paper, the reaction is assumed as "neat", no solvent, and that through engineering and operational procedures the reaction exotherm is manageable.

The batch slurry base case is a 1000 gallon reactor, operated at 1000 psig, made of 316L stainless steel, with heating and cooling, and fitted with a sintered metal catalyst filtration and recycle system. Manufacturing area total investment in year 2000 US dollars is $1,390,000. The batch slurry base case has a thruput of 8000 pounds per day (ppd) with a 24 hr cycle time, operating 6000 hr per year. This 6000 hr includes cleaning time, and the somewhat high downtime is the typical high maintenance due to seals, catalyst filtration, etc. Production annually is 2,000,000 lbs. A 4,000,000 ppy plant has a 2000 gallon reactor: a 16,000,000 ppy plant, a 8,000 gallon reactor, which is likely the practical limit for a 1000 psig vessel with slurry mixing.

The analysis from this point depends greatly on the understanding of investment scaling with the power rule. The classic equation is

C2=C1 (V2/V1) exp alpha

where "C" is plant investment cost and "V" is plant production volume or capacity. If you know subscript 1 cost and volume, you can calculate subscript 2 cost at a different capacity. When the exponent is 0.6, we have the "six-tenths" rule. In 1988 I wrote a paper discussing investment scaling (Ref 3). This paper argued that the investment scaling exponent alpha, in the equation, has two limiting solutions, zero and one, and that the scaling exponent is a continuum between those values, not a single value. I also set a base case or center point as the normal capacity, where the exponent is 0.5.

Applying this methodology to the batch slurry plant, refer to log-log Plot I and Table 1.

Table 1 Batch Slurry Plant

Annual Capacity million ppy	MFG Invest $million	MFG Invest $/lb
0.0625	0.58	9.32
0.125	0.58	4.66
0.25	0.61	2.44
0.5	0.76	1.53
1	1.00	1.00
2	1.39	0.70
4	2.00	0.50
8	3.04	0.38
16	5.30	0.33
32	10.60	0.33

Plot I-Capital Investment vs Capacity

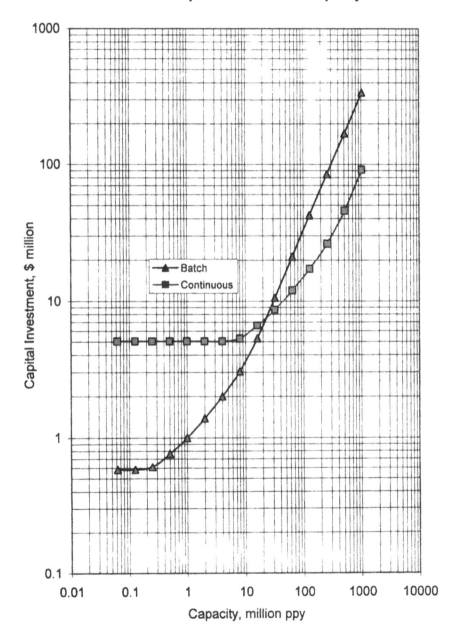

The center point is 2 million ppy. As one builds batch slurry plants of lower capacity, eventually the plant total cost does not reduce any more with a smaller reactor, filter, etc. and alpha equals zero and the investment levels out at about $600,000. As one builds larger batch slurry plants, eventually the equipment reaches its maximum size and one will build a second line, and alpha equals one at 16,000,000 ppy. Inverting the thought and converting the numbers into $ of investment per pound annual capacity, this is also shown on Table 1. Now on log-log Plot II, the left-hand slope is one; the right hand slope is zero.

The continuous fixed bed case has a much higher base case or center capacity, 64,000,000 ppy. This plant has an operating time of 7785 hours/yr., higher that the batch slurry plant since the maintenance is less and there is essentially no catalyst filtration. Pressure is 1000 psig, material of construction is 316L stainless steel, and the fixed bed design is either a large single bed, or primary and secondary fixed beds, with typical space velocity of 2 hr(-1) for the 100 molecular weight product requiring one mole of hydrogen per mole. This unit has hydrogen recycle and purge along with heat management using either product recycle or interstage cooling. The center point investment for 64,000,000 ppy is $12,000,000. Using the same scaling methodology as the batch plant, Table 2 and log-log Plots I and II show data.

Table 2   Continuous Fixed Bed Plant

Annual Capacity million ppy	MFG Invest $million	MFG Invest $/lb
1	5.04	5.04
2	5.04	2.52
4	5.04	1.26
8	5.28	0.66
16	6.60	0.41
32	8.64	0.27
64	12.00	0.19
128	17.28	0.14
256	26.28	0.1
512	45.72	0.09
1024	91.44	0.09

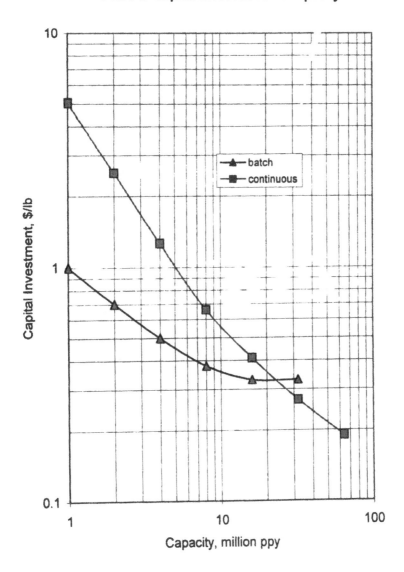

Note that the break-even investment is at about 25,000,000 ppy. Also note that at lower capacities, the batch plants are significantly less costly. The opposite is true for large capacities. But what about annual operating cost plus profit?

The batch and slurry plants have typically a minimum of two operators. From experience, I have developed an equation to calculate total operators for 24-hour, 365 days per year operation, based on annual capacity in millions of ppy. For each significant reaction step, the equation is number of operators equals 15 times (capacity/100) exp 0.2. Capacity is in million ppy. To develop a complete operating cost, the detailed basis is shown on Table 3.

Table 3   Cost and Profit Basis

No Shifts	4
Hr/yr-operator	2000
Op $/hr wages	20
Benefits, %wages	50
Op Supv, %W+B	20
Op Overhead, %W+B	30
Supplies, %W+B	10
Depr, book % invest	10
Maint W+B, % invest	6 *
Maint Matl, % invest	4 *
Maint Overhead, %W+B	30
Maint Supv,%W+B	20
No Tech Prof-see other tables	
Tech wages/yr	70
Tech benefits, %W	50
Local taxes, %invest	1
Plant Adm, %invest	2
QC, % Op W+B	25
QC Overhead, % W+B	30
Corp expense, % above	8
Pretax margin, %sales	20

*For continuous, 4%L & 3%M
Op= operating
W= wages
B= benefits

Note that 10% straight-line depreciation is included. For profit a 20% pretax margin on sales is used.

Often batch plants change product every week, coming back to that product each quarter. A cost including the cleaning time is obtained by a 7/6 adjustment to the straight time cost. This scenario also results in about 10 different products over the year. Catalyst expense is shown on Plot III, a declining cost with increasing annual volume, representing operating improvements and lower prices with higher purchase volumes. Utilities cost declines with increasing annual volume. And hydrogen cost per unit of production decreases with increasing annual production due to a declining purchase price as purchased quantity per year increases (shown on Plot IV), and reduced physical losses with increasing volume (shown on Table 4). Combining all these elements, for a total cost plus profit, see Table 5.

Table 4 H2 operating losses vs capacity

Capacity Million ppy	% losses
2	20
4	13
8	8
16	6
32	5.5
64	5

Plot III- Catalyst Expense vs Capacity

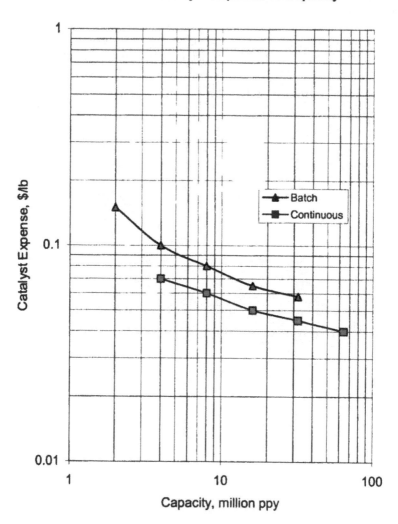

Plot IV-H2 Price vs. Capacity

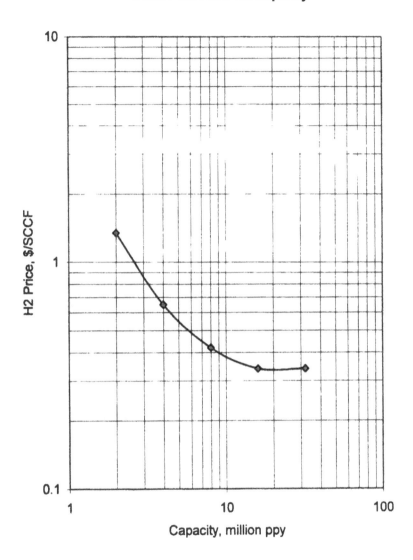

Table 5   Batch Case

Capacity million ppy	2	4	8	16	32
ISBL cap invest, $ million	1.39	2	3.04	5.3	10.6
No. operators	6.9	7.9	9.1	10.4	11.9
No. tech professionals	0.5	0.5	0.5	1	1
Op Cost, $1000/yr					
Operators labor & benefits	412	473	543	624	717
Op supervision	82	95	109	125	143
Op overhead	123	142	163	187	215
Op supplies	41	47	54	62	72
Maint labor	83	120	182	318	636
Maint overhead	25	36	55	95	191
Maint supervision	17	24	36	64	127
Maint material	56	80	122	212	424
Technical	53	53	53	105	105
Local taxes	14	20	30	53	106
Depr-book	139	200	304	530	1060
Plant adm	28	40	61	106	212
Quality control	103	118	136	156	179
Corp-Adm/Sales/R&D	94	116	148	211	335
Margin	317	391	499	712	1130
Sales at 100% capacity	1587	1954	2494	3560	5652
Sales Price, $/lb	0.793	0.488	0.312	0.223	0.177
Sales Price incl clean	0.926	0.570	0.364	0.260	0.206
Catalyst	0.15	0.1	0.08	0.065	0.057
Utilities	0.02	0.015	0.01	0.01	0.01
H2	0.062	0.028	0.018	0.014	0.014
Total	1.158	0.713	0.472	0.349	0.287
Capital Invest, $/lb	0.70	0.50	0.38	0.33	0.33

ISBL= inside battery limits

As a side comment, you can convert these costs to $/hr operating fee if pricing for toll manufacturing. Taking the 4,000,000 ppy case, the 2000 gal reactor, and rolling up cleaning into the price, the hourly fee is about $375.

The continuous case employs the same operator equation. See the detailed operating cost plus profit basis on Table 3. Maintenance percentages are lower due to fewer seals and no catalyst filtration.   This continuous plant typically makes product on the same catalyst for two months in order to get reasonable catalyst expense, so two different products are assumed as the production during the two months, and a 6 day cleanout and turnaround is assumed. So one makes a 60/54 adjustment to the straight time cost to get full cost including cleaning and trunaround, and there are a total of 10 products made each year. The campaigns are longer than typical of a batch plant. Campaigns are about $1/10^{th}$ of the annual capacity. Catalyst cost is lower than slurry, see Plot II. Utilities are less than batch, and the hydrogen cost and losses are assumed to be the same as the batch plant, but in reality the losses are somewhat higher in the batch plant. See Table 6 for the summary of annual cost plus profit for this continuous case.

Plotting both cost plus profit cases on log-log paper (See Plot V) the break-even is about 10,000,000 ppy. The batch slurry plant reactor size would be about 5,000 gallons. Since this is an operating cost plus 20% pretax margin comparison, the lower capital investment for the batch case would result in a break-even of about 15,000,000 ppy when a rate of return calculation was done. Good advice from Grant and Ireson, a classic text in the area: "Often it is a good idea to compute the prospective rate of return on an investment rather than merely to find out whether the investment meets a given standard of attractiveness." (Ref 4) So it is no wonder that most reasonably sized hydrogenation plants are batch slurry. Plus many firms that build these hydrogenation plants are investment adverse, so the lower capital investment plants are logically preferred, especially when one includes some of the noneconomic considerations briefly discussed next.

Table 6   Continuous Case

Capacity million ppy	4	8	16	32	64
ISBL cap invest, $ million	5.04	5.28	6.6	8.64	12
No. operators	7.9	9.1	10.4	11.9	13.7
No. tech professionals	0.5	0.5	1	1	1
Op Cost, $1000/yr					
Operators labor & benefits	473	543	624	717	823
Op supervision	95	109	125	143	165
Op overhead	142	163	187	215	247
Op supplies	47	54	62	72	82
Maint labor	202	211	264	346	480
Maint overhead	60	63	79	104	144
Maint supervision	40	42	53	69	96
Maint material	151	158	198	259	360
technical	53	53	105	105	105
Local taxes	50	53	66	86	120
Depr-book	504	528	660	864	1200
Plant adm	101	106	132	173	240
Quality control	118	136	156	179	206
Corp-Adm/Sales/R&D	163	178	217	267	341
Margin	550	599	732	900	1152
Sales at 100% capacity	2749	2995	3660	4498	5762
Sales Price, $/lb	0.687	0.374	0.229	0.141	0.090
Sales Price incl clean, $/lb	0.763	0.416	0.254	0.156	0.100
Catalyst	0.07	0.06	0.05	0.045	0.04
Utilities	0.01	0.0075	0.0075	0.005	0.005
H2	0.028	0.018	0.014	0.014	0.014
Total	0.871	0.502	0.326	0.220	0.159
Capital Invest, $/lb	1.26	0.66	0.41	0.27	0.19
Invest, $million at 64,000,000 ppy	12				

Plot V-Cost Plus Profit vs Capacity

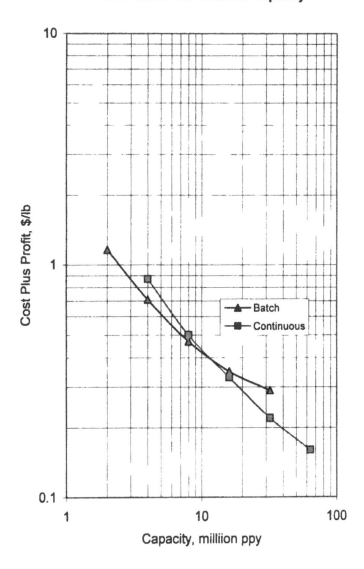

Most lab work is done in batch slurry, so the scale-up to commercial is often believed easier for batch slurry. The catalyst is understood, and it is a direct scale-up most of the time. But in a batch slurry plant, the catalyst filtration is the key to operability. There are often great difficulties with filtration (Ref 5). My personal experience is that filtration is the most troublesome scaleup problem in the plant after one has apparently successful results in small lab slurry reactors. Catalyst attrition often results in long filtration times, filter media changes, etc. Even with a heavy sponge metal catalyst that does settle, the fines must be filtered, often with problems. Supported base metals are often also difficult to filter. The attrition, losses and activity decline of supported precious metals on carbon is another troublesome area.

Fixed beds essentially avoid most catalyst filtration problems. And with a very costly catalyst, fixed bed does result in lower catalyst cost and higher precious metal recovery with a well-designed system. But there are significant technical challenges in the mass and heat transfer area, issues that many firms are not willing to solve. And once adequate pilot data is developed for fixed bed, often a space velocity of less than 1 hr(-1), say 0.25 to 0.5, results in very large fixed beds, at high investment.

Another consideration is by-product chemistry, influencing product quality. This can often be influenced by the type of reactor selected, especially when considering the reactant or product concentrations and back mixing associated with the reactor selection. Other factors such as removing heat of reaction or the industrial hygiene issues associated with catalyst handling can influence the choice of operating mode.

Batch slurry plants have significant flexibility, able to process one truckload of feed in 3-5 days time, able to change catalyst and products frequently. Batch slurry plants are truly multiproduct. The fixed bed plants require runs of 500,000 to 2,500,000 lb., so campaigns are relatively large. To carry inventory from large campaigns may result in a financing burden that can be unacceptable.

A continuous stirred tank reactor (CSTR) conducting slurry hydrogenation is an extension of the batch case and has similarities with the continuous fixed bed case, with additional catalyst filtration equipment, more instrumentation, and often hydrogen recycle. It was deemed that this case was in between the two considered cases, and not different enough to merit separate treatment.

At first I believed that a semi-continuous unit was a separate case, but as the analysis was developed, it was an extension of the batch slurry case and was also not different enough to merit treating it as a separate case. Sometimes a batch unit is called semi-continuous as one reactant other than hydrogen is fed over a period of time, incremental feed. A batch unit can also be made into a semi-continuous by adding off-line continuous filtration with a lower pressure hold tank and two filters.

In conclusion, given the qualitative pros and cons, and the economic considerations, the industry will continue to build many more batch slurry hydrogenation plants compared to continuous fixed bed plants.

References

1) A K S Murthy, "Design and Scaleup of Slurry-Hydrogenation Systems," Chemical Engineering, September 1999, pages 94-107.
2) A K S Murthy, S R Schmidt, J M Solomon, Alan Stout, Applied Hydrogenation Technology, Course sponsored by Center for Professional Advancement, November 17-19, 1997.
3) J D Super, "A New, Theoretically Sound Manufacturing Investment Scaling Method", AIChE Symposium Series No. 285, Volume 87, 1991, pages 82-86.
4) E L Grant, W G Ireson, and R S Leavenworth, Principles of Engineering Economy, 7th Ed, Wiley, 1982, page 117
5) J Concordia, " Catalyst Handling in Batch Hydrogenation", Chemical Engineering Progress, June 1980, pages 35-40.

Safety and Economics Considerations in the Design of an Industrial Hydrogenation Facility

Bruce D. Nacker[a], Thomas R. Kowar[a], Mike G. Scaros[a],
Susan R. Cyganiak[b], Mike Grolmes[c], Vasilios H. Iskos[a],
Andy Longinow[d], Christopher N. Robb[a],
Kim Mniszewski[e], and John W. Venitz[b]

[a]Department of Chemical Sciences, Searle/Monsanto
[b]Environmental Health and Safety, Searle /Monsanto
[c]Consultant, Centaurus Technology Inc.
[d] Consultant, Wiss, Janney, Elstner Associates, Inc.
[e] Consultant, Triodyne Fire and Explosion Engineers, Inc.

Chemical Sciences, Searle/Monsanto
4901 Searle Parkway
Skokie IL 60077

Abstract

The issues and considerations involved in the design of a prospective, new construction hydrogenation facility are described. The planned facility includes a 1000-gallon stainless steel hydrogenator, associated catalyst handling equipment and the building to house the hydrogenation equipment. The iterative design process is based on a hydrogenation facility described in the literature. The primary objective is to design an intrinsically safe facility but many important environmental and siting issues also require consideration. A hurricane-stable building with complete spill containment capability is planned. The potential for future expansion to 2000-gallon scale equipment is also evaluated for this cost effective hydrogenation facility.

Introduction

Hydrogenation plays an important role in the manufacture of pharmaceuticals and fine chemicals. These are specialized reactions that operate on various scales and the designs of facilities for the conduct of hydrogenation reactions are not widely documented.

The process being developed by Searle/Monsanto for the synthesis of a new pharmaceutical product involved the use of a hydrogenolysis reaction. The

51

plant that was scheduled for the production of this new product did not have hydrogenation reaction equipment. Therefore, the design of a new hydrogenation facility was undertaken (1).

The hydrogenolysis reaction was to be conducted in a stainless steel hydrogenator under a hydrogen pressure of 60 psig and at a temperature of 50 °C. These conditions were considered moderate with respect to high-pressure reactions. However, the design process was governed by Searle/Monsanto high standards of safety to ensure the protection of personnel and property. Environmental and economic considerations were also of high priority. Finally, all processing was to be conducted under Good Manufacturing Practice (GMP) conditions.

Production plans required the initial use of a 1000-gallon stainless steel hydrogenator in the new facility and the capability to accommodate the addition of a second 2000-gallon hydrogenator at some future date.

The design process was based on a hydrogenation laboratory described in the literature. The elements of that design were adapted to the design of a manufacturing-scale hydrogenation facility. The design process was iterative in nature and the initial literature design was transitioned through two distinct phases before the final design was achieved.

Chemical Process

The hydrogenation reaction involved the reduction of the aromatic hydroxybenzamidine 1 to the corresponding benzamidine 2 employing 5% Pd/C as the catalyst, five equivalents of glacial acetic acid, and water as a solvent at 60 psig hydrogen pressure and 50 °C.

The general procedure called for the hydroxybenzamide to be charged to the reactor followed by the catalyst (5%Pd/C) and deionized water containing five equivalents of glacial acetic acid. The reactor was purged sequentially with nitrogen and hydrogen. The reactor was stirred under a constant hydrogen pressure of 60 psig and 50 °C for 4 h (2).

The reaction is initially a slurry but as the hydrogenation proceeds the resulting water soluble benzamidine acetic acid salt is formed. The reaction completion was confirmed by HPLC analysis. The catalyst was removed by hot filtration (50 °C) through a layer of powdered cellulose and the catalyst was washed with a small amount of hot deionized water.

Building Design Issues

Safety

Safety governed all design considerations. The proper protection of personnel and property was addressed at each step of the design process.

The safety issues inherent with the transport, storage, and use of hydrogen gas were addressed as the first priority in the facility design. Hydrogen gas is extremely flammable and it reacts explosively with oxygen in the atmosphere.

Hydrogen is stored and transported under pressure. Delivery of high pressure compressed hydrogen (2800 psig) using a hydrogen tube trailer was anticipated. Therefore, a protected hydrogen storage and delivery area was required. A fire-protected concrete-walled parking area for the tube trailer was planned. This area needed to be accessible to ground transportation and had to be located at a distance inherently safe with respect to other elements on the plant site. Automatically sealing break-away couplings were to be used for connection of the tube trailer to the stationary piping.

The piping and valving for transport of the hydrogen from the storage area to the hydrogenation reactor needed to be routed appropriately and had to meet exceptional quality standards. The components utilized in the piping system were planned in order to comply with the American Petroleum Institute (API 706) and the American Society of Mechanical Engineering (ASME Section I, IV, and VIII) codes (3).

The hydrogen supply piping was to be protected from regulator failure by the use of overpressure relief vents equipped with flame stacks for hydrogen flare-off.

The 1000-gallon hydrogenation reaction vessel was designed as a 300 psig ASME code vessel. The pressure rating was based on reactor content detonation calculations and included a generous safety margin. The overpressure relief line was to be vented to an inerted 2000-gallon vessel rated for 50 psig and for the containment of flammables (4).

The worst case scenario considered in the design process was the release of the entire hydrogen contents of a 2000-gallon reactor at 60 psig pressure into the building followed by exposure to an ignition source. Modeling calculations of the pressures and temperatures resulting from such an ignition at various points inside and outside of the hydrogenation building drove the development of the design.

Environmental

The new facility was to be designed for the chemical process described above and for future chemical processes that might involve the use of various organic solvents and chemical process intermediates. The new facility had to accommodate potential spills of these materials in a way that would prevent contamination of the environment.

Further, the design had to consider the accidental spill of hydrogenation catalyst (precious metal) and prevention of soil and sewer contamination by these catalysts. Spill containment capacity had to be sized for both 1000-gallon and 2000-gallon reactors.

To address these spill issues, a secondary sump was planned for the floor of the building. The sump would connect to the process sewers only by means of a sump pump that would discharge through a filter element allowing for recovery of the precious metals and preventing their discharge to the environment.

Appropriate process and polish condensers to minimize solvent discharges to the atmosphere were planned for the facility. The planned use of a containment tank to capture effluent from the overpressure relief line minimized the potential for solvent release to the environment.

Site

The location of the new hydrogenation facility had to be consistent with the long-term site plan. Therefore, the distance of the facility from current and future buildings, roads, chemical storage and chemical transport had to be evaluated.

Specifically, the following issues had to be considered for the new facility:

- distance to the tank farm
- distance to the hydrogen supply
- restricted access to the building by normal traffic
- run length of the substrate slurry feed
- run length of the catalyst slurry feed

Civil

The plant was located on a Caribbean island and, therefore, the planned facility had to be constructed to withstand external hurricane-force winds and the internal pressure generated by the potential ignition of hydrogen gas within the building. The building was to be designed to conform to the National Fire Protection Association (NFPA 68) code and also was required to withstand an internal pressure buildup of 1.5 psi.

The tropical temperatures suggested the use of open surfaces in the building to ensure adequate ventilation and temperature control.

The subsoil limestone base exhibited extreme porosity that influenced both the structural aspects of the building and the potential environmental impact in the event of a spill.

Quality

All operations in the new facility were to be conducted under GMP conditions. All chemical charging and catalyst loading and unloading had to be conducted within the confines of the structure.

All open surfaces of the building had to be screened to protect against insects and wild life.

Economics

The design of the new hydrogenation facility was to address all the above issues at minimum cost. The final design was the simplest in concept and, accordingly, was available at the lowest cost.

Building Design

Description of Design Iterations

The initial design proposed by an engineering consulting firm consisted of a four-sided reinforced monolithic poured concrete structure, with explosion venting through roof-mounted blowout panels. According to NFPA 68 standards, this type of four-sided concrete building would be required to withstand 2 psig of internal pressure from a worst case hydrogen explosion.

The burn injury threshold distance is the lower limit for skin blistering upon exposure to a hydrogen explosion fireball. The calculated (5,6) burn injury threshold distance was used to judge the safe distance from the building for a human being in the event of the worst case explosion scenario. In the case of this initial vented roof design, the burn injury threshold distance was estimated to be 20 feet around the entire building.

This design was lacking in any natural ventilation that might mitigate an potential explosion, and the multiple blowout roof panels posed potential maintenance problems. Furthermore, the costs associated with building a structure to withstand 2 psig of pressure and the schedule required for the construction of the custom-fabricated roof encouraged the investigation of other design alternatives.

A number of excellent literature articles on the design of hydrogenation/high pressure facilities were reviewed (7,8,9,10). The following narrative describes the considerations driving the iterative design process that began with these literature examples and concludes with a presentation of the details the final hydrogenation facility design planned for the Searle plant.

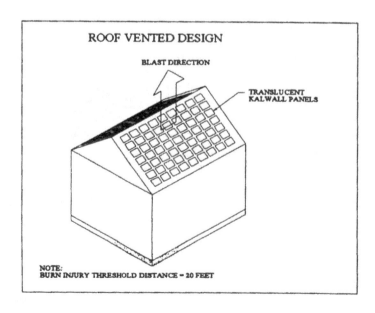

Design Iteration 1: Open-sided structure with external deflection bunker

The first design iteration was based on the design concepts presented for a laboratory hydrogenation facility (8). This proposal consisted of a four-sided building with one open vertical wall, outside of which was a concrete bunker to trap shrapnel from a blast. The concrete bunker was later replaced with an earthen bunker in order to reduce cost.

Conceptually, this basic design proposal was evolved by a number of modifications. First, a sloped roof was added to the building so that the natural progression of leaked hydrogen accumulating near the roof would be towards the open end of the building. This provided a simple means for minimizing the accumulation of adventitious hydrogen. The installation of a blast mat across the open end of the building was included in order to trap and direct shrapnel.

The burn injury threshold distance was estimated to be 52 feet on the open side of the building which required that a significant land area around the proposed structure be sequestered and limited to use as a buffer zone around the building.

The open wall design resulted in a calculated hydrogen explosion pressure of 1.5 psig. This was expected to substantially reduce the cost of the building walls compared to the initial four-walled design. However, cost analyses and schedule requirements showed that this building could not be built within budget and schedule constraints.

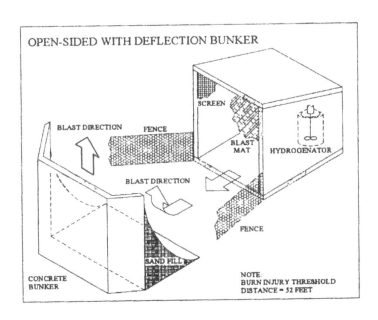

Design Iteration 2: Umbrella Design

This creative design was based on a hydrogenation building with four vertical concrete walls and a structural steel-supported gable roof located six feet above the top of the concrete walls. The location of the roof was likened to an umbrella. The area of the open space between the top of the building and the roof was calculated to be of sufficient capacity to allow pressure relief in the event of an explosion (6).

The most significant advantage of the umbrella design is that it allows hot gas from an explosion to vent above ground while eliminating the need for integral roof blast panels or deflector walls. This design would dissipate the blast fireball in a 360 degree pattern surrounding the structure and would necessitate personnel protection on all sides of the structure.

The other advantage of the umbrella design is that it provides for natural convection through the roof vent and prevents hydrogen gas accumulation.

The burn injury threshold distance for the umbrella design was estimated to be 28 feet around the entire building. This distance was significantly lower than the open side structure.

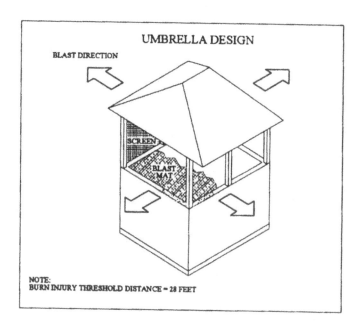

Disadvantages of the umbrella design were the additional cost for steel to support the elevated roof and additional construction time required to build the innovative roof. Consequently, this design was also not pursued.

Design Iteration 3: Searle/Monsanto Final Design

The final design was derived from the considerations evaluated in the previous design versions. The final design consisted of a four-sided building with one open, vertical wall and with the omission of the bunker. This design was the simplest in concept and the least in cost.

The sloped roof for natural ventilation was maintained in this design. Although the structure would be intrinsically self-ventilating, future, potential use of heavier-than-air solvents dictated the need for additional ventilation. Forced air ventilation was included at the back of the structure.

A burn injury threshold distance of 52 feet was calculated for the final design. It was concluded that a fenced, unoccupied, pie-piece-shaped blast zone, extending 70 feet out from the building would provide more that adequate personnel protection in the event of a worst case explosion scenario.

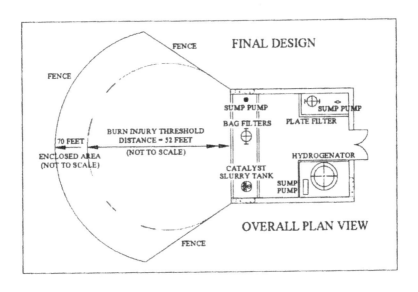

This final hydrogenation building design provided for a facility that was the safest and least expensive to build amongst the several designs that were evaluated. Land had to be isolated as a buffer zone to safely accommodate an unlikely potential explosion. This area would allow the fireball from the blast to dissipate it's energy load without personnel or equipment damage. Management accepted this proposal, and dedicated the additional land footprint for the buffer zone.

Conclusions

Three design iterations for a new hydrogenation facility were evaluated. All designs achieved the required safety, GMP, and environmental objectives. The final design was the simplest in concept and was consistent with the site plan. This design was the least expensive and its construction schedule was compatible with the product production plan. Unfortunately, because of unforeseen problems in the late stages of the drug development, the project was dropped and the facility was not built.

Acknowledgments
The authors greatly acknowledge Searle/Monsanto for permission to publish.

References

1. A. K. S. Murthy, *Chemical Engineering*, September, 1999

2. M. G. Scaros, P. K. Yonan and S. A. Laneman, *Heterogeneous Catalysis and Fine Chemicals IV*, **108**, H. U Blaser, A. Baiker and R. Prins (editors) 1997 Elsevier Science

3. Jim Laskonis, Lasweld Corporation, 1308 S. Crystal lake Rd., McHenry, IL 60050

4. Mike Grolmes, Consultant, Centaurus Technology Inc., 4590 Webb Road P. O. Box 598, Simpsonville, Kentucky

5. Andy Longinow, Project Mananger, Wiss, Janney, Elstner Associates, Inc., 330 Pfingsten Road, Northbrook, IL 60062-2095

6. Kim Mniszewskie, Project Consultant, Triodyne Fire and Explosion Engineers, Inc., 2907 Butterfield, Suite 120, Oak Brook, IL 60521

7. J. C. Bowen, *Ann. N. Y. Acad. Sci.*, **145**, p 169-177 (1967)

8. E. Lavagnino and J. Campbell, *Ann. N. Y. Acad. Sci.*, **214**, p 3-13 (1973)

9. D. W. Blackburn and H. E. Reiff, *Ann. N. Y. Acad. Sci.*, **145**, p 192-203 (1967)

10. M. A. Rebenstorf, *Ann. N. Y. Acad. Sci.*, **145**, p 178-191 (1967)

Effect of Carbon Activation on Deposition of Pd and Pt

Timothy J. McCarthy,[a] Baoshu Chen,[b] Misty L. Ernstberger,[b] and Francis P. Daly [c]

[a] Degussa-Hüls Corporation, P.O. Box 365
Piscataway, NJ 08855
[b] Degussa-Hüls Corporation, P.O. Box 649,
Calvert City, KY 42029
[c] Apyron Technologies, Inc., 4030-F Pleasantdale Rd,
Atlanta, GA 30340

Abstract

In this study precious metal powder catalysts were prepared on wood-based carbons that underwent two different activation processes: 1) with phosphoric acid and 2) with steam. Reduced and unreduced catalysts were prepared on these carbons using chloride and nitrate precious metal (Pt, Pd) salts. Mother liquors and washing solutions collected during the catalyst preparations were analyzed by inductively coupled plasma (ICP) to determine the amount of soluble precious metal (PM). Temperature-programmed desorption (TPD) was used to characterize the functional groups on the carbon surface. The results presented suggest that palladium chloride is only strongly adsorbed on the steam-activated carbon, but not on the phosphoric acid-activated carbon. For two unreduced catalysts leaching was observed with palladium chloride on the phosphoric acid-activated carbon support (0.277wt.% Pd loss), whereas, little or no leaching was observed on the steam-activated carbon (0.0051wt.% Pd loss). To a lesser extent the same effect was also observed with reduced-type catalysts. Changing the precious metal precursor from palladium chloride to palladium nitrate or platinum chloride reduced the PM loss during the catalyst preparation. Pretreatment of the phosphoric acid-activated carbon by water washing or thermal treatment in inert gas minimized the PM loss problem. A specific type of C-O functional group on the carbon surface is suspected to be responsible for PM loss during the catalyst preparation.

Introduction

Catalysts based on precious metals (PM) supported on activated carbons are frequently used in the fine chemical industry because such materials possess

63

unique properties compared to oxide-based supports such as silica or alumina. Activated carbons typically have high specific surface areas ($800\text{-}1200 \text{m}^2/\text{g}$), are stable in both acidic and basic media, and permit economic recovery of precious metals. A review of carbons as supports for industrial precious metal catalysts has recently been published (1). Carbon-supported precious metal catalysts are mainly used in liquid phase hydrogenation (2,3,4), dehydrogenation (5) or oxidation reactions in the fine chemical industry (6,7). Preparation of a carbon-supported precious metal catalyst requires detailed information on the surface chemistry of the support. Activated carbons contain variable amounts of oxygenated functional groups such as carboxylic and phenolic groups (8). The activation process in the manufacturing of the carbon determines the surface chemistry of the support that strongly influences the precious metal deposition reaction and subsequent metal dispersion and distribution. With Pd and Pt prices in excess of $300 per troy ounce and strict environmental limits in place, it is important that all of the PM be fixed onto the carbon support during the manufacturing process. Also, it is crucial that PM is stabilized on the carbon support to minimize PM leaching in the application.

In this study, catalysts were prepared on wood-based carbons that underwent two different activation processes: 1) chemical activation with phosphoric acid and 2) steam activation. Chemically activated carbons are manufactured by the simultaneous carbonization and activation of the raw material at high temperature ($600\text{-}800°C$) (9,10). The activating agent for the process, e.g. H_3PO_4, is incorporated into the raw material prior to heating. Steam activated carbons are manufactured from a pre-carbonized material, which is obtained by thermal decomposition of a carbonaceous precursor at $600\text{-}800°C$ in the absence or under controlled admission of air. The activation step is usually performed in the presence of steam and/or CO_2 at $800\text{-}1100°C$.

The objective of this study was to examine the affect of various catalyst preparation variables, including carbon surface chemistry, on PM leaching during the catalyst preparation process.

Experimental Section

Catalyst Preparation: Two activated carbons were used as supports in the current study. The phosphoric acid activated carbon is designated as carbon I and the steam activated carbon as carbon II. For each catalyst, a 50g sample was prepared using carbon slurry methods where the PM salts (H_2PdCl_4, $Pd(NO_3)_2$, or H_2PtCl_6) are adsorbed by the carbon followed by a pH adjustment to 8.5 at $70\text{-}90°C$ with a 20% solution of Na_2CO_3. The concentrations of PM metal in H_2PdCl_4, $Pd(NO_3)_2$, and H_2PtCl_6 solutions were 24% Pd, 9% Pd, and 29% Pt, respectively.

Samples were chemically reduced in the solution with different liquid reducing agents prior to filtration and drying. These samples are labeled "reduced." Table 1 lists several catalysts and their main variables. In most experiments carbon supports were used as received while the metal salt or preparation methodology was changed. In two experiments, however, carbon I was pre-washed with water at 90°C and dried under vacuum prior to the catalyst preparation (catalysts H and I). In another experiment carbon I was heat-treated at 990°C in nitrogen prior to the catalyst preparation (catalyst J).

Table 1 List of catalysts and their description.

Catalyst	Batch #	Variables
A	6MLE09	2.75wt.% Pd, carbon I, H_2PdCl_4 salt, reduced
B	6MLE10	2.75wt.% Pd, carbon II, H_2PdCl_4 salt, reduced
C	6MLE48	2.75wt.% Pd, carbon I, H_2PdCl_4 salt, unreduced
D	6MLE49	2.75wt.% Pd, carbon II, H_2PdCl_4 salt, unreduced
E	6MLE22	5wt.% Pt, carbon I, H_2PtCl_6 salt, reduced
F	7MLE01	5wt.% Pt, carbon I, H_2PtCl_6 salt, unreduced
G	6MLE46	2.75wt.% Pd, carbon I, $Pd(NO_3)_2$ salt, unreduced
H	7MLE46	2.75wt.% Pd, carbon I (pre-washed by water at 90°C), H_2PdCl_4 salt, reduced
I	7MLE47	2.75wt.% Pd, carbon I (pre-washed by water at 90°C), H_2PdCl_4 salt, unreduced
J	7MLE48	2.75wt.% Pd, carbon I (pre-heat-treated in N_2 at 990°C), H_2PdCl_4 salt, reduced

After the precious metal was deposited on the carbon, the slurry was filtered, and the solution was collected and labeled as "mother liquor." Then, the carbon cake was washed with a certain amount of DI water, and the solution was collected and labeled as "washings." Both "mother liquor" and "washings" were analyzed for precious metal content by inductively coupled plasma (ICP).

CO Chemisorption and Hydrogenation Activity Tests: Precious metal dispersions on the catalysts were measured by CO chemisorption at 25°C using an Altamira AMI-1 unit equipped with a thermal conductivity detector (TCD). All catalyst samples (50mg) were pre-reduced in situ at 300°C for one hour with H_2 prior to CO chemisorption. Cinnamic and crotonic acid hydrogenation rates were used to characterize the activity of Pd/C and Pt/C, respectively. A stirred glass reactor was used at 25°C and ambient pressure. The catalyst (200mg) was slurried with 40ml of ethanol and added to a H_2-flushed reactor containing an 80ml solution of cinnamic acid or crotonic acid in ethanol. Catalyst performance was determined by H_2 uptake measurements.

Temperature-programmed desorption: TPD experiments were conducted with a Zeton Altamira AMI-100 unit. The catalyst sample (100 – 200mg) was supported on a wad of quartz wool inside a 4 mm I.D. glass U-tube and purged at 30°C with helium. The sample temperature was raised from 30°C to 950°C at a rate of 30°C/min. Gas effluent from the reactor was analyzed with a quadrupole mass spectrometer.   Hydrogen, CO, and $10\%CO_2$/He gases were used to calibrate the mass spectrometer.

Results and Discussion

Carbon-supported precious metal (e.g. Pd or Pt) powder catalysts are widely used in the fine chemical industry. Catalytic performance is strongly dependent on the catalyst preparation methodology. It is crucial to obtain desired metal dispersion and distribution. Also, how well the precious metals deposit or bond to the carbon support is equally important, since it is directly related to catalyst stability, e.g. PM leaching during the preparation and in the application. Several factors can impact these catalyst characteristics, including carbon support, precious metal salts, and preparation methodology.

In this study ten catalyst samples were prepared by varying the carbon support, metal salts, pretreatment procedure for carbon, and the catalyst preparation method. The objective was to examine the effect of these variables on metal deposition while keeping the metal dispersion and distribution similar.   The focus was on the precious metal loss during the catalyst preparation.   Table 2 lists results for precious metal losses in both "mother liquor" and "washings" for these catalysts.

The dispersion for Pd and Pt were measured by CO chemisorption and were around 10% and 30%, respectively. The hydrogenation activities for Pd catalysts were determined with cinnamic acid hydrogenation, and were between 350 and 450ml H_2/g catalyst/min. Similarly, the hydrogenation activities for Pt catalysts were determined with crotonic acid hydrogenation, and were between 370 and 470ml H_2/g catalyst/min. Although these catalysts have similar dispersion and activity, the metal loss during the preparation is significantly different (Table 2). The effect of various preparation variables on PM loss is discussed in the following sections.

Table 2 Precious metal losses in mother liquor and washings for Pt/C and Pd/C catalyst preparations.

Catalyst	PM loss in mother liquor (wt.%)	PM loss in washings (wt.%)
A	0.013	0.49
B	<0.0053	<0.012
C	0.277	0.378
D	<0.0048	<0.0058
E	<0.0056	<0.0048
F	<0.0051	<0.0022
G	<0.0047	<0.0073
H	<0.0044	0.0275
I	0.0165	0.0832
J	Not detectable	Not detectable

Effect of Carbon Supports

By comparing catalyst A and B, it is clear that using carbon I as the support (catalyst A) resulted in significant metal losses in both "mother liquor" and "washings." The Pd losses in mother liquor and in washings were 0.013 and 0.49wt.%, respectively (Table 2). This catalyst was prepared with H_2PdCl_4 precursor, and was reduced chemically prior to filtration and washing. However, when carbon II was used (catalyst B) and other parameters were kept the same (metal salt, methodology) as for catalyst A, the metal loss was much lower in mother liquor and washing solution. Less than 0.0053 and 0.012wt.% of the input Pd metal was dissolved in mother liquor and washing solution, respectively (Table 2).

For unreduced catalysts, using carbon I as the support without the final reduction (catalyst C) resulted in even higher metal loss, and 0.277wt.% of input Pd metal was dissolved in the mother liquor (Table 2). During washing with DI water, a high concentration of Pd (0.378wt.% Pd loss) was also detected in the washings. When carbon II was used without reduction, no significant Pd loss was observed (catalyst D). Pd losses in the mother liquor and in washings were less than 0.0048 and 0.0058wt.%, respectively, (Table 2).

These results suggest that both recipes with or without metal reduction give similar metal losses. With H_2PdCl_4 as metal precursor, the use of carbon I resulted in much higher metal loss than carbon II. Note that carbon I was activated with H_3PO_4 and carbon II was activated with steam. The different activation methods may have caused the formation of different surface species, on which metal deposits and forms certain complexes or bonds. Cameron et al. (11) reported that the H_3PO_4-activated carbons had pronounced buffering capacity around a neutral pH, and this buffering capacity mainly came from the inorganic acidity/basicity, much of which can be removed by water or acid washing. Also, they indicated that additional buffering capacity was associated with oxygen groups. Thus, when H_2PdCl_4 is deposited on carbon I, it is possible that some complexes or bonds between the Pd compound and the surface species (or functional groups) on carbon are not stable and can be washed out or dissolved in water.

Effect of Metal Salts

To determine if varying metal salts can minimize the problem of metal loss, platinum chloride (H_2PtCl_6) instead of palladium chloride (H_2PdCl_4) was used to make catalysts on carbon I. Two recipes with or without a reduction step were followed to prepare 5wt.% Pt/carbon I catalysts (catalysts E and F). No significant Pt metal loss was observed for both mother liquors and washings (Table 2). Similar results were obtained when carbon II was used for the preparation of Pt/C catalysts (H_2PtCl_6 as precursor). No significant metal loss was observed in the mother liquor or washings. In other words, H_2PtCl_6 adsorbs strongly on both types of carbons (carbons I and II) while H_2PdCl_4 is only strongly adsorbed on steam activated carbons (carbon II).

When the $Pd(NO_3)_2$ precursor was used in place of H_2PdCl_4 to prepare 2.75%Pd/carbon I (catalyst G), a lower Pd loss was observed as reported for catalyst C. Less than 0.0047 and 0.0073 wt.% of added palladium was dissolved in mother liquor and washings, respectively (Table 2). Note that different metal salts have different redox potentials, and thus, can oxidize carbon to different levels (11). These results indicate that H_2PtCl_6 or $Pd(NO_3)_2$ can either change the carbon surface properties or bond to carbon more strongly than H_2PdCl_4.

Pretreatment of Carbon Support

Since using the steam-treated carbon (carbon II) did not result in a significant metal loss when palladium chloride was deposited on it (catalysts B and D), the pretreatment of carbon support is important for the stability of metals on the support. Two treatments were used on carbon I to minimize or even to eliminate the metal loss during the catalyst preparation. In one treatment, carbon I was washed with DI water at 90°C, followed by filtration and vacuum drying. Two catalysts were prepared using this pretreated carbon (catalysts H and I) with recipes identical to catalysts A and C, respectively. In another treatment, carbon I was heated to 990°C in flowing nitrogen. The same recipe as for catalyst A was used for the preparation of catalyst J.

After the hot-water treatment, Pd concentrations in both mother liquors and washings were reduced significantly during the preparation of catalysts H and I. By comparing catalyst A (untreated carbon I as the support) with catalyst H (water-washed carbon I as the support), Pd losses decreased from 0.013wt.% to less than 0.0044wt.% in the mother liquor, and from 0.49 to 0.0275wt.% in the washing solution (Table 2). Similar results were observed for unreduced catalysts (C and I). Pd loss decreased in the mother liquor from 0.227 to 0.0165wt.%, and in washings from 0.378 to 0.0832wt.%. Note that the H_3PO_4-activated carbon contains a significant amount of inorganic components, which can be removed by water washing (11). These results indicate that washing carbon I with hot water modified the carbon surface chemistry. Thus, the bond between Pd metal or Pd compounds and the carbon support becomes stronger.

When catalyst J was prepared using H_2PdCl_4 and carbon I, which was heat-treated at 990°C in flowing N_2, no detectable Pd loss was observed in either mother liquor or washing solution (Table 2). These results are expected since the high-temperature-treatment removes unstable surface species on carbon, such as oxygenated groups (12). The presence of some of these surface species is probably the cause of precious metal loss during the catalyst preparation.

Temperature-Programmed Desorption

To understand the effect of various treatments on the surface species of carbon, TPD was used to detect the evolution of H_2, CO, and CO_2 during the linear heat-up in inert gas. TPD experiments were carried out on carbons I and II and catalysts A through J.

TPD on carbon I (Figure 1) shows that two CO desorption peaks are present at temperatures of 700 and 900°C, respectively. A total of 1419μmol CO/g carbon was detected. At the similar temperature range between 700 and 1000°C, a large hydrogen desorption peak was observed with a peak

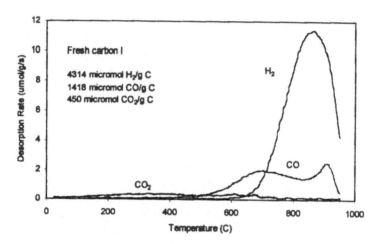

Figure 1 TPD on fresh carbon I.

temperature of 850°C. The total amount of H_2 is 4314µmol per gram carbon. This results in a H/CO molar ratio of 6. A small amount of CO_2 (450µmol/g) desorbed in a broad temperature range. The hydrogen desorption peak may be related to strongly adsorbed hydrogen on a variety of different reaction sites (13). The CO_2 peak might be from surface acidic complexes like carboxylic acids, anhydrides, and lactones, and similarly, the CO peaks may be from the decomposition of phenolic and quinonic groups (14). Two types of CO-yielding complexes may be present on the carbon I surface since two CO desorption peaks were observed. They are thermally stable and do not decompose until the temperature goes above 500°C. These surface groups may facilitate anchoring of the metal ions.

For comparison, TPD on carbon II showed only one CO desorption peak with a peak temperature at about 900°C (Figure 2). A significantly smaller amount of CO (590µmol/g carbon) was desorbed. The hydrogen desorption peak did not start until the temperature was above 850°C. A similar amount of CO_2 was evolved during TPD at the lower temperature (lower than 700°C). These results indicate that one type of surface species leads to the CO and H_2 desorption at temperatures above 600°C. The quantity of removable surface species on carbon II is significantly less than on carbon I.

TPD spectra on catalysts A and C are almost identical to that on fresh carbon I. Two CO desorption peaks were present, and the amount of CO desorbed was similar to that on carbon I alone. For catalysts B and D, which were prepared using carbon II, TPD spectra are similar to that on pure carbon II in terms of one CO desorption peak and the amount of CO desorbed. These results are consistent with PM loss data during the catalyst preparation using the two carbons. PM loss was much higher when carbon I was used as the support (catalysts A and C) than that using carbon II as the support (catalysts B and D), as shown in Table 2. Thus, one can hypothesize that the higher the amount of thermally removable surface species, the more PM loss may be expected.

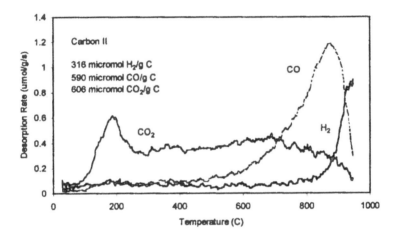

Figure 2 TPD on fresh carbon II.

This hypothesis can be supported by considering the experimental data on catalyst J, which was prepared using the heat-treated carbon I. After carbon I was treated at 990°C in flowing nitrogen, all removable surface species on the carbon desorbed. TPD on this heat-treated carbon did not show significant CO, CO_2, or H_2 peaks. It was reported that oxygenated groups on activated carbon were eliminated at temperatures higher than 700°C (12), and such thermal treatment created basic groups (pyrone-type basic groups) (15). When this carbon was used as the support for the catalyst preparation (catalyst J), no detectable Pd loss was observed in either mother liquor or washing solution (Table 2).

It is possible that the surface species of carbon I are modified when different metal salts are used. Note that PM losses are reduced significantly during the preparation of catalysts E, F, and G when H_2PtCl_6 and $Pd(NO_3)_2$

were used as precursors to replace H_2PdCl_4 (Table 2). Note that these metal salts have different oxidation potentials (10). Interestingly, TPD spectra on catalysts E, F, and G only showed one CO peak with a peak temperature at 700°C. The second CO desorption peak (at about 900°C), which was observed on carbon I alone and catalysts A and C, disappeared. For example, TPD spectra on catalyst F showed only one CO desorption peak between 500 and 900°C with a peak temperature at 700°C (Figure 3). A total amount of 1795μmol CO/g catalyst was obtained.

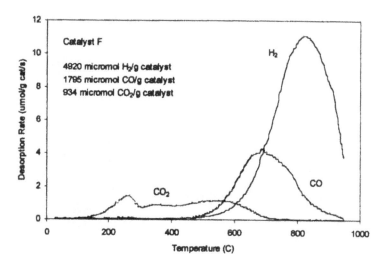

Figure 3 TPD on the catalyst F (5 wt.%Pt/carbon I, unreduced).

These results indicate that a specific type of surface species is responsible for Pd loss during the preparation of catalysts A and C. When H_2PdCl_4 was used as a precursor, some type of complex or bond between Pd compounds and this species was formed. Apparently, this complex or bond is not stable in the mother liquor, and is easily dissolved in an aqueous solution. Thus, Pd loss resulted during filtration and washing. The use of metal salts other than H_2PdCl_4 either removed or modified this type of surface species due to their different oxidation potentials, and thus less PM loss was observed.

When carbon I was pre-washed with DI water at 90°C, the surface species that might cause the Pd loss was partially removed. As shown in Figure 4, TPD on washed carbon I produced 1255μmol CO/g carbon, which is less than that on unwashed (fresh) carbon I (1419μmol CO/g carbon). Most interestingly, the CO desorption peak at 900°C is significantly less when compared to the TPD spectra on fresh carbon I (Figure 1). The reduction in the amount of CO desorbed is primarily due to the loss of the second CO desorption peak.

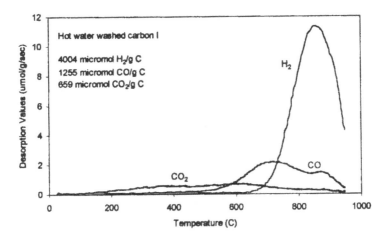

Figure 4 TPD on water-washed carbon I.

Similar TPD spectra were obtained on catalysts H and I, which were prepared using washed carbon I. The second CO desorption peak (900°C) is much smaller than that on catalysts A and C, which were prepared using fresh carbon I. Note that Pd loss for catalysts H and I are significantly lower than those for catalysts A and C. These results further support the premise that one type of species (out of two types) on the carbon surface is responsible for PM loss during the catalyst preparation.

Conclusions

Carbon-supported precious metal powder catalysts are widely used in the fine chemical industry. Stabilizing precious metals on carbon supports is important for both the catalyst manufacturer and catalyst user. Our studies have shown that precious metal loss during catalyst preparation is significantly impacted by moieties on the carbon surface. Removal of such moieties can be achieved by using either a different activation procedure (changing carbon support), varying the PM precursor or employing a carbon pretreatment step, e.g., water washing and thermal treatment.

References

1. E. Auer, A. Freund, J. Pietsch and T. Tacke, *Appl. Catal.* **173**, 259 (1998).
2. M. Freifelder, In *Practical Catalytic Hydrogenation: Techniques and Applications*, John Wiley & Sons, Inc., New York, 1971.
3. R. N. Rylander, In *Hydrogenation Methods*, Academic Press, London, 1985.
4. R L. Augustine, *Catal. Today* **37**, 419 (1997).
5. J.A.A. van den Tillaart, B.F.M. Kuster and G.B. Marin, *Appl. Catal.* **120**, 127 (1994).
6. B.M. Despevroux, K. Deller and E. Peldszus, In *New Developments in Selective Oxidation*, G. Centi and F. Trifiró, eds., Elsevier Science Publishers B.V., Amsterdam, 1990, pp. 159-168.
7. C. Brönnimann, Z. Bodnar, P. Hug, T. Mallat and A. Baiker, *J. Catal.* **150**, 199 (1994).
8. K. Kinoshita, In *Carbon: Electrochemical and Physicochemical Properties*, Wiley, New York, 1988.
9. R.C. Bansal, J.-B. Donnet and F. Stoeckli, In *Active Carbon*, Marcel Dekker, New York, 1988.
10. F. Rodríquez-Reinoso, M. Molina-Sabio, *Carbon*, **30**, 1111 (1992).
11. D.S. Cameron, S.J. Cooper, I.L. Dodgson, B. Harrison, and J.W. Jenkins, *Catal. Today*, **7**, 113 (1990).
12. C. Pinel, E. Landrivon, H. Lini, and P. Gallezot, *J. Catal.*, **182**, 515 (1999).
13. B.R. Puri, R.C. Bansal, *Carbon*, **1**, 451 (1964).
14. A. Dandekar, R.T.K Baker, and M.A. Vannice, Carbon, **36**, 1821 (1998).
15. R. Schlögl, In *Handbook of Heterogeneous Catalysis*, (G. Ertl, H. Knözinger, J. Weitkamp, Eds.), Wiley-VCH, Weinheim, 1997. Vol. 1.

The Flexibility of the Chemical and Physical Properties of Metalyst™ for the Optimal Performance of Fixed Bed Reactions on Activated Base Metal Surfaces

Daniel J. Ostgard, Monika Berweiler, Stefan Röder, Konrad Möbus, Andreas Freund, and Peter Panster

Degussa-Huels AG, Silicas and Chemical Catalysts Division
Rodenbacher Chausee 4
P.O. Box 1345
D-63403 Hanau (Wolfgang), Germany

Abstract

Although many applications use activated base metal catalysts (ABMC) as powders in slurry reactors, the use of fixed bed ABMC is gaining wider commercial interest and acceptance. The fixed bed approach allows for a continuous reaction that can be readily controlled by the reaction conditions as well as the physical and chemical properties of the catalyst. Degussa-Huels's Metalyst™ delivers the advantages of an ABMC while offering the flexibility one needs to take full advantage of their fixed bed catalytic system. This paper describes how the catalytic performance of Metalyst™ for the hydrogenation of nitrobenzene, glucose, acetone, and furfural is effected by parameters such as its activation depth, the structure of the preactivated alloy, its chemical composition, and bulk density.

Introduction

The Metalyst™ product line is the result of Degussa-Huels's fixed bed ABMC patented (1,2) technology. Metalyst™ has the distinct advantage of providing increased pore volume and catalytic activity without sacrificing mechanical stability when compared to activated granules of base metals alloyed with aluminum (1). Metalyst™ alpha, Metalyst™ beta, and Metalyst™ gamma are the codes used for the activated Ni, Co, and Cu versions respectively. The chemical properties of Metalyst™ such as the presence of promoters enhance certain reactions by modifying its structure, the adsorption of the substrate on the activated surface, and hydrogen content. Additionally, the use of rapidly cooled alloys (RCA) for the production of Metalyst™ has improved its porosity and decreased its bulk density, thereby making it even more attractive for the hydrogenation of unsaturated organic compounds. This paper discusses in detail the effects of various compositions and structures of Metalyst™ on key reactions such as the hydrogenation of nitro- and carbonyl moieties.

Experimental Section

The specific interface density (S_V) describes the fineness of the phase structure of an alloy and is defined by the following equation (2):

$$S_v = \frac{4}{\pi} \bullet \frac{(\text{Perimeter of the phase})}{(\text{Area of the phase})} \quad [\mu m^{-1}]$$

The larger the S_V, the smaller the corresponding phase domains. The S_V was determined by preparing a transverse section from a granule of the catalyst alloy and examining it under a microscope where the different phases appear as different shades of gray. The structure of the alloy was automatically analyzed with a PC-supported image analysis system that differentiated between the various shades of gray and performed the above-mentioned calculation. The Ni/Al alloys used here were analyzed by energy dispersive X-ray analysis and found to have Al_3Ni_2, Al_3Ni, and $Al-Al_3Ni$ eutectic phases.

In accordance with the literature (1,2), the preparation of these catalysts started with a homogeneous mixture of the desired alloy and shaping aids that was tableted, calcined, activated in caustic solution, washed, and stored under water. Since Metalyst™ is preactivated, activation in the reactor isn't necessary. The crush strength of Metalyst™ tablets is approximately 300 N, thus eliminating catalyst attrition during reactor loading and the reaction itself.

The amount of activated metal was determined using temperature programmed oxidation (TPO). In the case of nickel, each activated nickel atom can take up one oxygen atom as a result of oxidation. To perform TPO, about 5 to 10 grams of the water-moist activated catalyst was dried in a stream of nitrogen flowing 10 l/h at 120°C for a period of 17 hours. The furnace was then carefully cooled to 20°C. After reaching a constant reactor temperature, the pure nitrogen was switched to a 4% oxygen in nitrogen mixture passing over the catalyst at the rate of 10 l/h while the temperature was ramped at the rate of 6°C/min to the end temperature of about 640°C. The oxygen content was measured during the experiment with a "Oxynos 100" paramagnetic detector and the consumed amount of oxygen was determined from the area of the curve.

The hydrogenation of nitrobenzene was carried out in a 0.5 liter autoclave stirred at 1000 rpm while containing 200 grams of a 50 wt.% ethanolic nitrobenzene solution, 10 grams of 4 mm catalyst tablets placed in a basket located at the optimum mixing zone of the reactor, and 40 bar of hydrogen pressure while ramping the temperature to 150°C. During the experiment, the initial reaction temperature and the rate of hydrogen uptake were determined.

The rate of acetone hydrogenation was determined in the trickle phase with a 120-ml tubular reactor using 20 ml of 3-mm catalyst tablets. This hydrogenation was performed at 5 bar hydrogen pressure, 75°C, a hydrogen-to-acetone ratio of 5-to-1, and at LHSV ranging from 1.0 to 4.0 h^{-1}. Acetone conversion and selectivity was determined by GC.

On the lab scale, glucose hydrogenations were carried out at 130°C over 5 hours in a 0.5 liter autoclave stirred at 1000 rpm while containing 330 grams of a 25 wt.% aqueous glucose solution, 10 grams of 4 mm catalyst tablets placed in a basket located in the optimum mixing zone of the reactor, and 200 bar of hydrogen pressure. After the experiment, glucose conversion and sorbitol selectivity were determined by HPLC. On a pilot scale, the hydrogenation of glucose was carried out over 40 l of 4-mm catalyst tablets in a fixed bed reactor at 200 bar and temperatures ranging from 70 to 120°C with from 20 to 40% aqueous glucose solutions. The LHSV for these pilot tests was 0.125 h^{-1}.

Furfural hydrogenations were carried out over 40 ml of 3 mm catalyst tablets in a fixed bed reactor at 3 bar, 170°C, a hydrogen-to-furfural ratio of 5-to-1, and a LHSV of 0.5 h^{-1} for the duration of 4 hours.

Results and Discussion

From Figures 1 and 2, one can clearly see that the irregularly shaped particles of the rapidly cooled alloy (RCA) used here consist of much smaller phase domains in comparison to the particles of the slowly cooled alloy (SCA). On a micro-scale, the finer dendritic structures of these smaller Ni/Al phases give a catalyst with increased surface area and meso/micro-porosity after activation with caustic solutions. During the calcination of Metalyst™ alpha (Ni) precursor tablets, some of the smaller phase domains sinter slightly to produce strong metallic bridges between the alloy particles, thereby increasing the catalyst's mechanical strength and allowing for the optimization or elimination of binders that may have a detrimental effect on catalyst activity. On a macro-scale, the irregular shaped particles can be pressed together to form a tablet with higher macro-porosity and consequently a lower bulk density.

Figures 3 and 4 show that Metalyst™ consists of shell activated tablets where the aluminum content is depleted in the active catalyst layer during caustic activation. Table 1 displays the differences of the precursor Metalyst™ alpha tablets made out of either a RCA or a SCA before and after the identical

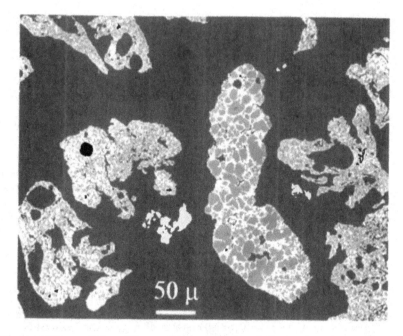

Figure 1 The particles of the rapidly cooled alloy (RCA). Magnification 200:1.

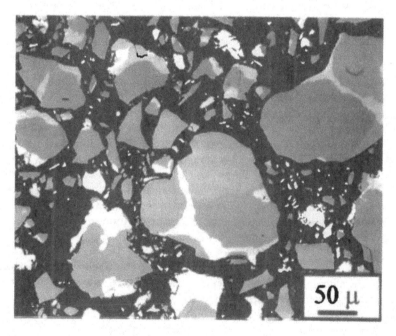

Figure 2 The particles of the slowly cooled alloy (SCA). Magnification 200:1.

Figure 3 The distribution of Ni before and after the activation of Metalyst^{TM}.

Figure 4 The distribution of Al before and after the activation of Metalyst^{TM}.

activation procedure. In comparison to the SCA, the higher porosity of the RCA allowed for a deeper activation profile and consequently the formation of more activated metal as determined by TPO. This resulted in a more active catalyst that starts to hydrogenate nitrobenzene at 16°C lower than the SCA analog. Figure 5 displays the effect of activation depth (as monitored by TPO) and the presence of promoters on the hydrogenation of nitrobenzene performed in an autoclave. There is a strong correlation between the catalyst's nitrobenzene activity and its TPO value. Interestingly, Metalyst^{TM} alpha tablets made from either the RCA or the SCA fit on the same curve where the SCA covers the lower TPO values and the deeper activated RCA spans the higher TPO oxygen uptakes. The slope of this curve tends to decrease as the TPO value increases indicating that the activity per activated Ni atom for the RCA is slightly lower than that of the SCA. Part of this decreased activity is due to increased mass transport limitations in the deeper activated tablets, and the other part is due to

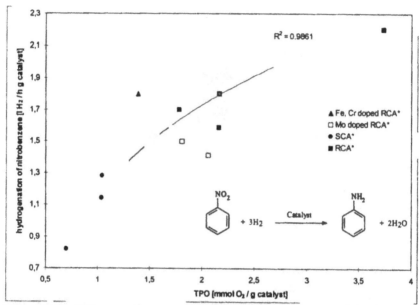

Figure 5 The influence of promoters on the correlation between TPO values and nitrobenzene hydrogenation activity as performed in an autoclave (SCA = slowly cooled alloy and RCA = rapidly cooled alloy).

Table 1 The Effects of the Alloy on the Properties and Performance of the Activated Metalyst™ Tablets.

Alloy Properties			Properties of the Resulting Metalyst™			Nitrobenzene Data	
% Ni/Al	Cooling Rate	S_V	Density g/ml	TPO mmol O_2/g	Activation Depth, mm	Initial °C	Activity ml H_2/min·g
53/47	Slow	0.08	2.23	1.04	0.20	131	1.14
50/50	Fast	1.60	1.65	2.16	0.47	115	1.59

the higher residual Al content in the activated RCA when activated by the procedure used here. It has recently been found that alloys with higher S_V values for the Al_2Ni_3 phase activate under mild conditions to higher residual Al contents and this slightly decreases the activity per activated metal atom (3). Therefore, the greatest advantage of using a RCA over a SCA comes not so much from the type of activated metal sites, but from increasing the amount of them without degrading the catalyst's stability. Promoters also change the activity per activated Ni atom. In the case of molybdenum, the surface is partial-

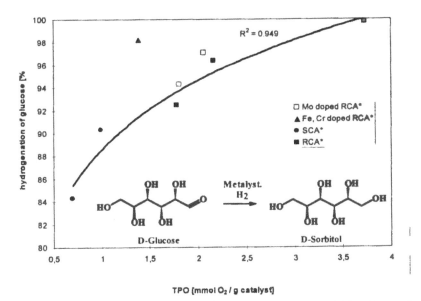

Figure 6 The influence of promoters on the correlation between TPO values and glucose hydrogenation activity as performed in an autoclave (SCA = slowly cooled alloy and RCA = rapidly cooled alloy).

-ly blocked and this Mo decorated activated Ni is less active for nitrobenzene hydrogenation. Doping the catalyst with chromium and iron increases its nitrobenzene activity, which suggests the development of improved sites where factors such as adsorption, desorption, increased catalyst surface area, and the availability of hydrogen could have been improved.

In agreement with the nitrobenzene results, the catalysts with higher TPO values were also more active for glucose hydrogenation as measured in an autoclave (see Figure 6). The slope of this curve also decreases with increasing TPO values as was discussed for nitrobenzene hydrogenation. Therefore, increased mass transport limitations and residual aluminum content also play their respective roles in glucose hydrogenation. As could be predicted, the glucose activity per TPO site of the chromium and iron doped Metalyst™ alpha (Ni) was much higher than those measured for the undoped catalysts. It has been shown previously that the use of Cr/Fe doping increases the activity of activated nickel catalysts (4,5,6) by improving factors such as surface area and coordination of the substrate to the catalytic surface. Unlike nitrobenzene, promoting Metalyst™ with molybdenum improved its glucose hydrogenation ability indicating that molybdenum may play a special coordination role in the adsorption of glucose. The positive influence of molybdenum in the hydroge-

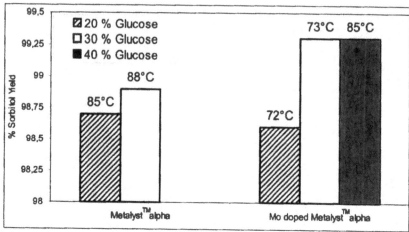

Figure 7 The influence of doping Metalyst™ alpha (Ni) with Mo on the reaction temperature and yield of sorbitol during the hydrogenation of glucose in a 40 liter fixed bed reactor.

nation of sugars was also noticed earlier for powdered activated nickel catalysts (6,7). While the glucose activities of these catalysts varied with their TPO values and promotion (see Figure 6), their sorbitol selectivities remained relatively constant with values of 98.5% or higher.

To further study the effect of Mo doping on the performance of Metalyst™ alpha (Ni) for the hydrogenation of glucose, pilot reaction studies were performed on 40 l of catalyst with a commercial glucose feed. Figure 7 clearly shows that the Mo doped version reaches roughly equivalent or higher sorbitol yields at substantially lower temperatures for each glucose concentration in comparison to the undoped catalyst. Thus, the use of Mo doping lowers the activation energy of glucose hydrogenation by changing its reaction mechanism. According to Gallezot et al. (6), Mo exists in a low-valent state on the activated Ni surface and acts as a Lewis adsorption site for the oxygen atom of the carbonyl group that then becomes more readily hydrogenated.

Acetone hydrogenations to isopropyl alcohol (IPA) were performed over Metalyst™ beta (Co), gamma (Cu), and versions of alpha (Ni) as seen in Figures 8, 9, and 10 and Table 2. Table 2 shows that the catalysts used here have a broad range of TPO values where the Co catalyst has the lowest value followed by the Ni catalyst made out of a SCA. All of the Ni catalysts made out of RCA, regardless of doping, have roughly the same TPO values, thus their comparisons could be made without having to consider large differences in their activation levels. The catalyst with the highest TPO level is the Metalyst™

Table 2 The Properties of the Metalyst™ Catalysts Used in the Fixed Bed Study of Acetone Hydrogenation.

Catalyst	Metalyst™ Type	Active Metal	Promoters	TPO Value mmol O_2/g
A^1	alpha	Ni	none	1.693
B^2	alpha	Ni	none	0.854
C^1	alpha	Ni	Mo	1.673
D^1	alpha	Ni	Cr/Fe	1.773
E	beta	Co	none	0.555
F	gamma	Cu	none	2.131

[1]These catalysts were made from a RCA.
[2]This catalyst was made from a SCA

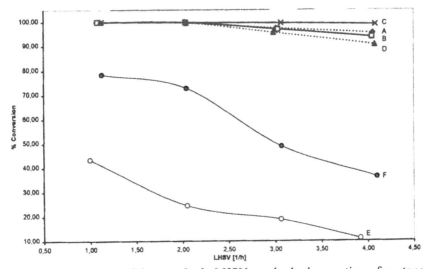

Figure 8 The effect of the reaction's LHSV on the hydrogenation of acetone over 20 ml of the catalysts described in Table 2.

gamma made out of Cu indicating that it is the deepest activated. Figure 8 shows the effect of the reaction's LHSV on acetone hydrogenation over the same volume of the various catalysts. On a per volume basis, the Mo doped Ni catalyst out performs the others followed by the sequence of the undoped Ni catalyst made out of a RCA, the undoped Ni catalyst made out of a SCA, and finally the Cr/Fe doped Ni catalyst. All of the Ni based catalysts had conversions higher than 90%, even at a LHSV of 4.0 h^{-1}. The hydrogenation of acetone over the Ni based catalysts were heavily diffusion limited at LHSV of 1 and 2, and it wasn't until the LHSV of 3 was reached that the mass-transfer-

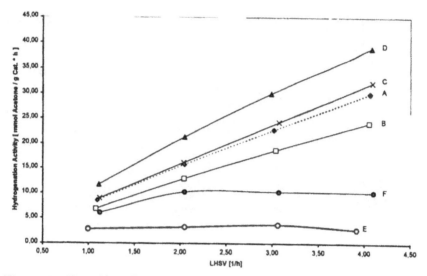

Figure 9 The effect of the reaction's LHSV on the millimoles of acetone hydrogenated per gram of catalyst per hour for the catalysts of Table 2.

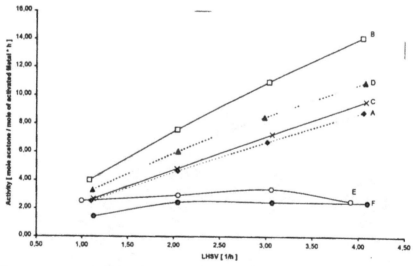

Figure 10 The effect of the reaction's LHSV on the moles of acetone hydrogenated per mole of activated metal per hour for the catalysts of Table 2.

limited-component started to decline. On a volumetric basis, the Co catalyst had the lowest activity and the Cu catalyst had conversions between those of the Ni catalysts and that of Co. Figure 9 displays this data when it is calculated on a per weight basis, and as expected, the performance rating changes due to the different bulk densities. On a per weight basis, the Cr/Fe doped Ni catalyst performs the best followed in order by the Mo doped Ni, the Ni catalyst made out of a RCA, and its analog made out of a SCA. The performance of the Cu catalyst, on a per weight basis, still lies between that of the Ni ones and the poorly performing Co. It is interesting to note that the activities per gram of catalyst for the Co and Cu catalysts are relatively constant after LHSV values higher than 2.0 indicating that the reaction at the activated Cu or Co surface is the limiting factor and the increased availability of reactants does not enhance the hydrogenation rate any further. When calculating these activities on a per mole of activated metal basis, as measured by TPO (see Figure 10), the performance rating changes again where the Ni catalyst made out of the SCA performs the best followed in order by the Cr/Fe doped Ni catalyst, the Mo doped Ni catalyst, and the straight Ni catalyst made out of the RCA. This activity per mole of activated metal follows that of the other studies previously mentioned due to the higher residual Al content of the activated RCA (3). Additionally, the increased mass transport limitations for the deeper activated RCA should also decrease the activity per mole of activated metal. On a per mole of activated metal basis, the Co catalyst was found to be slightly more active than Cu, and this is in agreement with the hydrogen content of these activated metals where the rating was found to be Ni >>> Co > Cu. Hence, the overall activity of the undoped catalysts per gram of active metal appears to be dependent on the hydrogen content of that metal, and the hydrogen contents of activated Co and Cu catalysts are so low that they can no longer increase the rate of acetone hydrogenation above the LHSV of 2.0.

The IPA selectivities of all of the Ni based catalysts were greater than 99.9%, while Cu displayed an IPA selectivity of 99.0% and that for Co dropped as low as 98.8%. As seen in Figure 11, side-reactions during the hydrogenation of acetone can produce diisopropyl ether (DIPE) or condensation products such as 4-hydroxy-4-methyl-2-pentanone. The later can lose water to form 4-methyl-3-pentene-2-one which can be hydrogenated to 4-methyl-2-pentanone and further hydrogenated to 4-methyl-2-pentanol. The 4-hydroxy-4-methyl-2-pentanone can also be hydrogenated to 2,4-dihydroxy-2-methylpentane, however this was not detected in any of our reaction mixtures. The side reactions on Co were found to form only the 4-hydroxy-4-methyl-2-pentanone, while the side reactions over Cu formed roughly an equal mixture of the dehydrated and deeper hydrogenated condensation products 4-methyl-2-pentanone and 4-methyl-2-pentanol. Considering the higher activity per active metal atom and the higher hydrogen content of activated Co, one could have expected this to be different. Obviously the adsorption of 4-hydroxy-4-methyl-2-pentanone is stronger on Cu than Co allowing it to react further in spite of the lower hydrogen content of Cu.

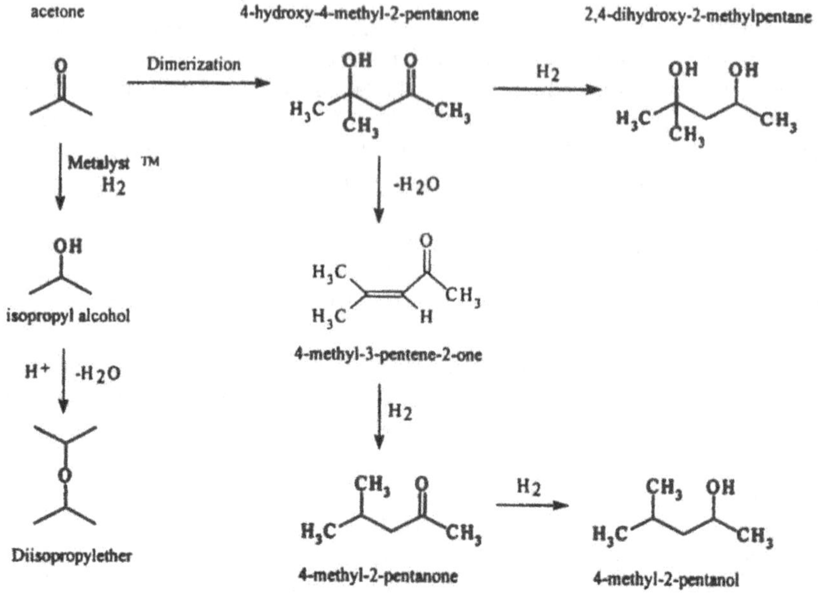

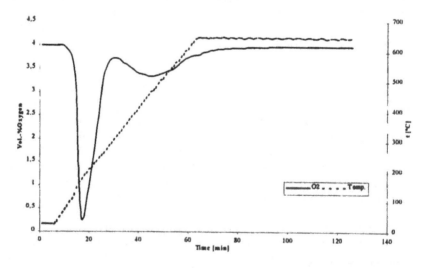

acetone
4-hydroxy-4-methyl-2-pentanone
2,4-dihydroxy-2-methylpentane
Dimerization
H2
Metalyst TM
H2
-H2O
isopropyl alcohol
4-methyl-3-pentene-2-one
H+
-H2O
H2
Diisopropylether
H2
4-methyl-2-pentanone
4-methyl-2-pentanol

Figure 11 The hydrogenation of acetone and its possible side products.

Time [min]
O2
Temp.

Figure 12 The TPO characterizaton of the activated Ni catalyst A in Table 2.

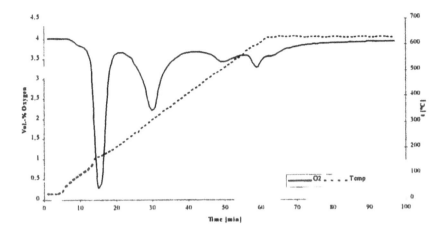

Figure 13 The TPO characterizaton of the activated Cu catalyst F in Table 2.

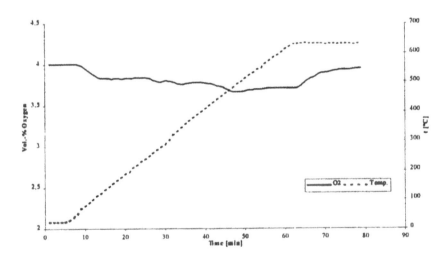

Figure 14 The TPO characterizaton of the activated Co catalyst E in Table 2.

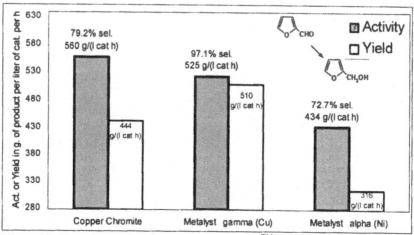

Figure 15 The enhanced performance of Metalyst^{TM} gamma (Cu) for the selective hydrogenation of furfural in comparison to Metalyst^{TM} alpha (Ni) and copper chromite.

Upon inspection of the TPO curves for activated Ni (Figure 12), Cu (Figure 13), and Co (Figure 14), one sees that the oxygen consumption of the Ni TPO curve has the largest low temperature contribution of all the metals displayed here. The Cu TPO curve displays slightly more contribution from various well-defined high temperature peaks than that of Ni, and the Co TPO curve shows a continuous oxidation starting from 65°C and going to ~640°C where the majority of this occurs at higher temperatures. A much closer analysis of the TPO curves than that given in Figures 13 and 14 showed that at the temperature used here for acetone hydrogenation (75°C), the interaction of Cu to oxygen is stronger than that for Co. The ability of Cu to interact strongly with oxygen, and most probably oxygen containing moieties (e.g., 4-hydroxyl-4-methyl-2-pentanone), combined with Cu's low hydrogen content make it ideal for the selective hydrogenation of olefinic carbonyl compounds to their respective olefinic alcohols. Figure 15 brings this point across by displaying the increased yield of the unsaturated alcohol from the hydrogenation of furfural over Metalyst^{TM} gamma (Cu) in comparison to Metalyst^{TM} alpha (Ni) and a commercially available copper chromite catalyst.

Conclusions

Metalyst^{TM} catalysts made from RCA were found to have lower bulk densities and increased pore volumes in comparison to those made from the SCA. The activity per mole of activated Ni was found to be slightly better for the SCA than the RCA when activated by the procedure used here. However, the increased amount of activated metal on both a weight and volume basis for the Metalyst™ tablets made out of the RCA, in comparison to those made out of SCA, was more than enough to overcome its slightly lower activity per mole of activated metal. Thus it is the increased amount of activated metal (without the loss of mechanical strength) and not the type of activated metal site that enhances Metalyst™ made out of RCA. As expected, the addition of promoters changed the catalyst's activity per activated metal atom and the Cr/Fe combination was found to be the best overall promoter system. The addition of Mo had a detrimental effect on the hydrogenation of nitrobenzene, while it was found to improve the hydrogenation of both glucose and acetone, thereby demonstrating the coordination-effect this Mo species has on carbonyl compounds. The hydrogen content rating of Ni >>> Co > Cu mirrored that of their acetone hydrogenation activities when normalized per mole of activated metal, thus showing the importance of catalyst hydrogen content for this reaction. Although the combination of low hydrogen content and strong interactions with oxygen containing compounds hurts the selectivity of acetone hydrogenation on Metalyst™ gamma (Cu), it was found to be advantageous for the hydrogenation of furfural to the corresponding unsaturated alcohol over the same catalyst.

Acknowledgements

We thank Degussa-Huels for permission to publish this work.

References

1. P. Schuetz, R. Burmeister, B. Despeyroux, H. Moesinger, H. Krause, and K. Deller, U.S. Patent 5,536,694, to Degussa-Huels (1996).

2. A. Freund, M. Berweiler, B. Bender, and B. Kempf, German Patent DE 19721898 A1, to Degussa-Huels (1998).

3. S. Knies, M. Berweiler, P. Panster, H.E. Exner, and D.J. Ostgard, *accepted for publication in "Studies of Surface Science and Catalysis" along with the other presentations from the 12th International Congress of Catalysis in Granada, Spain.*

4. S. Montgomery, Chem Ind. (Marcel Dekker), 5, (*Catal. Org React.*), 383 (1981).

5. S.N. Thomas-Pryor, T.A. Manz, Z. Liu, T.A. Koch, S.K. Sengupta, and W.N. Delgass, Chem Ind. (Marcel Dekker), 17, (*Catal. Org React.*), 195 (1998).

6. P. Gallezot, P.J. Cerino, B. Blanc, G. Fléche, and P. Fuentes, *J. Catal.*, 146, 93 (1994).

7. R. Albert, A. Strätz, and G. Vollheim, Chem Ind. (Marcel Dekker), 5, (*Catal. Org React.*), 421 (1981).

Advanced Fixed-Bed Hydrogenation of Unsaturated Fatty Acids

T. Tacke[a], I. Beul[b], C. Rehren[b], P. Panster[b], and H. Buchold[c]

[a]Degussa-Hüls Corp., Applied Research and Development Chemical Catalysts, P. O. Box 649, Calvert City, KY 42029, USA
[b]Degussa-Hüls AG, Research and Applied Technology Chemical Catalysts and Zeolites, P. O. Box 1345, 63403 Hanau, Germany
[c]Lurgi Öl Gas Chemie GmbH, Lurgiallee5, 60295 Frankfurt a. M., Germany

Abstract

Activated carbon supported palladium catalysts are very attractive for the continuous fixed-bed hydrogenation of fatty acids. The development process which has been carried out has resulted in the preparation of a very active and selective catalyst with a long lifetime. Important parameters for the development of a superior catalyst are the selection of support material and catalyst reduction method. In addition to the catalyst development, various reaction conditions were investigated in order to develop the concept of a commercial process. A particular activated carbon supported palladium fixed-bed catalyst retains a high activity after more than 4300 hours on stream in a long-term test run which was carried out in order to simulate a multi-step continuous reactor cascade. The new fixed-bed process offers advantages over the currently applied discontinuous stirred-tank reactor process with supported nickel catalysts. These advantages include i) no nickel contamination of the product, ii) significantly reduced catalyst poisoning by sulphur, iii) no need for the separation of catalyst powders, and iv) no catalyst sensitivity against water. A more than 20 times higher catalyst productivity has been determined for our fixed-bed process compared to the currently applied discontinuous process. Also, a comparison of both processes revealed economic advantage for our continuous fixed-bed process.

Introduction

Free fatty acids can be produced by i) oxidation of paraffins, ii) oxo synthesis, iii) hydrolysis of nitriles, or iv) splitting of fats and oils. The pressure splitting of fats and oils is very important due to the use of renewable raw materials. Of the more than 100 million metric tons of fats and oils produced worldwide,

91

approximately 15% is used for the manufacture of free fatty acids and their derivatives (1). Various processing techniques are applied for the production of fatty acids and their derivatives: distillation, fractionation, catalytic hydrogenation, spray cooling, and esterification.

The state-of-the-art-technology for the hydrogenation of fats and oils, fatty acid esters, and free fatty acids is characterized by discontinuous operation and the use of a powdered nickel on silica or nickel on kieselguhr catalyst. Since the melting point of the hydrogenated products increases, the hydrogenation is also expressed as hardening. In 1997 approx. 1.2 million metric tons of free fatty acids were hydrogenated to form completely hydrogenated free fatty acids for oleochemical applications (2). The current commercial batch fatty acid hardening process with nickel on silica or nickel on kieselguhr as a catalyst has some disadvantages: discontinuous operation, low space-time-yields, and high variable costs (e. g. man-power, energy, filtration). The nickel on kieselguhr catalyst also causes some problems in that the catalyst is deactivated through formation of nickel soaps (3), the costs of product purification are high (distillation of free fatty acids), and disposal of nickel residues is becoming more and more difficult. In addition, it is necessary to remove the residual moisture under elevated temperature and vacuum (4.0 – 5.5 KPa, 0.58 – 0.80 psig), since water would damage the reduced nickel catalyst (formation of nickel hydroxide).

Due to the limitations of batch-processing and the disadvantages of the nickel catalyst, various attempts have been started in the past to overcome the problems. Over the last twenty years various companies (Ruhrchemie (4), Henkel/Degussa (5), Henkel (6, 7), J. Brown/Davy Process Technology, and Lurgi/Degussa(8)) were involved in the development of continuous processes operating with palladium based fixed-bed catalysts. In the mid 90s Henkel's continuous fixed-bed process came on stream in Düsseldorf-Holthausen, Germany. Activated carbon or titania based palladium catalysts are highly active for the hardening of free fatty acids and are stable under reaction conditions even in the presence of water impurities. However, the catalyst/process performance and, in particular, catalyst life-time still needs to be improved for economic reasons.

Experimental

A. Catalyst preparation and screening

Various activated carbon supported palladium catalysts were prepared based on different methods and recipes. In particular, the following parameters have been investigated in detail: various activated carbon supports, precious metal precursors, precious metal impregnation step (e.g., degree of pore volume impregnation), precious metal fixation step, precious metal reduction step, and precious metal content. Catalysts were tested under screening conditions in a continuous multi-purpose laboratory scale apparatus (T_{max} 400°C, p_{max} 40.0 MPa) as described previously (8). In addition, the palladium dispersion of various catalysts was characterized by dynamic CO-pulse chemisorption measurements and is expressed in ml CO/g catalyst. In particular, the trickle bed hydrogenation of fatty acids over the palladium catalysts was carried out at 2.5 MPa (25 bar, 362.6 psig) pressure with a continuous flow of hydrogen amounting to 40 litres/hour. During any particular screening the reactor was operated at two different temperatures (140°C and 190°C) and at two different space velocities (LHSV = 3 h^{-1} or 1 h^{-1}). The volume of catalyst was 30 ml for each screening and was replaced for each new screening test. The hydrogenated fatty acids were collected at suitable intervals and analysed for the level of hydrogenation by measuring the iodine value. The lower the iodine value, the greater the level of hydrogenation that has occurred. Typically, for a sample of the unhydrogenated fatty acid the iodine value was found to be 52 g iodine/100 g of feedstock. Values less than this were obtained for hydrogenated samples. The sulphur content of fatty acids applied during screening experiments was in the range of 24 to 35 ppm.

For comparison reasons fatty acid hydrogenation experiments were carried out in a 0.5 l discontinuous stirred-tank reactor with both a supported nickel catalyst as well with an activated carbon supported palladium catalyst.

B. Pilot plant catalyst testing

The flow diagram of the pilot plant trickle-bed reactor is shown in Figure 1. Hydrogen and the starting material are introduced from the top of the reactor. The hydrogen volume flow is regulated with a mass flow controller (Hitec). The fatty acid flow is regulated with a membrane pump. The product leaving the reactor is cooled to 115°C and then separated from unconverted hydrogen in a high-pressure separator. The volume of unconverted hydrogen is determined with a flow meter. Downstream, the liquid product is decompressed in a low-pressure separator from which it can be recycled into the reactor if desired. The

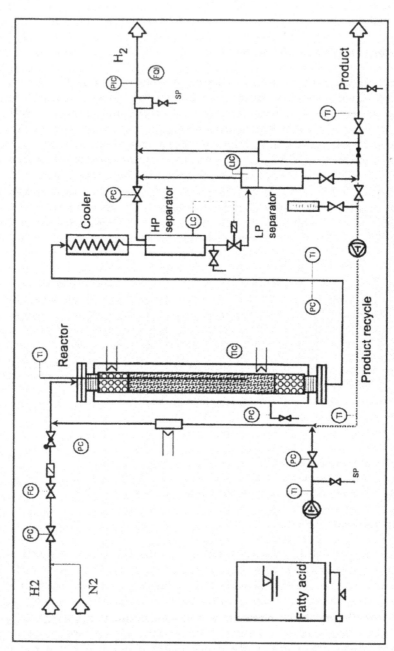

Figure 1 Laboratory reactor for the continuous hydrogenation of fatty acids

whole apparatus is kept at 90°C. The double shell reactor (38 x 3.5 x 1340 mm) is thermostated with high-pressure water. Inside the reactor a sieve was positioned 100 mm above the flange. Above the sieve the reactor was filled with a 100 mm layer of inert glass balls (6 mm), a 750 mm layer of catalyst (140 g = approx. 400 ml), and an additional layer of glass balls, acting as a static mixer and preheating zone. The reaction temperature is measured inside the catalyst bed with a thermocouple, which can be moved during the reaction if desired.

Results and Discussion

A. Catalyst preparation and screening test results

Various activated carbon supported 2 wt. % palladium catalysts were prepared on different activated carbon supports. The catalyst preparation method includes aqueous impregnation with a palladium precursor, catalyst reduction, washing, and drying. Palladium nitrate, as well as palladium chloride, are suitable palladium precursors. Commercially available activated carbon supports in the form of granules and extrudates and with particle sizes in the range of 1 to 5 mm were used. Specific surface areas and total pore volumes of these supports were in the range of 1000 to 1600 m^2/g and 0.3 to 1.4 ml/g, respectively. In Table 1, screening results for the trickle-bed hydrogenation of free fatty acids over various activated carbon catalysts at a liquid hourly space velocity (LHSV) of 3 h^{-1} and at two different temperature levels (140 and 190°C) are given. A lower iodine number of the hydrogenated fatty acid is a measure for a higher hydrogenation activity of the catalyst, since more C-C-double bonds in the fatty acid product have been saturated.

Table 1   CO numbers and iodine numbers for hydrogenated fatty acids at two different reaction temperatures for various catalysts

Catalyst	IV number at 140°C in [g I_2/100 g]	IV number at 190°C in [g I_2/100 g]	CO number [ml CO/g cat.]
A	11.4	2.3	2.59
B	13.0	5.0	1.57
C	34.9	13.9	1.30
D	43.2	28.0	0.70
E	48.4	30.9	1.00
F	51.0	38.3	0.67

Catalyst A on activated carbon extrudates (2.3 mm in diameter) was superior to all other tested activated carbon based catalysts and gave the lowest iodine numbers at both temperature levels. Table 1 also indicates that there is a relationship between the iodine number of the hydrogenated product and the palladium dispersion of a particular catalyst. Catalyst A has a very high CO number of 2.59 ml of CO/g of catalyst at a palladium content of 1.99%. The CO value for catalyst A correlates to a palladium dispersion of 60%. The activated carbon support for catalyst A is characterized by a specific surface area in the range of 1400 to 1600 m^2/g and a pore volume of more than 1.0 ml/g. Basically, in all catalysts investigated, palladium is distributed as a shell on the activated carbon support in order to use the expensive precious metal more efficiently.

Based on the standard catalyst preparation recipe some further investigations were made. In particular, the catalyst reduction method was investigated in more detail. It was found that a wet phase catalyst reduction with chemicals like hydrazine, sodium formate, and sodium hypophosphite gives a superior catalyst in terms of activity compared to gas phase catalyst reductions with either pure hydrogen or a hydrogen/nitrogen mixture. A wet phase catalyst reduction gives very active catalysts with CO numbers up to 2.59 ml CO/g catalyst, whereas in gas phase reductions less active catalysts with CO numbers in the range of 0.2 to 1.0 ml CO/g catalyst were found.

During catalyst screening it was necessary from time-to-time to change to a new container of fatty acids. Although using similar fatty acids, most often from the same lot number, sometimes different sulphur contents and iodine numbers were found. We observed that a lower iodine number can be achieved under the same reaction conditions when a purer fatty acid with a lower sulphur content was applied. Therefore, at any time when the fatty acid was changed, even when only a different container from the same lot was used, a reference catalyst was tested in the beginning.

Based on our screening test results, a catalyst was selected and scaled-up for the process development work.

B. Process development work and catalyst life-time tests

Various reaction conditions (liquid space velocity, temperature, hydrogen pressure, and hydrogen to feedstock ratio) were investigated in order to develop the concept of a commercial plant. In all pilot plant test runs a scale-up catalyst based on catalyst recipe A was used, which was found to be superior compared to other catalysts according to our screening test results. In screening tests it has been found that a lower sulphur content of the feedstock has an impact on the catalyst performance in terms of a lower iodine number. Therefore, various fatty acid qualtities have been hydrogenated over a mid to long-term period of time in order to investigate catalyst poisoning effects and in order to generate catalyst life-time data. Among the different fatty acid qualities we hydrogenated were very pure fatty acids like distilled palm kernel fatty acids and less pure fatty acids like undistilled tallow fatty acids. As reaction conditions the following parameters have been investigated in more detail:

- Reaction temerature   100-200°C
- Pressure   1.0 – 5.0 MPa
- LHSV   0.2 – 3.0 h^{-1}
- Hydrogen excess   10-200% (mole H$_2$/mole feedstock)
- Catalyst   Catalyst A (2 wt. % Pd/C on 2.3 mm extr.)
- Catalyst amount   140 g
- Catalyst volume   400 ml

A slight excess of hydrogen was found to be sufficient to get good hydrogenation results. During the experiments no major drop in selectivity, determined by any decrease in acid value, was observed. It turned out that the measurement of the catalyst bed temperature in the reactor is a very good tool to determine catalyst deactivation along the catalyst bed. As an example, after 518 hours the peak for the highest catalyst bed temperature in one of our pilot plant hydrogenaton runs shifted only from 20% to 27% of the entire catalyst bed length. The mass balance with regard to sulphur shows that after 100 to 200 hours on stream the sulphur content of the product seems to be in the same order of magnitude as for the feedstock. After that time the catalyst is still very active. It seems to be that sulphur is not the main catalyst deactivation mechanism as we thought based on earlier fatty acid hardening experiments with precious metal powder and fixed-bed catalysts. In fact, our catalyst seems to be very sulphur resistant.

A final experiment was carried out as a catalyst life-time experiment with catalyst A. A distilled split tallow fatty acid from Peter Greven Fett-Chemie GmbH + Co. KG, Germany, was used. The typical properties of the starting material are given below:

- IV (iodine value):   55.6 g iodine/100 g
- Acid value: 203.6 mg KOH/g
- Sulphur content: 9 – 26 ppm (determined by AAS)
- Phosphorus content: 1 ppm
- Iron content: 10 ppm
- Water content: 0.03 wt. %
- Density (60°C):   0.860 g/ccm

Despite the high sulphur content of the starting material the catalyst was found to retain a high activity after more than 4300 hours on stream. Various experiments were carried out in order to simulate process designs including a multi-step continuous reactor cascade. The applied raw material and the desired final iodine number determine the most economic process design and reaction conditions. One of various feasible process designs is shown in Figure 2.

The substrate, blended with pre-hardened material, is introduced into the reactor with a LHSV of 1.5 1/h. One half of the obtained product, having an IV of 5 g iodine/100 g product (Run 1), is then used for the initial blending. The other part is once again dosed into the reactor at a LHSV of 0.75 1/h leading to an IV of approximately 1 g iodine/100 g product (Run 2). A further hardening step leads, if desired, to iodine values of 0.5 g iodine/100 g product (Run 3). By running this experiment in the described mode it was possible to simulate a two to three fixed-bed reactor cascade. The corresponding iodine values as a function of time on stream for a certain run time are shown in Figure 3.

After 4300 hours on stream we determined a catalyst consumption of only 0.126 kg of catalyst (2.52 g palladium) per metric ton of fatty acid. Compared to a commercially applied supported nickel catalyst operated in a discontinuous stirred-tank reactor, we found at least a 23 times lower catalyst consumption for the fixed-bed process as shown in Table 3. At that time, the fixed-bed catalyst retained a high activity. Both processes operate under similar process conditions in terms of temperature (150-200°C) and pressure (2.5 MPa). The same fatty acid feedstock was applied in the determination of catalyst productivity numbers. In addition, we found at least a 23 times lower catalyst consumption for the fixed-bed process compared to the discontinuous stirred-tank reactor process when operating with an activated carbon supported palladium catalyst.

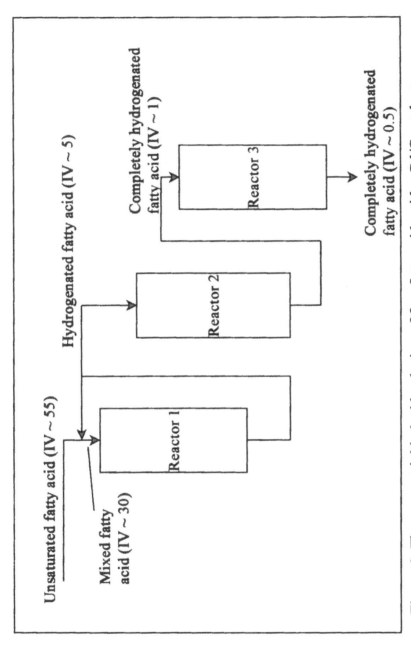

Figure 2 Three-step trickle-bed hardening of free fatty acids with a Pd/C catalyst as one of various feasible process designs

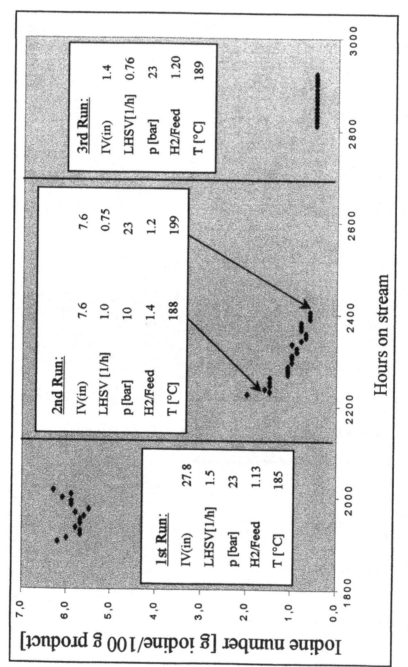

Figure 3 Results for a three-step trickle-bed hardening of free fatty acids with a Pd/C catalyst as one of various feasible process designs

Table 2 Catalyst consumption for the complete hardening of free fatty acids in different processes

Catalyst	Process	Catalyst consumption [kg cat./kg fatty acid]
22% Ni/silica (powder)	Stirred-tank reactor (discontinuous)	3 - 4
5% Pd/C (powder)	Stirred-tank reactor (discontinuous)	3 - 4
2% Pd/C (fixed-bed)	Trickle-bed reactor (continuous)	< 0.126

C. Characterization of fresh and used hydrogenation catalysts

Fresh and used catalysts were investigated by XRF and element analysis. No loss of palladium was determined for the used catalyst. On the used catalysts we found Fe, S, P, and Ni in small amounts (< 0.1 wt. %) and in smaller traces Mn, Cr, Zn, and Sn.

Conclusion

This study shows that activated carbon supported palladium catalysts are very attractive for the continuous fixed-bed hydrogenation of fatty acids. The development process which has been carried out has resulted in the preparation of a very active and selective catalyst with a long life time, both on a laboratory and a pilot production scale, for the hydrogenation of fatty acids. Important parameters for the development of a superior catalyst are the selection of support material and catalyst reduction method. High catalyst hydrogenation activities were found for catalysts with high palladium dispersion numbers, in particular for a catalyst with a palladium dispersion in the order of 60%.

Within the development work for a fixed-bed process for the continuous hardening of free fatty acids, various fatty acid qualities were tested. Despite the high sulphur content of the starting material, catalyst A was found to retain a high activity after more than 4300 hours on stream in a long-term test run. A variety of reaction conditions (space velocity, temperature, hydrogen pressure, and hydrogen to feedstock ratio) were investigated in order to develop the concept of a commercial process.

The new fixed-bed process offers advantages over the currently applied discontinuous stirred-tank reactor process with supported nickel catalysts. These advantages include:

- High throughput of raw materials
- Low catalyst consumption
- Stable fixed-bed catalyst
- No nickel contamination of the product
- Significantly reduced catalyst poisoning by sulphur
- No need for the separation of catalyst powders
- No catalyst sensitiviy against water

Based on our results we estimated total catalyst costs per metric ton fatty acid for the new fixed-bed process and the currently applied stirred-tank reactor process with supported nickel catalyst. The comparison revealed economic advantage for our continuous fixed-bed process.

References

1. Lurgi brochure 'Fatty Acid Technology' 197e/3.91/30
2. Estimation, based on 'The World's Fatty Acid Industry 1998', Hewin International Inc., van Leyenberghlaan 158, P. O. Box, 7813 Amsterdam, The Netherlands
3. Fette, Seifen, Anstrichmittel, 78 (1976), 385; 79 (1976), 181, 465; 80 (1978), 1
4. Ruhrchemie (Canadian Patent No. 1 157 844, 1981)
5. Henkel/Degussa (DE 41 09 502 C2; EP 0 505 863 B1, 1991)
6. Henkel (DE 42 09 832 A1, EP 0 632 747 B1, 1992)
7. Gritz, E. V. W. in Handbook of Heterogeneous Catalysis, Ertl, G., Knözinger, H. (Eds.), VCH Verlagsgesellschaft mbH, Weinheim, 1997, p. 2221
8. H. Buchold, T. Tacke, I. Beul, C. Rehren and P. Panster, New palladium fixed-bed catalyst and process for the continuous hardening of unsaturated fatty acids, poster presentation, annual AOCS meeting, Orlando, FL, 1999

Modeling of Propylene Oxidation on Hydrophobic Catalyst in a Slurry and a Trickle Bed Reactor

Wugeng Liang and K. T. Chuang

Department of Chemical and Materials Engineering, University of Alberta, Edmonton, Alberta, Canada T6G 2G6

Abstract

The oxidation of propylene to acrylic acid over Pd-supported hydrophobic catalysts has been studied in a slurry and a trickle bed reactor. Hydrophobic catalyst was shown to have much higher activity than hydrophilic catalyst due to its advantage of low mass transfer resistance.
Key words: hydrophobic catalyst, propylene oxidation

Introduction

Palladium-catalyzed propylene oxidation to acrylic acid has advantages of low reaction temperature and a single-step process compared to the current commercial process using metal oxides catalysts. Low space-time yield is the main problem for the commercializating this process. Polymer-supported hydrophobic catalysts have been shown to have higher efficiency than hydrophilic catalysts for this process (1, ·2). In this work, propylene oxidation over hydrophobic catalyst was studied in slurry and trickle bed reactors. A comparison between the performances of hydrophilic and hydrophobic catalysts is discussed.

Experimental

In this work, a hydrophobic high surface area poly-styrene-divinylbenzene (SDB) has been used as the support. The catalysts were prepared by the impregnation of the SDB with Pd salt, which was described in detail elsewhere (1, 2). A 10% Pd catalyst was used for the experiment. The slurry and trickle bed reactors were made of a stainless-steel and were heated using an electrical heat jacket. The gas phase reactants and products were analyzed using an on-line gas chromatograph equipped with a TCD detector. The outlet gas flow rate was measured using a bubble flow meter. After each experiment the liquid product was analyzed using GC and GC-MS. More details about reactors were reported elsewhere (1, 2).

Results and Discussion

1. Mass Transfer Processes over Hydrophobic Catalyst

When the hydrophobic catalyst is contacted with water, there is a micro-layer of gas surrounding each particle of the catalyst (3, 4). In a slurry or a trickle bed reactor, there is a liquid film passing over and around the catalyst. However the presence of the gas layer formed around the catalyst particles ensures that the gas phase can contact the catalyst without resistance from gas-liquid mass transfer (Figure 1). Therefore the hydrophobic catalyst has the advantage of low mass transfer resistance compared to the hydrophlic catalyst. The comparison of the diffusion processes with hydrophobic and hydrophilic catalysts are shown in Figure 2.

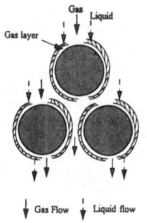

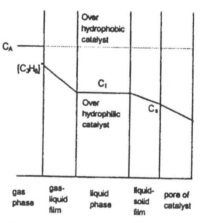

Figure 1. The gas-liquid-solid contacting over hydrophobic catalyst

Figure 2. Comparison of the mass transfer resistance between hydrophobic and hydrophilic catalyst

2. Reaction in the slurry reactor

Due to the high concentration of water compared to all other active components in the reaction system, and based on the reaction kinetics of propylene oxidation on Pd catalyst and the diffusion processes shown in Figure 2, the reaction rate under uniform catalyst particle distribution can be expressed as (5):

$$r_A = k\bar{\varepsilon}_s C_A \qquad (1)$$

The residence time of gas reactants in the slurry reactor is:

$$t = V\bar{\varepsilon}_g / U_g \qquad (2)$$

So, propylene conversion is:

$$x = 1 - \exp(-k\bar{\varepsilon}_s V \bar{\varepsilon}_g / U_g) \qquad (3)$$

When the hydrophilic catalysts are used for propylene oxidation in a slurry reactor, in contrast to hydrophobic catalysts, in order to have the reactants reach the active sites, the gas reactants must first dissolve in water, and the gas reactants must then diffuse in the liquid phase to reach the active sites in the catalyst pores. The reaction rate is:

$$r_A = \frac{k_l a' k \bar{\varepsilon}_s \eta}{H(K_l a' + k\bar{\varepsilon}_s)} C_A \qquad (4)$$

where η is the effectiveness factor of the catalyst, as determined using the Thiele modulus:

$$\eta = \frac{3}{\phi}(\frac{1}{\tanh\phi} - \frac{1}{\phi}) \quad (5) \quad \text{and} \quad \phi = (d/6)\sqrt{k/D_e} \qquad (6)$$

Comparing the mass transfer resistance for the two types of catalysts shown in Figure 2, it can be seen that the following features of the mass transfer processes cause the significant difference between the two catalysts.

(1): *Low solubility of propylene in water.* Under the operating conditions, the Henry Law constant (H) of propylene is 10.0 (6). For a hydrophobic catalyst the gaseous reactants can diffuse directly to the active sites in the pores of the catalyst, so that the reactants need not first dissolve into the liquid.

(2): *Lower intracrystalline diffusion coefficient in the liquid phase.* In the pore of the hydrophobic catalyst, the diffusivity of gaseous propylene is in the range of 10^{-3} cm^2/s, and the Thiele modulus under the reaction condition is:

$$\phi = (d/6)\sqrt{(k/D_e)} = 0.32 \qquad (7)$$

Therefore, η is 0.94. Over hydrophilic catalyst, the reactants diffuse into the pore of the catalyst in the liquid phase, where the diffusivity is in the range of 10^{-5} cm^2/s. Therefore, the catalyst effectiveness factor with the hydrophilic catalyst will be about 1/5 of that of hydrophobic catalyst, i.e., $\eta = 0.19$.

(3): *Low gas-liquid and liquid-solid mass transfer rates over hydrophilic catalyst.* Under the above operating conditions, $K_l a_g$ is 0.01 s^{-1}, and $k_c a_c$ is 0.50 s^{-1} with the hydrophilic catalyst (7). Such gas-liquid and liquid-solid mass transfer resistance does not exist for gas phase direct reaction over the hydrophobic catalyst.

Due to the above differences between hydrophobic and hydrophilic catalysts, the hydrophobic catalyst shows much higher activity than the hydrophilic catalyst. The comparison of the experimental results in the slurry reactor with the two types of catalysts is shown in Table 1. The predicted yield for the hydrophilic catalysts is obtained with Equations (4–7). The results show that the above explanation of the difference between the two types of catalysts in the slurry reactor is reasonable.

Table 1: Comparison between hydrophobic (O) and hydrophilic (I) catalysts

Catalyst	T (°C)	P (kPa)	Selecti- vity(%)	Y	Y Predicted by Eq(4 - 7)	References
10%Pd/C (I)	40	790	96.0	1.2	1.1	(8, 9)
10%Pd/C (I)	50	790	93.1	1.9	1.8	(8, 9)
10%Pd/C (I)	65	790	93.2	3.2	3.2	(8, 9)
5%Pd/C (I)	100	2450	67.1	11.1	10.0	(8, 9)
10%Pd/SDB (O)	112	997	95.0	31.0		This work
10%Pd/SDB (O)	112	2493		57.0		Predicted by Eq (3)

3. Reaction in the trickle bed reactor

In the trickle bed reactor, if the gas phase is assumed to be plug flow, then the mass conversion rate of propylene is expressed by:

$$\frac{dx}{dz} = \frac{1}{U_g C_{A0}} r_A \qquad (8)$$

where:

$$x = 1 - C_A / C_{A0} \qquad (9)$$

with the boundary condition:

$$z = 0; \quad x = 0 \qquad (10)$$

$$x = 1 - \exp(-k'z\bar{\varepsilon}_g / U_g) \qquad (11)$$

When hydrophilic catalysts are used for this reaction in the trickle bed reactor, due to other diffusion processes as indicated in Figure 2, the mass balance for the reactant propylene in the gas phase in the trickle bed reactor is:

$$U_g \frac{dC_g}{dz} + (K_l a_g)(C_g / H - C_l) = 0 \qquad (12)$$

And the mass balance for propylene in the liquid phase is:

$$U_l \frac{dC_l}{dz} - (K_l a_g)(C_g / H - C_l) + (k_c a_c)(C_l - C_s) = 0 \qquad (13)$$

In the above equations, the axial dispersions of gas phase and liquid phase are both neglected for simplicity.

Under steady state,

$$k_c a_c (C_l - C_s) = k' \eta C_s \qquad (14)$$

The analytical solution of the above equations can be obtained under isothermal condition. The reaction conversion is:

$$x = 1 - C/C_0 \qquad (15)$$

where:

$$\frac{C}{C_0} = \frac{1}{m_2 - m_1}[m_2 \exp(m_1 z) - m_1 \exp(m_2 z)] \qquad (16)$$

$$m_1 = -\frac{\beta}{2} + \frac{1}{2}(\beta^2 - 4\gamma)^{1/2} \text{ and } m_2 = -\frac{\beta}{2} - \frac{1}{2}(\beta^2 - 4\gamma)^{1/2} \qquad (17)$$

In the above equations, the constants α, β, γ are defined as follows:

$$\alpha = \frac{k_c a_c}{k_c a_c + k'\eta}; \quad \beta = \frac{U_g + U_l / H + k_c a_c (1 - \alpha) U_g /(K_l a_g)}{U_g U_l /(K_l a_g)};$$

$$\gamma = \frac{k_c a_c K_l a_g (1 - \alpha)}{H U_g U_l} \qquad (18)$$

Under the conditions F_{C3H6} = 180 mL/min, F_{O2} = 120 mL/min, P = 993 kPa, F_l = 17 mL/min and T_0 = 127 °C, and based on the above Henry Law Constant and mass transfer coefficient, quantitative comparison from the model predictions can be made using Equations (11) and (16), respectively, for the two types of catalysts in the trickle bed reactor (Figure 3). The results indicate that for the hydrophobic catalyst the activity is about 20 times higher than that for the hydrophilic one. The experimental results obtained are also shown in the same figure, and are consistent with the model predictions.

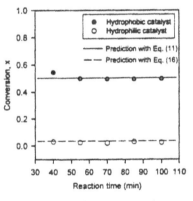

Figure 3. Propylene conversions over hydrophobic and hydrophilic catalysts in the trickle bed reactor

Conclusions

Based on the experimental results obtained in slurry and trickle bed reactors, the activity of the hydrophobic catalyst is approximately one order of magnitude higher than that of the hydrophilic catalyst. This advantage comes from the much less mass transfer resistance with hydrophobic catalyst.

Acknowledgment   The financial support from the Hydrocarbon Technology Inc. (HTI) and the National Science and Engineering Research Council of Canada (NSERC) is gratefully acknowledged.

Nomenclature

a_c = liquid-solid interface area; a_g = gas-liquid interface;
C_A = concentration of propylene in gas phase (mol/m³);
C_{A_0} = concentration of propylene in gas phase at the inlet of reactor (mol/m³);
C_l = concentration of propylene dissolved in liquid phase (mol/m³);
C_s = concentration of propylene at the surface of the catalyst;
d = diameter of the catalyst (cm); D_e = Effective diffusion coefficient (m²/s);
H = Henry Law constant; k = reaction rate constant in slurry reactor (1/s);
k' = reaction rate constant (1/s) in trickle bed reactor;
k_c = liquid-solid mass transfer resistance;
K_l = gas-liquid mass transfer resistance;
P = reaction pressure (kPa); r_A = reaction rate (mol/s);
R = radius of the catalyst particle (m); t = Residence time of reactants (min);
U_g = gas velocity in the reactor (cm/s); U_l = liquid velocity in the reactor (cm/s)
x = reaction conversion;
Y = yield of acrylic acid per gram of Pd per hour (g/g(Pd).h)
z = axial position in the reactor (cm)
Greek letters
ϕ = Thiele modulus; η = catalyst effectiveness factor
$\bar{\varepsilon}_g$ = average gas holdup in the reactor;
$\bar{\varepsilon}_s$ = average catalyst particle holdup in the reactor

Literature Cited

1. Chuang, K. T.; Fu, L. U. S. Patent 5,210,319, 1993.
2. Sood, S. One-Step Oxidation of Propylene to Acrylic Acid, Thesis of Ms. Sci. University of Alberta: Edmonton, Canada, 1995.
3. Enright, J. T.; Chuang, T. T. *Can. J. Chem. Eng.* **1978**. *56*, 246.
4. Butler, J. P. *Separation Sci. & Tech.* **1980**, *15*, 371.
5. Liang, W. G,; Chuang, K. T.; Zhou, B. *Dev. Chem. Eng. Min. Proc.* **1998**, *6*, 211.
6. Hancock, E. G. Propylene and Its Industrial Derivatives; Ernest Benn Ltd.: London, 1973; P. 78.
7. Smith, J. M. Chemical Engineering Kinetics, 3rd ed., McGraw - Hill: New York, 1981.
8. Lyons, J. E. *Catalysis Today* **1988**, *3*, 245.
9. Lyons, J. E.; Suld, G.; Hsu, C.-Y. Multiple Roles of Palladium in Liquid Phase Oxidation. In *Homogeneous and Heterogeneous Catalysis*; Yu, Y. et al., Eds.; VNU Science Press: Utrecht, The Netherlands, 1986.

The Safe Handling of Activated Nickel Catalysts: Heat-Induced Interactions Between the Catalyst, Ethyl Acetate, and Water

Daniel J. Ostgard[+], Monika Berweiler[+], Peter Panster[+], Rainer Müller*, and Felix Roessler*

*Hoffmann-La Roche Ltd., Research and Technology Development
VFH 214/8A, CH-4070 Basel, Switzerland

[+]Degussa-Huels AG, Silicas and Chemical Catalysts Division
Rodenbacher Chaussee 4, P.O. Box 1345, D-63403 Hanau-Wolfgang, Germany

Abstract

The fresh activated nickel catalyst (ANC) was found to react exothermically with ethyl acetate initially at 76°C and more violently after 116°C. The presence of water suppressed this interaction and heating the catalyst in only water generated the lowest exotherm measured in this study. The spent catalyst was not as sensitive to ethyl acetate as the fresh one, and treatments that partially cleaned its metal surface increased the global exothermicity of this reaction.

Introduction

ANC are usually generated by leaching most of the Al out of a 50% Ni/50% Al alloy. The resulting catalyst consists of a hydrogen rich porous 88-96% Ni/4-12% Al structure that is very effective for hydrogenation reactions. This catalyst system is pyrophoric and, as such, it is important for safety reasons to know how it behaves in the presence of solvents at various temperatures. This paper looks at the interactions of fresh and spent ANC in the presence of water and/or ethyl acetate as they are ramped through a heat and search program. This program slowly heats the catalyst suspension in a stepwise fashion and stops if an exotherm is detected so as to determine the properties of this potentially hazardous interaction.

Results and Discussion

Table 1 displays the interactions of the fresh catalyst in water and/or ethyl acetate suspensions during the heat and search experiments. If the aqueous catalyst slurry is heated in a closed reactor (figure 1), the resulting exotherm is much lower indicating that the build up of hydrogen pressure inhibits the

Table 1 Fresh catalyst interactions in water and/or ethyl acetate suspensions.

Exp.	Properties of the Catalyst Suspension	Temperature °C	Exotherm kJ/kg of Catalyst
A	105 g Catalyst	113	6.7
	68 g water	128	13.3
	in a closed system	144	3.3
		Sum =	23.3[a]
B	105 g Catalyst		
	68 g water	>75	>100
	in a vented system		
	(Max press. = 1.4 bar abs.)		
C	105 g Catalyst	82	1.9
	18 g water	90	7.5
	44.4 g ethyl acetate	101	5.7
	in a closed system	111	3.8
		120	11.4
		133	10.5
		Sum =	40.8
D	44 g Catalyst	76	13.0
	44.4 g ethyl acetate	96	4.0
	in a closed system	>116 - 201	181
		Sum =	198

[a] Performing the same experiment with only 45 g of ANC and 35 g of water resulted in an exotherm of 20 kJ/kg, thereby confirming these results.

additional evolution of hydrogen. In figure 1 there appears to be two main types of exothermic hydrogen evolution. The first emittance of hydrogen has the weakest exotherm and it appears to start at 113°C. This desorption most likely starts at the same temperature as that of the vented system (75°C, figure 2 / exp. B), but the developing hydrogen partial pressure suppressed it below the detection limit of the apparatus. The second desorption starts at 128°C and is far more exothermic than the first. This data fits well with that of Nicolau (1) where the highly exothermic evolution of hydrogen resulted from zero valent Al in the catalyst reacting with water to form hydrogen and aluminum oxide. The weaker desorption exotherm was found to occur in the same temperature range as the loss of interstitial hydrogen (1). Mars et al. (2) also found that the Al in ANC reacts "explosively" with water at temperatures around 150°C to form hydrogen.

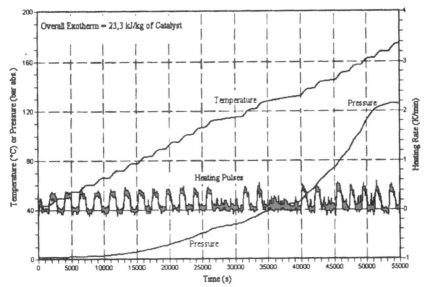

Figure 1 Experiment A of table 1.

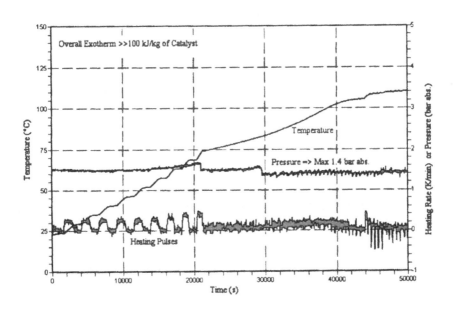

Figure 2 Experiment B of table 1.

Table 2 The exothermic reactions of the spent catalyst in ethyl acetate

Exp.	Post-Reaction Treatments	Temperature °C	Exotherm kJ/kg of Catalyst
E	The catalyst was washed with ethyl acetate after the reaction	84	3.3
		96	9.3
		104	6.3
		112	17
		128	41
		146	10.3
		155	5.7
		Sum =	92.9
F	None	77	8.3
		95	9.7
		103	7.3
		110	4.0
		117	8.0
		125	9.0
		139	38
		Sum =	84.3
G	The catalyst was purged with nitrogen in the reactor at 25°C and then washed with ethyl acetate after the reaction	71	7.3
		78	9.7
		113	117.7
		Sum =	134.7

Although our data suggests that the catalyst's Al starts to react with water at 128 instead of 150°C, the data are still in good agreement since the larger catalyst amounts used here improve the sensitivity of these measurements. Heating the fresh catalyst in the presence of only ethyl acetate (figure 3 / exp. D) generated the largest exotherm and gas pressure measured here. It has been found before that solvents such as methanol (3,4), ethanol, acetone, isopropanol, cyclohexane, benzene, dioxane, and amyl acetate (4) can decompose exothermically on ANC at the temperatures used here to generate methane. Hence, as suggested by the literature, the increased pressure in the presence of ethyl acetate may be due to its decomposition on the ANC to form methane. Experiment C clearly shows that adding water to the system moderates ethyl acetate decomposition by, most probably, competitive adsorption. Thus, the addition of a competitive adsorbate is a viable method to control the potentially dangerous decomposition of solvents on ANC during their use.

In comparison to the fresh catalyst, the spent catalyst (table 2) with or without post-reaction treatments is far less dangerous in the presence of ethyl acetate. This decrease in activity is due to carbonaceous residues blocking ethyl acetate's access to the ANC's surface. Washing the catalyst with ethyl acetate before performing the heat and search experiment slightly increases this exothermic decomposition by gaining additional access to unpoisoned active centers. One possible mechanism for this improved access could be the removal of soluble materials situated at the entrance of pores thus providing additional fresh uncoked catalytic surfaces for ethyl acetate decomposition. The very modest increase in ethyl acetate decomposition after washing fits the pore-mouth-blocking concept well, since the catalyst's pore volume is only 0.09 cc/g. Purging the spent catalyst with nitrogen while it is still in the reactor followed by an ethyl acetate wash boosted the catalyst's ability to decompose ethyl acetate far more than simple washing (figure 4). Therefore this surprising and reproducible treatment must remove more than readily soluble materials situated in the pores. For this kind of enhancement, there must be the chemical removal of coke from the catalyst's surface. Usually a reductive or an oxidative treatment is required for such removal, and a nitrogen purge is not considered to be either. One source of hydrogen could be from the catalyst itself. Although the topic of hydrogen forms in ANC is still in debate, the literature suggests (1, 5) that ANC contain both surface chemisorbed and interstitial hydrogen. While the core of ANC particles does not contain interstitial hydrogen, the existence of the ß-hydride phase has been found in the first couple layers from the surface (5). It has been further shown by Ceyer and coworkers (6, 7) that the nickel ß-hydride phase is more active than chemisorbed hydrogen making it ideal for coke removal. Due to the decreased hydrogen partial pressure during the nitrogen purge, the nickel ß-hydride phase formed during the previous high-pressure hydrogenation can start to desorb and on the way out it could remove coke from the surface by reduction. This would then provide more catalytic surface for the increased ethyl acetate decomposition observed during experiment E. While this is a proposed possibility, additional work is needed to clearly define this phenomenon.

Conclusions

Heating a fresh ANC in the presence of only ethyl acetate generates both a 198 kJ/kg exotherm and a higher gas pressure that starts slowly at 76°C followed by a rapid increase after 116°C. Adding water to the ANC and ethyl acetate mixture suppresses this reaction due to water's competitive adsorption. Hence, the use of a competitive adsorbate could allow for the safe removal of solvents when recycling ANC. Water worked well in this case, because heating the ANC in only water produced a much lower exotherm (23 kJ/kg). Spent

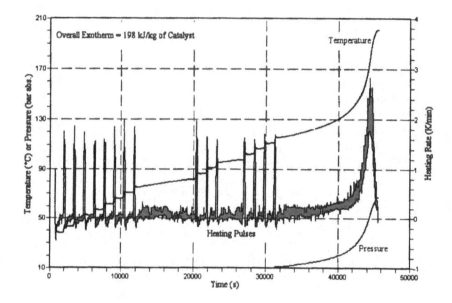

Figure 3 Experiment D of table 1.

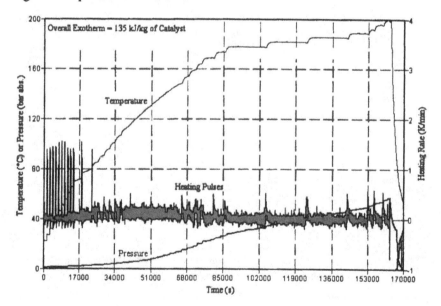

Figure 4 Experiment G of table 2.

catalysts were found to react less with ethyl acetate than fresh ones, and post-treatments that resulted in the partial cleaning of the surface increased ethyl acetate decomposition.

Experimental

The activated nickel catalyst (6.5 – 7.0% Al) used in these studies was provided by Degussa-Huels and the spent catalyst was generated prior to use by performing a high-pressure hydrogenation on it where ethylacetate was used as a solvent. The "heat and search" experiments were performed on the material combinations described in tables 1 and 2 in a 120 ml autoclave outfitted with pressure gauges, thermocouples, and a computer-controlled heating system. The data presented in the data tables were calculated from the temperature increases and the total heat capacity of the catalyst and the solvent system after taking into consideration the properties of the reactor. This data were then divided by the weight of the catalyst for the sake of comparison.

Acknowledgments

We thank Hoffman-La Roche and Degussa-Huels for the permission to publish this work.

References

1. I. Nicolau, and R. B. Anderson, *J. Catal.*, **68**, 339 (1981).
2. P. Mars, J. J. Scholten, and P. Zwietering, in *"Actes Cong. Int. Catal., 2nd, 1960"*, 1245 (1961).
3. O. Klais, in *"Hazards from Pressure: Exothermic Reactions, Unstable Substances, Pressure Relief, and Accidental Discharge"*, **102**, 25 (1987).
4. K. Hotta, K. Konishi, S. Kishida, and T. Kubomatsu, in *"The Proceedings of the 17th meeting of the Japanese Chemical Society"*, **40**, 1 (1966).
5. A. B. Fasman, Chem Ind. (Marcel Dekker), **75**, (*Catal. Org. React.*), 151 (1998).
6. A. D. Johnson, S. P. Daley, A.L. Utz, and S.T. Ceyer, *Science*, **257**, 223 (1992).
7. K. L. Haug, T. Bürgi, T. R. Trautman, and S.T. Ceyer, *J. Am. Chem. Soc.*, **120**, 8885 (1998).

The Effect of Surface Area on the Partial Oxidation of *n*-Butane over VPO Catalysts

S. H. Sookraj, D. Maripane, L. Breytenbach, K. Pillay and R. Ramatsebe

Sasol Technology (Pty) Ltd., Box 1, Sasolburg, 9570. South Africa

Abstract

The catalytic behaviour of vanadyl pyrophosphate in the partial oxidation of *n*-butane to maleic anhydride is reported. The effect of increasing surface area of the catalyst is investigated. The results obtained indicate that increasing the surface area does play a role in the selectivity and yield of maleic anhydride. However, there is a maximum surface area above which no further benefit with regards to selectivity and yield to maleic anhydride are obtained.

Introduction

The partial oxidation of *n*-butane to maleic anhydride (MA), over vanadium phosphorus mixed oxides was demonstrated by Bergman and Frisch in 1966 (1). During the 1960s' this process did not receive much attention because the benzene to MA route was well established and more cost effective. The decreasing cost of *n*-butane and the increasing environmental pressure to limit benzene emissions have gradually led to *n*-butane replacing benzene as feedstock for the production of MA. The MA yield in most commercial fixed bed reactors is generally between 50 to 60% and there exists an economic incentive to try and increase this yield (2); it is therefore important to obtain information that pertains to factors affecting the production of MA over these catalysts.

The vanadium phosphorus oxide (VPO) catalyst system represents the only example of a commercial catalyst used for the selective oxidation of an alkane, in this case *n*-butane to MA. The formation of MA requires the abstraction of 8 hydrogen atoms and the insertion of 3 oxygen atoms into the *n*-butane molecule. The reaction proceeds *via* a complex multistage mechanism, which is affected by the structural, redox and topological properties of the catalyst (3-5). It is generally accepted that the reaction proceeds entirely on the catalyst surface.

A series of VPO catalysts with different surface areas were prepared and tested for the partial oxidation of *n*-butane to MA. The results of these tests are reported in this paper.

Experimental

The catalysts are prepared in an organic medium, a solution of two or more alcohols is added to V_2O_5 and the mixture is refluxed with stirring for 7 hours, cooled to room temperature and stirred overnight. Ninety eight percent H_3PO_4 is added and the mixture is then stirred for 3 hours under reflux. The mixture is cooled and the precipitate removed by filtration. The precipitate is washed thoroughly with water and the blue solid is oven dried for 12 hours.

25 g of the catalyst precursor $VOHPO_4.\frac{1}{2}H_2O$, is calcined at a time. An inert gas like argon or nitrogen is used at a flow rate of 6 ml per minute. The temperature is steadily increased to 450°C over 8 hours. XRD studies are used to confirm that the precursor is converted to $(VO)_2P_2O_7$. The P/V ratio in the calcined catalysts was approximately 1.10. The surface area of the catalysts was determined after calcination.

The catalyst tests are carried out in a fixed bed micro-reactor at atmospheric pressure. The micro-reactor is loaded with 1.5 g of calcined catalyst. The feed to the reactor consists of 1%-butane in air. The operating temperature is varied between 380 and 420°C while the GHSV is varied between 1050 and 3500 h^{-1}. Selectivity and yield are calculated on a molar basis, yield is defined as (conversion x selectivity).

Results and Discussion

Previous studies using C4 and C5 hydrocarbons indicate that over promoted VPO catalysts there is a small improvement in MA yield as surface area increases (6). However the experimental data indicate that above a certain maximum surface area there are no further improvements in terms of MA yields. This paper attempts to establish if a relationship exists between surface area (in unpromoted VPO catalysts) and MA yields.

Figure 1 indicates that *n*-butane conversion increases with increasing surface area, reaches a maximum and then declines (at a temperature of 400°C with a GHSV of 1050 h^{-1}). The MA selectivity reaches a maximum and then levels off as surface area increases. The CO_2 selectivity also increases initially with increasing surface area and then levels off, while the CO selectivity initially increases and then decreases dramatically as the surface area is increased. Interestingly at higher surface areas the selectivity of CO_2 is higher than the selectivity of CO, a possible reason for this could be the presence of more active

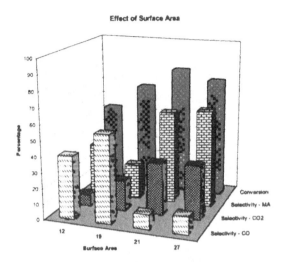

Figure 1 The effect of surface area on *n*-butane conversion, MA selectivity, CO selectivity and CO_2 selectivity [T = 400°C, GHSV = 1050 h^{-1}].

sites in the higher surface area catalysts that promote the secondary oxidation of MA to CO_2.

In a similar vein, the higher MA selectivities at higher surface area could be attributed to the presence of more sites that are active for MA formation. A probable explanation for the lower *n*-butane conversion at the highest surface area could be that less of the correct plane is exposed (7). The increase in CO_2 implies that a reduction in contact time with the catalyst may result in a decrease in the secondary oxidation of MA.

The effect of temperature and GHSV was then investigated over the two catalysts with the largest surface area, *viz.* 21 and 27 m^2/g. As expected *n*-butane conversion increases as temperature is increased, while the GHSV is held constant.. MA selectivity decreases while the total carbon oxide selectivity increases as the temperature is increased.

The results obtained when GHSV was increased (temperature was held at 400°C) indicate that as GHSV increases over the VPO catalysts tested, the conversion of *n*-butane decreases, the selectivity to both carbon oxides and MA

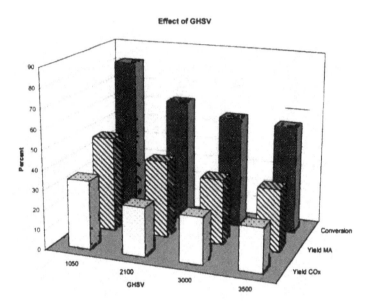

Figure 2 The effect of changing GHSV on catalyst with surface area of 21 m²/g [T = 400°C].

also decreases. Figure 2 illustrates the typical trends observed over the catalysts tested. The rate of decrease in selectivity to the carbon oxides is relatively slower than the rate of decrease in selectivity of MA. A similar trend is observed for the catalyst with the larger surface area.

In conclusion, it does appear that surface area of the VPO catalyst does play a role in the production of MA, although there does appear to be an optimum surface area beyond which there is no apparent gain in terms of MA yield.

References

1. R. L. Bergman, N. W. Frisch, US Patent 3 293 268, (1966)
2. Process Economics Program Report 46C, SRI International, October 1989
3. F. Cavani, F. Trifiro, *Catalysis*, 11, 247, (1994)

4. G. Centi, *Catal. Today*, **16**, 5, (1993)
5. G. Centi, J. R. Ebner, V. M. Franchetti, F. Trifiro, *Chem. Rev,*, **88**, 55, (1988), and references therein.
6. D. Engelbrecht, S. H. Sookraj, *Catal. Today*, **49**, 161, (1999)
7. G. Koyano, M. Misono, T. Okuhara, *J. Am. Chem. Soc.*, **120**(4), 767, (1998)

Reactivity of B/P/O-based Heterogeneous Catalysts in Gas-Phase *O*-Alkylation of Diphenols with Methanol

F. Cavani and T. Monti

Dipartimento di Chimica Industriale e dei Materiali, V.le Risorgimento 4, 40136 Bologna, Italy

Abstract
The gas-phase *O*-alkylation of catechol (1,2-dihydroxybenzene) with methanol over B/P/O catalysts, both bulk and supported on α-alumina, was studied with the aim of finding relationships between the acid properties of the catalysts and catalytic performance, in terms of activity and selectivity. Different catechol conversions and guaiacol (1-methoxyphenol, the product of mono-alkylation) selectivities were observed as functions of the B/P atomic ratio for the bulk samples. In some cases the initial activity however declined in a few hours, and the reason for deactivation was different depending on the catalyst B/P ratio. Supported catalysts gave more stable catalytic performance; moreover the latter was rather unaffected by the B/P ratio.

Introduction
Alkylation of aromatic substrates with olefins or alcohols constitutes an important class of reactions widely applied in both the petrochemical and fine-chemicals industries. Both heterogeneous and homogeneous acid-type materials are used to catalyze these reactions. The former systems however are preferred to the latter due to the fewer problems associated with toxicity, corrosiveness, inorganic wastes, need for stoichiometric amounts of base, and separation and recovery procedures. The heterogeneously-catalyzed vapour-phase alkylation of phenol and polyhydroxybenzenes at the oxygen atom (*O*-alkylation or etherification) with alcohols for the synthesis of alkylarylethers is one example, since interest exists for the replacement of traditional alkylating agents (dimethylsulphate, or dialkylcarbonate) with more environmentally friendly and more C-efficient reactants, such as alcohols, and for replacement of homogeneous catalysts.

Interest in diphenol alkylation has increased in recent years because practical uses have been found for the products of *O*-alkylation, especially when obtained with high purity. Alkylarylethers are important industrial chemicals and are extensively used as starting materials for the production of dyes, agrochemicals, antioxidants for oil and grease and as polymerization inhibitors.

Of particular interest is guaiacol (1-methoxyphenol), which is synthesised by methylation of catechol (1,2-dihydroxybenzene) with dimethylsulphate using sodium hydroxide as a homogeneous catalyst in the liquid phase, or by means of alkylation with dimethylcarbonate in aqueous solutions. The former methylating agent is corrosive and toxic, therefore special care is necessary in handling it, furthermore tedious procedures are required for the disposal of the waste water. In the methylation with dimethylcarbonate, where there are no environmental problems, only one carbon atom in three is utilised in the reaction, and this is even more a problem due to the relatively higher cost of the reactant. To avoid these disadvantages, vapour-phase methylation with methanol over heterogeneous catalysts has been attempted (1,2). In the vapour-phase methylation the reaction leads to different products (either O- or C-alkylated) (Figure 1), the relative amounts of which depend on the reaction conditions and on the catalyst acidity (3,4). Furthermore, frequently, coke is also formed during the reaction and consequently the catalyst deactivates with time.

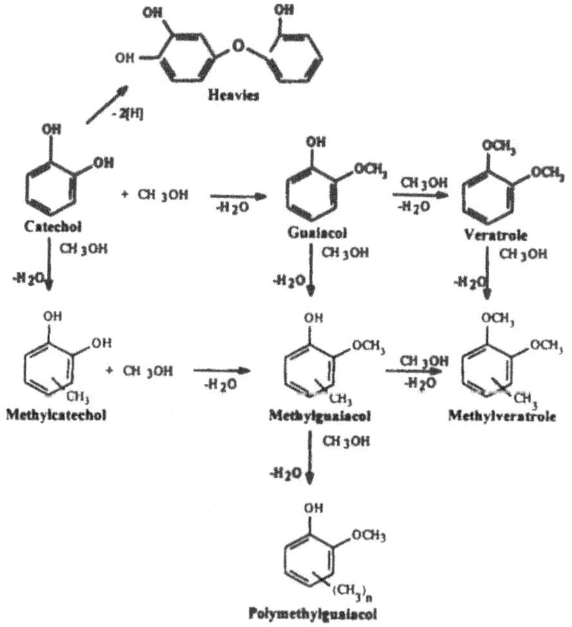

Figure 1 Reaction scheme of catechol methylation.

The catalytic systems described in the literature for this reaction include acidic materials such as zeolites, γ-aluminas and mixed oxides (e.g., boron phosphate and rare earth phosphates) (1-4). B/P/O systems in general are claimed as catalysts for dehydration of low weight alcohols and etherification of –OH groups in phenol (1,5,6). In these catalysts BPO_4 is the main component, where the boron and phosphorus atoms are tetrahedrally coordinated with oxygen to form a three-dimensional network of covalent bonded tetrahedra. The characteristics and reactivity of B/P/O catalysts however are mainly affected by their B/P atomic ratio.

The objective of the present investigation was to check the effect of the B/P ratio on the catalytic performance in the gas-phase O-alkylation of catechol with methanol, with the aim of finding relationships between acid properties of the catalysts and their reactivity. Both bulk and α alumina-supported catalysts were studied.

Experimental Section

Samples of bulk B/P/O catalysts (composition varying in the range of B/P atomic ratio = 0.6-1.4) were obtained by heating a mixture of H_3BO_3 and 85 wt.% H_3PO_4, taken in appropriate proportions, with continuous stirring in an aqueous solution. The resulting slurries were dried for 12h at 120°C and then calcined at 600°C for 4h. The surface area of the calcined powders was between 2 and 6 m^2/g. The supported catalysts were prepared by deposition of the active B/P/O phase onto preheated α-alumina (14 wt.% of active phase); after impregnation and drying, the samples were calcined at 600°C for 4h. For these samples the surface area after impregnation was between 6 and 8 m^2/g (the surface area of the support was 12 m^2/g). For the temperature-programmed-desorption experiments the catalyst (500 mg) was pretreated at 500°C under helium flow in a glass micro-catalytic tubular reactor for 10h. After cooling, the sample was exposed to ammonia flow for 30 minutes at 80°C, stabilised for three hours under helium flow and then heated at 500°C at the rate of 12°C/min. The effluents were analysed using a quadrupole mass spectrometer Balzers Prisma TM QMS 200. The catalytic reaction was carried out at 275°C and atmospheric pressure in a glass tubular fixed-bed reactor loaded with 6.0 g of bulk or supported catalyst. The premixed reaction solution, with a 1/1 weight ratio catechol/methanol was fed into the reactor using a micro-feed pump, together with N_2; the total flow rate was 35.05 ml/min. The products were condensed, identified and analysed by means of GC-MS. Before reaction, the catalysts were "activated" in a stream of methanol and N_2 at 275°C for 24h.

Results and Discussion

Figure 2 shows the catechol conversion of the unsupported catalysts with different B/P ratios as a function of the reaction time (time-on-stream, tos), from the beginning until the fourth hour of reaction. The distribution of products, after

1.5 hours tos and after 4.2 hours tos, is shown in Table 1. The most important by-products: methylcatechol, methylguaiacol, polymethylguaiacol and heavy compounds, are obtained from ring alkylation, and are probably due to activation of reactants on strong acidic surface sites. The formation of veratrol is instead obtained by consecutive O-methylation on guaiacol; veratrol is then ring-alkylated to yield methylveratrol.

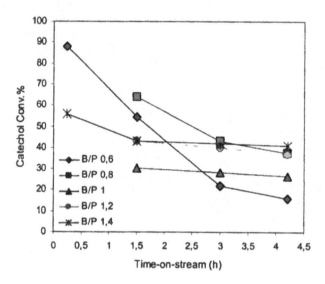

Figure 2 Catechol conversion for the bulk B/P/O catalysts with different B/P ratios, as a function of time-on-stream; T 275°C.

The catalysts showed very different initial catechol conversions, and different catalytic behaviours as functions of tos. The most active samples, those with excess phosphorus (B/P < 1.0), exhibited a progressive deactivation with a decrease in conversion from the initial 88% to 15% in a few hours for the catalyst having B/P = 0.6. The selectivity to guaiacol for these samples increased with increasing tos (Table 1), probably as a consequence of the decrease in conversion that reduced the contribution of consecutive reactions. In correspondence, the amounts of all the by-products decreased with tos.

The samples with excess boron (B/P > 1.0) showed a decrease in activity during the first hour of reaction (see sample with B/P=1.4 in Figure 2), but the catechol conversion then remained almost constant (around 40%). The selectivity to guaiacol (Table 1) was high (more than 90%) and constant after the first hour of reaction. The catalyst having B/P = 1.0 showed the lowest initial catechol conversion. The selectivity to guaiacol was constant with time, and in this case the most important by-products were veratrol, methylcatechol, methylguaiacol and heavy compounds.

Table 1 Distribution of products for the bulk B/P/O catalysts with different B/P ratios, at 1.5h (**I**) and 4.2h (**II**) tos.

B/P ratio	Sample	Selectivity %						
		Gc	Vr	MGc	MVr	MCt	PMGc	H
0.6	I	46.6	1.9	9.2	0.4	19.5	13.6	8.6
	II	70.7	1.1	2.7	0.7	9.5	6.3	9.0
0.8	I	57.0	2.3	11.6	0.4	4.6	17.7	6.4
	II	78.4	1.6	6.5	0.2	3.0	6.7	3.6
1.0	I	95.4	1.0	0.6	0.3	1.0	0	1.7
	II	94.2	0.6	0.9	0.4	1.9	0.04	1.9
1.2	I	93.6	1.6	2.9	0.06	0.3	0.8	0.7
	II	95.0	1.3	2.1	0.05	0.3	0.3	0.9
1.4	I	95.0	1.2	2.1	0.06	0.1	1.2	0.4
	II	95.6	0.9	1.7	0.02	0.1	0.9	0.7

Guaiacol (**Gc**), Veratrole (**Vr**), Methylguaiacol (**MGc**), Methylveratrole (**MVr**), Methylcatechol (**MCt**), Polymethylguaiacol (**PMGc**), Heavy products (**H**).

The different catalytic behaviours observed can be ascribed to the different surface acidity and to the different type of catalyst deactivation. Indeed, the ratio between the rates for C-alkylation and O-alkylation in phenol methylation was found to be mainly a function of the strength of the acid sites (7-11).

The FT-Ir spectra of these catalysts before reaction (Figure 3) showed all the fundamental frequencies of BPO_4 in the cristobalite structure (1094, 937, 573 and 627 cm^{-1}). The spectra showed, however, some additional bands the intensity of which depends on the composition of the samples. In the spectra of the catalysts containing excess boron, new bands appeared (1483, 1196 and 780-740 cm^{-1}), assigned to the fundamental vibrations of the trigonal BO_3 group, that indicate the presence of H_3BO_3 or B_2O_3. In the spectra of the samples with B/P < 1.0, the BPO_4 band at 1094 cm^{-1} was shifted toward higher wavenumbers, and another band (shoulder) at 1180 cm^{-1} (attributed to the P-O-P bond) appeared that may be taken as evidence of the presence of condensed phosphoric acid on the surface of these samples. In fact the only possible phase of the B/P mixed oxides is BPO_4, and therefore the excess P is in the form of phosphoric acid, probably with different degrees of condensation (from *ortho* to polyphosphoric acids). The distribution of the phosphoric acids depends on the water content and on the calcination temperature (3).

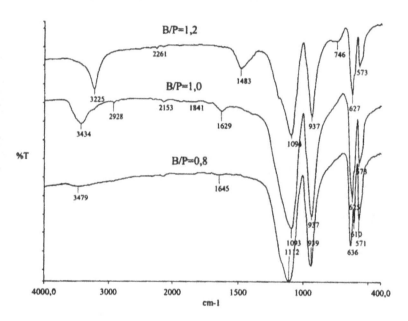

Figure 3 FT-Ir spectra (KBr disk technique) of the bulk B/P/O catalysts with different B/P ratios, after calcination.

Thermal-programmed-desorption (TPD) of NH_3 was studied to correlate the catalytic performance with the acid characteristics of these compounds. All the bulk samples gave (Figure 4) a large desorption peak at 170°C, due to weak acid sites, and another peak at 220°C. Moreover the catalyst containing excess B (B/P = 1.2) gave one further desorption peak at higher temperatures (about 270°C), due to the boric acid present on the surface. It is surprising that in samples with B/P < 1.0 the excess of P did not result in a high temperature desorption peak, under the operating conditions of the present study. This is in contradiction with the indications from catalytic tests, which suggest the probable presence of strong acid sites. Moreover the FT-Ir characterisation of the samples after TPD experiments showed that no residual ammonia was adsorbed on the catalyst after the TPD run. Therefore the presence of a considerable amount of strong acid sites in calcined, fresh catalysts could be excluded, even for samples with B/P < 1.0.

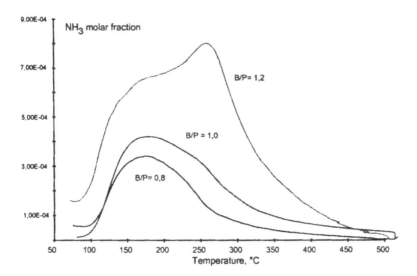

Figure 4 TPD spectra of ammonia adsorbed on bulk B/P/O catalysts with different B/P ratios, after calcination.

In order to gain further information concerning the presence of –OH surface groups, possibly responsible for the Brønsted-type acidity of these samples, FT-Ir tests have been done using the self-supported disk technique, in the transmission mode, treating the catalyst under vacuum at room temperature. The spectra in the –OH vibrations spectral range are reported in Figure 5, for three samples having different B/P atomic ratios.

Catalysts having B/P ≥ 1.0 showed a broad absorption centred around $3200 \ cm^{-1}$, much more intense in the case of the catalyst containing an excess of B. This absorption can be attributed to hydrogen–bonded –OH groups associated to the B atoms (14), even though a contribution of adsorbed water can not be excluded (it is known that complete dehydration of B/P/O occurs only after treatment at high temperatures (13,14)). In the case of the catalyst with B/P = 0.8, instead, the absorption was relatively weaker, and centred at lower wavenumbers (around $2900 \ cm^{-1}$), thus indicating the presence of a limited number of acid sites, relatively stronger than those present in catalysts having B/P ≥ 1.0 (15).

In order to obtain a better understanding of the different behaviours exhibited by the various catalysts, spent catalysts (unloaded after 4.2 hours tos) were characterized. The appearance of the unloaded samples with B/P ≥ 1.0 was the same as before reaction, both in colour and in particle shape, but those containing excess P looked like a sort of mud made of aggregated catalyst particles and coke (the catalyst was almost black). This explains the continuous deactivation exhibited by these catalysts, due to coke deposition which accumulated on the catalyst surface. However, this behaviour cannot be related to the presence of strong acid sites, as determined by TPD tests on calcined catalysts.

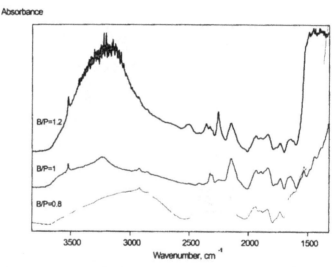

Figure 5 FT-Ir spectra of bulk B/P/O catalysts, (self-supported disk technique), under vacuum.

Figure 6 reports the FT-Ir spectra of samples after reaction. The fundamental bands relative to BPO_4 were unmodified with respect to fresh samples, but the bands attributed to B_2O_3 (or to H_3BO_3) in samples having excess B disappeared. This suggests that in these samples the higher initial activity, with respect to the B/P = 1.0 catalyst, was due to the contribution of free boric acid. The decrease in conversion which occurred in the first hour of reaction was due to the loss of excess B, in the form of volatile esters (these were effectively identified in the effluents). Stoichiometric BPO_4 was thus left, the reactivity of which was constant, at least for a few hours of reaction.

One possible hypothesis to explain the absence of strong acid sites in calcined catalysts having B/P < 1.0 is that the excess P is in the form of polycondensed phosphate groups (thus possessing a low number of strongly acid groups), and that these groups were hydrolyzed in the reaction environment (water is present, since it is a coproduct of the etherification reaction), with *in-situ* development of many P–OH groups. This was confirmed by the NH₃-TPD spectrum of the catalyst having B/P = 0.8, unloaded after reaction (Figure 7), which shows an intense desorption peak at 500°C that was not present in the same sample before reaction. It is possible that the P–OH groups generated during reaction are responsible for i) the high initial activity, ii) the progressive accumulation of coke on the catalyst, and iii) the high selectivity to the products of C-alkylation. A part of these groups was covered by coke during reaction, thus explaining the decrease in activity, but a part was still active in the spent catalyst, and was thus responsible for the NH₃ desorption peak at high temperature.

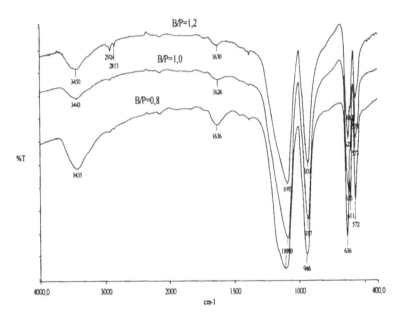

Figure 6 FT-Ir spectra (KBr disk technique) of the bulk B/P/O catalysts with different B/P ratios, after reaction.

The B/P/O active phase, with different B/P ratios, was then supported on α-alumina; the catalytic behaviour of these catalysts was quite different with respect to the unsupported ones. For all supported catalysts, catechol conversion was lower than for unsupported ones having the same B/P ratio, notwithstanding the higher specific surface area (Figure 8). This effect can not be attributed to the higher amount of active B/P/O phase loaded in the reactor in the case of bulk catalysts, since catalytic activity has reasonably to be attributed to the fraction of catalyst which is in contact with the gas phase, and therefore it rather is a function of the overall surface area. Also guaiacol selectivity was more stable with time-on-stream, regardless of the B/P ratio (Figure 8).

Another difference between bulk and supported catalysts concerns the deactivation rate, that for all supported catalysts was much lower than for bulk ones, regardless of the B/P ratio. In agreement with this result, supported catalysts having B/P < 1.0 gave a much lower formation of heavy by-products, and the spent catalysts were not covered with coke. Furthermore, the selectivity to guaiacol for supported samples with B/P < 1.0 was similar to that obtained with samples having B/P ≥ 1.0, and very different from that observed for bulk catalysts. Also in the case of supported catalysts having B/P > 1.0, excess B was soon lost during the first hour of reaction.

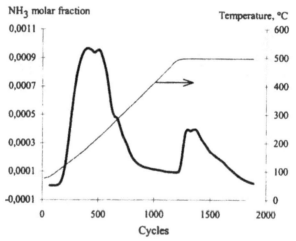

Figure 7 TPD spectra of ammonia adsorbed on bulk B/P/O catalysts with B/P = 0.8, after reaction.

Differences between bulk and supported catalysts can be attributed to a different distribution of surface acid sites, probably arising from the interaction between the active phase and the support. The data here reported indicate that stoichiometric BPO_4 possesses unique surface characteristics of mild acidity, which make it possible to obtain i) good catalytic activity in catechol etherification, ii) high selectivity in the reaction of O-methylation with a low degree of ring alkylation, and iii) low rate of catalyst deactivation. Moffat (12-14) proposed the existence of strained or bridged oxide sites in BPO_4, where partial positive charges are located on the B atom exposed on the surface. These sites might possess the weak Lewis acidity suitable to activate the –OH groups in reactants and catalyze their condensation, but not sufficiently high to form a carbocation and thus catalyze the ring alkylation. Also surface –OH groups exist, having different acidity depending on the B/P ratio, which may play a role in the reaction. It has been proposed that the B–O–P bridges in BPO_4 can be hydrolyzed in the presence of water, leading to the generation of B–OH and P–OH groups (14). One further aspect is that it is known that the hydrolysis of bulk BPO_4 under reaction conditions leads to the loss of boron and, sometimes, phosphorus, with a progressive decrease in performance for prolonged time-on-streams (1). This makes it necessary to cofeed volatile B compounds in order to prevent this deactivation phenomenon. In our case, however, it is likely that the hydrolysis of BPO_4 and the loss of the active components did not occur extensively under our conditions in our B/P = 1.0 catalyst, at least within the few hours of reaction of the experimental tests, since this would have led to a modification of the catalytic performance with time-on-stream.

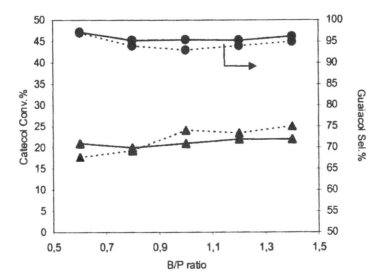

Figure 8 Catechol conversion (▲) and Guaiacol selectivity (●) of supported catalysts as functions of the B/P ratio. Results after 1.5h (—) and after 4.2h tos (---).

The presence of excess P leads to detrimental effects on both lifetime and selectivity. Excess B is instead soon lost from the catalyst in the form of volatile esters. B_2O_3 is characterized by a medium-strength acidity (12), as also determined by our TPD tests and FT-Ir measurements, which explains the initial high conversion.

Conclusions

B/P/O-based catalysts, both bulk and supported over α-alumina, have been tested as catalysts for the etherification of catechol with methanol to yield guaiacol. In bulk catalysts, the catalytic performance was affected by the B/P ratio; catalysts having B/P > 1.0 exhibited an initial high activity, which rapidly declined within the first hour of reaction due to the loss of excess boron in the form of volatile esters, and then remained almost constant with time-on-stream. Catalysts having B/P < 1.0 exhibited a continuous, rapid deactivation due to strong acid sites which developed on exposure to the reaction environment, and consequent build-up of coke. The most selective catalysts were those having B/P ≥ 1.0, due to the intrinsic mild acidity of BPO_4.

The reactivity of supported catalysts was instead less dependent upon the B/P ratio; moreover, they did not show any deactivation phenomenon during the first hours of reaction.

References
1.   S. Furusaki et al., Eur. Patent 420,756 A2, 1990; S. Umemura et al., JP 52152889, 1977; S. Nagai et al., JP 76108026, 1976; S. Nagai et al., 76122030, 1976, all assigned to Ube Ind.
2.   M. Brunelli et al., US Patent 4,654,446, 1987, assigned to Eniricerche S.p.A.
3.   S. Porchet, L. Kiwi-Minsker, R. Doepper and A. Renken, *Chem. Eng. Sci.*, **51**(11), 2933 (1996)
4.   S. Porchet, S. Su, R. Doepper and A. Renken, *Chem. Eng. Tech.*, **17**, 108 (1994)
5.   G.J. Hutchings, I.D. Hudson and D.G. Timms, Stud. Surf. Sci. Catal. (Elsevier Science), **88** (*Catalyst Deactivation 1994*), 663 (1994)
6.   J.B. Moffat and A.S. Riggs, *J. Catal.*, **42**, 388 (1976)
7.   M. Marczewski, J.P. Bodibo, G. Perot and M. Guisnet, *J. Molec. Catal.*, **50**, 211 (1989)
8.   R. Pierantozzi and A.F. Nordquist, *Appl. Catal.*, **21**, 263 (1986)
9.   V. Durgakumari, S. Narayanan and L. Guczi, *Catal. Lett.*, **5**, 377 (1990)
10. R. Tleimat-Manzalji, D. Bianchi and G.M. Pajonk, *Appl. Catal., A: General*, **101**, 339 (1993)
11. E. Santacesaria, D. Grasso, D. Gelosa and S. Carrà, *Appl. Catal.*, **64**, 83 (1990)
12. H. Miyata and J.B.Moffat, *J. Catal.*, **62**, 357 (1980)
13. J.B. Moffat and J.F. Neeleman, *J. Catal.*, **39**, 419 (1975)
14. J.B. Moffat and J.F. Neeleman, *J. Catal.*, **34**, 376 (1974)
15. G.C. Pimentel and A.L. McClellan, *The hydrogen bond*, W.H. Freeman, S. Francisco, 1960

Production of Aromatics Catalyzed by Commercial Sulfated Zirconia Solid Acids

J. H. Clark[a]*, G. L. Monks[b], D. J. Nightingale[a],
Peter M. Price[a] and J. F. White[c]

[a] *Department of Chemistry, University of York, Heslington, York YO10 5DD, UK.*
[b] *MEL Chemicals, PO BOX 6, Clifton Junction, Swinton Manchester M27 8LS, UK*
[c] *Engelhard Corporation, R&D, 23800 Mercantile Road Beachwood, Ohio 44122-5945, USA.*

Abstract

Commercial sulphated zirconia is a useful solid acid catalyst for Friedel-Crafts reactions. In benzoylations the catalyst has moderate activity which is sensitive to pretreatment conditions, and shows good reusability. In the alkylation of benzene using 1-dodecene, the catalyst activity and selectivity are very impressive and comparable to aluminium chloride while not having the problems of catalyst separation and aqueous waste.

Introduction

Lewis and Brønsted acid catalysis is widely used throughout organic chemistry for the production of many useful intermediates and products. One such example is Friedel-Crafts catalysis, which is commonly carried out on the industrial scale using homogeneous $AlCl_3$ or HF. Although reaction rates are good, major concerns exist over the use of $AlCl_3$ and, to a lesser extent, HF. These are problems associated with poor product selectivity, plant corrosion, catalyst separation and in the case of $AlCl_3$ formation of large quantities of aqueous aluminous waste, which, on an industrial scale, is environmentally unacceptable (1,2).

Solid acid catalysts can be useful alternatives to the traditional homogeneous materials. Their heterogeneous nature allows for simple separation at the end of the reaction, hence removing the necessity for an aqueous quench stage in the process. This is far more attractive from an environmental point of view. We have previously reported the use of supported $AlCl_3$ as an alkylation catalyst (3,4), however here we would like to report the use of sulfated zirconia as a potential alternative to $AlCl_3$. This commercially available catalyst is simple to prepare, thermally stable up to 700 °C and, owing to its heterogeneous nature, easily recoverable.

Experimental

Catalyst Preparation

40g of crushed (average particle size <100μm), commercially available sulfated zirconium hydroxide extrudate(grade EC0150E, supplied by MEL Chemicals/MEI and Engelhard Corporation) was placed in an open crucible and calcined in static air at a chosen temperature between 300 – 620 °C for 3 hours. After calcination, the sample was immediately placed in a desiccator and allowed to cool to room temperature over P_2O_5 and silica gel. In order to investigate how the cooling method utilised affected activity, a number of samples were cooled back to room temperature with the crucible left open to the atmosphere. After transferring to screw capped jars, the prepared catalysts were stored in a desiccator over P_2O_5 and silica gel. It was not necessary to pre-dry the catalysts again before use in the reactions.

Sample Characterisation

Surface area measurements were obtained on a Coulter SA 3100 machine using a 5 point BET method with nitrogen as adsorbate at –196 °C. Samples (~ 0.2g) were outgassed at 200 °C for 60 minutes before analysis.
Sulfate analysis was performed on a Leco 532-500 sulfur analyser.
Pyridine titrations were carried out as follows: 0.2 g of catalyst was weighed into a sample tube and the tube placed in a desiccator with pyridine in the bottom. After 18 hours, the sample tube was removed and transferred to a Schlenk line where it was heated under vacuum at 50 °C for 90 minutes to remove any excess pyridine. The catalyst was then ground with KBr and analysed by DRIFTS. Pre-drying the catalyst at 300 °C for 90 minutes prior to pyridine exposure gave identical spectra.

Reaction Procedures

Acylation – Sodium dried benzene (0.5 mol), benzoyl chloride (0.05 mol) and dodecane (0.013 mol, internal standard) were stirred at 85 °C under argon. After addition of the catalyst (10 g), samples were taken periodically and analysed by GC and GC-MS.

Alkylation – Catalyst (1.0 g) was stirred under argon in sodium dried benzene (0.25 mol) at 35 °C. Dodecene (0.025 mol) was added drop-wise using a syringe pump over 35 minutes, and a sample taken at 40 minutes and analysed by GC and GC-MS.

Results and Discussion

Physical Properties of Sulfated Zirconia

Surface Area and Sulfate Loading

Both surface area and sulfate loading of the catalysts have been found to exhibit a linear decrease with increasing calcination temperature (Figure 1 and 2).

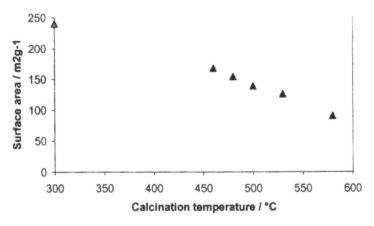

Figure 1 : Variation in surface area with calcination temperature (300 °C sample calcined for 26 hours)

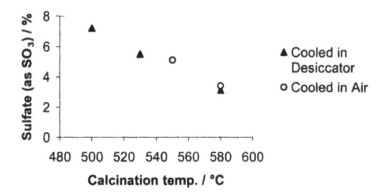

Figure 2 : Variation in sulfate loading with calcination temperature

Pyridine Titrations

Pyridine is a useful probe to identify the presence of Lewis or Brønsted acid sites in a material. The major band at 1541 cm^{-1} in Figure 3 indicates that the calcined sulfated zirconias are strongly Brønsted in nature. A relatively low concentration of Lewis acid sites (indicated by a band at 1440 cm^{-1}) are present. This is in contrast to a number of papers which report that sulfated zirconia contains a high percentage of Lewis sites (5,6). However, this is only true when the catalyst is calcined and then exposed to pyridine *in situ*. Our samples have been calcined, cooled in a desiccator or open to the air, stored and handled as they would be for use in a reaction. Presumably, even this limited exposure to the atmosphere significantly reduces the number of Lewis sites. Indeed other workers report that sulfated zirconia only exhibits Brønsted acidity (7).

Benzoylation of Benzene using Benzoyl Choride

The benzoylation of benzene using benzoyl chloride, a typical Lewis acid catalysed reaction, was chosen as a model reaction to investigate the activity of sulfated zirconia (Figure 4). Only a limited number of reports can be found in the literature utilising sulfated zirconia as a Friedel-Crafts acylation catalyst (7-11).

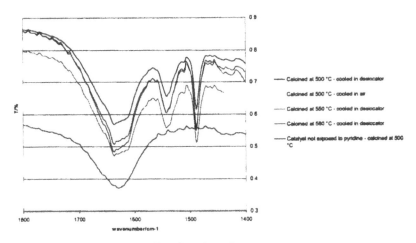

Figure 3 : DRIFTS of pyridine doped catalyst

Figure 4 : Benzoylation of benzene using benzoyl chloride

The effect of calcination temperature/cooling on catalyst activity

Activity of the catalyst is dependent on two important factors. The first is the calcination temperature used to prepare the catalyst. This is clearly shown in Figure 5 with maximum activity obtained at 500 °C. This value is somewhat lower than the optimal calcination temperatures found by Jia *et al.* of between 600-650 °C (9). Interestingly, it was found that catalyst calcined at 500 °C under argon was dark brown in colour in contrast to the usual cream colour obtained after calcining in static air. The activity of this catalyst was found to be

paticularly low (Figure 5). This proves the importance of oxygen in the
activation process presumably so as to keep sulfur in the +6 oxidation state

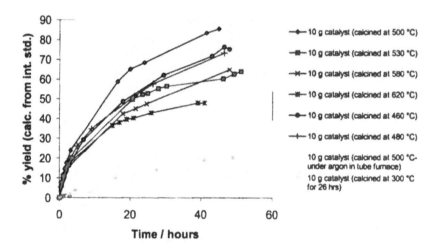

Figure 5 : Effect of calcination temperature on the benzoylation of benzene (0.5
mol) using benzoyl chloride (0.05 mol) at 85 °C

The second important factor is how the catalyst is cooled back to room
temperature after calcination. Figure 6 shows that cooling in a desiccator yields
a more active catalyst. Presumably, exposure to atmospheric moisture leads to a
decrease in the quantity and strength of Lewis acid sites, hence reducing the
activity of the catalyst. This is supported by the reduction in the band at 1440
cm⁻¹ characteristic of Lewis acid sites in the spectrum of the pyridine doped
material (Figure 3). It was also found that with both the desiccator and air
cooled catalysts, pre-drying at 300 °C for 1-2 hours before use failed to increase
catalytic activity perhaps because this is insufficient to remove strongly
adsorbed water.

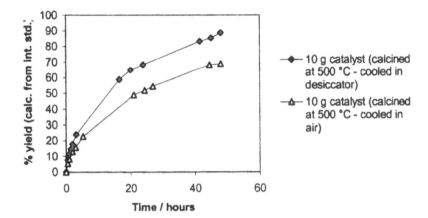

Figure 6 : Effect of cooling method on the benzoylation of benzene (0.5 mol) using benzoyl chloride (0.05 mol) at 85 °C

Reusing the catalyst is very simple. After reaction, the sulfated zirconia was allowed to settle, the liquor decanted off and the remaining catalyst washed with two 25 cm^3 portions of benzene (to remove any adsorbed benzophenone) prior to recharging with fresh reactants. The graph in Figure 7 shows that a significant decrease in the activity only occurs after the 4th use, and this is partially due to removal of catalyst from the system through sampling. This means the catalyst has a turnover number in excess of 20, much greater than the turnover number of AlCl$_3$ of 1 although the homogeneous reaction is much faster.

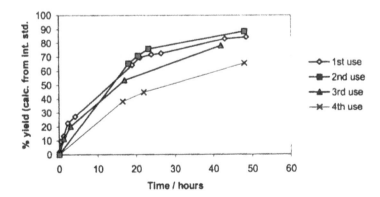

Figure 7 : Catalyst reuse in the benzoylation of benzene

Reaction of 4-chlorobenzoyl chloride with benzene occurs at about the same rate in the presence of sulfated zirconia (32% 4-chlorobenzophenone after 2.5h at 85°C).

Alkylation of Benzene using Dodecene and other alkenes

The second class of reaction we have studied is the industrially important alkylation of benzene, (Figure 8). Initial studies have used dodecene as the alkylating agent. This particular reaction is extensively used in the manufacture of detergent linear alkylbenzene sulfonates (LABS).

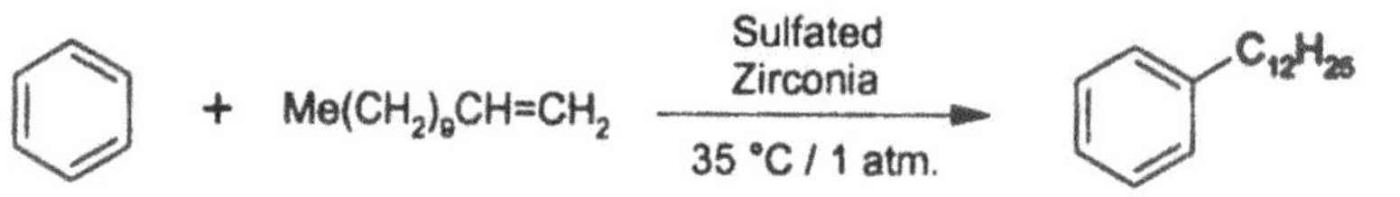

Figure 8 : Alkylation of benzene using dodecene

We have found that sulfated zirconia will rapidly catalyse the alkylation of benzene giving excellent selectivity to the mono-alkylated product. Moreover, selectivity to the industrially preferred 2-isomer is identical to that using $AlCl_3$ under the same conditions. Importantly however, the heterogeneous nature of sulfated zirconia allows simple separation of the catalyst after reaction, without the need for an aqueous quench step. This heterogeneous alkylation reaction can be compared to those using solid acid zeolites which only give good reaction rates under forcing conditions of temperature and pressure (12).

Catalyst	Dodecene / % GC Area	LAB / % GC Area	2-Isomer / % of LAB	Dialkylate, Oligomers / % GC Area
Sulfated Zirconia[a]	0	93	43	6
$AlCl_3$[b]	0	93	43	5

Table 9 : Alkylation of benzene (0.25 mol) with dodecene (0.025 mol) at 35 °C. Dodecene added over 35 mins, samples taken at 40 mins [a] 1g of sulfated zirconia, calcined at 480 °C. [b] 0.133g of $AlCl_3$ – equivalent to 1g of sulfated zirconia with a sulfate loading of 1.0 mmolg^{-1}

We have successfully extended the studies to the alkylation of benzene using other alkenes. 1-Hexene, 1-octene and 1-hexadecene all react with benzene in the presence of sulfated zirconia under identical conditions to those described above to give 91-94% monoalkylate.

We are currently studying the effects of benzene/alkene ratio and the use of internal alkenes and commercial alkene/alkane mixtures. We are also investigating the optimisation of the alkylation process including studies on catalyst reactivation and reuse.

Acknowledgments

We would like to thank MEL Chemicals Ltd./MEI and Engelhard Corporation for financially supporting this work and supplying catalysts, and the Royal Academy of Engineering/EPSRC for a Clean Technology Fellowship for JHC.

References

1. A.J. Butterworth, S.J. Tavener and S.J. Barlow, In *Chemistry of Waste Minimization*, J.H. Clark, Ed., Blackie Academic & Professional, Glasgow, p.522 (1995).
2. P.M. Price, J.H. Clark, K. Martin, D.J. Macquarrie and T.W. Bastock, *Org. Process Res. Dev.*, **2**, 221 (1998).
3. J.H. Clark, P.M. Price, K. Martin, D.J. Macquarrie and T.W. Bastock, *J. Chem. Research (S)*, 430 (1997).
4. J.H. Clark, K. Martin, A.J. Teasdale and S.J. Barlow, *Chem. Commun.*, 2037 (1995).
5. M.-T. Tran, N.S. Gnep, G. Szabo and M. Guisnet, *Appl. Catal. A: Gen.*, **171**, 207 (1998).
6. B.H. Davis, R.A. Keogh, S. Alerasool, D.J. Zalewski, D.E. Day and P.K. Doolin, *J. Catal.*, **183**, 45 (1999).
7. V. Quaschning, J. Deutsch, P. Druska, H.-J.Niclas and E. Kemnitz, *J. Catal.*, **177**, 164 (1998).
8. G. D. Yadav and A. A. Pujari, *Green Chem.*, **1**, 69 (1999).
9. C.-G. Jia, M.-Y. Huang and Y.-Y. Jiang, *Chin. J. Chem.*, **11**, 452 (1993).
10. K. Tanabe, T. Yamaguchi, K. Akiyama, A. Mitoh, K. Iwabuchi and K. Isogai, *Int. Congr. Catal., (Proc.) 8th*, **5**, V601 (1984).
11. K. Matsuzawa, *Prepr. Am. Cehm. Soc., Div. Pet. Chem.*, **42**, 734 (1997)
12. J.L.G. de Almeida, M. Dufaux, Y. Ben Taarit, C. Naccache, J. Oil Am. Chem. Soc., **71**, 675 (1994).

High Octane Fuel Ethers from Alkanol/Crude Acetone Streams via Inorganic Solid Acid Catalysis

John F. Knifton and P. Eugene Dai

Shell Chemical Company, P.O. Box 1380, Houston, TX 77251-1380

Abstract

This paper deals with the generation of high-octane aliphatic ethers - particularly methyl *tert*-butyl ether (MTBE), diisopropyl ether (DIPE), and isopropyl *tert*-butyl ether (IPTBE) - from readily available alkanol and crude acetone feed streams. Each etherification synthesis has been demonstrated using inorganic solid acid catalysis – particularly oxide-supported heteropoly acids and HF-treated clays in the case of MTBE production, and zeolite Beta, or transition-metal-modified β-zeolites, where DIPE or IPTBE are the intended products. Crude acetone hydrogenation to isopropanol (IPA) intermediate is effected using a nickel-rich, bulk-metal, catalyst.

Methyl *tert*-Butyl Ether

Methyl *tert*-butyl ether (MTBE) production in the U.S. now stands at ca. 290,000 barrels/day (1). The majority of this MTBE is generated through isobutene etherification with methanol (eq 1) over an acidic (sulfonated) resin catalyst (2), the isobutene being introduced as a mixed C-4 alkene/alkane stream, such as Raffinate-1, that typically has an isobutene content of 10-20%. Some MTBE is also produced commercially from *tert*-butanol – a by-product of propylene oxide manufacture. In this case, the *tert*-butanol may be reacted directly with methanol in the presence of an acid catalyst to give the desired MTBE (eq 2), with water as a coproduct.

For reaction (2) the choices of solid acid catalyst include: Hydrogen ion-exchange resins, such as Amberlyst 15 (3), and inorganic solid acids. We have demonstrated the successful application of three classes of inorganic solid acid catalysts for the generation of MTBE via synthesis route (2). These classes of catalyst include:

145

- Heteropoly acids on Group III and IV oxides (4).
- Hydrogen fluoride-treated montmorillonite clays (5).
- Mineral acid-activated montmorillonite clays (6).

$$\underset{CH_3}{\overset{CH_3}{>}}C = CH_2 \; + \; MeOH \; \longrightarrow \; CH_3 - \underset{CH_3}{\overset{CH_3}{C}} - O - Me \qquad (1)$$

$$CH_3 - \underset{CH_3}{\overset{CH_3}{C}} - OH \; + \; MeOH \; \longrightarrow \; CH_3 - \underset{CH_3}{\overset{CH_3}{C}} - O - Me \; + \; H_2O \qquad (2)$$

We focus here on catalyst screening in continuous reactor configurations, where we have identified certain intrinsic advantages to the use of the inorganic solid acids (versus hydrogen ion-exchange resins). The advantages of the inorganic solid acids – based upon their greater thermal and oxidative stabilities – include:

- They allow etherification to be conducted at higher operating temperatures (> 140°C) where rates of MTBE formation are extremely facile.
- They allow the optional production of pure isobutene as an attractive coproduct.
- They allow use of crude *tert*-butanol feed streams containing various oxygenated impurities, including peroxides.

Oxide-Supported Heteropoly Acids

MTBE syntheses from *tert*-butanol plus methanol has recently been described for 12-molybdophosphoric acid ($H_3PMo_{12}O_{40}$, designated Mo-P) and 12-tungstophosphoric acid (W-P) impregnated into high surface area titania, silica, and alumina supports (7). Each experimental series was conducted using a 25 cc capacity, plug-flow, continuous reactor over a range of etherification conditions. The feedstock was a 2:1 molar mixture of methanol and *tert*-butanol. We conclude (7) that:

- MTBE may be selectively generated from methanol/*tert*-butanol mixtures in up to 90 molar %, at moderate temperatures, using the 12-tungstophosphoric acid-on-titania and 12-molybdophosphoric acid-on-titania catalysts. These MTBE selectivities are substantially higher than those reported previously for heteropoly acids.

- *tert*-Butanol (tBA) conversion levels reaching 81% may be achieved at moderate feed rates (LHSV 1), and even at high throughputs (LHSV 4-8), the *tert*-butanol conversions exceed 70%.

- Increasing the etherification temperature generally leads to a greater isobutene (i-C$_4$) by-product make, so that by 180°C, with >80% tBA conversions per pass, the product effluent now comprises two-phase mixtures:
    1. A lighter, isobutene-MTBE product rich phase, and
    2. A heavier, aqueous methanol rich phase.

We also compared our published ^{31}P and ^{1}H MAS NMR data for the fresh W-P/TiO$_2$ catalysts (8) with data collected after their extended use in MTBE service under the etherification conditions typified *supra*. Generally, we find that these oxide-supported 12-tungstophosphoric acid catalysts show excellent stability during etherification service, however, there are some significant changes in their ^{31}P and ^{1}H NMR spectra. In particular, we find:

- Extended usage in MTBE service (4) generally leads to a qualitative decrease in total phosphorus signal. Some loss of surface-bonded heteropoly acid moiety is evident as a function of time on stream, and this is most noticeable at temperatures >160°C where there is product phase separation. These ^{31}P data are consistent with P-elemental analyses for the same recovered catalyst samples.

- In addition to the intact Keggin ion signal at ca. −15 ppm, the used catalyst, after >200 hrs service also shows resonances for a range of partially and highly fragmented clusters in the region −2 to 2 ppm. Continued MTBE service leads to brand new bands in the 4 ppm region due to organophosphate species. These spectral changes are illustrated in Figure 1 for a typical W-P/TiO$_2$ catalyst after 430 to 700 hours of high-temperature service (140–165°C). The P-OR, organophosphate, species on the titania surface were confirmed by Block decay and cross polarization experiments.

- ^{1}H Solid state, Block decay, MAS data for the same series of used catalysts generally exhibit a significant shift in the remaining acidic protons, to 3.15-3.22 ppm, and new resonances at 1.04-1.07 ppm due to the aliphatic protons of the coordinated organic, P-OR, species noted *supra*. A typical spectrum is illustrated in Figure 2. There is still some evidence for the highly acidic protons (8) at 11.6 ppm.

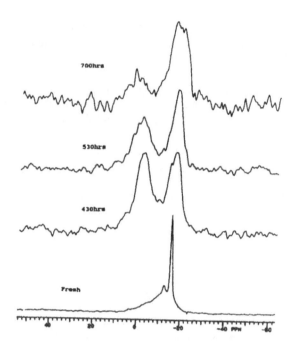

Figure 1. ^{31}P NMR spectra for fresh and used W-P/TiO$_2$ catalysts. Ref. 7.

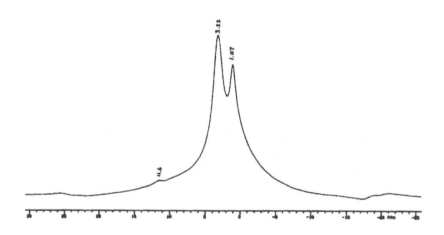

Figure 2. ^{1}H NMR specrtum of W-P/TiO$_2$ catalyst after 530 hrs of MTBE service. Ref. 7.

HF-Treated Clays

A clear illustration of the isobutene-MTBE product phase separation phenomena at higher etherification temperatures is provided here using a hydrogen fluoride-treated montmorillonite clay catalyst. This modified clay was prepared by treating a neutral montmorillonite with 48% hydrofluoric acid solution to a fluoride content of 1.2%. Feeding the plug-flow reactor with a 1.1:1 molar mix of methanol and *tert*-butanol at a series of etherification temperatures, we observe the onset of product phase separation at above 150°C and two distinct product phases in the operating temperature range 160-180°C. The compositions of each phase for two on-line samples taken at 160°C and 180°C are illustrated in Table 1. At both temperatures the lighter phase comprises isobutene and MTBE as the principal components and the heavier phase is mainly aqueous methanol. Estimated *tert*-butanol conversion levels at 160°C are ca. 84%; they exceed 90% at 180°C. Desired MTBE product is thereby concentrated in the lighter phase, while by-product water and recycle methanol are found primarily in the heavier phase.

In general, for both the HF-treated clays and oxide-supported heteropoly acid catalysts, a methanol-to-*tert*-butanol feed mix close to molar stoichiometry (i.e. containing very little excess MeOH), enhances the onset of MTBE product phase separation and leads to higher isobutene yields in the 160°-180°C etherification temperature regime.

The same HF-modified montmorillonite clay catalyst has also proven to have excellent performance in MTBE service over extended periods using crude *tert*-butanol feedstocks. The data in Figure 3 illustrate 60 days of MTBE service using a *tert*-butanol feedstock that also contained significant quantities of water, acetone, isopropanol, di-*tert*-butylperoxide, *tert*-butyl formate, plus traces of MTBE recycle. Etherification was conducted using a 2:1 molar mix of methanol and crude *tert*-butanol, conditions were 140°C, 20 bar, and LHSV 2. Over the 1200 hour period, tBA conversion levels remained close to 71% with MTBE molar selectivities in the range 65-71%. The remaining coproduct is then primarily isobutene (31-24% molar selectivity).

It is particularly noteworthy that neither the peroxide (di-*tert*-butylperoxide), nor the ester (*tert*-butyl formate), components in this crude TBA feedstock affected the HF/clay performance, since it is known that these materials can lead to rapid deactivation of hydrogen ion-exchange resins. Hydrogen fluoride treatment of the montmorillonite clays provides particularly active and stable acid sites – more so than with the mineral acids (6).

Table 1 MTBE from methanol/*tert*-butanol.

Catalyst	MeOH/TBA Molar Ratio	Temp. (C)	Feed Rate (LHSV)	Time On Stream (Days)	Sample	Product Composition (Wt. %) H2O	MeOH	isobutene	tBA	MTBE
HF/Clay	1.1:1				Feed		32.5		67.1	
		120	2	1	1	3.4	28.4	4.1	51.1	12.7
					2	3.8	30.1	4.1	49.4	12.3
		140		2	3	7.7	23.8	9.2	32.4	26.7
					4	8	24.1	9	32.2	26.5
		160		3	5a	9.9	22.1	25.8	11	30.9
						30.2	37.5	6.6	10.5	14.9
					6a	9	21.2	27.4	10.8	31.4
						30.7	38.1	6.3	10.1	14.5
		180		4	7a	0.8	7.8	68.9	2.8	19.4
						29.6	52.2	5.8	5.6	6.4
					8a	0.9	8.1	68.6	2.8	19.5
						29.7	52.3	5.7	5.6	6.3

a. Two-phase product

Diisopropyl Ether

Diisopropyl ether (DIPE) has been proposed as an alternative oxygenate to methyl *tert*-butyl ether (MTBE) and *tert*-amyl methyl ether (TAME) in gasoline fuel blending since it has similar physical and blending properties to MTBE and TAME (9-11). Mobil Corp. has announced a new etherification technology to produce DIPE, based upon propylene (a readily available refinery product) and water, that has the advantage of allowing refiners to produce an oxygenate without relying upon an "external" supply of alcohols (11,12).

We have investigated an alternative route to DIPE, starting from low-value, crude acetone streams. These liquid streams are generated in large volume as a result of propylene oxide/MTBE manufacture (13) and typically comprise 20-

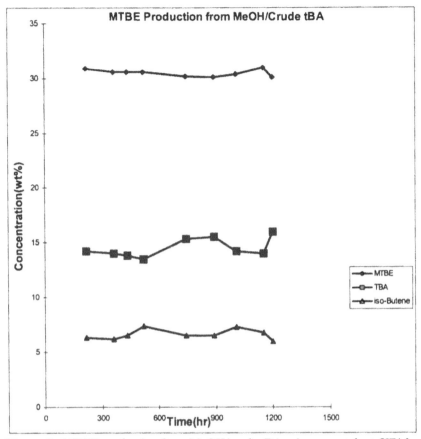

Figure 3. MTBE production from MeOH/crude tBA mixture; catalyst, HF/clay. Ref. 7.

80% acetone. In particular, during propylene oxide cogeneration, a large number of C-1 to C-4 oxygenates may be coproduced, including acetone, *tert*-butanol, formic acid, acetic acid, and their ester derivatives (13). We have now demonstrated that the acetone fraction may be selectively converted to DIPE in a two-step process involving initial hydrogenation to isopropanol (IPA), followed by IPA dehydration to DIPE (eq 3) (14).

Selective hydrogenation of said acetone stream in a continuous, upflow, reactor system packed with nickel, copper, chromium bulk-metal catalyst comprising ca. 72% Ni, at 160°C, provided near quantitative (99%) acetone conversion levels at LHSV of 0.5, with isopropanol as the major product fraction. Typical experimental data are summarized in Table 2. Selectivity to IPA was typically in the range 76-80 mole%. Catalyst activity may be sustained for extended periods without loss of performance – particularly in terms of acetone conversion levels and IPA selectivities. An added, critical feature of using this type nickel-rich, bulk-metal catalyst is that any allyl *tert*-butyl peroxide (ATBP) or *tert*-butyl hydroperoxide (HTBP) fractions present in this crude acetone feed are quantitatively converted to more innocuous alcohols, e.g. *tert*-butanol, under the hydrogenation conditions of Table 2 – without causing catalyst deactivation.

Table 2   Crude acetone hydrogenation to isopropanol.

			Composition(%)a					
Temp. (C)	LHSV	Sample	Acetone	IPA	MeOH	tBA	tBF	ATBP
		Feed	61.7	0.1	13.9	16.7	0.1	3.3
160	0.5	Product	0.8	48.3	15.8	30.8	<0.1	<0.1

a.   Designations: IPA, isopropanol; MeOH, methanol; tBA, *tert*-butanol; tBF, *tert*-butyl formate; ATBP, allyl *tert*-butyl peroxide

Dehydration of the IPA intermediate to diisopropyl ether (DIPE) has also been demonstrated in continuous, upflow, reactor systems using three classes of acidic, large-pore zeolites (14):

- Zeolite Beta.
- Transition metal-modified β-zeolites.
- Dealuminized Y-zeolite.

Typically, DIPE is generated in up to ca. 14% concentration in the crude liquid product under moderate dehydration conditions. Table 3 illustrates DIPE syntheses from the crude IPA stream of Table 2 using an acidic β-zeolite catalyst (80% Beta, 20% alumina binder in 1/16" diameter extruded form) over a reaction temperature range of 120-180°C. At 180°C, the estimated isopropanol conversion level is 67% and the crude liquid effluent comprises 13.9 wt% diisopropyl ether (see sample #4). Total ether production in this case is 31.4 wt%, with methyl isopropyl ether (MIPE) and dimethyl ether (DME) as significant components – together with C-4 olefin/dimer. Clearly these particular ethers, plus alkenes, could be avoided if one started with a purer acetone feed stream, or if the IPA intermediate was fractionated prior to dehydration. At lower operating temperatures (e.g. 120°C) with the same crude IPA feed stream, MTBE is the major product fraction – from methanol/tert-butanol etherification, but at 180°C, methyl tert-butyl ether is no longer equilibrium favored [see (15) and Table 1]. It is particularly noteworthy that the Beta-zeolite catalyst maintains dehydration activity and good DIPE selectivity with this crude hydrogenated acetone feedstock over extended periods (14).

Table 3 Isopropanol dehydration to diisopropyl ether.

Temp. (C)	Sample	Ethers					Composition(%)a					
		DIPE	MIPE	MTBE	IPTBE	DME	Acetone	IPA	MeOH	tBA	H2O	C4H8/C8H16
							0.8	48.3	15.8	30.8	5.7	
120	#1	0.9	0.6	8.1	1.3	0.2	1.3	46.4	12.4	9.3	10.5	7.8
140	#2	3	1.8	4.6	0.9	1.5	1.8	42	12.9	5.1	12.1	10.2
160	#3	9.1	6.6	1.2	0.2	2.8	2.3	29.3	10.9	1.8	16.5	10.1
180	#4	13.9	12.8	0.3	0.4	4	2.2	15.8	4.4	0.8	8	10.9

a. Designations: MIPE, methyl isopropyl ether; IPTBE, isopropyl tert-butyl ether; DME, dimethyl ether

We conclude then that both the large-pore Beta zeolite, as well as transition-metal modified β-zeolites (15), are very effective catalysts for sustainable DIPE production using crude isopropanol feedstocks. While the presence of alkyl peroxides, such as ATBP and HTBP, in the original crude acetone is not deleterious to the performance of these solid acid catalysts (14), the concentration of water in the feed should be closely controlled. As noted by others (12), an aqueous content in the crude IPA feed of >10% can significantly impact the level of DIPE productivity. On the other hand, the use of crude IPA containing both MeOH and *tert*-butanol may provide a dialkyl ether-rich mix of products – including DIPE, MIPE, MTBE, IPTBE and DME - in high total yield and effluent concentration (e.g. 31.4% concentration, see Table 3, sample #4). Here the DIPE:MIPE:MTBE:IPTBE:DME product molar ratios at 180°C are typically ca. 1: 1.3 : 0.02 : 0.02 : 0.6. Lower etherification temperatures lead to higher proportions of MTBE and IPTBE (see samples #1 and 2, Table 3, generated at 120° and 140°C, respectively). These changes in dialkyl ether product distribution with increase, or decrease, in etherification temperature do not appear to follow the thermodynamic data for these five ether syntheses (15). More likely, the increases in DIPE, MIPE, and DME concentrations over the temperature range 120-180°C are associated with kinetic factors and the competition for surface active etherification sites.

Optionally, these types of DIPE, MIPE, MTBE, IPTBE, and DME ether mixtures could themselves be employed as valuable blends of oxygenates for octane enhancement and gasoline blending (9).

Isopropyl *tert*-Butyl Ether

Of the fuel ethers seriously being considered in the US for reformulated gasoline programs, isopropyl *tert*-butyl ether (IPTBE) has the triple advantages (16) of:
- Highest octane blending values.
- Lowest oxygen content.
- Low vapor pressure.

However, until now the only route to IPTBE was through the acid-catalyzed etherification of isobutene (oftentimes in short supply) with isopropanol (17,18). We have developed an alternative route to IBTBE from crude acetone streams, making it a co-product of propylene oxide manufacture, and have demonstrated that IPTBE can be made in good yield via:
- Selective hydrogenation of the crude acetone by-product stream to give isopropanol (eq 4).
- Isopropanol etherification with the *tert*-butanol coproduct to yield IPTBE plus water (eq 5).

$$>\!=\!O \quad + \quad H_2 \quad \longrightarrow \quad >\!-\!OH \qquad (4)$$

$$>\!-\!OH \quad + \quad t\text{-}BuOH \quad \longrightarrow \quad t\text{-}Bu\!-\!O\!-\!< \quad + \quad H_2O \qquad (5)$$

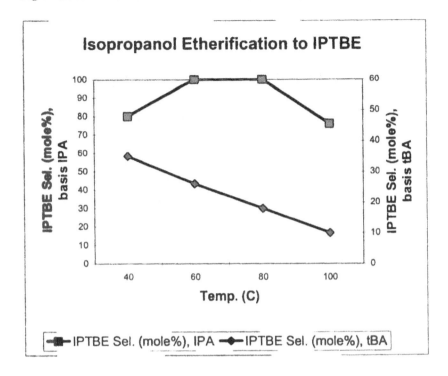

Figure 4. Crude IPA etherification to IPTBE with tBA. Ref. 20.

IPTBE is typically generated in near quantitative molar selectivities (basis IPA converted) in the crude liquid products under mild etherification conditions. Figure 4 illustrates IPTBE syntheses from the crude IPA stream of Table 2, plus added *tert*-butanol (IPA:tBA mole ratio 1:1.6), using an acidic β-zeolite catalyst (80% Beta, 20% alumina binder, in 1/16″ diameter extruded form), over a reactor temperature range of 40 to 100°C. At 60-80°C the estimated isopropanol conversion level is moderate (ca. 12%) and the crude liquid effluent comprises 7-8 wt% IPTBE. However, IPTBE molar selectivity basis tBA converted is below 50% with this crude feedstock due to competing tBA dehydration and oligomerization to isobutene and diisobutene, as well as the formation of smaller quantities of MTBE through etherification. It is significant, nevertheless, that the Beta-zeolite catalyst does maintain etherification activity and good IPTBA molar selectivity with this crude hydrogenated acetone feedstock for extended periods (19).

Liquid-phase isopropyl *tert*-butyl ether synthesis from IPA/tBA mixtures (eq 5) is essentially thermo neutral (20), in contrast to IPTBE from isobutene.

Isolating and purifying (99%) the intermediate isopropanol from Table 2, and running etherification with close to stoichiometric quantities of *tert*-butanol, provides measurably higher IPTBE selectivities over this temperature range and fewer competing reactions (20). However, etherifications conducted much above 100°C lead to the formation of diisopropyl ether as a dominant product (Table 3).

Conclusions

It has been demonstrated that MTBE, DIPE, and IPTBE can all be generated efficiently from low-cost feedstocks using inorganic solid acid catalysis.

References

1.   *Chemical Marketing Reporter*, pp. 5, March 29, 1999.
2.   *Chemical Economics Handbook* – SRI International, Report 543.7500 (May 1994).
3.   J. M. Adams, K. Martin, R. W. McCabe, and S. Murray, *Clay and Clay Minerals*, **34**, 597 (1986).
4.   J. F. Knifton, U.S. Patent 4,827,048 to Texaco Chemical Company (1989).
5.   J. F. Knifton, U.S. Patent 5,157,161 to Texaco Chemical Company (1992).
6.   J. F. Knifton, U.S. Patent 5,099,072 to Texaco Chemical Company (1992).
7.   J. F. Knifton and J. C. Edwards, *Applied Catal.*, **183**, 1 (1999) .
8.   J. C. Edwards, C. Y. Thiel, B. L. Benac, and J. F. Knifton, *Catal. Lett.*, **51**, 77 (1998).
9.   W. J. Piel, *Fuel Reformulation*, **2**, 34 (Nov./Dec. 1992).
10. F. P. Heese, M. E. Dry, and K. P. Moller, *Catal. Today*, **49**, 327 (1999).
11. A. Wood, *Chemical Week*, pp. 7 (April 15, 1992).
12. M. N. Harandi, W. O. Haag, H. Owen, and W. K. Bell, US Patent 5,144,086 to Mobil Oil Corporation (1992).
13. *Stanford Research Institute PERP Report 2E*, "Propylene Oxide" (August 1994).
14. J. F. Knifton and P. E. Dai, US Patent 5,430,198 to Texaco Chemical Inc. (1995).
15. J. F. Knifton, and P. E. Dai, *Catal. Lett.*, **57**, 193 (1999).
16. W. J. Piel, *Fuel Reformulation*, **4**, 28 (March/April 1994).
17. J. A. Linnekoski, A. O. I. Krause, A. Holman, M. Kjetsa, and K. Moljord, *Applied Catal.*, **174**, 1 (1998).
18. A. Calderon, J. Tejero, J. F. Izquierdo, M. Iborra, and F. Cunill, *Ind. Eng. Chem. Res.*, **36**, 896 (1997).
19. J. F. Knifton, E. L. Yeakey, and P. E. Dai, US Patent 5,449,838 to Texaco Chemical Company (1995).
20. J. F. Knifton, P. E. Dai, and J. M. Walsh, *Chem. Comm.*, 1521 (1999).

Vapour Phase Synthesis of Indole and Its Derivatives

M. Campanati[a], F. Donati[a], A. Vaccari[a], A. Valentini[a] and O. Piccolo[b]

[a]Dip Chimica Industr. e Materiali, Viale Risorgimento 4, 40136 Bologna (Italy)
[b]Chemi SpA, Via dei Lavoratori 54, 20092 Cinisello Balsamo MI (Italy)

Abstract

The vapour phase synthesis of indoles from anilines and diols was investigated using copper chromites and novel ZrO_2/SiO_2 catalysts. With the copper chromite catalysts an excess of aromatic amine was necessary to avoid polyalkylation products and the best results were obtained using a catalyst containing only promoters improving the physical properties (BaO, CaO and SiO_2), with operation at low temperature and LHSV values. The hourly productivity was improved by reducing the contact time, with a corresponding increase in the volume of carrier gas to be recycled. Both the excess of aniline derivatives and large volume of carrier gas required to operate at low contact time were avoided using novel ZrO_2/SiO_2 catalysts, able to operate with almost stoichiometric mixtures and using water as the main carrier gas. These catalysts showed catalytic results better than those reported in the literature, as well as a very good regenerability. This synthesis can be applied to a wide number of substrates (anilines and/or diols). In particular, using ethylene glycol the best yields in the corresponding indole were obtained when an alkyl group was located in the ortho-position to the amino group and the length of the chain was increased. Moreover, the differences in reactivity between aniline and alkylanilines were significantly smoothed by increasing the length of the diol chain.

Introduction

Nitrogen-containing heterocyclic compounds, especially indole or alkylindoles, are of considerable industrial interest, finding applications as intermediates in the production of pharmaceuticals, herbicides, fungicides, dyes, etc. (1-3). They are recovered from the biphenyl-indole fraction obtained by coal tar distillation, although today this source no longer seems able to cover completely the increasing market demand. Alternatively, they can be obtained in high yields by well-known liquid-phase reactions, which, however, present many drawbacks (4-8). In recent years increasing interest has been focused on reactions occurring in the vapour phase with heterogeneous catalysts (9-15) that exhibit many advantages in comparison to liquid phase syntheses (continuous production,

157

simplified product recovery, catalyst regenerability, absence of liquid waste stream, etc.). The patent literature is rich in data regarding the nature and composition of catalysts as well as the operating conditions for the synthesis of indole, while very little data has been reported on the synthesis of alkylindoles. Moreover, many different catalysts have been claimed, making it difficult to identify the reaction requirements and optimum catalyst properties. On the other hand, almost no data are avaiable in the open literature on the vapour phase synthesis of indoles nor on the possible reaction pathway.

Reported in this paper is a study on the vapour phase synthesis of a wide number of indoles, carried out taking into account the possible scale-up of the process. The synthesis was performed starting from aniline or alkylanilines and different types of diols (low price and widely available feedstocks), according to the following reaction:

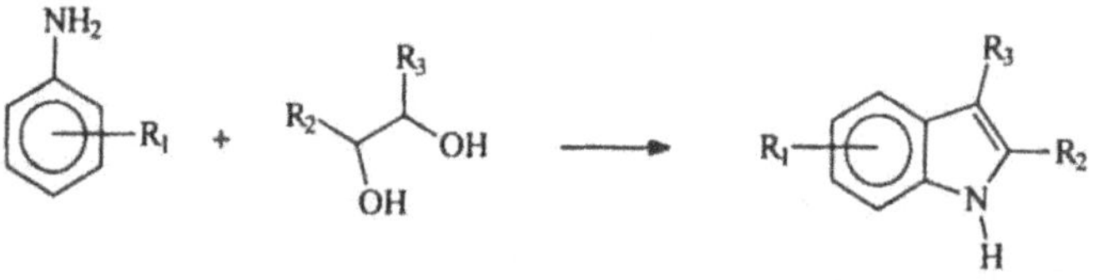

R_1 = H; -CH$_3$; -CH$_2$CH$_3$; -CH$_2$CH$_2$CH$_3$; -Cl
R_2 = H; -CH$_3$; -CH$_2$CH$_3$
R_3 = H; -CH$_3$

Experimental Section

The catalytic tests were carried out using 2.0 mL (ca.2.5 g, 425-850 μm particle size) of some commercial copper chromites [CAT 1 = Cu/Cr/Ba/Si/Ca (36:34:4:22:4 at. ratio %); CAT 2 = Cu/Cr/Mn/Ba (44:47:8:1 at. ratio %); CAT 3 = Cu/Cr (61:39 at. ratio %)] or of two home-made copper chromites (16), already tested in selective hydrogenation reactions (17,18) [CAT 4 = Cu/Cr/Mg (40:50:10 at. ratio %); CAT 5 = Cu/Cr/Co (40:50:10 at. ratio %)]. Alternatively, 4.0 mL (ca. 2.5 g, 425-850 μm particle size) of ZrO$_2$/SiO$_2$ mixed or supported catalysts (19) were used.

All the catalysts were previously activated in situ, increasing progressively the temperature in the range 573-603 K and using a 6 L/h flow of a H$_2$/N$_2$ (1:9 v/v) gas mixture. The tests were carried out in a fixed-bed glass microreactor (i.d. 7 mm, length 400 mm), placed in an electronically controlled oven and operating at atmospheric pressure. The isothermal axial temperature profile of the catalytic bed during the tests was determined using a 0.5 mm J-type thermocouple, sliding in a glass capillary tube. The organic feedstock was introduced by an Infors Precidor model 5003 infusion pump, while the gas composition and flow were controlled using Brook mass flow meters. After 1h of time-on-stream to reach stationary conditions, the products were condensed in two traps cooled at 268 K and collected in methanol, adding tridecane as an internal standard.

The analyses were carried out using a Perkin Elmer AutoSystem XL gas chromatograph, equipped with FID and a wide bore SE 54 column (Length 30 m, i.d. 0.53 mm, film width 0.8 μm). The products were tentatively identified by GC-MS using a Hewlett-Packard GCD 1800 system; the identifications were afterwards confirmed by comparing the experimental GC and GC-MS patterns with those obtained for pure reference compounds.

Results and Discussion

The synthesis of 7-ethylindole (ETI) from 2-ethylaniline (ETAN) and ethylene glycol (EG) was investigated, focusing attention on the role of catalyst composition and reaction parameters. The stability of ETI in the reaction conditions was first checked by heating in an autoclave for 4 h at 563 K under nitrogen, in the presence of 0.1 g of CAT 1; no significant formation of other compounds was observed.

The different copper chromites were investigated under two different atmospheres feeding an organic mixture containing an excess of ETAN. Table 1 shows that the best yields in ETI were obtained with CAT 1 (containing only promoters improving the physical properties), regardless of the hydrogen content in the reaction gas mixture. Moreover, high molecular weight compounds were not detected for any of the catalysts investigated. The increase in hydrogen content did not show any positive effect, whereas when pure nitrogen was used a significant worsening of the yield in ETI was observed, mainly due to a decrease

Table 1. Yield (%) in 7-ethylindole for the different copper chromite catalysts.[ETAN/EG = 10:1 mol/mol; T= 603 K; GHSV = 3,000 h^{-1}; LHSV = 0.1 h^{-1}; yield values referred to the ETAN fed (maximum = 10%)].

Sample	H_2/N_2 (1:1 v/v)	H_2/N_2 (1:9 v/v)
CAT 1	1.4	2.1
CAT 2	0.8	1.0
CAT 3	1.3	1.0
CAT 4	0.8	0.9
CAT 5	0.6	0.8

in conversion. Thus a reaction gas mixture containing a small percentage of hydrogen was adopted in all the following tests, considering also the advantages from the safety point of view.

On the basis of the previous results, CAT 1 was employed to define the role of the different reaction parameters. When the reaction temperature was increased, at first a significant increase in the catalytic activity up to 523 K ca. was observed while further increases in temperature worsened the yield as a consequence of the dramatic decrease in selectivity in ETI (Fig. 1).

In the following tests different ways to increase the hourly productivity were investigated. Increasing the flow rate of the organic mixture did not have any effect up to a LHSV value of 0.2 h^{-1}, after which the conversion decreased significantly, probably due to saturation of the catalyst active sites. The possibility to increase the ETI productivity by increasing the EG content in the organic mixture was also investigated and a decrease in selectivity in ETI, due to the formation of different polyalkylated indoles (Fig. 2) was obserevd. Thus, using chromite catalysts it is not possible to increase significantly the ETI productivity by increasing the flow rate of the organic mixture or moving its composition towards the stoichiometric one (ETAN/EG = 1:1 mol/mol), due to the saturation of the active sites and their high activity in the further alkylation of ETI. Finally, no significant difference in the yield in ETI was observed when GHSV was decreased from 3,000 to 180 h^{-1}, i.e. to the value reported in the patents [14,15].

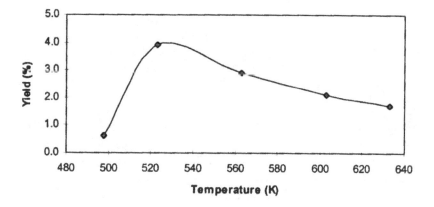

Figure 1. Yield (%) in 7-ethylindole as a function of the reaction temperature. [CAT 1 = 2 mL; ETAN/EG = 10:1 mol/mol; H$_2$/N$_2$ = 1:9 v/v; GSHV = 3,000 h^{-1}; LHSV = 0.1 h^{-1}; yield values referred to the ETAN fed (maximum = 10%).]

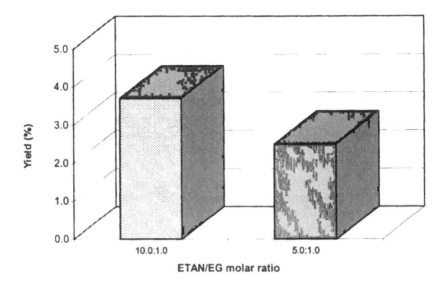

Figure 2. Yield (%) in 7-ethylindole as a function of the ETAN/EG molar ratio. [CAT 1 = 2 mL; T = 523 K; H_2/N_2 = 1:9 v/v; GSHV = 3,000 h^{-1}; LHSV = 0.2 h^{-1}; yield values referred to the ETAN fed (maximum = 10 and 20%, respectively).]

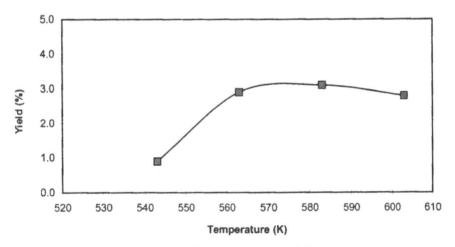

Figure 3. Yield (%) in 7-ethylindole as a function of the reaction temperature. [ZrO_2/SiO_2 mixture (4:2 v/v) = 4 mL; ETAN/EG = 10:1 mol/mol; H_2/H_2O = 3:7 v/v; GSHV = 2,000 h^{-1}; LHSV = 1.6 h^{-1}; yield values referred to the ETAN fed (maximum = 10%).]

In order to overcome the constraints of chromite catalysts, novel ZrO_2/SiO_2 catalysts were investigated [19]. ZrO_2 was chosen, because of its hydro-dehydrogenating and weakly acid properties [20-22]. ZrO_2 favours the activation of EG to 2-hydroxyacetaldehyde, while the acid sites of SiO_2 favour its attack on ETAN and the following formation of ETI. On the other hand, the acidity of the SiO_2 [23] is not so strong as to catalyse the transformation of EG to crotonaldehyde and, consequently, the alternative synthesis of 2-methyl-8-ethylquinoline from ETAN and EG, as already reported using an acid-treated commercial clay [24]. In the first tests, a mixture of ZrO_2 and SiO_2 was used (4:2 v/v), which showed an improvement in the yield in ETI in comparison to copper chromites, regardless of the reaction temperature, with maximum activity at 563-583 K (Fig. 3). Furthermore, with this catalyst it was possible to operate at very high LHSV values.

Unlike the copper chromites, using the ZrO_2/SiO_2 mixture it was possible to decrease the ETAN/EG molar ratio significantly, without any evidence of the formation of polyalkylated indoles. N-(2-hydroxyethyl)-2-ethylaniline (NEE) and N-vinyl-2-ethylaniline (NVE) were the main by-products observed (Table 2). In this way, it was possible to increase the hourly productivity of ETI, reaching the maximum value for a ratio ETAN/EG = 1:1 (mol/mol). NEE was the main product formed using a two-layer catalytic bed (ZrO_2 + SiO_2, and maintaining the previous volumetric ratio) or only ZrO_2 calcined at low temperature, thus showing the role of acidity in the synthesis of ETI and the requirement of an intimate interdispersion of acid and hydrogenating active sites.

Another relevant advantage of this catalyst is that it allowed one to replace nitrogen with steam in the gas mixture, although the percentage of hydrogen is

Table 2. Yield in 7-ethylindole and the two main by-products N-(2-hydroxyethyl)-2-ethylaniline (NEE) and N-vinyl-2-ethylaniline (NVE) as a function of the ETAN/EG molar ratio [ZrO_2/SiO_2 mixture (4:2 v/v) = 4 mL; H_2/H_2O = 3:7 (v/v); T= 563 K; GHSV = 2,000 h^{-1}; LHSV = 1.6 h^{-1}; yield values referred to the ETAN fed].

ETAN/EG (mol/mol)	ETI	NEE	NVE
10:1	2.9	0.0	0.0
4:1	4.5	0.0	0.0
1:1	8.5	0.5	< 0.1
1:2	9.7	4.7	< 0.1
1:10	12.1	12.3	7.5

higher. This allows operation at high values of GHSV, with a corresponding decrease in reactor volume, without increasing the gas volume above that which can be economically recycled. The steam can be condensed after reaction, so it is easily separated from the organic mixture and recycled after vapourization.

Moreover, using steam as the inert gas, an increase was observed in the yield in ETI, associated with the disappearance of NVE. Although a detailed description of the reaction pathway for the synthesis of ETI is outside the scope of the present study and will be reported in a following paper, it can be anticipated that a key intermediate in the synthesis of ETI may be identified as N-(2-ethylphenyl)-2-aminoacetaldehyde (A) (Fig. 4), which may cyclize to ETI or be reversibly hydrogenated to NEE (which in turn forms NVE by dehydration) or polymerize to high molecular weight compounds, which are responsible for catalyst deactivation with time-on-stream (Fig. 5).

In comparison to the ZrO_2/SiO_2 mixture (4:2 v/v, corresponding to 50:50 w/w), significant improvements in catalytic performances were achieved by preparing the ZrO_2-based catalysts by incipient wetness impregnation of the same SiO_2 with a solution of zirconium (IV) acetylacetonate in acetic acid, notwithstanding the lower ZrO_2-contents. For these catalysts, the yield in ETI increased with decreasing ZrO_2 up to 5% (w/w), while a further decrease to 2% (w/w) gave rise to worsening of the yield (Fig. 6). Thus, these data support the previous hypothesis on the role of both acid and dehydrogenating sites in the synthesis of ETI and the requirement of a their optimum interdispersion to have good catalytic performances.

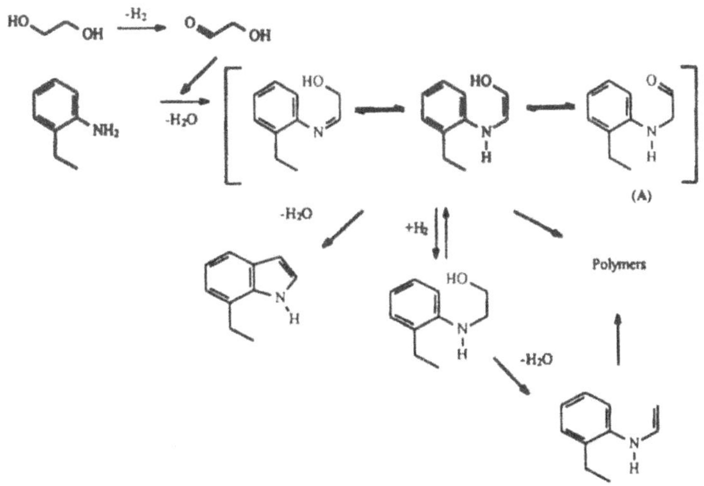

Figure 4. Proposed reaction pathway for the vapour phase synthesis of 7-ethylindole from 2-ethylaniline and ethylene glycol.

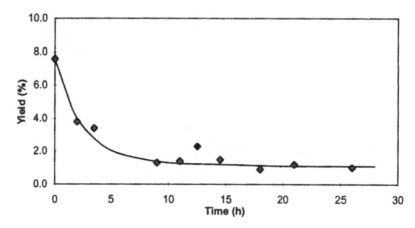

Figure 5. Yield (%) in 7-ethylindole as a function of the time-on-stream. [ZrO_2/SiO_2 mixture (4:2 v/v) = 4 mL; ETAN/EG = 1:1 mol/mol; H_2/H_2O = 2:8 (v/v); T = 583 K; GSHV = 2,900 h^{-1}; LHSV = 1.6 h^{-1}; yield values referred to the ETAN fed.]

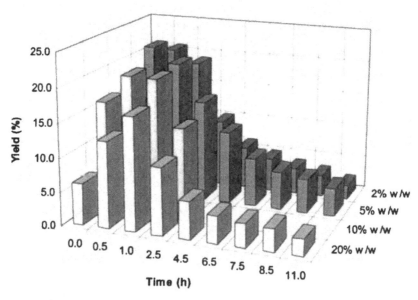

Figure 6. Yield (%) in 7-ethylindole as a function of the ZrO_2-content for the catalyst prepared by incipient wetness impregnation. [Catalyst = 4 mL; ETAN/EG = 1:1 mol/mol; H_2/H_2O = 2:8 v/v; T = 583 K; GSHV = 2,900 h^{-1}; LHSV = 1.6 h^{-1}; yield values referred to the ETAN fed.]

Taking into account the catalyst deactivation with time-on-stream, an investigation was carried out on the possibility of recovering the initial activity by calcination in situ using a O_2/N_2 mixture, and progressively increasing the temperature up to 773K and the O_2-content up to 20% (v/v), followed by a new activation step using a H_2/N_2 (1:9 v/v) mixture (see Experimental Section). Figure 6 shows the good regenerability of the ZrO_2/SiO_2 (5:95 w/w) catalyst, with almost complete recovery of the original conversion value also after three catalytic cycles, and a decrease in selectivity in ETI, due to a higher formation of the two main by-products NEE and NVE.

Because of the interesting performances obtained in the synthesis of ETI using the ZrO_2/SiO_2 (5:95 w/w) catalyst, the possibility of extending the reaction to the synthesis of a wide number of indoles was investigated. However, feeding aniline (AN), a yield in indole 50 % lower than that in ETI, notwithstanding the lower steric hindrance in the cyclization step (Table 3). Moreover, this result was not significantly modified by increasing the temperature to 603 K or the AN/EG ratio to 10 [14,15].

Based on the hypothesis that the lower activity of AN was attributable to the

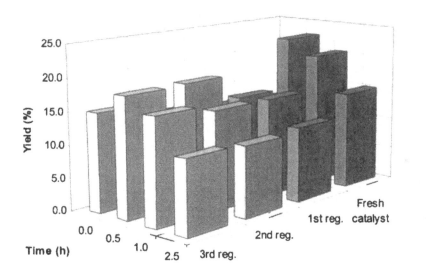

Figure 7. Yield (%) in 7-ethylindole as a function of the regeneration cycles. [ZrO_2/SiO_2 (5:95 w/w) prepared by incipient wetness impregnation = 4 mL; ETAN/EG = 1:1 mol/mol; H_2/H_2O = 2:8 v/v; T = 583 K; GSHV = 2900 h^{-1}; LHSV = 1.6 h^{-1}; yield values referred to the ETAN fed.]

absence of an activation effect by the alkyl chain in the ortho-position to the amino group, 2-methylaniline (METAN) was then employed, but only a slight increase in the yield in 7-methylindole was found. On the contrary, remarkable improvements were detected with further increases in the chain length, i.e. feeding 2-ethylaniline (ETAN) or, mainly, 2-propylaniline (PRAN). These results cannot be explained simply on the basis of mesomeric or inductive effects, thus it may be hypothesized that interaction occurs between the terminal methyl group of a sufficiently long alkyl chain and the carbonyl group of the reaction intermediate (A), favoring the synthesis of the corresponding indole in comparison to that of the by-products. On the other hand, the very low activity of the 2-chloroaniline, can be attributed to the deactivation effect of an electron attracting substituent on the aromatic amine.

The differences in reactivity between aniline and alkylanilines were significantly smoothed by increasing the length of the diol chain, while also improving yields in the corresponding indoles (Table 3). Furthermore, with both 1,2-propylene glycol and 1,2-butandiol no deactivation was observed with time-on-stream. These data are in agreement with the reaction pathway proposed above, considering the increase in stability in the diols as a function of the chain length and the possible stabilization effect on the carbonyl group of the intermediate due to inductive effects by the alkyl chain. However, this latter effect is limited by the steric hindrance of the diol, such as observed by feeding 1,2-hexanediol, for which a significant worsening of the yield in 3-butylindole was detected.

Table 3. Yield (%) in the different indoles as a function of the composition of the organic mixture fed [ZrO_2/SiO_2 (5:95 w/w) prepared by incipient wetness impregnation = 4 mL; Aromatic amine/glycol = 1:1 mol/mol; H_2/H_2O = 2:8 (v/v); T = 583 K; GSHV = 2900 h^{-1}; LHSV = 1.6 h^{-1}; yield values referred to the aromatic amine fed].

Amine	Glycol	Product	Yield
Aniline	Ethylene glycol	Indole	8.1
2-Methyl-aniline	"	7-Methylindole	12.4
2-Ethyl-aniline	"	7-Ethylindole	20.5
2-Propyl-aniline	"	7-Propylindole	25.2
2-Chloroaniline	"	7-Chloroindole	0.8
Aniline	1,2-Propylene glycol	3-Methylindole	12.0
"	1,2-Butanediol	3-Ethylindole	20.4
"	1,2-Hexanediol	3-Butylindole	11.8

Conclusions

The vapour phase synthesis of indole and its derivatives using heterogeneous catalysts represents a novel example of an economic and environmentally friendly process, with significant advantages in comparison to those operating in the liquid phase with homogeneous catalysts. Using well known copper chromites, the best results were obtained with a catalyst containing promoters affecting only the physical properties and operating at low temperature and LHSV values, while an excess of aniline derivatives was necessary to avoid polyalkylation products. Furthermore, it was shown that the hourly productivity can be improved by reducing the contact time, but with a corresponding increase in volume of the carrier gas to be recycled.

The problems of the excess of aniline derivatives and large volume of carrier gas required to operate at low contact time were both overcome by developing novel ZrO_2-based catalysts, able to operate with an almost stoichiometric ratio between the aromatic amine and the diol and using water as the main carrier gas. Water can be easily separated by condensation and recycled to the reactor, thus avoiding any environmental constraints. The ZrO_2/SiO_2 catalysts, mainly those prepared by incipient wetness impregnation, showed industrially useful results, better than those reported in the current literature. Furthermore, these catalysts exhibited good regenerability, recovering the initial activity almost completely.

This synthesis can be applied to a wide number of substrates (anilines and/or diols). In particular, using ethylene glycol the best yields in the corresponding indoles were obtained when an alkyl group was located in the ortho-position to the amino group and the length of the chain was increased, in opposition to the theory of steric occupancy. Moreover, the differences in reactivity between aniline and alkylanilines were significantly smoothed by increasing the length of the diol chain, although with a limit due to the steric hindrance of the diol.

Acknowledgments

Financial support from the National Research Council [CNR, Rome (I)] and CHEMI SpA [Cinisello Balsamo MI (I)] are gratefully acknowledged. Thanks are due to Engelhard and Süd-Chemie for providing the commercial catalysts.

References

1.   R.J. Sundberg, in Kirk-Othmer Encyclopedia of Chemical Technology, J.I. Kroschwitz and M. Howe-Grant, eds., Vol. 14, John Wiley-Interscience, New York, 1995, pp 161-173.

2.   G. Collin and H. Höke, in Ullmann's Encyclopedia of Industrial Chemistry, F.T.

Campbell, R. Pfefferkorn and J.F. Rounsaville, eds, Vol. A14, VCH, Weinheim, 1989, p. 167-170.

3.   H.G. Frank and J.W. Stadelhofer, Industrial Aromatic Chemistry, Springer-Verlag, Berlin, 1988, pp. 417-418.

4.   B. Robinson, *Chem. Rev.* **63**, 373 (1963).

5.   R.L. Augustine, A.J. Gustavsen, S.F. Wanat, I.C. Pattison, K.S. Houghton and G. Koleter, *J. Org. Chem.* **38**, 3004 (1973).

6.   Y. Tsuji, S. Kotachi, K.T. Huh and Y. Watanabe, *J. Org. Chem.* **55**, 580 (1990)

7.   M.C. Fagnola, I. Candiani, G. Visentin, W. Cabri, F. Zarini, N. Mongelli and A. Badeschi, *Tetrahedron Lett.* **38 (13)**, 2307 (1997).

8.   G. Bacolini, *Topics Heterocycl. Syst. Synth. React. Prop.* **1**, 103 (1996).

9.   Ube Industries Ltd, Jpn. Kokai Tokkyo Koho 81,36,452 (1979).

10.   T. Honda, F. Matsuda, T. Kiyoura and K. Terada, Eur. Pat. Appl. 69,242 to Mitsui Toatsu Chemicals Inc. (1981).

11.   P. Hardt, Eur. Pat. Appl. 120,221 to Lonza AG (1983).

12.   A. Kudoh, H. Tadatoshi, M. Kontani, T. Tsuda and S. Kiyono, German Pat. 32,22,153 to Mitsui Toatsu Chemicals Inc. (1983).

13.   A. Kudoh, T. Honda, M. Kotani, K. Terada, T. Tsuda and S. Kiyono, German Pat. 33,24,092 to Mitsui Toatsu Chemicals Inc. (1983).

14.   F. Matsuda and T. Kato, US Pat. 4,376,205 to Mitsui Toatsu Chemicals Inc. (1983).

15.   F. Matsuda and T. Kato, US Pat. 4,436,917 to Mitsui Toatsu Chemicals Inc. (1984).

16.   M. Piemontese, F. Trifirò, A. Vaccari, E. Foresti and M. Gazzano, in Preparation of Catalysts V, G. Poncelet, P.A. Jacobs, P. Grange and B. Delmon, eds., Elsevier, Amsterdam, 1991, pp 49-58.

17.   F. Trifirò, A. Vaccari, G. Braca and A.M. Raspolli Galletti, in Catalysis of Organic Reactions, M.G. Scaros and M.L. Prunier, eds, Marcel Dekker, New York, 1994, pp 475-478.

18.   G.L. Castiglioni, C. Fumagalli, R. Lancia and A. Vaccari, in Catalysis of Organic Reactions, E.R. Malz, ed., Marcel Dekker, New York, 1996, pp 65-74.

19.   O. Piccolo, A. Vaccari, M. Campanati and P. Massardo, Italian Pat. MI98A 02,307 to CHEMI SpA (1998).

20.   A. Cimino, D. Cordischi, S. De Rossi, G. Ferraris, D. Gazzoli, V. Indovina, G. Minelli, M. Occhiuzzi and M. Valigi, *J. Catal.* **127**, 744 (1991).

21.   G.A.M. Hussein, N. Sheppard, M.I. Zaki and R.B. Fahim, *J. Chem. Soc. Farady Trans.* **87**, 2661 (1991)

22.   K. Tanabe and T. Yamaguchi, *Catal. Today* **20**, 185 (1994).

23.   A.P. Legrand, ed., The Surface Properties of Silicas, John Wiley, & Sons, Chichester, 1998,

24.   M. Campanati, P. Savini, A. Tagliani, A. Vaccari and O. Piccolo, *Catal. Lett.* **47**, 247 (1997).

Carbons as Catalysts

Francis J. Waller, John B. Appleby and Stephen C. Webb

Air Products and Chemicals, Inc.
7201 Hamilton Boulevard,
Allentown, PA 18195

Abstract

Carbons are most frequently used as catalyst supports or adsorbents. Some carbons are sufficiently acidic or can be treated with nitric acid to be made acidic to catalyze the hydrolysis of esters. The carbons are also more thermally robust than acidic sulfonic ion-exchange resins and can routinely operate at temperatures greater than 100 °C. The nature of the acidic group(s) on carbon responsible for the catalytic activity for ester hydrolysis do not catalyze the dehydration of methanol to dimethyl ether at temperatures up to 200 °C.

Introduction

Ester hydrolysis, like the reverse reaction of esterfication, reaches equilibrium under reaction conditions of reactant concentration, temperature and pressure.

$$RCO_2R_1 + H_2O \rightleftharpoons RCO_2H + R_1OH$$

For liquid-phase hydrolysis of esters, there are only a few solid acid catalysts which are acceptable for activity, stability and lifetime. Acidic ion-exchange resins such as the Amberlite 100 series or the Amberlyst series where the functional group is sulfonic acid are often used at temperatures of 100 °C or less. Higher temperatures lead to thermal instability of the catalyst. Other solid acid catalysts such as zeolites, heteropolyacids, and zirconium phosphonate have been reported in the literature for the hydrolysis of esters (1).

Thermodynamics for the hydrolysis of esters suggest that higher temperatures will lead to a more favorable equilibrium constant. Therefore, in order to hydrolyze esters at these higher temperatures, a catalyst not only must have the acceptable features of the acidic ion-exchange resins, but also the higher temperature tolerant catalyst must not catalyze the formation of ethers. The ethers are obtained by the reaction of the hydrolyzed alcohol with itself.

$$2R_1OH \longrightarrow R_1OR_1 + H_2O$$

The hydrolysis of methyl acetate has been studied and reported for a strongly acidic cation-exchange resins containing sulfonic acid groups. When Amberlite IR-120 catalyst was evaluated in the batch mode, the hydrolysis of methyl acetate was found to have an activation energy of 73kJ/mol between the temperature range of 40-55 °C (2).

Oxidizing agents such as nitric acid have been used to introduce surface oxygen functionality on carbons. A recent study has compared the effect of different oxidizing agent treatments (3). Nitric acid treatment increased the carboxyl groups 70 and 200% for an activated granular charcoal and an activated charcoal cloth respectively. In addition, the nitric acid oxidation caused the surface area of both carbons to decrease the most when compared to the other oxidants.

Carbons are unique and versatile adsorbents. Reports in the literature suggest that carbons can also promote a variety of surface reactions. Several examples are selected from the literature in order to demonstrate the role of the carbon as a catalyst.

Recently, graphite has been reported to catalyze Friedel-Crafts acylation (4). In particular, anisole and benzoyl bromide reacted to form p-methoxybenzophenone in 89% yield in refluxing benzene. Without the graphite, the reaction was not observed and an "active charcoal" gave only a trace of product.

The thermal cracking of dimethyl ethylidene dicarbamate to N-vinyl-O-methyl carbamate can be carried out under partial vacuum at 190-200 °C in the presence of a carbon catalyst (5). The more catalytically active catalyst was a DARCO carbon which was acid washed.

$$CH_3CH(NHCO_2Me)_2 \longrightarrow CH_2 = CHNHCO_2Me + H_2NCO_2Me$$

The catalytic oxidation of sulfur dioxide by activated carbon has been studied at 25 °C in water (6). Under the experimental conditions employed, the overall reaction can be written as

$$HSO_3^- + 1/2O_2 \longrightarrow SO_4^{-2} + H^+$$

An activated carbon catalyst is also responsible for the selective oxidation of hydrogen sulfide to elemental sulfur (7). This reaction has been studied over the temperature range of 25-240 °C. The oxidation is negligible in the absence of a carbon.

$$H_2S + 1/2O_2 \longrightarrow H_2O + 1/2 S_2$$

This work will focus on the hydrolysis of methyl acetate, a co-product often made when polyvinyl acetate is transesterified with methanol to polyvinyl alcohol (8).

Experimental Section

Carbon treatment [*care should be exercised when heating HNO_3 with carbon*]

Method A: A 5 gram sample of carbon was placed in a 250 mL beaker. 25 mL of concentrated HNO_3 was slowly added to the carbon. The carbon was then allowed to sit for approximately 15 minutes. The beaker was then placed on a hot plate and the mixture was reacted by gentle boiling to dryness at 80-100 °C for about one hour. The carbon was then washed with distilled water to neutral pH and dried at 110 °C for 12 hours.

Method B: In some cases, 25 mL of 0.4M Cu(II) acetate solution was added following the HNO_3 but before boiling

Method C: A 5 gram sample of carbon was placed in a 250 mL beaker. 25 ml. of concentrated HNO_3 was slowly added to the carbon. The carbon was allowed to sit for approximately 15 minutes. The carbon was then rinsed with distilled water to neutral pH and dried at 110 °C for 12 hours.

Reactor

Catalyst evaluation was done using a packed bed reactor system. The reactor tube was 0.5" o.d. with a 0.049" wall and surrounded by a tube furnace. A single phase liquid feed was pumped through the catalyst bed by an ISCO pump. Pressure, sufficient to maintain the system in liquid state, was maintained using a back-pressure regulator. The effluent from the reactor tube was cooled using a refrigerator bath. Samples were analyzed on a HP 5890 Gas Chromatograph. Organic products were determined using a flame-ionization detector, while water was determined on a thermal conductivity detector.

Reactor feed solutions

Synthetic feed solutions representing a 10% or 30% converted methyl acetate solution were prepared by blending the appropriate amounts of methyl acetate, water, acetic acid and methanol together to obtain a single phase. Both solutions were based on an initial ratio of water/methyl acetate of 2.

Contact time

Contact time is defined here to mean the volume of catalyst charged to the reactor divided by the volumetric flow rate of the feed solution at reaction conditions.

Error limits for GC analysis

The % conversion reported in the tables for acetic acid and methanol are ±1% based on all the experimental data available during this study.

Conversion calculation

The equilibrium expression for the hydrolysis of methyl acetate is expressed in the following equation. If x represents the amount

$$K = \frac{[HOAc]\,[MeOH]}{[MeOAc]\,[H_2O]}$$

in moles of methyl acetate reacted, then the equilibrium expression is reduced to

$$K = [x]^2/[MeOAc-x]\,[H_2O-x]$$

The value of x is obtained by GC analysis of the samples withdrawn during the hydrolysis experiment. Molar response factors were used to determine the molar amounts of methanol, acetic acid and methyl acetate within the samples withdrawn at reaction time, t. The GC derived expression $[HOAc_t]/([HOAc_t] + [MeOAc_t])$ when multiplied by the concentration of MeOAc before 10 or 30% converted gives the value of x. For the two solutions utilized in this study, the initial mole fraction of water and methyl acetate is 0.666 and 0.334 respectively. Similarly, x can also be determined by the GC derived expression $[MeOH]_t/([MeOH]_t + [MeOAc]_t)$. Both expressions can be used to calculate x, conversion, or K. In this example K is a unitless number.

Results and Discussion

In this study, an Amberlyst series of resin was evaluated in the packed bed reactor for ester hydrolysis activity. The hydrolysis reaction was compared at two temperatures, 100 and 150 °C. A synthetic blend representing a composition of methyl acetate, methanol, acetic acid and water at 10% converted methyl acetate was premixed and pumped continuously as one phase.

Table 1 Methyl Acetate Hydrolysis with Amberlyst[1]

Temp.(° C)	Press. (psig)	Contact Time (min)[2]	% Conv.[3]	DME/MeOH
100	100	19	36	0.003
100	100	75	36	0.010
150	150	10	40	0.116
150	150	37	42	0.155

1. Amberlyst 131; 2.0 grams of catalyst.
2. 10% converted MeOAc feed solution: mole fraction of water, methyl acetate, methanol, and acetic acid is 0.633, 0.300, 0.033, and 0.034 respectively.
3. Conversion based upon acetic acid; samples for GC analysis collected after 8 hours.

The conversions based on acetic acid at 100 °C and contact times of 19 and 75 minutes suggest the hydrolysis reaction is at equilibrium at a resident time of 19 minutes or less. At 150 °C, the reaction is at equilibrium at a residence time of 37 minutes. In all examples, dimethyl ether (DME) is produced through the dehydration of methanol catalyzed by the strong sulfonic acid group of the Amberlyst series.

Two carbons were compared for methyl acetate hydrolysis activity and the results are summarized in Table 2. These carbons were not pretreated with nitric acid, however, both were dried under air at 110 °C for twelve hours before being used as catalysts. The activated carbon from Barneby and Sutcliffe was sieved to 60-100 mesh particles. A carbon, OL from Calgon, had a particle size of 20-30 mesh. Both were evaluated at a reaction temperature of 150 °C and 200 psig. The AC carbon catalyzed the conversion of methyl acetate in a 10% converted methyl acetate feed solution to 42% conversion at a contact time between 48 to 95 minutes. In a similar manner, the OL carbon catalyzed the conversion of methyl acetate in a 30% converted methyl acetate feed solution to 40%

Table 2 Methyl Acetate Hydrolysis with Untreated Carbons

Catalyst	% Feed	Contact Time (min)[1]	% Conv[2]	% Conv[3]
AC Carbon[4]	10	24	27	27
		48	40	38
		95	42	40
OL Carbon[5]	30[6]	26	38	37
		52	40	39
		104	40	39

1. Reaction temperature is 150 °C at 200 psig.
2. Conversion is based on acetic acid; DME was not detected by GC.
3. Conversion is based on methanol.
4. AC Carbon from Barneby and Sutcliffe; 60-100 mesh; 4 grams of catalyst.
5. OL Carbon from Calgon, 20-35 mesh; 4 grams of catalyst.
6. Mole fraction of water, methyl acetate, methanol and acetic acid is 0.567, 0.233, 0.1, and 0.1 respectively.

conversion at a contact time between 26 to 52 minutes. With each catalyst, no dimethyl ether was observed. The conversion calculation based on either acetic acid or methanol determined by GC at each contact time is in close agreement.

The effect of a nitric acid pretreatment at room temperature is shown in Table 3. The OL carbon from Calgon was selected and evaluated for methyl acetate conversion using a synthetic 30% converted methyl acetate feed solution. The carbon when not treated with nitric acid required a contact time between 26 to 52 minutes to reach 40% conversion. When the same carbon is treated with nitric acid, a 42% conversion was reached at a contact time of 12 minutes or less. Dimethyl ether was not observed in either experiment.

Table 3 Methyl Acetate Hydrolysis with Untreated and Treated Carbon

Catalyst	Contact Time (min)[1]	% Conv.[2]
OL Carbon[3,4]	26	38
	52	40
OL Carbon[3,5]	12	42
	24	42
	47	42

1. Reaction temperature is 150 °C at 200 psig.
2. Conversion is based on acetic acid; DME was not detected by GC.
3. 30% converted methyl acetate feed solution; 20-35 mesh; 4 grams of catalyst.
4. Untreated carbon.
5. Method C treatment.

The effect of nitric acid pretreatment was further examined by monitoring the hydrolysis activity of a carbon for 17 days of continuous operation. Table 4 summarizes these results. At 150 °C and a contact time of 94 minutes (~1.5 hours) the conversion of methyl acetate was 40-41% based on acetic acid analysis. Under these conditions, the catalyst bed experienced approximately 272 volume changes of a synthetic 10% converted methyl acetate feed solution. If any residual nitric acid was present either on or in the carbon, the nitric acid should have been washed off the catalyst. Within experimental error the treated and untreated carbon, compare Table 1, behave similarly.

Table 4 Methyl Acetate Hydrolysis for Seventeen Day Operation with Treated Carbon[1]

Catalyst	Time-on-Stream (hrs)[2,3,4]	% Conv.[5]	% Conv.[6]
AC Carbon[7]	120	40	39
	192	41	40
	312	41	40
	408	41	39

1. Method B treatment.
2. Reaction temperature is 150 °C at 200 psig.
3. Residence time is 94 minutes.
4. 10% converted MeOAc feed solution.
5. Conversion based upon acetic acid.
6. Conversion based upon methanol.
7. Barneby and Sutcliffe; 60-100 mesh; four grams of catalyst.

A pretreated carbon was also evaluated in the liquid phase at 200 °C. The results presented in Table 5 can be compared with Table 3 at 150°C. The catalyst at 200 °C hydrolyzed both a 10% and 30% converted methyl acetate feed solution. The treated carbon required between 23-47 minutes to obtain a methyl acetate conversion of 43% when feeding a 10% converted methyl acetate feed solution but the untreated carbon required only 12 minutes to reach a methyl acetate conversion of 43% when starting with 30% converted methyl acetate feed solution. In both examples, DME was not detected by GC.

Table 5 Methyl Acetate Hydrolysis at 200 °C with an Untreated and Treated Carbon

Catalyst	% Feed	Contact Time (min)[2]	% Conv.[3]
OL Carbon[1,4]	10	12	40
		23	41
		47	43
OL Carbon[5]	30	12	43
		23	43
		47	43

1. Method C treatment.
2. Reaction temperature is 200 °C at 750 psig.
3. Conversion is based on acetic acid; DME was not detected by GC.
4. Calgon; 20-35 mesh; 4 grams of catalyst.
5. Untreated carbon.

Lastly, a carbon was pretreated with hot nitric acid and evaluated at 100 °C. It was found not to be as active as the Amberlyst 131 shown in Table 1.

Table 6 Methyl Acetate Hydrolysis at 100 °C[1] with Treated[2] Carbon

Catalyst	Contact Time (min)[3]	% Conv.[4]
Peat carbon	12	21
	24	24

1. Reaction pressure of 100 psig.
2. Method A treatment.
3. 10% converted MeOAc feed solution; 2 grams of catalyst.
4. Conversion based upon acetic acid; DME was not detected by GC.

Longer contact time may be required to measure the same % conversion with the 10% converted methyl acetate feed solution. However, the carbon was catalytically active. It did not reach equilibrium conversion at the contact time evaluated.

Conclusion

The common carbons are carbon black, charcoal, graphite. Depending upon the method used to manufacture these carbons, their surface area, pore structure and surface functional groups are different. Many of the functional groups present can be phenolic, carboxylic acid, quinone, ketones and lactones. Some of these functional groups impart acidic characteristics. The introduction of the acidic functional groups occurs during the manufacturing process or by treatment of the carbons with oxidizing agents (3). It has been found that carbons can act as catalysts to hydrolyze methyl esters to the corresponding organic acid and alcohol. Since carbons are more thermally tolerant than organic polymers, the carbons can operate at higher reaction temperatures. In addition, the carbon surface can be oxidized with nitric acid to increase the carbon's hydrolysis activity. The most likely functional group that behaves as a catalyst is postulated to be a carboxylic acid. A carboxylic acid group is a weak organic acid and not sufficiently acidic to catalyze the formation of DME from methanol. Any residual nitrate esters would be hydrolyzed under catalyst preparation or ester hydrolysis conditions. However, a stable acidic carbon-nitrogen species cannot be completely ruled out (9).

Acknowledgments

We thank Mrs. T. Hoppe for word processing the manuscript and Air Products and Chemicals, Inc. for permission to publish this work.

References

1. Y. Izumi, *Catalysis Today* **33**, 371 (1997).
2. T. Mizota, S. Tsuneda, K. Saito and T. Sugo, *Ind. Eng. Chem. Res.* **33**, 2215 (1994).
3. B. K. Pradhan and N. K. Sandle, *Carbon* **37**, 1323 (1999).
4. M. Kodomari, Y. Suzuki and K. Yoshida, *Chem. Commun.*, 1567 (1997).
5. F. J. Waller, US Patent 5,233,077 to Air Products and Chemicals, Inc. (1993).

6. F. R. Pinero, D. C. Amoros and E. Morallon, *J. Chem. Educ.* **76,** 958 (1999).

7. R. C. Bansal, J. B. Donnet and F. Stoeckli, In Active Carbon, Marcel Dekker, Inc., New York, 1988; pp 413-423.

8. J. B. Appleby, F. J. Waller and S. C. Webb, US Patent 5,872,289 to Air Products and Chemicals, Inc. (1999).

9. H. Teng and E. M. Suuberg, *J. Phys. Chem.* **97**, 478 (1993).

New Coupling and Isomerization Reactions Catalyzed by a Protic Superacid

Alexei V. Iretskii, Sheldon C. Sherman, and Mark G. White

School of Chemical Engineering, Georgia Institute of Technology, Atlanta, GA 30332-0100

Abstract

A supposed protonation of a sp^3 or sp^2-carbon respectively in cycloalkanes (cyclohexane and methylcyclopentane) or alkylphenyls (dimethylbiphenyls) leads to oxidative coupling of paraffins, but promotes isomerization of substituted benzenes. Cyclohexane and methylcyclopentane are converted to a mixture of isomeric dimethyldecahydronaphthalenes in the presence of a superacid at mild reaction conditions (80-165°C). This reaction has been characterized for different reaction times, acid/substrate ratio, and different reaction temperatures. The coupling reaction is not observed for other substrates such as cyclopentane, cycloheptane, and methylcyclohexane. The isomerization of dimethylbiphenyl proceeds rapidly to 3,3'- and 3,4'-dimethylbiphenyl in superacid at 100°C. The kinetics were measured as a function of temperature, and acid/substrate ratio.

Introduction

Some of the prior art for the synthesis of substituted decahydronapthalene (also known as decalin) involves homogeneous catalysts. For example, the synthesis was described in the patent literature for coupling cycloparaffins with cycloparaffin halides (*e. g.*, cyclohexylchloride) in the presence of AlCl₃ complex (1). This preparation gives dimethyldecalin in 40% yield. A related patent (2) describes the synthesis of 2,6-dimethyldecalin from either C₁₂ dicyclic napthenic isomers, methylcyclopentane, or cyclohexane using a catalyst of HF and TaF₅ or HF and NbF₅ at 25°C under a hydrogen partial pressure of 2 atmospheres for a duration of 95 h. The yield of the title compound was 12.2 wt%. Unlike other liquid-catalyst systems, this catalyst does not need a promoter (halide, *vide supra*) or initiator (olefin, *vide infra*); however a strong Lewis acid, co-catalyst is required. Omae, *et al.* described the synthesis (3), of dimethyldecalin from a mixture of methyl cyclopentane and C₆ naphthenic hydrocarbons at 40-80 °C. The catalyst was HF/BF₃. In a typical run for 6 hours, the yield of dimethyl decalin was 35% at a selectivity of 66%. In a subsequent patent, these researchers described the influence of adding an olefin, such as propylene, to the reaction mixture as an initiator (4). Bushick and Suld also

181

reported the effect of this olefin additive in a series of US and foreign patents (5). They describe the dimerization of C_6 naphthenes at -20 to 80°C in the presence of propylene and an HF/BF_3 catalyst. The yield of dimethyldecalins was 40% of which about 30 wt% was the 2,6-isomer. The synthesis has been reported in the literature for dimethyl and tetramethyldecalins (6).

This literature shows that strong acids promote the oxidative coupling, e.g. the hydrocarbon build-up reactions to form decalins from cycloparaffins. However, some of these homogeneous catalysts require either a substrate functionalized with a halogen or an olefin initiator. We examined this chemistry with the goal to develop a system that does not require either a functionalized substrate or an initiator. We also wanted a catalyst that did not use a strong Lewis acid such as $AlCl_3$. Our experience with sulfated zirconia systems shows that poisoning of the surface was unavoidable even with simple substrates such as toluene and CO (7). Finally, it was desired to use a strong liquid acid that could be regenerated easily in a continuous operation concept, such as trifluoromethanesulfonic acid (8).

Different transformations occur when a strong acid interacts with substituted aromatic hydrocarbons. Under mild reaction conditions the preferential reaction is the isomerization. Literature on the catalysis by superacids shows that some difunctionalized arenes can be isomerized preferentially to the *meta* isomer, for example xylene (9). The authors explained these results by an intramolecular process whereby the substrate formed carbo-cations with the HF/BF_3 superacid in equilibrium with the substrate. The relative stability of the carbo-cations defined the distribution of isomers in a separate equilibrium. When water was added to stop the reaction, the distribution of isomers is governed by the distribution of isomeric carbo-cations. The authors arrived at this conclusion based on simple arguments of arene cation stability. They reported in the same publication data for the isomerization of trimethyl- and tetramethylbenzenes that supported this hypothesis. Olah showed that di-isopropylbenzene was converted quantitatively to the *meta*-di-isopropylbenzene isomer when a superacid was the catalyst (10). However, they invoked an intermolecular process to explain their results since the disproportionation products were also observed. These data suggested that other difunctionalized arenes may be isomerized preferentially to the 3,3'- and 3,4'-isomers. In this paper, we extend this approach and report the isomerization of disubstituted biphenyls, such as dimethylbiphenyls (dmbp) as a second objective.

Experimental Section

We completed two different sets of reactions: coupling of cycloalkanes and isomerization of dimethylbiphenyl. Some reactions were completed in Pyrex glassware or in Fischer-Porter tubes under reflux and inert atmospheres with vigorous stirring for reaction temperatures $\leq$ 165°C. Other reactions were completed in an autoclave having a volume of 75 mL, System MAWP 1000 (Autoclave Engineers, Erie PA) constructed of Hastelloy-C under a pressure of inert N_2 (100-200 psig). The temperature of the glass reactor was maintained to within 1 C using an oil-bath when the temperatures were less than 165°C. For temperatures greater than 165 C, we used a heated autoclave (*vide supra*). The temperature of this reactor was controlled to the nearest 1°C by an automatic temperature controller (Eurotherm).

All reactions were completed under Ar or N_2 using standard Schlenk technique. Care was exercised in transferring the triflic acid to minimize exposure to the atmosphere. Typically, 10-15 mmol of triflic acid were taken with the required volume of substrate in a 100 Fischer-Porter tube. The contents were stirred so that the observed concentration versus time data were not a function of the stirring speed, thus ensuring that the kinetic reaction rates were controlling the overall reaction process. The contents of the reactor were refluxed for reaction times up to 5 h. Small aliquots of the reaction mixture were withdrawn at frequent intervals (45 minutes) to establish the reaction kinetics. Samples were introduced into ice water to quench the reaction. These samples were diluted with acetone (100 μL per 1.75 mL of acetone) and analyzed on a HP 5890 Series II GC/HP 5972 MS. The partitioning agent was a Supelco SPB-5 column (30 m x 0.25 mm x 0.5 μm).

The substrates were obtained from Aldrich and used without further purification (cyclopentane, methylcyclopentane, cyclohexane, methylcyclo-hexane, ethylcyclohexane, methylcycloheptane, 3,3'-dimethylbiphenyl, and 4,4'-dimethylbiphenyl). Trifluoromethanesulfonic acid, CF_3SO_3H, (< 0.1 wt% H_2O) was obtained from Alfa and used without further purification.

Results and Discussion -- Coupling of cycloalkanes

Reactivity of substrates. We tested the following substrates for reactivity in triflic acid at 165°C: cyclopentane, methylcyclopentane, cyclohexane, methylcyclohexane, and cycloheptane. Equimolar amounts of the substrate and acid were mixed in an isothermal, batch reactor for 3 hours before the reaction was quenched by adding water. The yields of dimethyldecalin were 2-2.5% when the substrate was either methylcyclopentane or cyclohexane. No yields of dimethyldecalin were observed when the substrate was cyclopentane, however the reaction of either methylcyclohexane or cycloheptane with triflic acid formed only decalin with other products showing an m/e = 194. As a final test of the relative reactivity of methylcyclopentane and cyclohexane, an equimolar mixture of the two substrates was combined with an equimolar amount of triflic acid at 165°C for 3 hours. The yield of dimethyldecalin was 3.2%.

Effect of acid/substrate ratio. Additional tests were completed for the reaction of methylcyclopentane at 165°C to determine the effect of changing the acid/substrate ratio upon yields of dimethyldecalin (Table 1).

Table 1 Effect of acid/substrate ratio on the yields of dimethyldecalin from coupling methylcyclopentane at 165°C.

Acid/substrate	0.25	0.33	0.5	1	2	3	4
Yields, %							
DiMe-decalin	1.2	1.4	1.9	2.5	1.6	1.8	0
Others[A]	0	0	0	0	1.8	2.9	1.0

[A]Includes: decalin, dimethyl- and other substituted adamantanes

At this temperature, additional products were observed when the acid/substrate exceeded unity. These additional products, suggest that further loss of H_2 led to a formation of a more condensed cycloparaffins: substituted adamantanes. An additional experiment was completed at an acid/substrate ratio of 4 for which the reaction temperature was lowered to 145°C. Dimethyldecalin (2.8%) was observed along with substituted adamantanes (2.5%). When the reaction temperature was 200°C and the acid/substrate ratio was 0.5 mol/mol, no dimethyldecalin was observed, but the yield of substituted adamantanes was 8.8%.

Effect of temperature. We examined the effect of reaction temperature upon the yields of dimethyldecalin for the coupling of methylcyclopentane at an acid/substrate ratios of 0.25 to 1 (Table 2). The reaction temperature was changed from 80 to 200°C in separate tests lasting 3 h each.

Table 2 Effect of temperature and acid/substrate ratio on the yields of dimethyldecalin from coupling methylcyclopentane

Temperature, C	80	115	145	165	165	165	200
Acid/substrate	1	0.5	0.25	0.25	0.5	1	0.5
Yields, %							
DiMe-decalin	0.6	1.0	1.0	1.2	1.9	2.5	0
Others[A]	0	0	0	0	0	0	8.8

[A]substituted adamantanes

The only product observed was dimethyldecalin when acid/substrate ratios ≤ 1 and the reaction temperatures were less than 165 C.   However, when the temperature was increased to 200 C large amounts of other products was observed (predominantly substituted adamantanes) even though the acid/substrate was 0.5.   Thus, the combined effects of temperature and acid/substrate ratio (*vide supra*) appear to control the selectivity to the formation of dimethyldecalin and/or substituted adamantanes.

Effect of reaction time.   The reaction progress was followed as a function of time for the coupling of methylcyclopentane when the acid/substrate was 1 and the reaction temperature was 165°C. (Figure 1).

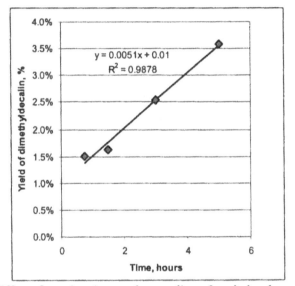

Figure 1 Effect of reaction time on the coupling of methylcyclopentane

These data confirm zero order kinetics which is expected since the conversion of substrate is less than 10%.

Mechanistic considerations.   The possible pathway for coupling C_6-naphthenic hydrocarbon in presence of trifluoromethanesulfonic acid is somewhat similar to the transformation of cycloparaffins in a 1:2 mixture of $AlBr_3$ - acyl halides (11). The Akhrem *et al.* catalyst (RCOX-AlBr₃) is thought to abstract a hydride from a cycloalkane moiety thus forming a carbocation as the active intermediate. We speculate that triflic acid generates carbocations through protonation of a sp^3-carbon followed by loss of dihydrogen (12). These authors showed that a carbocation was formed and that stoichiometric amounts of hydrogen were eliminated by the action of triflic acid upon a substrate having a tertiary carbon (isobutane). For methylcyclopentane that process is shown below:

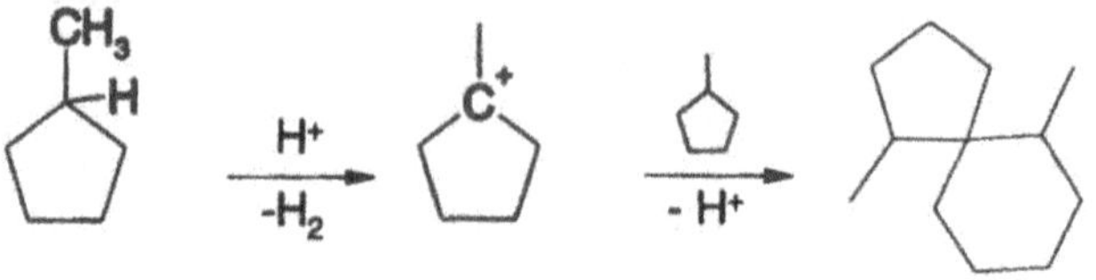

Subsequent insertion into C-C bond of another MCP followed by loss of the proton may lead to a formation of isomeric *spiro* intermediates – dimethylspirodecanes. The further transformations depend upon the reaction conditions. When these conditions are relatively mild (t < 145 °C, 1 mol or less of HOTf per mol of substrate) the obvious rearrangement involving an α-hydrogen occurs:

The result is the formation of dimethyldecahydronaphthalenes. A different reaction may happen when reaction conditions are harsher. The numerous rearrangements may have occurred, including migration of methyl groups and the additional evolvement of hydrogen. The final products are dimethyl- and other substituted adamantanes.

We speculate that the formation of relatively stable carbocations in triflic acid media may be favored in those cycloparaffins that show at least one isomer containing a tertiary carbon atom. The nature of the substrate directs the reaction pathway.

Results and Discussion -- Isomerization of dimethylbiphenyl

Effect of acid/substrate ratio at room temperature. These data (Table 3) show that the rate of approach to chemical equilibrium was accelerated by the use of excess acid in neat substrate (4,4'-dmbp). Notice that the distribution of isomers contains no 2,X'-dmbp isomers, and that 3,4'-dmbp composition was the greatest of the remaining three isomers. At the highest acid/substrate ratio, the yield of 4,4'-dmbp is very low (1-3%) with the 3,4'-dmbp showing the highest yield (57%) and the balance was 3,3'-dmbp (30%).

Table 3 Effect of acid/substrate ratio on the isomerization of 4,4'-dimethylbiphenyl at room temperature.

Time Hours	HOTF/substrate = 10			20			50		
	4,4'-	3,3'-	3,4'-	4,4'-	3,3,'-	3,4'-	4,4'-	3,3'-	3,4'-
0	100	0	0	100	0	0	100	0	0
2	72	2	26	8	7	85	4	8	88
18	28	18	54	4	29	68	1	27	73
43	7	40	53	14	28	58	3	30	67
67	7	36	57	7	33	60	--	--	--

Effect of acid/substrate at 100°C. We wished to explore the effects of changing the acid/substrate ratio using a different isomer of dmbp (3,3'-dmbp) as the substrate and using a different temperature (100°C). The effect of changing the acid to substrate ratio was elaborated further for the isomerization of 3,3'-dmbp at 100°C using the protocol described above. The acid/substrate ratio in these studies was 10 and 25 mol/mol. The results (Table 4) show that the concentrations of isomers do not change for reaction times greater than 10 minutes at either acid/substrate ratio. The distribution of isomers was 2,X'-dmbp = 0; 3,3'-dmbp = 37-39%; 3,4'-dmbp = 49-55%; and 4,4'-dmbp = 5-9%. This distribution of isomers obtained by reacting 3,3'-dmbp was very similar to the distribution (36%/56%/7% for 3,3'-/3,4'-/4,4'-dmbp) obtained from reacting the 4,4'-dmbp at a similar acid/substrate ratio and 25°C for 67 h. The similar final product composition observed at long reaction times when the substrate was either 3,3'- or 4,4'-dmbp suggest that the system approached equilibrium (Figure 2).

Table 4 Effect of acid/substrate ratio on the isomerization of 3,3'-dimethylbiphenyl at 100°C.

Time min.	HOTF/substrate = 10			25		
	4,4'-	3,3'-	3,4'-	4,4'-	3,3,'-	3,4'-
0	0	100	0	0	100	0
16	8	39	49	5	38	56
30	8	39	49	6	38	56
60	8	39	49	6	37	56
120	8	39	49	6	38	56

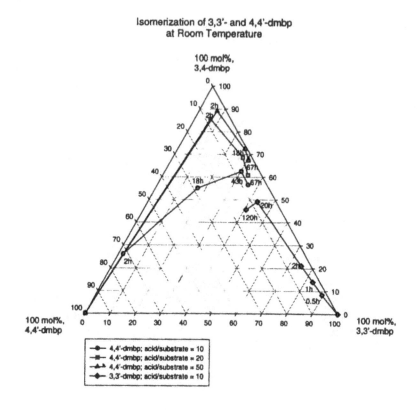

Figure 2 Effect of reaction time on the isomerization of 3,3'- and 4,4'-dmbp at constant temperature and acid/substrate ratio.

<u>Effect of temperature at constant acid/substrate ratio.</u> The effect of temperature was demonstrated for the isomerization of 3,3'-dmbp using an acid/substrate ratio of 10 mol/mol. The protocol is as before and the temperatures used were 30, 50, and 100°C (Table 5). The reaction time could be reduced to less than 10 minutes by using a temperature of 100°C. The distribution of isomers does not appear to change much with increasing temperatures: 2,X'-dmbp = 0%; 3,3'-dmbp = 31-39%; 3,4'-dmbp = 49-52%; 4,4'-dmbp = 8-9%. The effect of temperature was confirmed for the isomerization of 4,4'-dmbp using the same protocol and an acid/substrate ratio of 10 (Table 6). The reaction temperatures were 25, 30, and 50°C. The final concentration of isomers were 31-37%/52-67%/2-8% for 3,3'-/3,4'-/4,4'-dmbp at these conditions. The distribution of isomers (8%/37%/52%) at the longest reaction time, 67 h, and 50°C was similar

to that observed when 3,3'-dmbp was isomerized at 100°C and for 2 h (9%/39%/49%).

Table 5 Effect of temperature on the isomerization of 3,3'-dimethylbiphenyl at an acid/substrate ratio of 10.

Time	Temperature = 30°C			50°C			100°C		
Hours	4,4'-	3,3'-	3,4'-	4,4'-	3,3,'-	3,4'-	4,4'-	3,3'-	3,4'-
0	0	100	0	0	100	0	0	100	0
0.17	<1	96	3	3	68	29	8	39	49
0.5	1	90	9	8	44	48	8	39	49
1	2	84	14	9	42	49	8	39	49
2	3	76	21	9	39	52	9	39	49
24	5	52	43	--	--	--	--	--	--
94	8	31	51	--	--	--	--	--	--

Table 6 Effect of temperature on the isomerization of 4,4'-dimethylbiphenyl at an acid/substrate ratio of 10.

Time	Temperature = 25°C			30°C			100°C		
Hours	4,4'-	3,3'-	3,4'-	4,4'-	3,3,'-	3,4'-	4,4'-	3,3'-	3,4'-
0	100	0	0	100	0	0	100	0	0
2	4	7	89	11	12	77	7	36	57
18	1	27	72	7	30	63	6	37	57
43	2	30	68	7	35	58	7	37	56
67	2	31	67	7	37	55	8	37	52

Effect of reaction time. The effect of reaction time on the isomerization of 3,3'- and 4,4'-dmbp was examined in separate experiments at room temperature (Fig. 2). The progress of the reaction was followed in separate tests to determine the reaction network. For example, starting with 4,4'-dmbp, the yields of 3,4'-dmbp increased rapidly compared to the yields of 3,3'-dmbp at every acid/substrate ratio. For acid/substrate = 20 and 50, the maximum yield of 3,4'-dmbp was 80-90% although the yield of 3,3'-dmbp was less than 10% when the reaction times were less than 2 hours. At a lower acid/substrate ratio (10), the reactivity is low and the concentration of 3,3'-dmbp builds at a rate still lower than the accumulation rate of 3,4'-dmbp (e. g., at 18 h, the yields of 3,3'- and 3,4'-dmbp are 18% and 55%, respectively). Contrast this behavior to that observed when the substrate was 3,3'-dmbp (Fig. 2). Here, the concentration of 3,4'-dmbp increases rapidly although the yield of 4,4'-dmbp increases to less than 10% in the first 20 h of reaction. Thus, it would appear that an intermediate involving 3,4'-dmbp is favored.

<u>Mechanistic considerations</u>. We propose the following reaction network for the reaction of 3,3'- and 4,4'-dmbp:

$$3,3'\text{-dmbp} \Leftrightarrow 3,4'\text{-dmbp} \Leftrightarrow 4,4'\text{-dmbp}$$

This pathway does not involve a direct interconversion between 3,3'- and 4,4'-dmbp. Similar product distributions may be obtained starting from either 3,3'- or 4,4'-dmbp. The reaction rate to form 3,4'-dmbp from 4,4'-dmbp appears to be faster than the rate to form 3,4'-dmbp from 3,3'-dmbp suggesting that some intermediate involving 4,4'-dmbp is the least stable of the intermediates.

The earlier researchers (9) speculated that methyl-substituted arenes were protonated by HF/BF_3 to form a cation complex. Moreover, they assumed that the product distribution was determined by the free energies of the carbocation complex. They used conventional models of arene, cation-complexes to argue that the *meta*-substituted, arene, carbo-cation was most stable species for the cases of xylene, trimethyl and tetramethyl-benzene. We calculated (13) the free energies of formation for neutral and carbocation complexes at the reaction conditions described in the literature using a simple molecular force-field calculation (AM-1) to determine the enthalpy of formation. From these calculations we predicted the equilibrium isomer distributions for the carbocations that agreed with the literature data and expanded it to the isomerization of disubstituted biphenyls. The agreement was good between the calculated equilibrium compositions with experimental data for these substituted arenes. The success of the AM-1 software to predict these equilibrium distributions suggests that the carbocation model is warranted in simulating the chemistry of these systems.

The reaction mechanism for this isomerization is given below:

$$\text{Isomer}_1 + H^+A^- \Leftrightarrow [\text{Isomer}_1\text{-H}]^+A^- \quad \text{Eq. (1), reversible}$$

$$[\text{Isomer}_1\text{-H}]^+A^- \rightarrow [\text{Isomer}_2\text{-H}]^+A^- \quad \text{Eq. (2), slow}$$

$$[\text{Isomer}_2\text{-H}]^+A^- + H_2O \rightarrow \text{Isomer}_2 + [H_3O]^+A^- \quad \text{Eq. (3), fast}$$

Step 1 is the reversible protonation of the arene by triflic acid and step 2, the "methyl shift", is the slow step of this process. After the reaction is completed then water is added,to quench the reaction, step 3. The addition of water to the products is very fast and produces an acid weaker than triflic acid. We showed how small amounts of water compromised the reactivity of another reaction catalyzed by triflic acid (14). Therefore, we believe that the addition of water 1) releases Isomer$_2$ from the acid, and 2) prevents the back-isomerization of Isomer$_2$ into Isomer$_1$. This kinetic trapping is the key to synthesis of the thermodynamically-unfavored neutral isomer.

Conclusions

Triflic acid catalyzes the coupling of cycloalkanes and the isomerization of substituted arenes without the need of a co-catalyst such as a Lewis acid. We speculate that triflic acid forms carbocations from these substrates and that these carbocations are the active intermediates. Active cycloalkanes for the coupling reaction either show a tertiary carbon or can be isomerized into another substrate that does show a tertiary carbon. Active arenes are those that show ring hydrogens which permit access of triflic acid to the ring.

Acknowledgements

We acknowledge the generous support from the KoSa Corporation (Spartanburg, SC) and thank Mr. W. B. Hillock for his aid in completing the cycloalkane coupling reactions.

References

1. Ming Chow, J. D. Fellman, U. S. Pat. (1993).
2. D. A. McCaulay, D. A., U. S. Patent 4,300,008, to Amoco Oil Comp. (1981)
3. I. Omae, T. Urasaki, and T. Shima, Jap. Pat. JP49,001,548 (1974).
4. I. Omae, T. Urasaki, and T. Shima, Jap. Pat. JP49,001,549 (1974).
5. R. D. Bushick, G. Mills and G. Suld, DE 1,927,374 (1969); US 3,509,223 (1970); GB 1,222,034 (1971); NL 6,908,683 (1969); BE-734,350 (1969); FR 2,013,344, to Sun Oil Company (1970)
6. A. Schneider, U. S. Pat. 3,346,656, to Sun Oil Company (1967).
7. D. Sood, and Mark G. White, "Characterization of Sulfated Metal Oxide Catalysis by Reaction with *iso*-Propyl Amine," 1995 Pacifichem, Honolulu, HI, December 1995; US 5,679,867: "Carbonylation via Solid Acid Catalysis", (1997).
8. S. I. Hommeltoft, Eur. Pat. 687658A1 (29.05.95), and US Pat. 5,759,357, to Haldor-Topsoe (1998).
9. D. A. McCaulay, and A. P. Lien, *J. Am. Chem. Soc.*, **1952**, *74*, 6246.
10. G. A. Olah, US Pat. 4,547,606 (1985).
11. I. S. Akhrem, A. V. Orlinkov, E. I. Mysov, and M. E. Vol'pin. *Tetrahedron Letters*, **1981**, *22*, 3891.
12. G. A. Olah, G. K. S. Prakash, R. E. Williams, L. D. Field, and K. Wade. *Hypercarbon Chemistry*. Wiley-Interscience: New York, **1987**; J. Sommer, and J. Bukala. *Acc. Chem. Res.*, **1993**, *26*, 370.

13. S. C. Sherman, A. V. Iretskii, M. G. White, C. Gumienny, L. M. Tolbert and
    D. A. Schiraldi, "Acid-promoted activation of arenes. I. Isomerization of
    substituted benzenes and biphenyls", submitted to *JACS*.
14. B. Xu, D. S. Sood, L. T. Gelbaum, and M. G. White, *J. Catal.*, **186**, 345-352
    (1999).

Alkylation of Phenol with Tert-Butyl Alcohol Catalyzed by Zeolite HY Catalyst

Kui Zhang*, Fei He, and Genhui Xu

State Key Laboratory of C₁ Chemical Technology, Tianjin University, Tianjin, 300072 P.R.China

Abstract

The tert butylation of phenol was investigated over zeolite HY catalyst using tert-butyl alcohol as the alkylating agent in a down-flow tubular reactor at atmospheric pressure. The important variables affecting the activity and selectivity of zeolite HY such as reaction temperature, space velocity and mole ratios of tert-butyl alcohol to phenol were studied.

Introduction

Alkylations of phenol using tert-butyl alcohol (TBA) have been studied extensively owing to industrial interest in the production of antioxidants, ultraviolet absorbers and heat stabilizers for polymeric materials. The catalysts to date include liquid acids[1-3], metal oxides[4-6], aluminum salt catalysts[7,8]and cation exchange resins[9-13]. Although cation exchange resin catalysts have some advantage over other catalysts, e.g. reducing equipment corrosion and no environmental pollution, they still possess some disadvantages. Their activity and the selectivity are not up to the mark[9-13]. Changing over to zeolites instead of using environmentally hazardous catalysts is more acceptable, because of their activity, selectivity and re-usability. Phenol tert-butylation has not been studied in detail over zeolite catalysts, except for a few papers reporting the reaction over zeolite and SiO₂-Al₂O₃ catalyst in gas or liquid phase[14-16]. Corma et al.[14] studied the influence of acid strength of zeolite HNaY catalyst on the tert-butylation of phenol with tert-butyl alcohol at 303, 318 and 353K in CCl₄ solutions. Their work reveals that C-alkylation requires stronger acidity than O-alkylation. Kijiya et al.[16] studied the effect of acidity of various catalysts (SiO₂-Al₂O₃, Al₂O₃, Zeolite X) on the reaction by using a continuous flow reactor under normal pressure at 455-523K. Our previous work[17] reveals that zeolite Hβ is an effective catalyst for tert-butylation of phenol giving high

---

* Correspondence to : Dr. Kui Zhang, Energy & Global Change, ABB Corporate Research Ltd., CH-5405 Baden-Dattwil, Switzerland. Tel. (+41) 56 48 670 81, email: kui.zhang@ch.abb.com

phenol conversions and high p-TBP selectivity with optimum reaction
temperature ca. 418K.

The objective of the present study is to demonstrate the feasibility of tert
buylation of phenol over zeolite HY. The important variables affecting the
activity and selectivity of zeolite HY such as reaction temperature, space
velocity and mole ratios of tert-butyl alcohol to phenol were studied. We found
that a large pore zeolite such as HY has a potential application in the production
of tert butyl phenols with high activity and 2,4-DTBP selectivity.

Results and discussions

Effect of reaction time

Fig.1 shows the relative activity and product selectivity of zeolite HY catalyst
as a function of time for the reaction of phenol alkylation with tert-butyl alcohol
at 418K, WHSV(h^{-1}) of 0.67(based on phenol) and molar ratio of tert-butyl

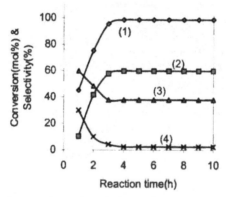

Fig. 1 Phenol conversion and product selectivity versus reaction time over
zeolite HY
(1) Phenol conversion; (2) Selectivity to 2,4-DTBP; (3) Selectivity to p-TBP;
(4) Selectivity to o-TBP
(Reaction temperature=418K; TBA:Phenol(molar ratio)=2.5:1;WHSV(h-
1)=0.67(based on phenol)

alcohol to phenol of 2.5:1. As the figure shows, after 3 or 4h, the catalyst
activity and product selectivity reach an equilibrium level; therefore, all the
experimental data were obtained after the stabilization of the activity.

Effect of reaction temperature

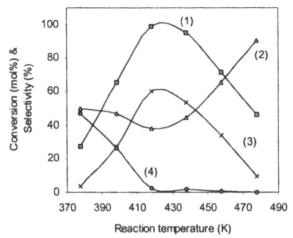

Fig. 2 Phenol conversion and product selectivity versus reaction
temerature over   HY. (1) Phenol conversion; (2) Selectivity to p-TBP; (3)
Selectivity to 2,4-DTBP; (4) Selectivity to o-TBP.
TBA:Phenol (molar ratio)=2.5:1; WHSV (h-1)=0.67(based on phenol)

In the alkylation of phenol with tert butyl alcohol over zeolite HY catalysts,
o-TBP, p-TBP and 2,4-DTBP and hydrocarbons were found as the predominant
reaction products. In addition to the phenol alkylation (which yields desired
products), undesired reactions such as the oligomerization of C_4 hydrocarbons to
produce C_8 olefins and C_{12} olefins occurred as well. Fig.2 clearly shows the
phenol conversion and desired product selectivity at various reaction
temperatures over zeolite HY. The suitable reaction temperature range for
phenol alkylation with TBA was from 398 to 438K. At temperatures lower than
418K, the conversion of phenol increased with increasing temperature, and
o-TBP and p-TBP were the main products in alkyl phenols. In contrast, at
reaction temperatures above 418K, phenol conversion decreased with increasing
temperature. The selectivity to p-TBP increased while that of o-TBP and
2,4-DTBP decreased at higher reaction temperatures. A moderate reaction
temperature (418K) enhanced the selectivity to 2,4-DTBP. Considering the
phenol conversion and product (2,4-DTBP) distribution, the proper reaction
temperature is 418K. At this temperature, the selectivity to 2,4-DTBP is the
highest at high phenol conversions.

According to the results from GC-MS, we found no phenyl ethers and meta-tert
butyl phenol (m-TBP) formed at even high phenol conversions under our

reaction conditions. This is different from the result obtained in CCl₄ solutions at lower reaction temperature (303, 318, 353K) reported previously[14].

Effect of weight hourly space velocity (WHSV(h⁻¹))

The influence of WHSV(h⁻¹)on the activity and selectivity of zeolite HY catalyst was studied at 418K and a tert-butyl alcohol/ phenol molar ratio of 2.5:1. Fig. 3 shows the conversion of phenol and product distribution for the reaction at WHSV (h⁻¹) ranging from 0.42 to 2.00(based on phenol). It is clear from Fig. 3 that both phenol conversion and product selectivity to o-TBP, p-TBP and 2,4-DTBP were almost unchanged when the WHSV (h⁻¹) was less than 1.66. The phenol conversion and selectivity to 2,4-DTBP decreased approximately 8% at space velocities greater than 1.66 due to shorter contact times.

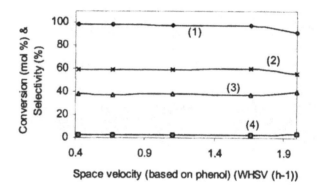

Fig. 3 Phenol conversion and product selectivity versus space velocity over zeolite HY. (1) Phenol conversion; (2) Selectivity to 2,4-DTBP; (3) Selectivity to p-TBP, (4) Selectivity to o-TBP
(Reaction temperature=418K; TBA:Phenol (molar ratio)=2.5:1)

Effect of tert-butyl alcohol / phenol(molar ratio)

The effect of varying the tert-butyl alcohol / phenol molar ratio on the activity and selectivity of zeolite HY at 418K and a WHSV(h⁻¹) of 0.67(based on phenol) is shown in Fig. 4. The results show that the conversion of phenol and selectivity to 2,4-DTBP increased with increasing molar ratio. The product selectivity to 2,4-DTBP reached 71.56% at 99.3% phenol conversion and a molar ratio of 3:1.

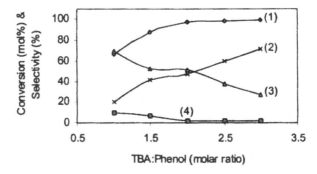

Fig. 4 Phenol conversion and product selectivity versus molar ratio over zeolite HY. (1) Phenol conversion; (2) Selectivity to 2,4-DTBP, (3) Selectivity to p-TBP, (4) Selectivity to o-TBP
(Reaction temperature=418K; WHSV (h-1)=0.67 (based on phenol)

Experimental

Zeolite NaY was obtained from the Catalyst Plant of Nankai University. Zeolite HY was prepared from the sodium form by ammonium exchange with 1M NH_4Cl solution at 363K for 4 hours. The operation was repeated 4 times and the solid washed with distilled water until it became Cl^- free. The product was dried overnight at 423K overnight and then calcined at 823K for 4 hours. The degree of H^+ exchange was 90%. Zeolite HY was then pelletized and crushed into 20-30 mesh particles without binder. All reagents were analytically pure (A.R grade) and used without further treatment.

The catalytic activity measurements were carried out at atmospheric pressure in a tubular, down flow stainless steel reactor (i.d. 4 mm) using 0.5g of the catalyst by changing the reagent feed ratio, the temperature and space velocity. The catalyst was activated in air at 773K for 2 hours prior to catalytic runs. The reactor was then cooled under a nitrogen flow ca. 50ml/min to the desired reaction temperature and the feed mixture was injected from the top using a syringe pump. The products were collected in a cold trap (273K), quantified by gas chromatography(GC) using a SE-30 column and confirmed by gas chromatography-mass spectrometry(GC-MS). The activities and selectivities considered for comparison of the behavior of different catalysts were those obtained after 4 to 6h. Product distribution were always reported in mol-%.

Conclusions

Zeolite HY is recommended to the reaction in order to obtain higher 2,4-DTBP selectivity with high phenol conversions. In the alkylation of phenol with tert butyl alcohol over zeolite HY catalyst, the suitable reaction temperature range is from 398 to 438K. Lower reactant molar ratios are beneficial to p-TBP and o-TBP, while higher of which are helpful to produce 2,4-DTBP. In order to enhance the selectivity to 2,4-DTBP at high phenol conversions, an optimum reaction temperature (418K), higher reactant molar ratio and suitable WHSY (h^{-1}) from 0.42 to 1.66(based on phenol) are recommended.

Acknowledgments

The authors acknowledge the financial support of the National Natural Science Foundation of China (Grant No. 29133071), the Postdoctoral Foundation of China and the Open Foundation of the State Key Laboratory of C_1 Chemical Technology.

References

1 Tsevktov,O.N.; Kovenev,K.D. Int.J.Chem.Eng.,6(1966)328.
2 USP 4,414,233
3 Ger(East) DD 267,250
4 Korenskii,V.I. ;Kolenko, I.P.; Skobeleva, V.D., Zh.Prikl, Khim(Leningrad), 57(9)(1984) 2016-20
5 Sartori Giovanni; Bigi Franca; Casiraghi Giovanni; Casnati Giuseppe; Chiesi Lorella and Arduini Arturo, Chem. Ind.(London).,(22)(1985)762-3
6 JP 85,178,836
7 Koshchii, V.A.; Kozlikovskii, Ya,B.; Matyusha, A.A. Zh. org. Khim.,24(7) (1988) 1508-12
8 JP 6100,036
9 Chandra,K, Gokul; Sharma, M.M. Catal. Lett., 19(4)(1993)309-17
10 JP 5852,233
11 Rajadhyaksha, Rajeev A; Chaudhari, Dilip D. Ind.Eng. Chem. Res. 26(7) (1987)1276-80
12 Braz. Pedidv PIBR 8002,607
13 Ger. offen. DE 3,443,736
14 Corma, Avelino; Garcia, Hermenegido; Primo, Jaime. J. Chem. Research(S), (1) (1988) 40-1
15 Peirou Xu; Bing Feng; Shaozhou Chen, Huadong Huagong xueyuan xuebao, 14(4) (1988)476-80
16 Tomiyasu Kijiya and Susumu Okazaki , Nippon Kagaku Kaishi, 8 (1978) 1071-1077

17   Kui Zhang; Changhua Huang; Huaibin Zhang; Shouhe Xiang; Shangyuan
     Liu; Dong Xu; Hexuan Li, Appl. Catal., 166(1998)89-95

Lithium Hydroxide Modified Sponge Catalysts for Control of Primary Amine Selectivity in Nitrile Hydrogenations

Thomas A. Johnson[a] and Douglas P. Freyberger[b]

[a]Consultant for Process Development Chemistry, 5361 Baldwin Lane, Orefield, PA 18069, USA
[b]Air Products and Chemicals, Inc., Allentown PA 18195, USA

Abstract

Some of the largest commercially produced primary amines are manufactured by catalytic hydrogenation of nitriles using sponge metal catalysts. The larger the market volume for the amine, the more important the technology used to control selectivity becomes to remain a viable producer. We've found that controlling the selectivity to the primary amine using lithium hydroxide modified sponge cobalt in backmix reactors, batch, semi-batch or continuous, at moderate pressures and temperatures provides an excellent means of minimizing by-products without sacrificing productivity. LiOH modified sponge cobalt was found to recycle in batch processing without loss of selectivity for primary amines. In continuous backmix processing LiOH modified sponge cobalt catalyst retained selectivity through numerous reactor turnovers compared to LiOH modified sponge nickel. NaOH and KOH modified catalysts tended to agglomerate under similar conditions. Procedures using a semi-batch system are provided for selecting optimum catalysts for nitrile hydrogenation, measuring the catalysts activity and its ability to resist poisoning by nitriles. This paper presents a practical approach to selecting the best selectivity control for the commercial production of primary amines and demonstrates that chemical additives alone are not enough to allow one to obtain the best possible control over selectivity and in fact, the mode of operation and reaction conditions are also important in the optimization process.

Introduction

It has long been known that depending on the catalyst and conditions nitriles may be transformed through catalytic hydrogenation into primary, secondary or tertiary amines. Generally a mixture of the amines is formed. For commercial reasons only one of these products is usually desired. In many cases only the primary amine is of interest, e.g., dimethylaminopropylamine (DMAPA), an important intermediate for producing soft soaps, and hexamethylenediamine (HMDA), a Nylon® intermediate. These products are produced at the multi-million and billion pound per year level, respectively. Industry's challenge is to produce these products in high selectivity because at these high volumes even a few tenths of a percent represents a very large co-product disposal problem and at the percent level, the byproducts become almost unmanageable from an ecological standpoint unless there is a commercial outlet for the co-products. This is often not the case. The purpose of this work was to identify optimum technology for controlling selectivity to the primary amine during nitrile hydrogenation in backmix reactors, whether batch, semi-batch or continuous equipment.

Background

The most common and least expensive catalysts for making primary amines from nitriles are supported and sponge nickel or cobalt. With these catalysts in a batch mode of operation one generally generates up to 20% secondary amine if no moderating agents are added to control the selectivity. Figure 1

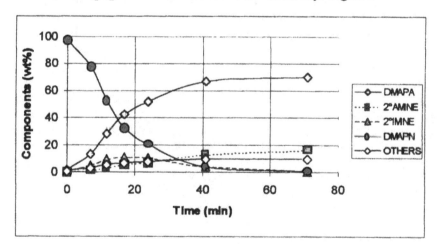

Figure 1   Batch Hydrogenation of DMAPN with Sponge Nickel. Catalyst, sponge nickel A4000, 0.92%; Water, 8%; Temperature, 100°C; Pressure, 850 psig.

shows a batch hydrogenation of dimethylaminopropionitrile (DMAPN) using sponge nickel without use of selectivity control additives. This reaction produced 18% of the secondary amine. Unless the secondary amine was needed for a second, smaller application, this process would be unsuitable for commercial scale-up because of the low yield of the primary amine. More selective process technology is clearly needed for this hydrogenation reaction to be economically viable.

The use of alkaline substances to enhance the selectivity to primary amine during the hydrogenation of nitriles has been known since 1939 when reported by E. J. Schwoegler and H. Adkins (1). Schwoegler and Adkins predicted and then demonstrated that adding sufficient ammonia to the hydrogenation of a nitrile would strongly inhibit the formation of secondary amine and thus greatly improve the selectivity to primary amine. This topic has been covered in several reviews (2, 3) of nitrile hydrogenation.

Ammonia has been commonly used to control selectivity and directs it to primary amine. In 1923 von Braun (4) proposed a path for the hydrogenation of nitriles to amines. This was later modified by Greenfield (5) to incorporate tertiary amines through the intermediacy of an enamine and again by Dallons (6) to incorporate surface species. This scheme is shown in Figure 2 and shows a surface bound primary imine (1°imine) species that is attacked by a primary amine (1°amine) which then expels ammonia in a reversible step during the formation of the secondary imine (2°imine). Hydrogenation of the 2°imine generates the 2°amine. The formation of the 2°amine is for all practical purposes identical to reductive amination of an aldehyde. However as Dallons (6) points out, the 1°imine and the adduct in Figure 1 are species that are never seen by gas chromatography and are thus most likely bound to the catalyst surface. It is quite logical then for addition of ammonia to minimize the concentration of the 2°imine, generate a second adduct where R' is H and drive this set of reactions to produce almost exclusively 1°amine.

The first US patent taking advantage of this finding was issued to H. P. Young and C. W. Christensen (7) in 1942. Here fatty nitriles were hydrogenated in the presence of aqueous ammonia to improve the selectivity to primary amines. Either water or a mixture of water and ammonia improved the selectivity to primary amine over anhydrous ammonia according to these authors. This indicated that the operative component controlling selectivity was hydroxide ion, not ammonia. In addition this was the first disclosure that sodium ethylate improved selectivity to primary amine during hydrogenation of a nitrile. The inventors rationalized that ethoxide would behave like hydroxide.

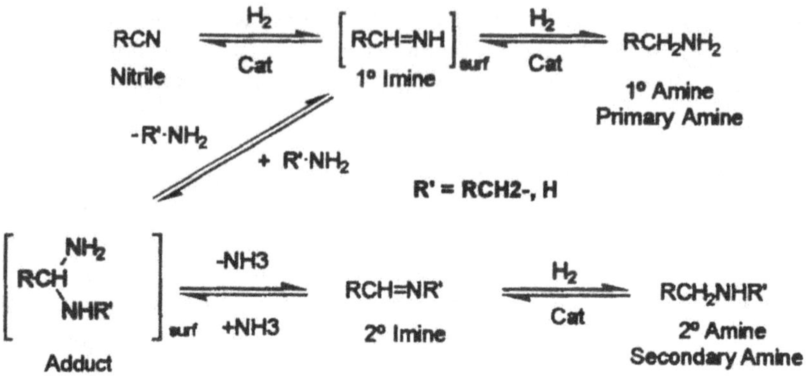

Figure 2 - Nitrile Hydrogenation Pathway

In 1948 M. Grunfeld (8) disclosed that selectivity to primary amine could be controlled by using alkali metal hydroxides, barium hydroxide or quaternary ammonium hydroxides during the hydrogenation of fatty nitriles and of dinitriles such as adiponitrile. In 1974 Bartalini and Giuggioli (9) were issued a US patent on a process which used continuous backmix technology to produce amines from nitriles and in which the selectivity to primary amine was controlled with addition of alkali metal hydroxide. Here adiponitrile was converted over Raney® nickel into hexamethylenediamine in high selectivity in a gas lift reactor, a type of continuous backmix reactor system.

Since that time numerous citations have been made using alkali metal hydroxides to control selectivity while hydrogenating nitriles. The reaction pathway proposed in Figure 2 however does not readily accommodate alkali or alkaline earth metal hydroxides for controlling selectivity. In 1998, Thomas-Pryor et al (10) shed some light on the possible reasons that these alkaline substances are effective. Adsorption studies of butylamine on Raney® nickel catalysts in the liquid phase showed that as much as 50 mmoles of butylamine is adsorbed onto the surface of 1 g of catalyst suspended in methanol at room temperature. When the suspension was treated with NaOH (0.1 to 0.4 mM), the amount of butylamine adsorbed dropped by about 50%. Hydrogenation reactions run under similar concentrations of NaOH, hydrogen and butyronitrile in methanol as in the adsorption generated roughly 8% 2°amine when no NaOH was present but only about 0.9% when it was present. What is certain is that NaOH dramatically decreased the amount of butylamine adsorbed on the catalyst surface and decreased the amount of 2°amine formed. What isn't as certain is how many different kinds of sites for butylamine adsorption were present and which types were being eliminated when NaOH was added. It does however look like the majority of sites that were important for forming the 2°imine were the ones being eliminated by the alkali and that those responsible for hydrogenation were not since the rate of hydrogenation was not greatly affected.

Also shown in the Thomas-Pryor work (10) was that the amount of adsorbed 1°amine on a sodium hydroxide treated catalyst surface decreased by almost 50% when hydrogen replaced the bubbling nitrogen passing through the stirred slurry in methanol. This shows that hydrogen may compete quite strongly for the catalyst surface in the presence of 1°amine. They unfortunately did not report what happens when no alkali was present.

Batch Hydrogenation of DMAPN to DMAPA

Our initial process for manufacture of DMAPA used a sponge nickel catalyst and ammonia to moderate secondary amine production. This method of selectivity control however has some significant drawbacks, e.g., cost of ammonia recovery or disposal, low productivity because of dilution with ammonia, low reaction rate because of low hydrogen partial pressure and an increase in cycle time caused by charging and venting ammonia. A significant improvement in rate and productivity came with the substitution of lithium hydroxide for ammonia. Table 1 shows the results of 8 consecutive plant runs using lithium hydroxide with chromium promoted sponge nickel to control secondary amine under conditions similar to the run depicted in Figure 1. LiOH was only charged for the first use of catalyst. However the presence of LiOH kept secondary amine from rising above 8.5% over 8 uses of the catalyst. As the number of uses increased, the time required to convert the DMAPN also increased. Each time the catalyst was reused, a little more LiOH appeared to be lost and the selectivity to DMAPA diminished. However adding more LiOH•H$_2$O to used catalyst resulted in the catalyst agglomerating into balls and a severely diminished reaction rate.

Dimethylamine (DMA) and mono-n-propylamine (MNPA) were the main light by-products generated during the hydrogenation using sponge nickel and LiOH. These light by-products were the result of the retro-Michael

Table 1 - Sponge Nickel and LiOH[a]

Normalized Weight Percent (Water-free Basis)

Catalyst Use	Reaction Time (min)	2°Amine (wt%)
1	270	2.0
2	225	4.3
3	270	5.8
4	285	6.1
5	405	4.9
6	330	6.2
7	330	7.8
8	600	8.5

[a]0.8% sponge nickel (A4000, Activated Metals, Inc.), 0.4% LiOH•H_2O, 2% water.

reaction and their presence during the hydrogenation resulted in the formation of a variety of secondary heavy by-products. Figure 3 shows the most likely reaction pathways to the primary and secondary by-products observed. Although this process provided a reasonable yield of DMAPA, it also produced from 1000 to 1500 ppm of N,N,N'N'-tetramethyl-1,3-propanediamine (TMPDA) which was inseparable from DMAPA by distillation. Since a large new DMAPA market had developed in recent years in which levels of TMPDA over 300 ppm was a problem, our process was incapable of producing a product of high enough purity. In this new application, DMAPA was an intermediate for the preparation of a betaine used in soft soaps made by the reaction of DMAPA with coconut fatty acid and then sodium chloroacetate. TMPDA does not form an amide with the fatty acid but does generate a bis-betaine by reacting with 2 moles of sodium chloroacetate. The bis-betaine imparted undesirable properties, primarily cloudiness, to the product. To be a viable supplier we needed to produce a higher purity product without raising the price thus adding a purification step was not even an option.

Kiel and Bauer, in an European Patent (11), claimed that DMAPA could be made essentially free of 1,3-propanediamine (PDA) by using sponge cobalt or nickel and a small amount of either calcium or magnesium oxide and ammonia to control selectivity to primary amine. The only example presented used sponge cobalt. In addition the patent taught that this could be done at 160-180°C at 2200 psig with batch processing. This patent did not report the selectivity or yield of DMAPA but showed that < 50 ppm of PDA was formed

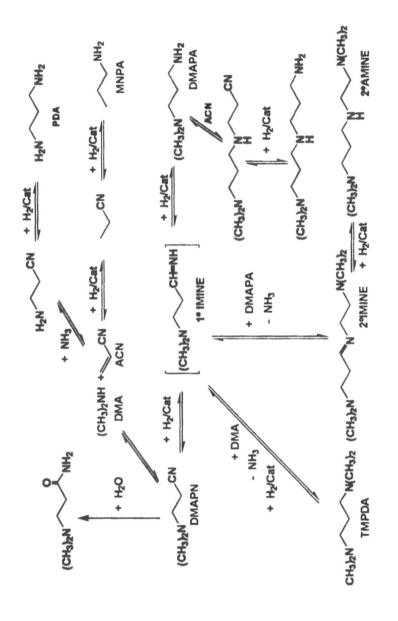

Figure 3 By-products Formed during Hydrogenation of DMAPN

under these conditions. Our experience at these high temperatures was that large amounts of retro-Michael reaction by-products formed and consequently a poor yield of DMAPA resulted. Not only was it unusual to operate at such a high temperature, presumably without making substantial amounts of by-product but to make DMAPA with < 50 ppm of PDA with so much ammonia present appeared to be a remarkable achievement. We repeated the patent example but at 1200 psig and 160°C using an unpromoted sponge cobalt and found that even at the lower pressure, the claims with respect to PDA were valid. However, even though the 2°amine content was only 3.4%, the DMAPA yield was only 90%.

We were intrigued by the use of sponge cobalt and decided to try it using our current plant operating conditions but with LiOH for selectivity control since reverting to ammonia would be a step backwards for us. This combination of conditions, catalyst and LiOH produced very high quality DMAPA but surprisingly produced only 90 ppm of TMPDA, no PDA and only 1.1% 2°amine. We then tried Cr and Ni promoted sponge cobalt. Table 2 shows the results of this combination through 9 recycles of the catalyst. As with sponge nickel, LiOH was only added to the first charge. We lost very little selectivity on recycling the catalyst. In these experiments the time for reaction was extended intentionally to between 4 and 5 hours through temperature and pressure programming. When the catalyst was fresh, the run was started at a low temperature. As the run progressed the temperature was increased. The initial runs were also conducted at a lower pressure. As the catalyst aged the initial temperature was increased as was the pressure. Extending the reaction time in this fashion was done to match the heat removal capability of the plant in which the process would eventually be operated to see if anything detrimental with respect to product quality was caused by slowing the reaction down via pressure and/or temperature programming. We noted that as the catalyst aged during the first 6 uses, there was a steady increase in amount of secondary amine and a corresponding decrease in DMAPA assay. Increasing the pressure in the 7th use brought the assay back up and the secondary amine dramatically decreased. It seems reasonable that increasing the pressure would have this kind of affect since the hydrogen concentration in solution should rise as pressure is increased and increase the probability of the adsorbed primary imine intermediate reacting with hydrogen over it reacting with primary amine. In the light of Thomas-Pryor's work (10), this suggests that at higher pressure hydrogen displaces 1°amine from the catalyst surface thus increasing the rate of hydrogenation   of   the   1°imine   over   addition   of   1°amine.

Table 2   DMAPN Hydrogenation Over Sponge Cobalt 2724 (W. R. Grace) Modified With LiOH (9.6 mmoles/g catalyst)

Catalyst Use	1	2	3	4	5	6	7	8	9
Temp (°C)	70-100	80-100	80-100	80-100	85-100	90-100	100	100	100
Press (psig)	400	400	400	400	400	400	750	750	750
RxTime (min)	290	220	295	280	295	340	265	240	275
Composition (wt%)									
DMA	0.12	0.17	0.18	0.21	0.33	0.92	0.87	0.93	0.94
MNPA	0.10	0.16	0.20	0.23	0.27	0.30	0.23	0.25	0.27
DMAPA	98.55	98.10	97.82	97.63	97.42	96.81	97.45	97.41	97.34
DMAPN	0.03	0.03	0.01	0.01	0.03	0.01	0.02	0.04	0.04
TMPDA (ppm)	40	38	59	99	-	33	83	100	96
2° AMINE	0.78	1.01	1.25	1.28	1.33	1.32	0.88	0.83	0.85
OTHER	0.43	0.52	0.54	0.63	0.60	0.53	0.45	0.54	0.47

Catalyst = 0.52 wt%, LiOH•H$_2$O = 0.21 wt%, Water = 3.0 wt%, Stirrer Speed = 1200 rpm

During the 1st through 6th use DMA and MNPA , retro-Michael by-products, definitely increased as the average temperature increased. From the 7th to 9th use, when the temperature remained at 100°C and the pressure was at 700 psig, the amount of these by-products remained relatively constant but quite high. The DMA and MNPA concentration did not decrease by the increase in pressure as did the 2°amine content.

The reaction that is being retarded by LiOH is the reductive amination of adsorbed 1°imine by amines. This is the path that leads to 2°amine and by-products such as TMPDA. However while 2°amine decreased by raising the pressure, TMPDA increased. At first this appears to be a discrepancy but, on further examination one notes that the average temperature also increased and with it the concentration of DMA, by a factor of about 3, and thus it is this increase that resulted in the increase in TMPDA.

Another effect of temperature however was that it improved the reaction rate as the catalyst deactivated. Thus there is a trade off between reaction rate and the amount of primary and secondary retro-Michael by-products. An example of this is shown in Table 3 where the 11th use of the catalyst was run at 140°C but at 400 psig. Here the reaction was over in about 135 minutes but the assay of DMAPA was significantly lower as a result of higher by-product formation across the board. Note, the water content in this experiment was intentionally kept lower to minimize hydrolysis of the nitrile group at this high temperature and minimize retro-Michael addition chemistry since water catalyzes the Michael addition, acting as a proton transfer agent (12).

Table 3   Effect of High Reaction Temperature

		Composition	Wt%
Catalyst Use	11	DMA	0.95
Temp (°C)	140	MNPA	0.46
Press (psig)	400-440	DMAPA	96.59
Water (wt%)	0.3	DMAPN	0.38
LiOH•H_2O (wt%)	0.21	TMPDA (ppm)	243
Rx Time (min)	135	2° AMINE	0.94
		OTHER	0.57

These experiments indicate that operating at high temperature will result in generating more by-products and lower hydrogen pressures will tend to generate more secondary amine. However operating at low temperature has a penalty in productivity. It would be a major improvement if a way could be found to operate at higher temperatures without being penalized by a lower yield. In fact there are ways of doing this but one needs to move away from batch processing to either semi-batch or continuous backmix technologies. These options are discussed later.

The use of LiOH with sponge cobalt was much more effective in controlling both 2°amine and TMPDA compared to sponge nickel with LiOH. LiOH appeared to not only control selectivity more effectively but also was more permanent, i.e., selectivity did not deteriorate appreciably with use. This is supported by plant data in Table 4. The data shows that even after 15 uses of the catalyst, the initial charge of LiOH was still very effective in controlling the 2°amine. The TMPDA level wasn't tracked on a batch by batch basis but the tank car of distilled product was found to have <250 ppm, well within the customer's specification and 4 to 6 times less than when LiOH modified sponge nickel was used..

Table 4   Manufacture of DMAPA Using Sponge Cobalt and LiOH[a]

Catalyst Use	Reaction Time (hr)	2°Amine (wt%)
1	3.50	1.32
2	5.25	1.54
3	5.50	1.15
4	3.75	1.11
5	3.75	1.73
6	4.25	1.39
7	4.25	1.20
8	4.75	1.34
9	4.75	1.32
10	4.25	1.58
11	4.75	1.89
12	5.25	1.79
13	3.75	1.81
14	3.50	1.86
15	3.75	1.67

a) 0.2% LiOH•H_2O, 0.4 wt% sponge cobalt

Is it lithium ion or hydroxide that imparts the special selectivity that we see in these hydrogenations? To answer that question lithium acetate and lithium carbonate were substituted for LiOH at a level of 12 mmole Li/g catalyst in two separate batch hydrogenations. The same result was obtained in both cases. Neither the acetate nor carbonate was any more effective in controlling selectivity than if nothing was added. It was also found that if lithium hydroxide hydrate is added as a solid, it took about 3 recycles of the catalyst/LiOH to see the full effect on selectivity. To get immediate selectivity response LiOH must be added as a solution in water. It is possible to isolate LiOH modified sponge cobalt catalyst that is essentially free of water. In this case it makes a difference as to what the liquid media is being used during the addition of the aqueous solution of LiOH. While exploring this feature we found that not all solvents can be used for the pretreatment. Table 5 shows the results of a number of batch hydrogenations in which different solvents were used to prepare the LiOH modified sponge cobalt. In these experiments the about 2 grams of catalyst was slurried in 100 ml of the solvent and then about 1 gram of lithium hydroxide hydrate dissolved in about 10 ml of water was added while stirring. After a few minutes of stirring the slurry was filtered and the recovered catalyst was re-suspended in another 100 ml of the solvent and filtered again. Finally the catalyst was washed into the autoclave using about 430 g of DMAPN containing 2% water. The hydrogenation of DMAPN was then conducted. In each case the product was analyzed for 2°amine content.

Table 5 shows that solvents and solvent mixtures in which LiOH is soluble, e.g., water, methanol, ethanol and/or mixtures of same, are not useful for preparing the LiOH modified catalyst for isolation. Furthermore when solvents in which water is immiscible, e.g., toluene, are used, much of the LiOH remains in the aqueous phase and is removed by this technique of isolation. This study shows that although LiOH is apparently strongly adsorbed onto the surface, polar solvents like water, methanol/water and ethanol/water can desorb it.

The most effective modified catalyst is made when the LiOH precipitates in a very fine particulate form and adsorbs onto the sponge cobalt as occurs in DMAPA, DMAPN, DMF and THF. In this form, the catalyst gives excellent selectivity control and may be recycled many times. If the solvent or solvent mixture dissolves LiOH, it can still be used but the LiOH modification needs to be made *in situ* but one should expect the selectivity control to dissipate or be completely removed on subsequent recycles.

Table 5 Hydrogenation of DMAPN with Co-2724/LiOH Prepared in Different Solvents

Solvent	LiOH•H₂O/ Catalyst	Amount DMAPN Charged	DMAPN Remaining	2°Amine
	mmoles/g	g	wt%	wt%
Water	11.7	426.6	0.042	11.41
Methanol	11.8	434.4	0.027	11.50
Ethanol	11.1	430.0	0.046	8.87
1-Propanol	12.2	456.0	0.002	2.35
Toluene	12.3	434.0	0.013	4.40
THF[a]	11.3	439.2	0.103	0.95
DMF[b]	11.2	436.0	0.067	0.70
DMAPA	2.95	429.5	0.115	0.66
DMAPA	11.9	432.8	0.200	0.69

[a] THF = tetrahydrofuran
[b] DMF = dimethylformamide

When the LiOH modified cobalt catalyst was prepared in methanol or DMAPA as indicated above and the two filtrates were analyzed for Li ion by ICP/AES, essentially all of the Li ion was found in the first filtrate when methanol was used but none when DMAPA was the medium. When the same experiment was repeated using NaOH in DMAPA, only a small fraction of the sodium hydroxide was found in the filtrate but the catalyst agglomerated into millimeter sized balls and also stuck to the side of the glassware.

The effectiveness of three alkali metal hydroxides is presented in Table 6. In these experiments the catalysts were modified with the alkali hydroxide in the same manner as when different solvents were used, but in all cases DMAPA was the solvent. This was done so the behavior of the catalyst could be observed when the alkali solution was added. In runs 1 to 4, amounts of LiOH varying from zero to 12.2 mmoles/g of catalyst were added to Cr and Ni promoted sponge cobalt. With no LiOH the run took 125 minutes but 9.3% of secondary amine was produced while at 3 mmoles/g only 0.6% was made and at 12.2 mmole/g only 0.4% secondary amine was made. The reaction time for LiOH modified catalysts was essentially the same as when no LiOH had been added. When NaOH was substituted for LiOH, a level of 0.8 mmoles/g reduced the secondary amine to 5.5% without increasing the reaction time. However at

Table 6   Effect of Alkali Metal Hydroxides on Activity and Selectivity
2 wt% Water, Temperature–100°C, Pressure–750 psig, 1200 RPM

Run	Catalyst	Catalyst Amount	Alkali Metal Hydroxide MOH	MOH/ Catalyst	Amount DMAPN	Reaction Time	DMAPN remaining	2° AMINE
		g	M	mmoles/g	g	min	wt%	wt%
1	Co 2724	2.04	none	0.00	433.7	125	0.12	9.34
2	Co 2724	2.20	lithium	1.42	431.8	135	0.02	6.27
3	Co 2724	2.11	lithium	2.95	429.0	140	0.12	0.60
4	Co 2724	1.81	lithium	12.16	429.1	120	0.01	0.46
5	Co 2724	2.22	sodium	0.81	427.2	130	0.26	5.77
6	Co 2724	2.06	sodium	8.65	428	140	86.0	0.42
7	Co 2700	1.85	sodium	9.54	427.8	185	88.0	0.23
8	Co 2700	2.19	sodium	5.80	426.7	260	86.9	0.24
9	Co 2724	1.95	potassium	5.24	412.5	155	87.4	0.43
10	Co 8000[a] A-	1.96	sodium	8.85	429.9	120	96.2	0.29
11	Ni 2800	2.51	sodium	6.93	432.4	120	94.7	0.33

a) Activated Metal and Chemicals, Inc.   All other catalyst are W. R. Grace & Co. products.

8 mmoles/g, although the secondary amine content of was only 0.4% after 140 minutes, 86% of the DMAPN remained unreacted.   Similar results were obtained with KOH and with NaOH using unpromoted sponge cobalt and sponge nickel. The explanation for this appears to be that NaOH and KOH above about 1 mmole/g cause the sponge catalysts to agglomerate and become sticky. When these modified catalysts were transferred to the autoclave with DMAPN they were already agglomerated. Upon opening the autoclave for cleaning after the hydrogenation, the majority of the catalyst was found stuck to the walls, stirrer and cooling surfaces when more than about 1 mmole of either NaOH or KOH was used. With LiOH, even loadings as high as 22 mmoles/g of catalyst gave very good results with respect to rate and selectivity to primary amine without any appreciable amount of the catalyst sticking to the autoclave surfaces.

These results appear to contradict those obtained by Thomas-Pryor (10) where reasonable reaction rates were obtained along with high selectivity to 1°amine. The difference is that methanol was used as a solvent in these cases which apparently keeps the catalyst from agglomerating. If a solvent is used commercially a significant loss in productivity will occur since the equipment contains solvent instead of product. Furthermore a significant cost penalty is suffered for losses on handling, lower productivity of equipment because of the time required for charging and distillation of the solvent.

The reason alkali metal hydroxides alter the selectivity of nickel and cobalt catalyst is not well understood. A reasonable explanation is that the alkali metal hydroxide neutralizes acidic sites on the surface and the primary amine is simply not adsorbed as strongly. This allows hydrogen to cover more surface sites and thus competes better. An attempt to examine the surface of a used LiOH modified catalyst by XPS was a frustrating exercise. If we washed the recovered catalyst to remove the crude DMAPA, we took the chance of also removing the LiOH. If we didn't wash it, but just removed as much organic as possible by heating in the vacuum we probably would leave a lot of organic material on the surface. However we chose the latter. What we found was a high proportion of the surface was covered by Li atoms (12 atom %) and in fact, no cobalt, aluminum, nickel or chromium was detected. As expected the remainder of the surface was covered by carbon and nitrogen in an atom ratio of 5-6/1 for C/N. This ratio was much too rich in carbon to be DMAPA or polyacrylonitrile. We only learned what we had known from bulk analysis that lithium was definitely on the surface of the catalyst.   When the LiOH modified catalyst was analyzed by ICP-AES, the Li/Co weight ratio was determined to be 0.106 vs. 0.986 calculated from the amount of catalyst (1.98 g) and LiOH•H$_2$O (1.027 g) used in preparation. Essentially all the LiOH was on the catalyst.

Semi-batch Hydrogenation

Semi-batch operation involves the incremental feeding of a reducible substrate, in our case, a nitrile to a slurry of catalyst in a solvent or in the hydrogenation product. In a typical semi-batch experiment hydrogen is supplied from an external ballast of sufficient capacity to satisfy the stoichiometry of the reaction and at an elevated pressure with respect to the desired operating pressure of the vessel. When operating in this mode, the amount of nitrile present in the reactor at any time will be determined by the global hydrogenation rate and the nitrile feed rate. Ideally when the feed rate is low with respect to the hydrogenation rate, the amount of nitrile will be at a very low steady state concentration. Conversely at a high feed rate, the steady state nitrile concentration will be relatively higher. When the nitrile feed is stopped, the hydrogenation again reverts to a batch operation and the residual nitrile is converted to product. When the steady state concentration is very low and the feed is stopped, the batch hydrogenation only proceeds for a few seconds before the residual nitrile is consumed. Conversely if the steady state nitrile concentration is high, it may take tens of minutes to convert the remaining nitrile to product. If the catalyst deactivates during the hydrogenation a steady state nitrile concentration is never reached because the global rate is constantly decreasing but the nitrile feed rate is constant. Thus an upward drift in the steady state nitrile concentration with time is an indication of catalyst deactivation.

Semi-batch hydrogenation is an excellent tool for comparing different catalysts and for determining the effect of additives, e.g., different alkali metal hydroxides, on the catalyst behavior. Figure 4 shows the results of a typical experiment for determining catalyst performance. In this experiment DMAPN containing 1.7 wt% water was pumped into a 2 liter reactor containing 2.15 g of sponge cobalt 2724, 32 g of water and 400 g of a pure grade of DMAPA. The reactor was at 140°C, 750 psig hydrogen pressure and stirred at 1200 rpm. As seen in Figure 4, DMAPN was fed in initially at nominally 2.5 g/min and the feed rate was periodically increased to a maximum of about 10 g/min. The steady state concentration of DMAPN in the bulk is shown above the trace of the feed rate. Note how it increased as the feed rate was increased. The rate of hydrogen pressure drop shown is from a pressurized ballast of 1 gallon capacity and is a measure of the rate of hydrogen as well as nitrile consumption. Figure 4 shows that for each change in feed rate there is a commensurate change in hydrogen consumption rate. The spikes in the hydrogen rate curves just before a change in feed rate is caused by taking two samples. The first sample clears

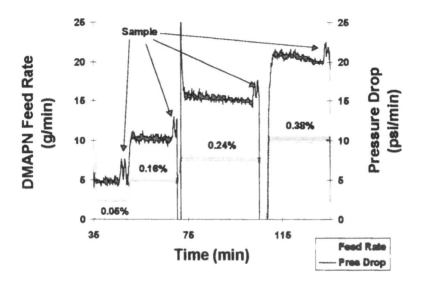

Figure 4   Semi-batch Hydrogenation of DMAPN

the sampling line and the second sample is analyzed by gc for composition, including the steady state DMAPN concentration. The feed pump is not turned off during sampling since we want to know what the steady state DMAPN concentration is while feeding at a particular rate. Before the last change in feed rate the reactor contained about 1200 ml so the volume was adjusted back to the initial level of about 500 ml before proceeding with the last feed rate change.

Figure 5 shows the results of the above experiment and a subsequent one where higher feed rates were used. The feed rate data was normalized to 1 gram of catalyst and converted to g DMAPN/g catalyst/hour or units of $g \, g^{-1} hr^{-1}$ or simply hr^{-1}. What this tells you is the performance, as measured by the residual nitrile concentration, you will obtain from a given catalyst loading and feed rate. For instance, if you have a 1000 gallon reactor containing 100 lbs of catalyst and you want to keep the DMAPN concentration in the product below 0.25%, you could feed at a normalized rate of 165 hr^{-1} or 16,500 lb DMAPN/100 lb catalyst/hr. This assumes that your commercial reactor is no more mass transfer limited than the lab reactor and can remove the heat of reaction at the chosen feed rate.

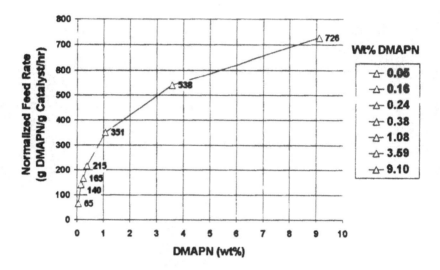

Figure 5 Normalized Feed Rate vs Residual DMAPN for Semi-batch
Hydrogenation. Temperature 140°C; Pressure 750 psig; Stirrer
Speed 1200 rpm.

This test method was used to measure the relative performance of
several commercial sponge catalysts. Figure 6 shows the results of these
comparative tests. These tests were performed in the same 2 liter autoclave at
140°C, 750 psig hydrogen pressure, 1200 rpm and with between 0.8 and 2.2 g
of catalyst and 12 to 13 mmoles of LiOH/g of catalyst. The results were
normalized to 1 g of catalyst. Sponge cobalt Co-2724 (Co:Ni:Cr:
Mo:Al::87.0:2.75:2.75:0:7.5; by wt.) performed much better in these tests than
other sponge catalysts, particularly at the higher feed rates. A-8036 was
promoted   by   nickel   and   molybdenum   (Co:Ni:Cr:Mo:Al::
80.46:8.43:0.008:6.62:4.47) while A-8046 was promoted by nickel and
chromium (Co:Ni:Cr:Mo:Al::83.46:7.93:3.23:0.10:5.29) and both performed
similarly.   Co-A8000   was   essentially   not   promoted
(Co:Ni:Cr:Mo:Al::.92.19:0.44:0.009:0.0:7.36). The latter catalyst showed the
poorest performance, particularly at high feed rates. Sponge nickel, Ni-A4000,
a chromium promoted catalyst performed almost identically to Co-A8036. At

low feed rates, all the catalysts performed reasonably well, although Co 2724 was better than the rest.

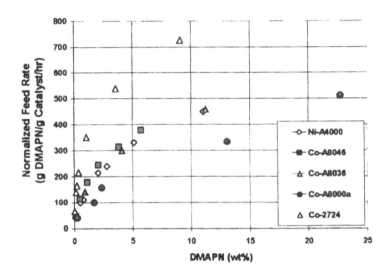

Figure 6   Comparison of Sponge Metal Catalyst Modified with LiOH (Ni-A4000, Co-A8046, Co-A8036 and Co-A8000 are products of Activated Metals and Chemicals, Inc. Co-2724 is a product of W. R. Grace and Co.)

In practice it is unlikely that anyone would operate a semi-batch reactor at rates much higher than 200 hr^{-1} because the amount of unreacted nitrile becomes excessive even with the best catalyst and the amount of heat removal will begin to tax most commercial reactor systems. So why run the test at such high feed rates? The test was run at the high feed rates to try to determine which of the catalysts was most resistant to poisoning by the nitrile feed. Figure 7 shows what happens when a sponge catalyst is exposed to high feed rates and then returned to a low feed rate. If the high concentration of nitrile had no effect on the performance of the catalyst, the same low steady state concentration of nitrile would have again resulted when the feed rate was returned to the former setting. If not, the catalyst will have been partially deactivated by the feed and consequently will not perform as well as it originally did with respect to rate.

Figure 7 shows two examples of how the catalyst performance deteriorates as a result of exposing it to high concentrations of nitrile. In both cases the feed rate was increased until the steady state nitrile concentration was

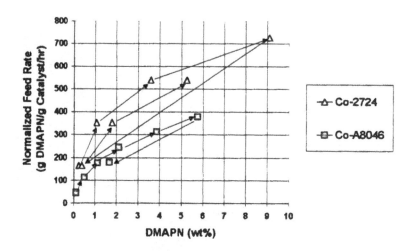

Figure 7   Assessment of Resistance to Activity Loss from Nitrile Contact

over 5% and then the feed backed off to an earlier feed rate. With Co-2724 after exposing the catalyst to over 9 wt% DMAPN the feed rate was returned to the initial rate of 165 hr^{-1}. The first time at 165 hr^{-1} the steady state DMAPN concentration was 0.23% but on returning, it increased to 0.28%, i.e., less nitrile converted. Now moving back up to 335 hr^{-1}, instead of 1.0% DMAPN, the steady state concentration was now 1.8%, etc. Catalyst Co-A8046 was only subjected to 5.8% DMAPN but when the feed rate was dropped to 185 hr^{-1}, the steady state DMAPN concentration had risen from 1.1 to 1.7%. If we look closely at the hydrogen uptake rate data in Figure 6 we also see rate deterioration, particularly when the feed rate was at the two highest levels. The DMAPN feed rate was constant but the hydrogen pressure drop, which is proportional to the nitrile hydrogenation rate, was definitely not. Although I had been calling the DMAPN concentration a steady state concentration, it surely wasn't, in fact it slowly increased as the catalyst activity deteriorated. This test is particularly useful when trying to chose between two catalysts which have about the same activity.

The lesson from this is that to get the maximum performance out of your catalyst, you need to operate at the lowest steady state DMAPN concentration that makes economic sense. In these experiments we saw almost no perceptible hydrogen uptake rate deterioration at normalized rates of about 100 hr^{-1}. This is a very substantial production rate in most cases and should give one excellent economics. This data also says, running nitrile hydrogenations in a batch mode will always give you more catalyst deactivation than operating semi-batch because you start out with a reactor full of nitrile and expose the catalyst to very high nitrile concentrations for extended periods. However, with careful handling and judicious choice of reaction temperature and pressure, it is possible to reuse the sponge cobalt catalysts for tens of cycles.

Continuous Nitrile Hydrogenation

The only difference between a semi-batch and continuous mode of operation is that products are removed at the same rate as starting materials are added. If a semi-batch experiment indicates that at a given normalized feed rate that the nitrile concentration should be some number for a given catalyst, the same concentration of nitrile should be obtained in a continuous reactor. For example if sponge nickel A4000 is used as the catalyst, Figure 6 indicates that at a normalized feed rate of about 100 hr^{-1}, the DMAPN concentration should be below 1%. Likewise one would expect the DMAPN concentration to be below 0.1% at 100 hr^{-1} if sponge cobalt 2724 is used. In fact this is seen when these catalysts are used under these conditions in a continuous stirred tank reactor (CSTR). Figure 8 shows the A4000 nickel catalyst performance in a CSTR. The DMAPN concentration initially rose to 1.5% as the water concentration in the bulk decreased from about 8% to 0.5%. When 1.1% water was added to the nitrile feedstock, the nitrile concentration in the bulk decreased to 0.5% and remained there for the next 4 hours. More water was added to the feed, increasing it to 4.5%. This brought the nitrile concentration in the bulk down to 0.2%. This shows that the reaction rate is greatly influenced by the amount of water in the system. However, as predicted by the semi-batch experiment which had about 2% water present, the nitrile concentration in the bulk at a normalized feed rate of less than 100 hr^{-1}, in this case 63.8 hr^{-1}, was less than 1%. The consequence of having increasing amounts of water in this experiment is shown in Figure 9. By-products increased from <0.2% to about 0.7% as a result of the increase in water concentration. As mentioned before, water promotes the formation of by-products, both by hydrolysis of nitrile and those arising from the retro-Michael addition reaction. Thus a compromise must be made between higher rate of reaction or higher yield. It also makes it necessary to use the same amount of water in the semi-batch test to measure catalyst performance to get meaningful results.

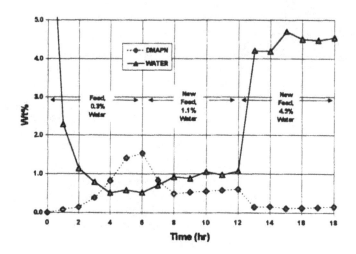

Figure 8 CSTR Hydrogenation of DMAPN. Catalyst, sponge nickel A4000, 4,7g; LiOH•H$_2$O, 2.0g; Volume, 350ml; Feed Rate, 300g/hr (63.8 hr^{-1}); Temperature, 140°C; Pressure, 850 psig

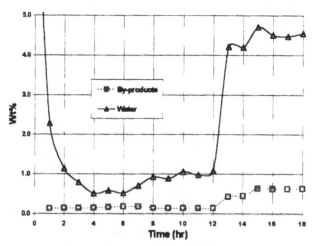

Figure 9 Effect of Water on By-product Make

In a second CSTR run, shown in Figure 10, using sponge nickel and LiOH, the water concentration was kept in the range of 4-6% throughout the experiment. It was quite obvious that the LiOH hydroxide was being lost as the run progressed by following the amount of 2°amine being formed as it rose from < 0.1% to about 1.6% over 24 hours on-stream. It was also noticed that the DMAPN concentration continually rose as well, rising by a factor of about 6 over the same period. However it wasn't certain that the catalyst activity really deteriorated since the water concentration also dropped as the experiment progressed and this could be responsible for the higher DMAPN concentration observed.

Under similar conditions when sponge cobalt was the catalyst, LiOH provided a much better control over selectivity than it did with sponge nickel.. In addition with sponge cobalt very little catalyst activity was lost over time on-stream. This is illustrated in Figure 11. In this experiment neither the 2°amine or nitrile concentration rose above 0.1% over a 12 hour period on-stream. Less water and more catalyst was used and the normalized feed rate was lower than in the comparable

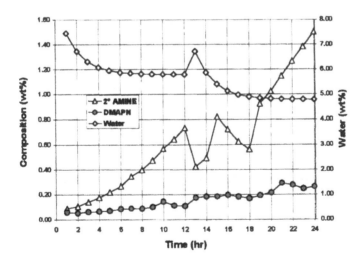

Figure 10   Selectivity Loss with Time with LiOH Treated Sponge Nickel. Catalyst, A4000, 4.7g; LiOH•H$_2$O, 2.0 g; Volume, 500 ml, Feed Rate, 300 g/hr (63.8 hr^{-1}); Temperature, 140°C; Pressure, 850 psig.

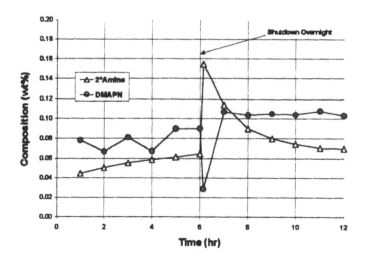

Figure 11   Stable Selectivity in CSTR Using LiOH Treated Sponge Cobalt. Catalyst, Cobalt 2724, 6.15g; LiOH•H₂O, 2.06 g; Volume, 500 ml; Feed Rate, 300 g/hr (49 hr⁻¹); Temperature, 140°C; Pressure, 850 psig.

experiment using LiOH modified sponge. There was no indication that the 2°amine concentration was increasing with time. In addition to this outstanding performance, even though the feed contained about 2% DMA, TMPDA fluctuated between 20 and 100 ppm but held steady at 20 ppm over the last 2 hours on-stream.

These experiments show that the combination of sponge cobalt and LiOH has special properties not demonstrated by sponge nickel with LiOH or of either sponge nickel or cobalt with other alkali metal hydroxides. LiOH apparently is more tightly bound to sponge Co and is thus more permanently attached. This makes it unnecessary to add additional LiOH to recycled catalyst in batch processes or to continually add LiOH to a continuous backmix process. However in the latter case more LiOH would have to be added if more catalyst is added to compensate for aging.

Adiponitrile Hydrogenation

To demonstrate that this technology was applicable for hydrogenation of other nitriles, adiponitrile was hydrogenated in a semi-batch reactor system. Initially we ran at 100°C and 750 psig at a normalized rate of 17.6 hr⁻¹in a 2 liter autoclave starting with a 400 g heel of pure hexamethylenediamine (HMDA), 8 g of water and 6.4 g of sponge cobalt modified with LiOH (10.05 mmoles/g). A total of 1400 g of adiponitrile (ADN) containing 2% water was hydrogenated at 100°C, adjusting the level back to about 500 ml when the level reached about

1500 ml to avoid overfilling and to simulate continuous operation. The temperature was then adjusted to 140°C and the normalized feed rate increased to 55 hr^{-1}. A total of 4100 g of adiponitrile was fed at this temperature, thus turning over the heel about 10 times. Table 7 shows the quality of the crude product run under these conditions. Note that since only 3.5 turnovers of the heel had been achieved while at 100°C, the concentration of impurities shown are only about 90% of steady state whereas the data at 140°C represents steady state.

At 100°C the steady state aminocapronitrile concentration was 0.09 wt%. At 140°C it was reduced to 0.009 wt% even though the normalized feed rate was 3 time higher. On the basis of nitrile content, i.e., equivalents of nitrile/g catalyst/hr, these rates are similar to the feed rates used in the DMAPN hydrogenation. Thus a 17.6 hr^{-1} adiponitrile feed rate is equivalent to a 32 hr^{-1}

Table 7 Crude Hexamethylenediamine Assay from the Hydrogenation of Adiponitrile over LiOH Modified Sponge Cobalt

Temperature (°C)	100	140
Component	Area%	
n-Pentylamine	0.007	0.032
n-Hexylamine	0.019	0.142
Cyclohexylamine	0.002	0.018
Hexamethyleneimine	0.017	0.058
1,2-Diaminocyclohexane	0.072	0.412
1,2-Diaminocyclohexane	0.064	0.180
Hexamethylenediamine	99.123	97.857
6-Aminocapronitrile	0.090	0.008
Adiponitrile	0.024	
Amide of Mass 172	0.014	0.012
Bishexamethylenetriamine	0.549	1.196
Unknowns	0.020	0.084

rate for DMAPN and a rate of 54 hr^{-1} for ADN to 98 hr^{-1} for DMAPN. Although the conversion of nitrile was very good at these rate and temperatures, the amount of by- products formed at the two temperatures varied greatly. At 140°C greater than twice as much secondary amines, hexamethyleneimine and bishexamethylenetriamine, over 4 times more cis-plus trans-1,2-diaminocyclohexane and about 7 times more n-hexylamine was formed. The overall loss in hexamethylenediamine assay was 1.26% by going to the higher temperature. The assay at 100°C was certainly more attractive.

When one compares the productivity at the two temperatures and feed rates, it becomes more difficult to select between these options. Again assuming the commercial equipment is no more mass transfer limited than the lab reactor and that it can remove the heat of reaction, a 2000 gallon autoclave configured as a CSTR can produce enormous amounts of product at these temperatures, rates and catalyst loadings. Table 8 shows these hypothetical numbers. Operating at 140°C and the higher feed rate increases the productivity of the equipment by a factor of 3 while increasing by-product make by 8. This makes the decision very difficult for the producer unless there is a commercial outlet for the by-products.

Table 8   Productivity of Crude HMDA from a 2000 Gallon CSTR

Temp	Catalyst Loading	Normalized Rate	Crude HMDA Production	By-Products
°C	lb	lb/lb/hr	lb/day	lb/day
100	200	17.6	86,768	876
100	600	17.6	260,304	2,629
100	1000	17.6	433,840	4,382
140	200	55.0	276,467	7,089
140	600	55.0	829,400	21,267
140	1000	55.0	1,382,333	35,444

Conclusions

LiOH hydroxide modified sponge cobalt catalysts give exceptional primary amine selectivity control in nitrile hydrogenations without sacrificing productivity, particularly when used in semi-batch and continuous backmix equipment. To obtain maximum performance from the modified sponge cobalt catalyst a medium in which LiOH is essentially insoluble is required. Addition of small amounts of water is essential for the hydrogenation reaction to attain and sustain high rates. When semi-batch or continuous backmix equipment is used with the LiOH modified sponge cobalt catalyst, higher temperatures than used in a batch operation may be used without loss of primary amine selectivity. In fact the amount of secondary amine is reduced by a factor of from 5 to 10 times over a batch operation at much lower temperatures. Being able to operate at higher temperatures also allows one to obtain much higher productivity. In batch hydrogenations LiOH modified sponge cobalt also performs exceptionally well in controlling selectivity to primary amines and also allows the catalyst to be recycled tens of times. The primary reason for the outstanding recycle performance is lithium hydroxide's ability to adsorb strongly to the surface of sponge cobalt without causing it to agglomerate.

Acknowledgements

The authors wish to thank Air Products and Chemicals, Inc. for permission to publish this work which was the basis for a US patent (13). We also wish to thank Mr. Richard Schmaldinst for operating the laboratory CSTR. Thanks are also due to Dr. Reinaldo Machado and Dr. Gamini Vedage for may helpful discussions concerning engineering, chemistry and catalysis. Dr. Dale Willcox was extremely helpful in identifying many of the by-products encountered in these studies by GC-MS and we are indebted to him for his contributions.

Reactivity and Surface Analysis Studies on the Deactivation of Raney™ Ni During Adiponitrile Hydrogenation

Alan M. Allgeier* and Michael W. Duch

E.I. duPont de Nemours Co., Experimental Station, Wilmington, DE 19880
E-mail: alan.m.allgeier@usa.dupont.com

Abstract

The heterogeneous catalyst, Raney™ nickel, deactivates during the hydrogenation of adiponitrile. The present study shows that the deactivation process is general to α, ω-dinitriles of varying length and also occurs for 6-aminocapronitrile but does not occur with mononitriles such as butyronitrile. In contrast to a previously reported mechanism for Ni catalyst deactivation in acetonitrile hydrogenation, these reactivity trends implicate deposition of oligomeric secondary amines and thus blocking of active sites as the mechanism of deactivation. Electron spectroscopy for chemical analysis (ESCA) reveals an increase in C and N on deactivated samples compared to nondeactivated samples and supports the conclusions drawn from reactivity studies.

Introduction

Raney™ Nickel is a prominent catalyst for the hydrogenation of unsaturated organic compounds (1). Of special industrial significance is the conversion of nitriles to primary amines (2), for which Raney™ Nickel is used commercially in the production of hexamethylene diamine (HMD) from adiponitrile (ADN). The mechanism of nitrile hydrogenation has been suggested to proceed through an intermediate aldimine (RCH=NH), which is very reactive and is readily hydrogenated to the primary amine, Scheme (2a,3). Instead of this hydrogenation step the aldimine can also react with a primary amine to form an aminal, which subsequently loses ammonia forming a secondary imine (Schiff base) (2a,3). Hydrogenation of secondary imines yields secondary amines, which are well-known byproducts of nitrile hydrogenation (4). Recently Huang and Sachtler have suggested an alternative mechanism, which does not invoke the aldimine but provides for H atom transfer to the carbon atom of the nitrile in an initial mechanistic step leading to the adsorption complex, RCH2-N=M. This complex reacts further to form a primary amine. The Huang/Sachtler mechanism was presented on the basis of convincing H/D exchange reactions and nitrile deuteration experiments (5). Either mechanism acknowledges and explains the formation of secondary amines during nitrile hydrogenation. In the

Scheme

case of α,ω-dinitriles, secondary amine formation can occur in an intramolecular fashion leading to cyclic products or an intermolecular fashion leading to dimeric or oligomeric products (4). In practice nitrile hydrogenations are generally carried out in alcohol solvents and/or in the presence of a base such as aqueous metal hydroxides or ammonia to increase selectivity to primary amine products avoiding secondary amines (2a-d). Additionally, basic additives serve to prevent or delay the rapid deactivation of Raney™ Ni, which occurs in the hydrogenation of ADN in alcohol solvent. As will be further discussed here, this is in contrast to behavior observed in butyronitrile hydrogenation, for which Raney™ Ni does not rapidly deactivate. The goal of this investigation was to better understand this intriguing deactivation process by a threefold approach: 1) Examine the reactivity trends in a series of nitrile molecules; 2) Compare the behavior of Raney™ Ni toward deactivation in various reaction media and 3) Utilize surface analytical techniques for further understanding of the deactivation mechanisms. This study was carried out in light of a previous investigation in the literature of the deactivation of nickel catalysts used in acetonitrile hydrogenation, which implicated growth of a nickel carbide surface as the source of deactivation (6). Additionally, reference has been made in the patent literature to the deactivation of nickel catalysts during ADN hydrogenation but no data was offered to support a mechanism of deactivation (7).

Experimental

Succinonitrile, glutaronitrile, butyronitrile and hexanenitrile were obtained from Aldrich Chemical Co.; adiponitrile, 6-aminocapronitrile and 1,12-dodecane dinitrile were prepared at DuPont. All nitriles were used as received. Doped Raney™ Ni was obtained from W.R. Grace (Raney™ 2400) and was supplied in its active state. (*Caution:* In its active state Raney™ Ni is supplied as an aqueous slurry (~43% solids) and becomes a pyrophoric material upon evaporation of the water of slurry.) It is prepared by the leaching away of Al from an Al and Ni containing alloy, leaving a porous, high surface area catalyst, which is predominantly composed of Ni with approximately 2%(w/w) Fe and 2% Cr promoters and approximately 8% residual Al. The surface area of Raney™ 2400 is 140 m^2/g (BET analysis) and the active surface area is 52 m^2/g (CO adsorption) and is based on a ratio of 1:1, CO: Ni (8).

Hydrogenation reactions were carried out in a stirred 100 cc or 160 cc Parr Mini Reactor, using appropriate caution to prevent accidents associated with high pressure and temperature. Ammonia was transferred via distillation into the reactor or by liquid transfer under pressure. Nitrile hydrogenation reactions were carried out under hydrogen at 75 °C and 500 psi total pressure, except when using ammonia for which the total pressure was 900 psi. Generally the initial reaction solutions were 6.2M in nitrile and carried out with a ratio of 0.38 mol nitrile / g Raney™ Ni slurry.

Electron spectroscopy for chemical analysis (ESCA) was conducted using a Physical Electronics PHI 5600ci spectrometer operated with a magnesium anode at 300 watts power (15 kV and 20 mA). High resolution spectra shown in the figures were recorded at an analyzer pass energy of 23.5 eV and a take-off angle of 45 degrees. Spectra also were recorded at 58.7 eV pass energy with higher count rates for quantifying spectral intensities and reporting atomic concentration data. An analyzer slit of dimensions 0.8 mm x 1.2 mm was used to image the sample analysis area. Spectra were scanned at 0.1 eV step intervals and 100 msec step time. An electron flood gun was used to minimize charging effects. The base pressure in the analysis chamber was typically below 5 x 10^{-8} Torr. Number of scans used to accumulate spectra for each element varied, depending on signal intensity. All spectral peak positions were referenced relative to the carbon 1s (C 1s) level at 284.6 eV. The linearity of the electron binding energy scale was checked by recording the Au 4f7/2 line at 84.0 eV and the Cu 2p3/2 line at 932.5 eV. Raney™ Ni samples were analyzed in their oxidized states. These samples were prepared by filtration of small quantities (< 2g) of catalyst on a sintered glass frit, washing with methanol and passivation in air for 30 min. (*Caution:* pyrophoric catalyst, passivation conducted on small quantities to limit the heat of oxidation.) The powder catalyst samples were prepared for analysis by tamping samples onto double-coated tape (3M Scotch@ # 665). PHI MultiPak@ software version 5.0 was used for data analysis.

Results and Discussion

The hydrogenation of adiponitrile (ADN) with Raney™ Ni 2400 in the absence of base leads to the rapid deactivation of the catalyst. For instance, when a 3.1

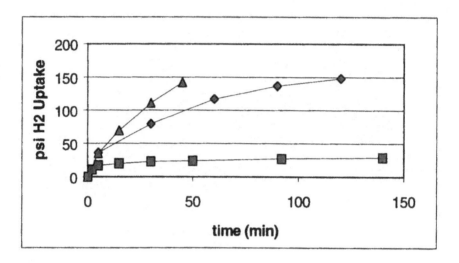

Figure 1. Comparison of the reaction profiles for nitrile hydrogenation. All reactions were carried out with a ratio of 0.38 mol nitrile / g Raney™ Ni slurry. ■) ADN without NaOH, ◆) ADN with NaOH additive, ▲) BN without NaOH.

M solution of ADN in methanol (i.e. 6.2 M in nitrile) is submitted to hydrogenation with a 20:1 ratio of ADN to aqueous Raney™ Ni slurry (~47:1 ADN:solid catalyst) under hydrogen at 500 psi, the catalyst becomes completely inactive before the reaction proceeds to 20% conversion, Figure 1. In contrast, hydrogenation of a 6.2 M solution of butyronitrile (BN) under similar conditions proceeds to complete conversion. Additionally, when a sample of Raney™ Ni is deactivated during ADN hydrogenation in methanol solvent and recovered and washed with methanol and water or, alternatively, mixed with 7% NaOH in 1:1 methanol/water for 30 min, it does not regain activity for ADN hydrogenation in the presence of caustic or, even for BN hydrogenation.

It may be hypothesized that ADN has a very strong two-point adsorption to Raney™ Ni, which could prevent complete hydrogenation, while BN with a one-point adsorption undergoes complete reaction. To evaluate this hypothesis Raney™ Ni was mixed with a 40%(w/w) solution of ADN in methanol at room temperature for 30 minutes; the ratio of ADN to Raney™ Ni slurry was 5:1. After this mixing period the catalyst was collected, washed with excess methanol and successfully used for BN hydrogenation. While these experiments do not provide thermodynamic data, characterizing the adsorption of ADN, they do indicate that factors other than a strong ADN adsorption are responsible for deactivation.

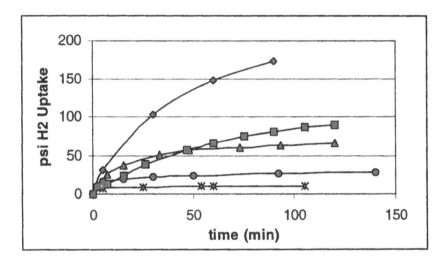

Figure 2. Comparison of the reaction profiles of dinitriles and mononitrile hydrogenation. All reactions were carried out with a ratio of 0.38 mol nitrile / g Raney™ Ni slurry in methanol solvent. ◆) HN(C6), ■) succinonitrile (SN) (C4), ▲) glutaronitrile (GN) (C5), ●) ADN(C6), *) C12 DN.

Obvious differences between ADN and BN include difunctionality versus monofunctionality and chain length differences. The hydrogenation of hexanenitrile (HN), a C_6 molecule, nearly equivalent in size to ADN, exhibits non-deactivating behavior, similar to BN, Figure 2. This eliminates a chain length dependent explanation for the contrast in behavior of ADN and BN. Furthermore, 1,12-dodecanedinitrile (C12 DN), a very large difunctional molecule, exhibits deactivation and suggests a correlation to α,ω-dinitriles. This correlation is substantiated by the examination of a series of difunctional molecules in comparison to hexanenitrile, Figure 2. All dinitriles molecules appear to result in deactivation of Raney™ Ni upon hydrogenation, however the rates of deactivation are not the same. It should be noted that a study comparing the hydrogenation behaviors of a series of dinitrile molecules of varying lengths has been conducted for the purpose of evaluating kinetics (9). There was no mention of deactivation behavior. Unlike the present study, however, the reactions were conducted in ethanol and the corresponding diamine and the ratio of the reaction components was not indicated.

After this intriguing revelation of the behavior of dinitriles, another difunctional molecule was investigated, 6-aminocapronitrile (ACN). Having a single nitrile group and similar chain length to hexanenitrile, one may expect

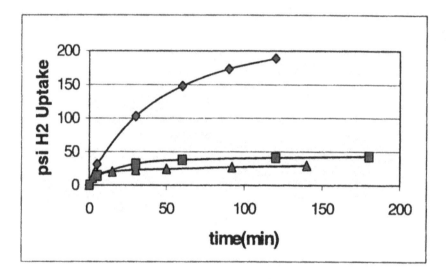

Figure 3. Comparison of the reaction profiles for the hydrogenation of a dinitrile, a mononitrile and an aminenitrile of similar length. All reactions were carried out with a ratio of 0.38 mol nitrile / g Raney[TM] Ni slurry in methanol solvent. ◆) HN, ■) ACN, ▲) ADN.

this substrate to fully hydrogenate without deactivating Raney[TM] Ni. On the contrary, it exhibits deactivation behavior nearly identical to ADN, Figure 3. Despite being a mononitrile, ACN bears the amine functionality, which can react in an amine condensation reaction to form a secondary amine product. This mechanism of byproduct formation, referred to in the introduction, provides the key to understanding the varying behavior of Raney[TM] Ni in hydrogenation of dinitriles as compared to mononitriles.

As α,ω-difunctional molecules, ACN, ADN and the other dinitriles of this study are not limited to forming only dimeric secondary amines, rather these condensation reactions can continue, yielding polymeric secondary amines, equation 1. As the concentration and length of these polyamines increases, the polymers eventually form an impenetrable layer, which is deposited on the surface of the catalyst and prevents adsorption and hydrogenation of nitriles remaining in solution.

$$NC\text{-}(CH_2)_4\text{-}CN \qquad H_2$$
$$+ \qquad \text{-----------}> \quad \text{-HN-}[(CH_2)_6NH\text{-}]_n\text{-} + NH_3 \qquad \textbf{(eq. 1)}$$
$$NC\text{-}(CH_2)_5\text{-}NH_2$$

Consistent with this deactivation mechanism, the use of ammonia as solvent prevents or greatly slows this deactivation mechanism by shifting the equilibrium back towards monomeric products. When methanol is replaced with ammonia in the above reactions they proceeded to nearly complete conversion in 240 min., though a gradual and uncharacteristic reduction in the rate over the course of the reaction suggested that slow deactivation was operative. It is further significant to note the trend exhibited in Figure 2. As dinitrile chain length increases the rate of deactivation also appears to increase. Note, the succinonitrile hydrogenation reaction reaches much higher conversion than that of the longer dinitriles especially C12 DN, which deactivates the catalyst before reaching 6% conversion. This is consistent with the suggestion that larger molecules require fewer condensation reactions to achieve a critical mass of polymer, which will be insoluble in the reaction medium and precipitate, covering the surface.

Other entries in the literature have commented on the deactivation of nickel catalysts during nitrile hydrogenation. A patent application prepared by BASF implicates oligomerization of amines as contributing to the deactivation of nickel containing catalysts for ADN hydrogenation but offers no supporting evidence (7). For gas phase hydrogenation of acetonitrile by Ni/SiO_2, a deactivation mechanism has been suggested, which includes the formation of nickel carbide (6). This results from "cracking" decomposition of acetonitrile. The reactions were conducted in a flow reactor at atmospheric pressure and 125 °C. The paper further comments that increasing hydrogen pressure decreases the rate of deactivation (6). Additional work on the reaction of methylamine over evaporated nickel surfaces corroborates the formation of carbonaceous species from the cracking of C-N bonds (10). The presence of such a mechanism is not refuted by the current experiments, however the contribution of a cracking mechanism is likely limited or negligible. The quoted mechanism would apply equally to dinitriles and mononitriles and not sufficiently explain the observed differences in the deactivation behavior of BN and ADN, for instance.

Additionally as indicated in the literature (6), the cracking mechanism is limited by increases in hydrogen pressure and decreases in temperature. Reactions in the current study were conducted at lower temperature and much higher pressure (75 °C and 500 psi) than the acetonitrile hydrogenation experiments in the literature (125 °C and atmospheric) (6).

Electron spectroscopy for chemical analysis (ESCA) provides corroborating evidence for the formation of insoluble polyamines as a mechanism of Raney™ Ni deactivation. Several samples of catalyst were analyzed by ESCA after being filtered on a sintered glass frit, oxidized and washed with methanol. These samples include: Raney™ Ni recovered from ADN hydrogenation in methanol solvent (deactivated), a fresh Raney™ Ni sample, which had not been exposed to a reaction mixture (control) and Raney™ Ni samples, recovered from ADN hydrogenation under non-deactivating conditions (spent in caustic and spent in NH_3). The "deactivated" Raney™ Ni sample showed a build-up of nitrogen on the surface to 4.6% (atom %, average of 4 experiments, standard deviation = 0.5%), Table 1, Figure 4. A corresponding build-up of carbon was noted on the deactivated samples, though the significance of the measured concentrations are more difficult to evaluate as a result of adventitious carbon build-up in the sample preparation. Nonetheless, comparison of "deactivated" Raney™ Ni to the control samples clearly indicated a difference in the quantity of carbon deposited at the surface and is consistent with deposition of polyamines on the catalyst. This deduction is further supported by the absence of detectable quantities of nitrogen on the surface of the "control" samples.

Table 1. ESCA results for Raney™ Ni samples

Sample	no. of exper.	atom % C	std dev	atom % N	std dev
Deactivated	4	60.8	4.3	4.6	0.5
Control	5	22.9	2.9	0	0
Spent in caustic	1	28.2		0	
Spent in NH_3	1	22.2		0	

It is important to realize that the mechanism delineated here occurs for Raney™ Ni / ADN reactions carried out in methanol and in the absence of base. When a base such as caustic (NaOH) or NH_3 is present these batch reactions do proceed to high conversion under the current conditions. For instance, a reaction of 30 g ADN, 2.5 g Raney™ Ni slurry, 45 g methanol and 0.15 g NaOH reaches 92% conversion in 35 minutes with no sign of deactivation. When the "spent in caustic" catalyst was recovered from this reaction and examined by

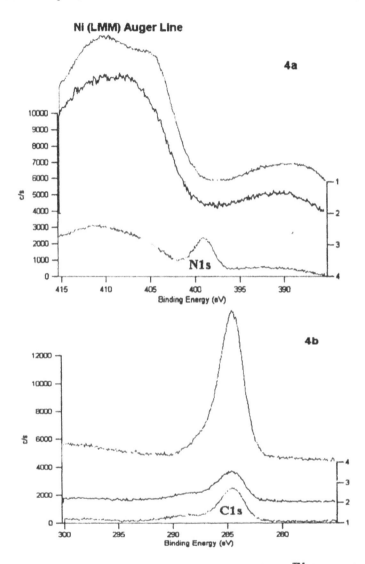

Figure 4. Comparison of ESCA spectra for various Raney™ Ni catalyst samples. a) the nitrogen 1s signal. b) the carbon 1s signal. Sample 1 = Raney™ Ni control sample, Sample 2 = Raney™ Ni recovered from an ADN hydrogenation reaction in the presence of ammonia, Sample 3 = Raney™ Ni recovered from an ADN hydrogenation reaction in the presence of NaOH, Sample 4 = deactivated Raney™ Ni recovered from an ADN hydrogenation reaction in the absence of base. Note the order of presentation is reversed for 4a and 4b so as not to obscure the data.

ESCA no detectable nitrogen was observed on the surface, Table 1, Figure 4. Similarly, a catalyst recovered from an ADN hydrogenation reaction conducted with NH_3 in place of the methanol solvent went to high conversion with no sign of deactivation and yielded a "spent in NH_3" catalyst with no trace of nitrogen present on the surface as detected by ESCA, Table 1.

Ammonia can suppress the formation of secondary amines by shifting the equilibrium reactions back toward primary amines (2a,b), equation 2. In the

$$RCH_2NH_2 + RCH=NH \quad \Longleftrightarrow \quad RCH=NCH_2R \quad + \quad NH_3 \qquad (eq. 2)$$

case of hydrogenation of α,ω-dinitriles with NH_3 this shift in equilibrium prevents polymerization and would suppress deactivation by blockage of catalytic sites. The influence of caustic, however, must be different than that of NH_3 with regard to reversing equation 2. It has been suggested that hydroxyl ions compete with butylamine for adsorption sites in the selective hydrogenation of butyronitrile over promoted Raney[TM] Ni catalysts (11). In the hydrogenation of dinitriles with caustic, reported herein, similar competitive adsorption effects also would prevent deposition and growth of polyamines. If competitive adsorption is significant with HO⁻, it does not decrease the rate of reaction as observed with NH_3, under these conditions. Notably, very little NaOH (0.2%(w/w)) is required to maintain activity in this system, whereas NH_3 must be present in very high proportions (e.g. 60% (w/w)). It is certainly true that NH_3 in the presence of aqueous Raney™ slurry will lead to increased HO⁻ concentration, which may lead to increased catalyst lifetime. Still, that HO⁻ alone is not sufficient to explain all of the reaction data and in particular, the decreased rate observed with NH_3. Even in the presence of NH_3 or caustic, Raney™ Ni eventually deactivates in ADN hydrogenation (2c, 7). Perhaps the mechanism delineated here is the operative mechanism even with basic additives and is merely slowed down under these conditions.

Conclusions

Raney™ Ni deactivates during dinitrile hydrogenation reactions in the absence of base as a result of deposition of polyamines, which physically block the surface. These polyamines are the product of condensation reactions, which occur in the hydrogenation process and their formation seems to be general for all dinitriles. The mononitriles evaluated here did not lead to catalyst deactivation, with the exception of 6-aminocapronitrile, which is susceptible to the same polymerization reactions observed for dinitriles. ESCA supports these conclusions, revealing a build-up of N and C on the deactivated catalysts in comparison to control samples or catalysts recovered from caustic or ammonia containing reactions. When ammonia or caustic are incorporated into the reaction mixture the batch reactions proceed to full conversion. Presumably ammonia and caustic prevent the formation and deposition of the polyamines.

Acknowledgements

We acknowledge E.I. duPont de Nemours Co. for supporting this work. Jeanette Woodward, Andrew Vu, Tom Dixon and Joe Norvell are acknowledged for excellent technical assistance and Dr. Theodore Koch for insightful discussions.

References

1. S.R. Montgomery, Chem. Ind. (Marcel Dekker), **5**, (*Catal. Org. React.*) 383 (1984).
2. a) C. De Bellefon and P. Fouilloux, *Catal. Rev. - Sci. Eng.* **36**, 459 (1994). b) J. Volf and J. Pasek, In *Catalytic Hydrogenation*; L. Cerveny, ed. Studies in Surface Science and Catalysis; Elsevier: New York, 1986, Vol. 27, Chapter 4. And for example: c) C.R. Campbell and C.E. Cutchens, U.S. Patent 4 359 585 (1982). d) S.B. Ziemecki. U.S. Patent 5 151 543 (1992). e) W. Schnurr, G. Voit, K. Flick, and R. Fischer, DE Patent 196 30 788 (1997). f) G. Cordier, P. Fouillouz, N. Laurain, J.F. Spindler, FR Patent 2 722 784 (1994). g) H.J.M. Bosman, and F.H.A.M.J. Vandenbooren, U.S. Patent 5 574 181 (1996). h) H. Chabert, U.S. Patent 3 862 911 (1975). i) K. Weissermel, H.-J. Arpe, *Industrial Organic Chemistry*, Second Edition; VCH: New York, Chapter 10 (1993).
3. J. von Braun, and G. Glassing, *Chem. Ber.* **36**, 1988 (1923).
4. A.Y. Lazaris, E.N. Zil'berman, E.V. Lunicheva and A.M. Vedin, *J. Appl. Chem. USSR,* 1076 (1965) ; *Zhurnal Prikladnoi Khimii,* **38**(5), 1097 (1965).
5. a) Y. Huang, and W.M.H. Sachtler, *J. Phys. Chem B,* 6558, (1998). b) Y. Huang and W.M.H. Sachtler, *Appl. Catal. C.,* **182**, 365 (1999). c) Y. Huang and W.M.H. Sachtler, *J. Catal.,* **184**, 247 (1999).
6. M.J.F.M. Verhaak, A.J. van Dillen, and J.W. Geus, *J. Catal.* **143**, 187 (1993).
7. W. Schnurr, G. Voit, K. Flick, J.-P. Melder, R. Fischer, and W. Harder, DE Patent 196 14 154 (1997).
8. a) Analysis supplied by W.R. Grace. b) S.R. Schmidt, Chem. Ind. (Marcel Dekker), **62**, (*Catal. Org. React.*) 45 (1995).
9. M. Joucla, P. Marion, P. Grenouillet, and J. Jenck, Chem. Ind. (Marcel Dekker), **53**, (*Catal. Org. React.*) 127 (1994).
10. J.R. Anderson and N.J. Clark, *J. Catal.* **5**, 250 (1966).
11. Thomas-Pryor, S.N.; Manz, T.A.; Liu, Z.; Koch, T.A.; Sengupta, S.K.; Delgass, W.N. Chem. Ind. (Marcel Dekker), **75**, (*Catal. Org. React.*) 195 (1998).

Catalytic hydrogenation of benzonitrile over Raney nickel. Influence of reaction parameters on reaction rates and selectivities

O. G. Degischer[a,b*], F. Roessler[a], P. Rys[b]

[a] F. Hoffmann-La Roche AG, Bau 214 Lab 13
4303 Kaiseraugst, Switzerland
[b] ETH Zürich, 8092 Zürich, Switzerland

Abstract

The liquid phase hydrogenation of aromatic nitriles has significant potential applications in the pharmaceutical and specialty chemicals industry. In contrast to aliphatic nitriles (1, 2, 3, 4), the systematic investigation of the hydrogenation of aromatic nitriles is not equally well documented in the literature (13, 14, 15). We therefore have chosen benzonitrile as a model substance and *Raney* Nickel as catalyst for our studies.

In order to establish a sound basis for the analysis of the experimental data, the phase equilibria of the system $MeOH/NH_3/H_2$ were measured. To exclude the influence of gas/liquid mass transfer, classical catalyst loading *vs.* reaction rate plots as well as $k_L a$ values were determined. (5, 6)

From the experimental kinetic and thermodynamic data a reaction model was developed for the system benzonitrile, methanol, *Raney* nickel, ammonia and hydrogen.

Introduction

The synthesis of primary amines by catalytic hydrogenation of nitriles is an important organic reaction, especially in polymer industry (production of hexa-methylenediamine from adiponitrile) and for the production of fine chemicals. Since the reaction affords secondary and tertiary amines as by-products, often the selectivity of the reaction is the most important criterion.

A mechanism for this reaction was first proposed by *Braun et al.* (7) (see Fig. 1). The addition of one hydrogen molecule to the nitrile triple bond leads to the reactive C=NH aldimine (primary aldime), which is either reduced to the primary amine or reacts further with another primary amine to afford the unstable aminal. This intermediate loses ammonia to form the C=N-R aldimine (secondary aldimine) whose further hydrogenation on the catalyst leads to the secondary amine. If the molecule contains a hydrogen atom at the position α to the nitrile function, the condensation of a secondary amine with a primary aldime leads to an unstable enamine (1, 17). Under the hydrogenation conditions this enamine reacts to the tertiary amine.

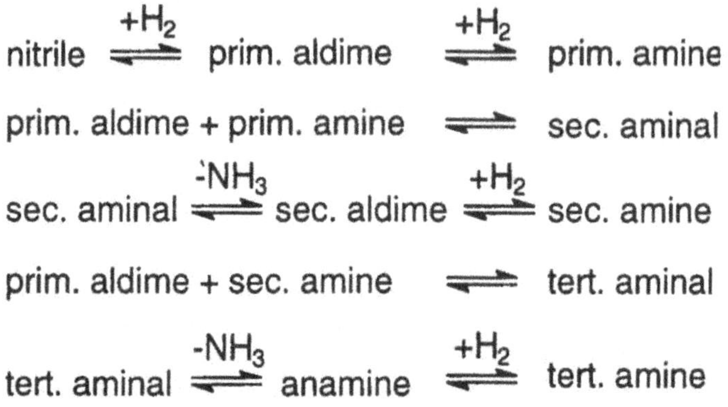

$$\text{nitrile} \xrightleftharpoons{+H_2} \text{prim. aldime} \xrightleftharpoons{+H_2} \text{prim. amine}$$

$$\text{prim. aldime + prim. amine} \rightleftharpoons \text{sec. aminal}$$

$$\text{sec. aminal} \xrightleftharpoons{-NH_3} \text{sec. aldime} \xrightleftharpoons{+H_2} \text{sec. amine}$$

$$\text{prim. aldime + sec. amine} \rightleftharpoons \text{tert. aminal}$$

$$\text{tert. aminal} \xrightleftharpoons{-NH_3} \text{anamine} \xrightleftharpoons{+H_2} \text{tert. amine}$$

Figure 1   Reaction system for the reaction of hydrogen with nitriles containing an α-hydrogen atom (7).

The rate of nitrile reduction and the product distribution are affected primarily by the nature of the catalyst used (1, 12, 16, 17): cobalt and nickel are preferred for the production of primary amines, copper and rhodium catalysts are mainly used for secondary amines, while tertiary amines can best be prepared on platinum and palladium catalysts. In contrast to aliphatic nitriles (1, 2, 3, 4), the systematic investigation of the hydrogenation of aromatic nitriles is not equally well documented in the literature (15). We therefore have chosen benzonitrile as a model substance and *Raney* nickel as the catalyst for our mechanistic studies.

Results

Reaction Pathway

The components benzonitrile (BN), benzylamine (BA), dibenzylimine (DBI) and dibenzylamine were detected by gas chromatography of the reaction mixture. Tribenzylamine was not found in the reaction mixture as the tertiary aminal cannot loose ammonia according to *Volf* and *Pasek* (1). This is strong evidence for the reaction system shown in fig. 2.

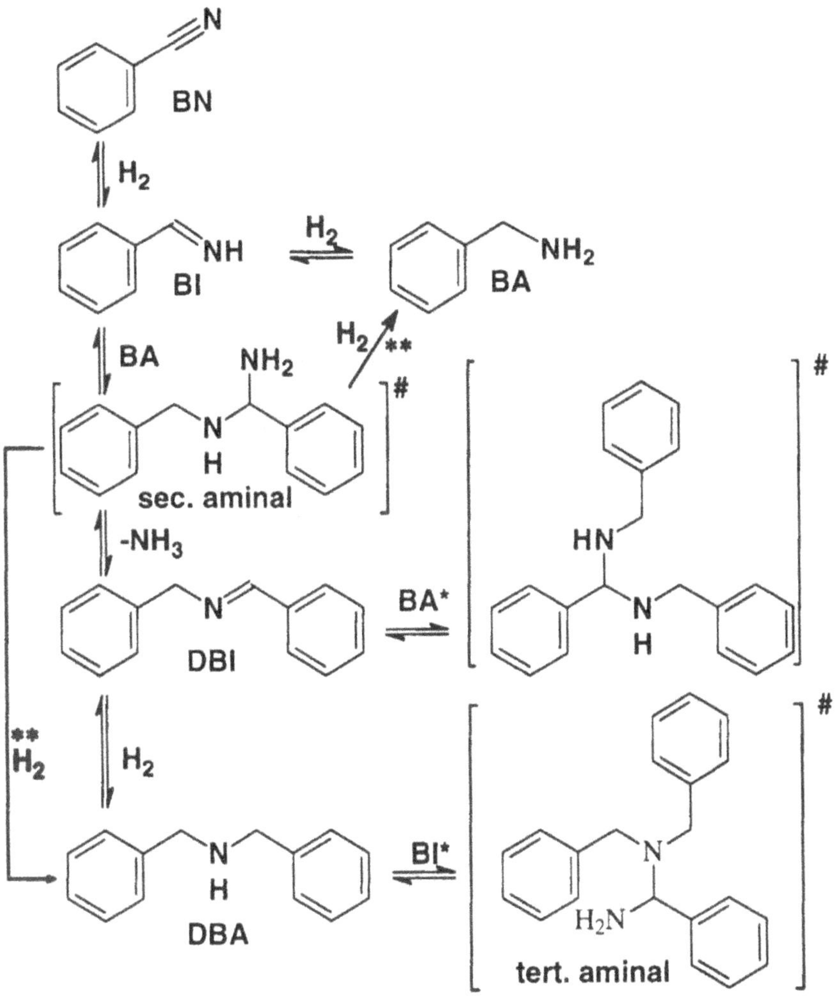

Figure 2 Reaction scheme for the hydrogenation of benzonitrile (1),
transient intermediate and species not observed, * reaction pathway not
observed, **hydrogenolysis mechanism

Influence of Agitation

For the evaluation of the effect of stirring on the reaction rate the impeller speed
was varied from 600 to 1200 rpm. The hydrogenation rate was independent from

the stirring speed above 790 rpm under the following reaction conditions: T = 100°C, 100 mL benzonitrile, 2 L methanol, hydrogen pressure 4 MPa, 5.7 g catalyst, which indicates the absence of limitation by gas-liquid mass transfer. All further hydrogenation experiments were therefore carried out at 795-800 rpm.

$k_L a$-Measurements

A second way to find out a potential limitation by gas-liquid mass transfer is to determine the $k_L a$-value of the system and to calculate the maximum possible hydrogen transfer rate. $k_L a$-values were measured with the dynamic physical absorption method described in the literature (5, 6). The saturation measurements were carried out without the catalyst. Calculations based on $k_L a$-values and on the ideal gas law corrected with the compressibility factor z (z = 1.0256 at 4 MPa) under standard catalyst loadings (catalyst/substrate = 5.7:100) showed a 10 times higher transport rate (ca. 27 mmol/s) than the normal reaction rate (ca. 2.8 mmol/s). Only at high catalyst loadings (>20% catalyst/substrate: >1:5) the reaction rate (16 mmol/s) reached nearly the rate of the hydrogen transport.

Catalyst Loading

When the catalyst loading was increased from 4% to 20% (wt/wt based on benzonitrile) the reaction rate showed a linear behavior. This indicates the absence of gas-liquid mass transport effects. The catalyst loading had no effect at all on the product distribution, *i.e.* on the selectivity.

Phase Equilibria

The ammonia concentration influences significantly the formation of the primary amine. The ammonia concentration in the solvent is essential in the hydrogenation of the nitrile. Because of the large differences in their physical properties (*i.e.* boiling points, critical temperatures) between ammonia and methanol, the phase equilibria for the ternary system methanol/ammonia/hydrogen was measured. The ammonia concentration in the liquid phase changes only very slightly, in contrast to the gas phase where the concentration of ammonia rises from 2 to 12% as a function of the temperature. This means that almost all ammonia is dissolved in the liquid phase and a simplified calculation - *i.e.* the ammonia only in liquid phase - will be appropriate.

Effect of the Temperature

The overall reaction rate is increased with higher reaction temperatures. A classical *Arrhenius* plot shows an activation energy for the overall reaction of 38 kJ/ mol nitrile (see Fig. 3). Aliphatic nitriles like valeronitrile have an activation

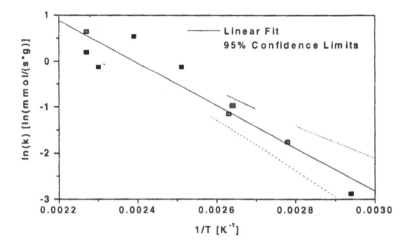

Figure 3 *Arrhenius* plot of benzonitrile. Reaction conditions: 100 mL benzonitrile, solvent 2 L methanol, hydrogen pressure 4 MPa, catalyst loading ca. 5.7 g, temperature: 60-160°C, stirrer speed 800 rpm

energy of 58 kJ/mol. In contrast to the hydrogenation of valeronitrile (3) or other aliphatic nitriles (1) where the selectivity drops with increased reaction temperature, the selectivity of the reduction of benzonitrile to the primary benzylamine increases with increasing reaction temperature in the range from 60 to 150°C (see Fig. 4). At higher temperatures (>150°C) this effect is less distinct. The observed effect of temperature on selectivity may be explained in a simple way by differences in activation energies for the reactions BI $\longrightarrow$ BA, DBI $\longrightarrow$ DBA and BA + BI $\longrightarrow$ DBI + NH$_3$ (kinetic control) as well as temperature dependent shift of the equilibrium BA $\Longleftrightarrow$ DBI + NH$_3$ (thermodynamic control). In addition, kinetics and thermodynamics of the adsorption and desorption of the various species on the catalyst surface could also play a role, as do mass transport (k_s-values and pore diffusion).

Modifiers

A variety of modifiers are reported to have a favorable effect on the formation of the primary amine. Many modifiers of *Raney* nickel are known, particulary LiOH (10) or NaOH (11). We have tested the influence of tertiary amines on the hydrogenation of nitriles. With triethylamine (Et$_3$N) as a modifier the selectivity for the benzylamine formation is increased. An increase of the Et$_3$N concentration decreases the reaction rate. The maximum amount of the intermediate diben-

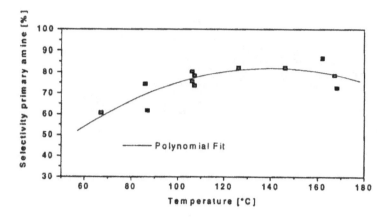

Figure 4 Selectivity as a function of temperature. Reaction conditions: 100 mL benzonitrile, solvent 2 L methanol, hydrogen pressure 4 MPa, temperature: 60-160°C, stirrer speed 800 rpm

zylimine did not change with Et_3N concentration (45-50% compared with 48-53% without Et_3N) (see Tab. 1).

Table 1 The effect of Et_3N on the hydrogenation of benzonitrile with *Raney* nickel. Reaction conditions: 100 mL benzonitrile, 5.7 g Ni_R, temperature 100°C, hydrogen pressure 4 MPa, stirrer speed 800 rpm, 2 L methanol.

Et_3N concentration [wt/vol%]	BA^a [%]	DBI^a [%]	DBA^a [%]	initial reac. rate [mmol/(s*g)]	max.DBI^b [%]
0	73.5	0.2	26.0	0.52	49.9
0.08	73.8	0.6	25.1	0.58	44.7
0.16	76.3	6.9	16.4	0.43	49.4
0.36	76.7	9.8	12.5	0.35	47.7
0.54	75.7	8.3	14.9	0.41	49.6
0.73	76.9	9.3	12.7	0.39	47.8

a. concentration at 120 minutes reaction time
b. maximum observed DBI concentration

Solvent

The solvent employed in the reaction system has a marked effect on the selectivity and the reaction rate. The results are summarized in Table 2. The best selectiv-

Table 2   The effect of the solvent on the hydrogenation of benzonitrile with *Raney* nickel. Reaction conditions: 100 mL benzonitrile, 5.7 g Ni_R, temperature 100°C, hydrogen pressure 4 MPa, stirrer speed 800 rpm, 2 L solvent.

solvent	polarity E_T (8) [kcal/mol]	H_2-solubility (9) 31 bar, 18°C [mol/l]	BA [%]	initial reac. rate [mmol*s^{-1}*g^{-1}]
methanol	55.5	0.129 (18)	75.5	0.38
ethanol	51.9	0.116 (19)	50.9	1.13
1-propanol	50.7	0.1 (19)	50.0	not measured
iso-propanol	48.6		47.2	0.90
1-butanol	50.2	0.096 (19)	49.9	0.96
	50.2		51.1	0.74
iso-butanol			46.8	0.51
tert-butanol	43.9		40.3	0.83

ity was reached with methanol (ca.70%, T=100°C, 2 L solvent, 100 mL BN, 5.7 g catalyst, stirrer speed 800 rpm, hydrogen pressure 4 MPa). With higher alcohols the selectivity was lower (ca. 50%) under the same reaction conditions, but the reaction rate was more than 2 times faster. This effect on the hydrogenation rate may occur because of the competative adsorption, *i.e.* methanol, a small and polar molecule, is adsorbed more strongly on the catalyst surface than the other molecules. The difference in selectivity and reaction rate with the higher alcohols may be due to a change in the solubility of hydrogen in the solvent. The differences in selectivity with solvents may also be due to changes in the equilibria pointed out in fig. 2 as well as the influence of solvent molecules on the kinetics and thermodynamics of adsorption and desorption.

Effect of the Hydrogen Pressure

The hydrogen pressure was varied from 0.97 MPa to 8 MPa to evaluate its influence on the hydrogenation of benzonitrile. At lower pressures (0.9 to 2 MPa) the initial reaction rate increased with increasing hydrogen pressure. But above a pressure of ca. 2 MPa, the reaction order with respect to the hydrogen pressure was zero. The selectivity for the primary amine was found to decrease with

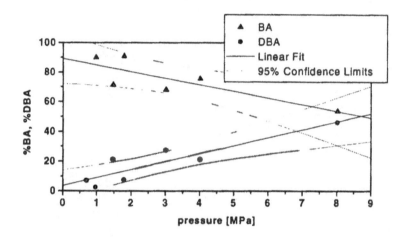

Figure 5 Effect of the hydrogen pressure on the selectivity of hydrogenation of benzonitrile. Reaction conditions: T = 100°C, 100 mL benzonitrile, solvent 2 L methanol, stirrer speed 800 rpm

higher hydrogen pressures (see Fig. 5). If the concentration of the dibenzylimine is plotted *vs.* the conversion of benzonitrile (see Fig. 6), we notice a significant effect of the hydrogen pressure on the course of the reaction: The maximum concentration of the dibenzylimine decreased with increasing hydrogen pressure. It seems that the reaction of BI with BA is very fast compared with the hydrogenation of BI and that the hydrogen coverage of the catalyst surface influences the individual hydrogenation reactions BI ⟶ BA and DBI ⟶ DBA differently, *i.e.* the reaction rate of some species is more hydrogen pressure sensitive than the other.

Addition of Ammonia

The classical way to increase the selectivity with respect to the primary amine is to add ammonia to the solvent. Experiments were carried out with ammonia con-

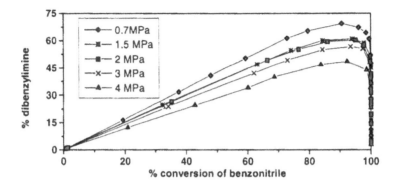

Figure 6 Effect of the hydrogen pressure on the observed amount of diben-zylimine on the extent of the conversion of benzonitrile. Reaction conditions: T = 100°C, 100 mL benzonitrile, 2 L methanol, stirrer speed 800 rpm.

centrations between 0 and 11% (wt% relative to solvent). To ensure constant hydrogen partial pressure in these experiments with varying ammonia concentra-tion, the total pressure was varied appropriately to compensate for the varying ammonia pressure. The maximum dibenzylimine concentration observed in the course of the hydrogenation could be reduced to 20% with an ammonia concen-tration of 11%, compared with 50% in the blank experiment, *i.e.* without ammo-nia. This resulted in a higher selectivity with respect to the primary amine at a total conversion of the nitrile and all intermediates (see Fig. 7). The hydrogena-tion rate was only slightly decreased by the ammonia. The increase in selectivity can be explained by the higher ammonia concentration in the liquid phase, *i.e.* the equilibrium between sec. aminal and DBI can be kept on the side of the sec. ami-nal, which can react back to the primary imine and then further be hydrogenated to the primary amine.

Kinetic Rate Laws

Initial Reaction Rate

The hydrogen consumption as a function of time in the first part (0-60% conver-sion) of the reaction is linear, the average rate was taken from the hydrogen con-sumption at 5, 10, 20 and 40% of the total hydrogen consumed.

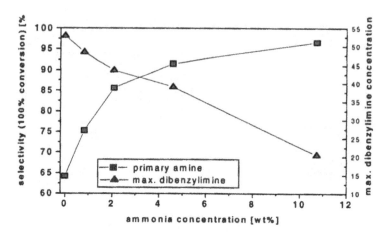

Figure 7   Selectivity as a function of ammonia concentration. Reaction condition: 100°C, 4 MPa hydrogen pressure, 2 L methanol, 5,7 g Ni$_R$, stirrer speed 800 rpm.

Power Law

A simplified model (see Fig. 8) of the reaction is used to describe the reaction *vs.*

$$BN \xrightarrow{\ k_1\ } BA$$

$$BN + BA \underset{k_3}{\overset{k_2}{\rightleftharpoons}} DBI\ (+\ NH_3)^*$$

$$DBI \xrightarrow{\ k_4\ } DBA$$

Figure 8   Scheme of the simplified model for the benzonitrile system, * not used in the model

time. The simplified model of the hydrogenation of benzonitrile to benzylamine takes into account the nitrile (BN), the amine (BA), the dibenzylimine (DBI) and the dibenzylamine (DBA) concentrations. Hydrogen is excluded from the system, because the hydrogen pressure was kept constant. The condensation reaction (BI + BA ⬌ sec. aminal) was simulated with the nitrile, amine and dibenzylimine concentrations, because no primary imine and aminal was detected. The experimental data were used to determine the values for k_1, k_2, k_3 and k_4 for the model in a least-square simultaneous parameters fitting of the differential equations. Fig. 9 shows a typical reaction profile with the calculated functions.

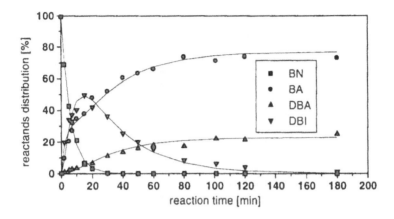

Figure 9   Typical reaction profile of the BN hydrogenation with experimental data and calculated functions for BN, BA, DBI and DBA. Reaction conditions: T = 100°C, 100 mL benzonitrile, solvent 2 L methanol, hydrogen pressure 4 MPa, 5,7 g Ni_R, stirrer speed 1000 rpm.

Although the proposed model is a strong simplification which does not describe the real mechanism, we calculated the k-values and activation energies. The *Arrhenius* plot is shown in fig. 10 and the activation energies are listed in tab. 3.

Experimental Section

Operating Conditions

The experiments are carried out in a 5 L stirred batch reactor. The autoclave is heated with an internal heat exchanger and cooled with water in the jacket. The reaction temperature is controlled within 2°C. The stirrer is a gas inducing impeller with a hollow shaft. The reactor operates at a constant pressure adjusting the hydrogen feed. The hydrogen consumption is calculated from the pressure variation recorded in the hydrogen feed tank. Samples of the liquid phase are taken during the reaction by using a capillary with a metal filter (porosity 2 μm). The samples are cooled in a mixture of methanol and dry ice to quench the reaction. The initial nitrile concentration varies from 0.19 kmol/m^3 to 2.42 kmol/m^3, the catalyst loading varies from 1.96 kg/m^3 to 9.65 kg/m^3, the temperature from 60°C to 170°C and the hydrogen pressure from 0.9 to 8 MPa.

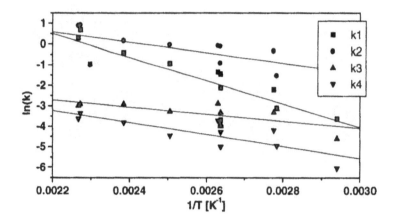

Figure 10 *Arrhenius* plot of k_1, k_2, k_3 and k_4 of the simplified model. Reaction conditions: T = 60-160°C, 100 mL benzonitrile, solvent 2 L methanol, hydrogen pressure 4 MPa, 5,7 g Ni_R, stirrer speed 800 rpm.

Table 3 Activation energies for the reaction of the model and the overall reaction, calculated from the slopes of figure 10 and figure 3

k_1	55.3 kJ/mol
k_2	28.9 kJ/mol
k_3	14.3 kJ/mol
k_4	24.3 kJ/mol
overall k	38.0 kJ/mol

For the determination of the phase equilibrium the samples are expanded in an evacuated vessel (0.5 L) to a pressure of 12 kPa to ensure that no methanol (vapor pressure (T=22°C): 14.4 kPa) will condense in the vessel. Afterwards a small overpressure with nitrogen is put onto the vessel to purge the diluted sample into a gas chromatograph with a TC detector.

Starting Materials

"Benzonitrile for synthesis" from *Merck* was used as substrate. *Pro Analysis* solvents were used. Hydrogen gas with a purity of 99.996% was applied. The catalyst used was a commercial *Raney* nickel from *Degussa* (Type B113Z). The catalyst properties of this manufacturer are an average area of 70 m^2/g and a hydrogenation activity of nitrobenzene of >22 mL H$_2$/(min*g cat).

Hydrogenation Procedure

The 5 L reactor described above was filled with 2 L of solvent or solvent mixture and the desired amounts of benzonitrile and triethylamine. The water wet catalyst was weighed and added. The reactor was closed and purged three times with nitrogen. Then the desired amount of ammonia was let into the reactor under stirring. The ammonia pressure vessel was mounted on a scale to determine the quantity of ammonia in the reactor. Then the autoclave was heated to the reaction temperature. Once the required temperature was reached, the stirrer was turned off and the autoclave was pressurized with hydrogen to the desired pressure with 3 feed tanks (1 L, 2 L and 3.3 L). The zero time of the experiment was taken when the stirring was started. At the same moment the 2 L and 3.3 L feed tank were closed and the pressure drop in the 1 L feed tank was recorded. For the analysis samples were taken from the sample outlet at regular time intervals without stopping the stirring.

Analytical Procedure

To improve the accuracy of the gas chromatographic analysis, a sample pretreatment was necessary. The samples have been evacuated and heated to 50°C to remove most of the methanol. After the dilution with n-hexane the analysis by gas chromatography was carried out on a *Supelco Simplicity-5* column (30 m x 0.32 mm x 0.25 μm film thickness) with a FID detector. Identified reaction products were benzylamine (BA), dibenzylimine (DBI) and dibenzylamine (DBA) together with unreacted benzonitrile (BN). The areas obtained from the gas chromatograms were multiplied with the substance specific correction factor (BN = 1, BA = 0.988, DBI and DBA = 1.23) to take into account the different responses.

References

1.   J. Volf, J. Pasek, *Stud. Surf. Sci. Catalysis*, **27**, 105-144 (1986)
2.   P. N. Rylander, Hydrogenation Methods, Academic Press, London, 1985, p. 94-103
3.   M. Besson, J. M. Bonnier, M. Joucla, *Bull. Soc. Chim. Fr.*, **127**, 5-12 (1990)

4.   M. Besson, J. M. Bonnier, D. Djaouadi, M. Joucla, *Bull. Soc. Chim. Fr.*, **127**, 13-19 (1990)

5.   A. Deimling, B.M. Karandikar, Y. T. Shah, N.L. Carr, *Chem. Eng. J.*, **29**, 127-140 (1984)

6.   R. V. Chaudhari, R. V. Gholap, G. Emig, H. Hofmann, *Can. J. Chem. Eng.*, **65**, 744-751 (1987)

7.   J. Braun, G. Blessing, F. Zobel, *Chem. Ber.*, **36**, 1988-2001 (1923)

8.   Liebigs Ann. Chem., **1**, 661, (1963)

9.   W. E. Pascoe, Chem. Ind. (Dekker), **47**, *(Catal. Org. React.)*, 93-104 (1992)

10.   T. A. Johnson, US Pat 5869653 to Air Products and Chemicals Inc., (1997)

11.   F. E. Herkes, Chem. Ind. (Dekker), **75**, *(Catal. Org. React.)*, 195-206 (1998)

12.   P. N. Rylander, L. Hasbrouck, I. Karpenko, *Annals of the New York Academy of Sciences*, **214**, 100-109 (1973)

13.   P. Tinapp, *Chem. Ber.*, **102**(8), 2770-2776 (1969)

14.   H. Greenfield, *Ind. Eng. Chem. Prod. Res. Dev.*, **15**(2), 156-158 (1976)

15.   R. A. Egli, *Helv. Chim. Acta*, **53**(4), 47-53 (1970)

16.   W. M. H. Sachtler, Y. Huang, *Appl. Catal. A*, **182**(2), 365-378 (1999)

17.   W. M. H. Sachtler, Y. Huang, *J. Catal.*, **184**(1), 247-261 (1999)

18.   E. Brunner, W. Hültenschmidt, G. Schlichthärle, *J. Chem. Thermodyn.*, **19**, 273-291 (1987)

19.   M. S. Wainwright, Th. Ahn, D. L. Trimm, *J. Chem. Eng. Data*, **32**, 22-24 (1987)

Catalytic Hydrogenation of Lignin Aromatics Using Ru-Arene Complexes

Terrance Y.H. Wong,[a] Russell Pratt,[a] Carolyn G. Leong,[a] Brian R. James,[a]
and Thomas Q. Hu[b]

[a]*Department of Chemistry, University of British Columbia, Vancouver, Canada
V6T 1Z1*
[b]*Vancouver Laboratory, Pulp and Paper Research Institute of Canada,
Vancouver, Canada V6S 2L9*

Abstract

Catalytic aromatic hydrogenation of 2-methoxy-4-propylphenol (1) and a
milled wood lignin (MWL) was achieved using various Ru systems (*e.g.*
$RuCl_3 \cdot 3H_2O$, $RuCl_2$(arene)(sulfonated phosphine), Ru_2Cl_4(arene)$_2$, and
[Ru_4H_4(arene)$_4$]Cl$_2$) with or without basic co-reagents in H_2O, ^{i}PrOH,
^{i}PrOH/H_2O or $ClCH_2CH_2Cl$/H_2O media at 80-100 °C under 50-60 atm H_2. For
1, complete conversions with selectivities for 2-methoxy-4-propylcyclohexanol
(1a) up to 91.3% are seen, while for MWL, conversions up to 64.9%
hydrogenated MWL (HMWL) can be achieved. For catalyst systems utilizing
water-soluble sulfonated phosphines, homogeneity was maintained, but the
systems were generally less active than the non-phosphine systems, which
involved active colloidal species that eventually precipitated as inactive metal.
Photo-accelerated yellowing studies revealed complete photostability for 1a and
a yellowing inhibition of at least 50% for HMWL.

Introduction

We are investigating catalytic hydrogenation of the aromatic rings of lignin as
an approach for the selective and effective inhibition of yellowing of
mechanical wood-pulps (1-3). Papers produced from these pulps, *i.e.* newsprint
and telephone directories, tend to yellow when exposed to light and/or heat, and
this phenomenon is caused by the photo-induced oxidative formation of
chromophores, such as *o*-quinones, in the lignocellulosic matrices (4-6). We
have shown previously that with the lignin model compounds, 2-methoxy-4-
propylphenol (1) and 4-propylphenol, aromatic hydrogenation could be
achieved using various Rh (*e.g.* RhCl$_3$, [RhCl(η^4-1,5-hexadiene)]$_2$, [RhCl(η^4-
1,5-cyclooctadiene)]$_2$) and Ru (*e.g.* [RuCl$_2$(η^6-C$_6$Me$_6$)]$_2$, Ru(η^6-C$_6$Me$_6$)(η^4-
C$_2$H$_4$), Ru(η^6-C$_6$Me$_6$)(η^4-diene), RuCl$_3$/(^{n}C$_8$H$_{17}$)$_3$N) systems in aqueous/organic

media at 20 - 100 °C under 1 - 50 atm H_2 (7); conversions to 99.7% and, for **1**, selectivities to 96.1% for 2-methoxy-4-propylcyclohexanol (**1a**) are observed. For a "milled-wood" lignin (MWL), the $RuCl_3/(^nC_8H_{17})_3N$ system was found most effective with ~38% hydrogenation of the aromatic rings achieved after 10 days at 80 °C under 50 atm H_2 in H_2O-^{i}PrOH-$MeOCH_2CH_2OH$. The systems involve colloidal catalysts as evidenced by inhibition with addition of mercury, and aggregation to inactive metal (7); the same criteria are used for the new systems described below.

In efforts to devise a purely aqueous catalyst system for eventual applications on wood-pulps, we are examining the water-soluble species, $[Ru_4H_4(\eta^6\text{-arene})_4]^{2+}$ and $[Ru_4H_6(\eta^6\text{-arene})_4]^{2+}$, reported by Süss-Fink's group (8,9), and related Ru(arene) complexes (e.g. $Ru_2Cl_4(\eta^6\text{-arene})_2$, $RuCl_2(\eta^6\text{-}$arene)(TPPMS) and $RuCl_2(\eta^6\text{-arene})$(TPPTS) where TPPMS = sodium diphenyl(m-sulfonatophenyl)phosphine and TPPTS = trisodium tri-m-sulfonatophenylphosphine). The present paper reports on the use of these species in the hydrogenation of **1**, other lignin model compounds (acetovanillone, 3-methoxy-4-hydroxycinnamaldehyde, and eugenol (4-allyl-2-methoxyphenol)), and MWL, as well as photo-accelerated yellowing studies on some of the hydrogenated products.

Experimental Section

The dinuclear complexes, $Ru_2Cl_4(C_6H_6)_2$ (of structure $(RuCl(\eta^6\text{-}C_6H_6))_2(\mu\text{-Cl})_2$ (10)) and the related $Ru_2Cl_4(p\text{-cymene})_2$ (10), $Ru_2Cl_4(C_6Me_6)_2$ (11), $Ru_2Cl_4(C_6H_2Me_4)_2$ (12), $Ru_2Cl_4(mesitylene)_2$ (13), and the tetranuclear $[Ru_4H_4(mesitylene)_4]Cl_2$ (8,9,14) were prepared according to literature methods while treatment with the water-soluble phosphines yields $RuCl_2(C_6Me_6)TPPMS$ (14), $RuCl_2(p\text{-cymene})TPPMS$ (14), and $RuCl_2(C_6H_6)TPPTS$ (15). Each of these compounds was well characterized and microanalytical data were acceptable. Catalytic hydrogenations of the aromatics and MWL (7) were carried out at 50 - 100 °C under 50 - 60 atm H_2 in various media (Tables 1 and 2). Catalyst and substrate concentrations (if not neat substrate) were in the ranges 0.30 - 10 mM and 15 - 600 mM, respectively. The identification of hydrogenated products was as described previously (7,16). Characterization of the hydrogenated lignin (HMWL) was achieved by ^1H NMR spectroscopy (17). Accelerated light-induced colour reversion (yellowing) experiments were carried out by irradiating Whatman filter paper sheets, impregnated with the test material, and subsequently measuring the sheets for the %ISO brightness and the CIE (Commission Internationale d'Eclairage) yellow coordinate (b*)

(18) at different time intervals using a Technibrite Micro TB-1C instrument (19). The molecular weight and polydispersity measurements for the HMWL were carried out at McGill University (Montreal, Quebec).

Results and Discussion

The hydrogenation reactions are shown in eqs. 1 and 2, and Tables 1 and 2 summarize data on conversions and selectivities. As described previously, hydrogenation of **1** gives rise usually to **1a**, largely as the *cis,cis*-stereoisomer, accompanied by hydrogenolysis to small amounts (<10%) of 4-propylcyclohexanol (**1b**) (7,16). For MWL, hydrogenation results in the appearance of cyclohexyl ^{1}H NMR signals in the δ 0.80 - 2.10 region (17).

$$(1)$$

$$(2)$$

We have already demonstrated that RuCl$_3$•3H$_2$O/3.5 TOA (TOA = tri-*n*-octylamine) catalyzes effectively the hydrogenation of **1**, presumably by amine-stabilized Ru(0) colloids (20), although these eventually aggregate and precipitate as inactive metal. Use of 7 equiv. TOA had no effect in maintaining homogeneity; however, when 1 equiv. of TPPMS was used in conjunction with TOA, a homogeneous system was observed, but with a smaller conversion (56.4% after 24 h) and lower selectivity (80.2%) to **1a** (Table 1).

The use of a water-soluble, "stabilizing" phosphine then prompted us to synthesize and study several RuCl$_2$(arene)TPPMS and RuCl$_2$(arene)TPPTS complexes and, during our studies, we learned that aqueous solutions of

Table 1 Conversion (%) and selectivity for **1a** in the hydrogenation of **1** using various Ru catalyst precursors usually at 50-90 °C under 50 atm H_2 for 24 h with [1]/[Ru] = 50.

Catalyst system (Ru conc. mM)	Temp. °C	Conv^n	Sel. for 1a
RuCl$_3$·3H$_2$O/7 TOA (2.0)^{a,b,c}	50	>99.9	87.8
RuCl$_3$·3H$_2$O/3.5 TOA/TPPMS (3.8)^a	50	56.4	80.2
RuCl$_3$·3H$_2$O/TPPMS (3.8)^a	80	0.0	-
RuCl$_2$(p-cymene)TPPMS (0.37)^a	80	4.0 (6.6)^d	55.8 (53.8)^d
RuCl$_2$(C$_6$H$_6$)TPPTS (0.31)^{a,e}	80	1.4	59.3
Ru$_2$Cl$_4$(C$_6$H$_6$)$_2$ (10)^{c,f,g,h}	90	82.2	63.9
Ru$_2$Cl$_4$(mesitylene)$_2$ (0.72)^{a,c}	80	36.7	82.6
Ru$_2$Cl$_4$(mesitylene)$_2$ (0.72)^{c,i}	80	0.7	90
[Ru$_4$H$_4$(mesitylene)$_4$]Cl$_2$ (1.0)^{a,c}	80	23.1	81.0
[Ru$_4$H$_4$(mesitylene)$_4$]Cl$_2$ (1.0)^{c,i}	80	4.0	60.4
Ru$_2$Cl$_4$(C$_6$Me$_6$)$_2$/10 Et$_3$N (1.0)^{c,j,k}	80	18.5	80.5
Ru$_2$Cl$_4$(p-cymene)$_2$/4 TOA (1.1)^{c,j,l}	80	6.5	91.3

^{a} In 20 mL ^{i}PrOH/H$_2$O (1:1 - 3:1). ^{b} [1]/[Ru] = 300. ^{c} Inactive metal deposited. ^{d} After 96 h. ^{e} After 72 h. ^{f} [1]/[Ru] = 12.5. ^{g} In H$_2$O. ^{h} Under 60 atm H$_2$. ^{i} In ClCH$_2$CH$_2$Cl/H$_2$O (10 mL: 10 mL). ^{j} In ^{i}PrOH. ^{k} No activity for (a) Ru$_2$Cl$_4$(C$_6$Me$_6$)$_2$ with or without added TOA or TPPMS in ^{i}PrOH, ^{i}PrOH/H$_2$O, or ^{i}PrOH/H$_2$O/CH$_3$OCH$_2$CH$_2$OH, or (b) Ru$_2$Cl$_4$(C$_6$H$_2$Me$_4$)$_2$ in ^{i}PrOH/H$_2$O. ^{l} 1-Methyl-4-isopropylcyclohexane generated.

$RuCl_2(C_6H_6)$TPPTS catalyze complete hydrogenation of benzene to cyclohexane after 2.5 h at 90 °C under 60 atm H_2 (15). We find that homogeneous hydrogenation of 1 was achieved using $RuCl_2(p$-cymene)TPPMS and $RuCl_2(C_6H_6)$TPPTS but conversions reached only 6.6% after 96 h, and selectivities (1a vs. 1b) were also lower (53 - 59%) (Table 1); $RuCl_2(C_6Me_6)$TPPMS effected no catalysis. Of note, hydrogenation of the allyl moiety of eugenol to give 1 using $RuCl_2(C_6H_6)$TPPTS is effected in a two-phase H_2O/substrate system, with 88.4% conversion after 21 h at 90 °C under 60 atm H_2.

The dinuclear arene complexes, $Ru_2Cl_4(C_6H_6)_2$, $Ru_2Cl_4(p$-cymene)$_2$, Ru_2Cl_4(mesitylene)$_2$, and $Ru_2Cl_4(C_6Me_6)_2$, when placed under H_2, form the water-soluble hydride clusters, $[Ru_4H_6$(arene)$_4]Cl_2$ (8,9), and catalytic hydrogenation of aromatics has been shown using $[Ru_4H_6(C_6H_6)_4]Cl_2$ (which can also be generated from $[Ru_4H_4(C_6H_6)_4]Cl_2$ (9)). Conditions for colloid-catalyzed effective conversion of 1 (up to 82.2%) were found with $Ru_2Cl_4(C_6H_6)_2$ and Ru_2Cl_4(mesitylene)$_2$ (and the corresponding $[Ru_4H_6$(mesitylene)$_4]Cl_2$), but no activity was found for $Ru_2Cl_4(C_6H_2Me_4)_2$ and $Ru_2Cl_4(C_6Me_6)_2$; conversions were always lower in the two-phase $ClCH_2CH_2Cl/H_2O$ system than in $^iPrOH/H_2O$ (Table 1). Unlike the analogous hydrido mesitylene clusters, $[Ru_4H_n(C_6Me_6)_4]Cl_2$ (n = 4,6), which can be generated in situ from $Ru_2Cl_4(C_6Me_6)_2$ under H_2 (14), does not effect conversion of 1. Bennett's group has noted that, in a two-phase aqueous/substrate system using $Ru_2Cl_4(C_6Me_6)_2$, catalytic hydrogenation of selected arenes can be achieved if Na_2CO_3 is present; this generates the water-soluble, air-stable hydride $[(C_6Me_6)Ru(\mu$-H)$_2(\mu$-Cl)Ru(C_6Me_6)]Cl (7,21). With 1, this species in $^iPrOH/H_2O$ affords a 37% conversion after 24 h at 80 °C under 50 atm H_2 with inactive metal eventually precipitated (22). In the present study, we also find that co-use of Et_3N with $Ru_2Cl_4(C_6Me_6)_2$ in ^iPrOH gives 18.5% conversion of 1; no conversions were observed when TOA or TPPMS was used as a co-reagent (Table 1, footnote k).

Monophosphines and amines (L) react with Ru_2Cl_4(arene)$_2$ to form $RuCl_2$(arene)L species (21); electronic (N vs. P) and steric (Et_3N vs. TOA) properties are certain to influence reactivity and the nature of the arene ligand is also important. For instance, a $Ru_2Cl_4(p$-cymene)$_2$/TOA system effects 6.5% conversion whereas no conversion is seen with $Ru_2Cl_4(C_6Me_6)_2$/TOA. The only effective homogeneous system is that involving $RuCl_3 \cdot 3H_2O$/3.5 TOA/TPPMS, and the TOA is essential; in the absence of this co-reagent, no conversion is evident (Table 1). The TOA is not acting as a colloid stabilizer here, and thus most likely plays the role of proton acceptor in generating a Ru-hydride (see above). The RuCl$_2$(arene)(sulfonated phosphine) systems are homogeneous but activity is poor, the *p*-cymene derivative being more active than the benzene and hexamethylbenzene derivatives. In the absence of phosphine, eventual aggregation of colloidal Ru(0) to generate inactive metal was always observed.

The mechanisms of the homogeneous and colloidal systems are not known, but the absence of cyclic diene and monoene products is consistent with the previously suggested "continuous substrate binding" mechanism, which accommodates formation of a *cis,cis* product (1). In the not very effective colloidal Ru$_2$Cl$_4$(*p*-cymene)$_2$/TOA system, a stoichiometric amount of 1-methyl-4-isopropylcyclohexane is generated from the *p*-cymene ligands, again implying successive hydrogen transfer to the coordinated aromatic.

[RuCl$_2$(mesitylene)]$_2$ and [Ru$_4$H$_4$(mesitylene)$_4$]Cl$_2$ were found effective for homogeneous hydrogenation of acetovanillone (eq. 3a) and 3-methoxy-4-hydroxycinnamaldehyde (eq. 3b) under conditions where the aromatic ring was not reduced. For acetovanillone, complete conversion to the alcohol takes place over 24 h at 80 °C under 50 atm H$_2$ using Ru$_2$Cl$_4$(mesitylene)$_2$ in mixed ^{i}PrOH/H$_2$O media (conditions as for the 7th entry in Table 1). For the cinnamaldehyde, conversions of 35 - 40% occur under the same conditions to generate mainly the saturated aldehyde and alcohol (eq. 3b), and three unidentified products (which, however, are not cyclohexyl derivatives); in ^{i}PrOH/H$_2$O, the selectivities for the alcohol (~45%) are greater than for the aldehyde (~30%), while in the two-phase ClCH$_2$CH$_2$Cl/H$_2$O system, greater selectivities are observed for the aldehyde (~55% *vs.* 20% for the alcohol). pH is probably important, especially for the [RuCl$_2$(mesitylene)]$_2$ system where formation of the Ru cluster generates HCl. A pH-dependent catalytic hydrogenation of cinnamaldehyde to the saturated alcohol via the saturated aldehyde or unsaturated alcohol was recently reported using a Ru$_2$Cl$_2$(TPPMS)/

$$R\text{-CO-Me} \longrightarrow R\text{-CH(OH)Me} \tag{3a}$$

$$R\text{-CH=CH=CHO} \longrightarrow R\text{-CH}_2\text{CH}_2\text{CHO} + R\text{-CH}_2\text{CH}_2\text{CH}_2\text{OH} \tag{3b}$$

R = HO— (ring) —

MeO

3 TPPMS system; depending on the pH, either $Ru(H)(Cl)(TPPMS)_3$ or $Ru(H)_2(TPPMS)_4$ was formed, the former preferentially catalyzing hydrogenation of the C=C moiety, and the latter, the C=O group (23).

Catalytic hydrogenation of MWL using $RuCl_3 \cdot 3H_2O/TOA$ was further examined in the three-solvent system (see Introduction). Under 50 atm H_2 and at 80 °C, a 47% conversion to HMWL was seen after 24 h (Table 2), the result correlating well with the previously data, assuming an approximately zero-order dependence on [MWL] (7). Increasing the catalyst concentration and reaction time had only a marginal effect while doubling the amount of TOA decreased conversion somewhat (Table 2); these systems (as with 1) precipitate inactive metal and the conversions will depend critically on the stage that metal

Table 2 Extent (%) of aromatic hydrogenation of a "milled-wood" lignin (10 mg) catalyzed by various Ru systems at 80 °C under 50 atm H_2 in ^{i}PrOH(5-13 mL)/H_2O(5-7 mL)/$CH_3OCH_2CH_2OH$(5 mL).

Catalyst system (Ru conc. mM)	Time h	Conversion (%)
$RuCl_3 \cdot 3H_2O/3.5$ TOA $(1.5)^a$	24	47
$RuCl_3 \cdot 3H_2O/3.5$ TOA $(3.0)^a$	24 (96)	56.4 (64.9)
$RuCl_3 \cdot 3H_2O/7$ TOA $(1.5)^a$	24	33
$RuCl_3 \cdot 3H_2O/3.5$ TOA/TPPMS $(1.9)^{b,c}$	24 (72)	21 (48)
$RuCl_2(p$-cymene)TPPMS $(1.5)^{c,d}$	48	16
$RuCl_2(p$-cymene)TPPMS/5 TOA $(0.8)^{b,c}$	72	14
$Ru_2Cl_4(C_6Me_6)_2/3$ Na_2CO_3 $(0.6)^e$	72	~5
$Ru_2Cl_4(p$-cymene)$_2/3$ Na_2CO_3 $(0.7)^a$	24	32
$Ru_2Cl_4(C_6Me_6)_2$ (2.2)	24	~5

^{a} Inactive metal deposited. ^{b} 20 mg MWL used. ^{c} In ^{i}PrOH(15 mL)/H_2O(5 mL). ^{d} 40 mg MWL used. ^{e} At 100 °C under 60 atm H_2.

is formed. As well as stabilizing the colloidal Ru(0), the TOA could also compete for coordination sites at the metal. Homogeneity was maintained when TPPMS (1 equiv.) was present either as a co-reagent or within a discrete complex but, as seen for 1, lower conversions than those obtained with the colloid systems were always found (Table 2). Some Ru-arene systems in the absence of TPPMS that appeared to be homogeneous were ineffective (~5% conversion), while a reasonably active Ru-p-cymene/Na$_2$CO$_3$ system gave metal deposition implying colloidal activity (Table 2).

Some accelerated light-induced colour reversion (yellowing) experiments were carried out. Removal of the aromaticity will eliminate completely the mechanism by which yellowing occurs (19). For example, irradiation of 1 (at 253 nm) for 45 h produces a brightness loss of ~26 points (Fig. 1) (and an increase of ~17 points in the CIE yellow coordinate (b*)), whereas no corresponding changes occur with the fully hydrogenated species 1a. Corresponding yellowing experiments for related aromatics such as 1,2-dimethoxy-4-propylbenzene and its cyclohexyl derivative again reveal complete photostability upon removal of the aromaticity [19].

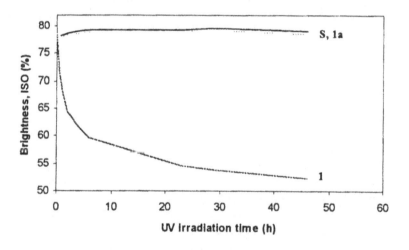

Figure 1   %ISO Brightness *vs.* UV irradiation time for paper sheets impregnated with MeOH (S), and compounds 2-methoxy-4-propylphenol (1) and 2-methoxy-4-propylcyclohexanol (1a) [19].

For MWL, we have shown previously that for a 16% hydrogenated sample (using the RuCl$_3$•3H$_2$O/TOA catalyst), a relative brightness loss reduction of up to 40% with an accompanying inhibition of yellowing by up to 25% can be achieved (22). In the present study, the results for a 52% hydrogenated MWL (using the $RuCl_3 \cdot 3H_2O$/TOA catalyst) are shown in Fig. 2. Exposure of MWL to UV radiation leads to a brightness drop of 9 points to ~67% ISO after 4 h with b* increasing 6 points to ~7. For HMWL, there is a

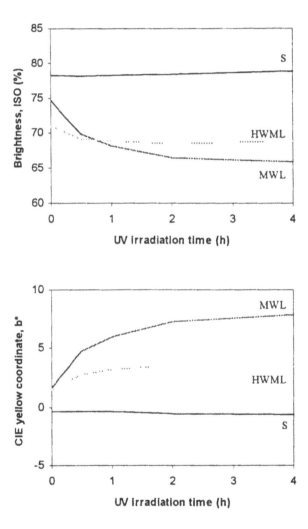

Figure 2 %ISO brightness and the CIE yellow coordinate (b*) vs. UV irradiation time for filter paper sheets impregnated with MeOH (S), MWL, and HMWL.

corresponding brightness loss of 2.5 points from an initial 71% ISO and an increase in b* from 1.5 to 3.5. Of note, the HMWL is visibly white (versus the yellow MWL); the initial low brightness and high b* readings result from the presence of trace Ru metal. Nevertheless, the b* data indicate a yellowing inhibition of at least 50% can be accomplished.

Figure 3 shows average molecular weight (M_n) and polydispersity (PD) data on HMWL (22). M_n decreases linearly from 6000 g/mol for MWL to 3000 g/mol for 37% hydrogenated MWL. Experimentally, we find that for MWL that is 52% hydrogenated, the product yield is only 28%. As seen for the model compound 1, hydrogenolysis takes place in addition to aromatic hydrogenation (eq. 1), and this is the likely cause for the decline in M_n. The polydispersity also decreases (Fig. 3) indicating increasing uniformity within the lignin polymeric structure. Whether this affects the tensile strength of thermomechanical pulps (TMP) remains to be seen, and studies on homogeneous catalyst systems with TMP are currently in progress. The principle has thus been established that hydrogenation of aromatic moieties in lignin leads to inhibition of yellowing; the aim is now to develop more effective homogeneous catalysts.

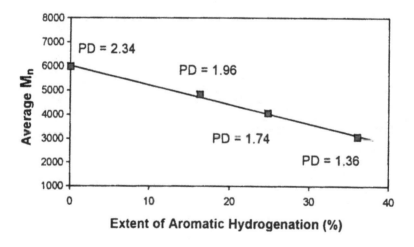

Figure 3 Average molecular weight (M_n) and polydispersity (PD) of HMWL.

Acknowledgments

We thank: the Natural Sciences and Engineering Research Council of Canada for financial support via the Mechanical and Chemimechanical Pulps Network and a Research Grant; Johnson Matthey Ltd. and Colonial Metals Inc. for loans of RuCl$_3$; and Prof. R.St.J. Manley of McGill University for performing the molecular weight determinations.

References

1.   B.R. James, Y. Wang and T.Q. Hu, *Chem. Ind.* (Marcel Dekker), **68**, 423 (1996).
2.   T.Q. Hu, B.R. James and C.-L. Lee, *J. Pulp Paper Sci.* **23**(4), J153 (1997).
3.   T.Q. Hu, B.R. James and C.-L. Lee, *J. Pulp Paper Sci.* **23**(5), J200 (1997).
4.   S.Y. Lin and K.P. Kringstad, *Norsk Skogindustri.* **25**(9), 252 (1971).
5.   J.A. Schmidt and C. Heitner, *J. Wood Chem. Technol.* **13**, 309 (1993).
6.   J.K.S. Wan, M.Y. Tse and C. Heitner, *J. Wood Chem. Technol.* **13**, 327 (1993).
7.   B.R. James, Y. Wang, C.S. Alexander and T.Q. Hu, *Chem. Ind.* (Marcel Dekker), **75**, (*Catal. Org. React.)* 233 (1998).
8.   G. Meister, G. Rheinwald, H. Stoeckli-Evans and G. Süss-Fink, *J. Chem. Soc., Dalton Trans.* 3215 (1994).
9.   L. Plasseraud and G. Süss-Fink, *J. Organomet. Chem.* **539**, 163 (1997).
10.   M.A. Bennett and A.K. Smith, *J. Chem. Soc., Dalton Trans.* 233 (1974).
11.   M.A. Bennett, T.-N. Huang, T.W. Matheson and A.K. Smith, *Inorg. Synth.* **21**, 74 (1982).
12.   M.A. Bennett, L.Y. Goh, I.J. McMahon, T.R.B. Mitchell, G.B. Robertson, T.W. Turney and W.A. Wickramasinghe, *Organometallics* **11**, 3069 (1992)
13.   J.W. Hull and W.L. Gladfelter, *Organometallics* **3**, 605 (1984).
14.   T.Y.H. Wong, R. Pratt, C.G. Leong, C.S. Alexander and B.R. James, unpublished data.
15.   D.J. Ellis, M.Sc. Dissertation, Imperial College, London, 1998.
16.   T.Q. Hu, B.R. James, S.J. Rettig and C.-L. Lee, *Can. J. Chem.* **75**, 1234 (1997).
17.   T.Q. Hu, B.R. James and Y. Wang, *J. Pulp Paper Sci.* **25**(9), 312 (1999).

18.   J.A. Bristow, *Tappi J.* 77(5), 174 (1994).
19.   T.Q. Hu and B.R. James, *J. Pulp Paper Sci.*, in press.
20.   F. Fache, S. Lehuede and M. Lemaire, *Tetrahedron Lett.* 36, 885 (1995).
21.   M.A. Bennett, T.-N. Huang and T.W. Turney, *J. Chem. Soc., Chem. Commun.* 213 (1979).
22.   C.S. Alexander, T.Y.H. Wong, Y. Wang, A. Nasiry, B.R. James and T.Q. Hu, Proc. 81st CSC Meeting, Whistler, BC, Canada, Paper 444.
23.   F. Joó, J. Kovács, A.Cs. Bényei and Á. Kathó, *Catalysis Today* 42, 441 (1998).

Reaction Pathways in the Catalytic Reductive N-Methylation of Polyamines

Richard P. Underwood and Richard V. C. Carr

Air Products and Chemicals, Inc.
7201 Hamilton Boulevard
Allentown, Pennsylvania 18195-1501

Abstract

Studies of the catalytic reductive N-methylation of polyamines have elucidated key aspects of the reaction mechanism. In investigations of the reductive permethylation of a homologous series of diamines, ranging from 1,2-ethanediamine to 1,6-hexanediamine, it was discovered that selectivity to by-product formamides was much greater for 1,3-propanediamine as compared to the other diamines. Detailed studies of 1,3-propanediamine and derivatives revealed the presence of intermediate cyclic aminals (hexahydropyrimidines) during N-methylation. Further probe reaction studies positively confirmed the intermediacy of the cyclic aminals in the formation of by-product formamides. Though certain details of the mechanism require further study, a generalized reaction scheme that accounts for products and by-products, as well as the differences in selectivity for ethyleneamine and propyleneamine substrates, has been developed.

Introduction

Catalytic reductive N-methylation of amines is a key synthetic methodology in the production of a variety of commercially important tertiary amines (1-7). A noteworthy application of this technology is in the production of polyurethane catalysts via permethylation of various polyamines. For these substrates, this chemistry affords N,N-dimethylamino- groups, which are particularly effective at catalyzing the reactions which occur in the production of polyurethane foam.

Catalytic N-methylation is typically carried out by reacting a primary or secondary amine with formaldehyde and hydrogen in the presence of a metal catalyst, most commonly Pd, Pt, or Ni. Though the details of the reaction mechanism have not previously been explored in detail, it has been generally assumed that the formation of an N-methyl group involves 2 sequential steps (8,9). Firstly, the amine nucleophilically adds to formaldehyde to form an N-methylol group, **1**. Secondly, for a primary amine substrate, the N-methylol moiety either dehydrates to the imine, **2**, and then is catalytically hydrogenated over the metal catalyst, or it undergoes direct hydrogenolysis.

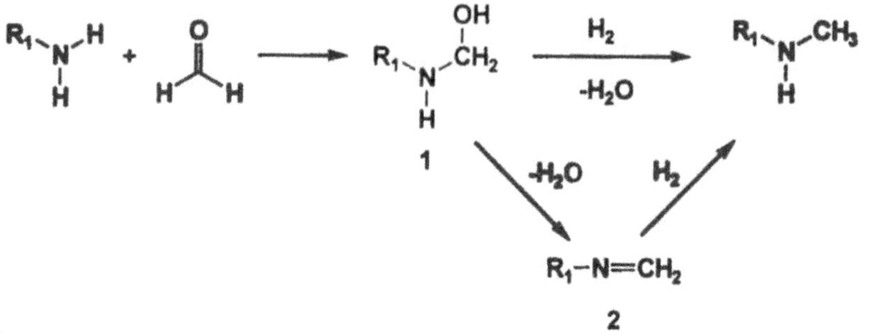

Since formation of an imine is precluded for secondary amine substrates, the N-methylol moiety undergoes direct catalytic hydrogenolysis, like that depicted above. For simple amines, the reaction tends to be quite selective to the desired N-methylated product. However, under certain conditions in N-methylation of polyamines, the undesired formation of N-formyl substituted amines (formamides) can be significant and problematic.

formamide by-product:

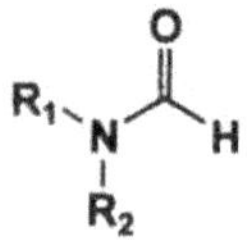

The present work was carried out to more fully understand the the factors which influence reaction selectivity and the mechanism of formation of the by-product formamides. Of particular interest was the effect of polyamine substrate structure on reaction selectivity.

Experimental

Reductive N-methylations were carried out at 7.8 atm and 121°C in an RC1 reaction calorimeter outfitted with a stainless-steel HP60 reaction vessel. The reactor was equipped with a draft-tube agitator to ensure good gas-liquid contacting. The absence of mass transfer limitations at the reaction conditions utilized was experimentally verified. The catalyst used was a commercial 5% Pd/carbon in the powder form. Hydrogen was supplied to the reactor via a 7.6 l ballast vessel equipped with a pressure transducer and thermocouple, which enabled the real-time determination of hydrogen uptake. Formaldehyde solution

(Methyl Formcel®), 55% formaldehyde with 35% methanol and 10% water, was supplied to the reactor via a diaphragm pump configured in an RD10 dosing control loop. Experiments were run semi-batch; catalyst and amine substrate were added to the reactor, heated to reaction temperature, and then formaldehyde was pumped in at a constant rate over several hours. Samples were taken from the reactor through a dip tube fitted with a sintered stainless-steel filter to separate catalyst from the sample.

Probe reaction experiments involving formaldehyde and some of the reaction intermediates were carried out in septum-capped vials immersed in a controlled temperature oil bath. Hydrogenations of some of the isolated reaction intermediates were done using a 0.3 l stainless-steel autoclave.

Reaction product analysis was done by capillary FID gas chromatography and structures were confirmed by GC/MS using both electron and chemical ionization.

Results and Discussion

Table 1 compares the selectivity to by-product formamides for the permethylation of a series of primary diamines, ranging in size from 1,2-ethanediamine (1,2-EDA) to 1,6-hexanediamine (1,6-HDA). With the exception of 1,3-propanediamine (1,3-PDA), the selectivity to formamides is very low for all of the other diamines and shows no apparent trend with the length of the bridging alkylene chain. The selectivity to formamides observed for 1,3-PDA, 55%, is strikingly higher than that for the other diamines in this series. The main by-product formamides produced from 1,3-PDA were N-formyl-N,N'N'-trimethyl-1,3-propanediamine, **3**, and N-formyl-N',N'-dimethyl-1,3-propanediamine, **4**.

The anomalously high selectivity to formamides for 1,3-PDA spurred further, detailed study of this substrate structure to gain additional insight into the reaction chemistry responsible.

Table 1 Selectivity in the permethylation of a series of diamines[a].

Diamine	Structure	Formamides Selectivity (GC area%)
1,2-EDA	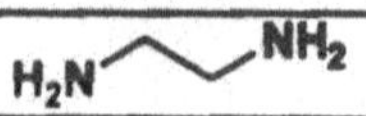 H_2N NH_2	0.9
1,3-PDA	 H_2N NH_2	55
1,4-BDA[b]	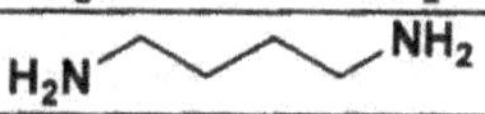 H_2N NH_2	0.8
2M-1,5-PDA[c]	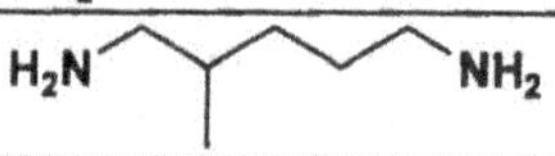 H_2N NH_2	1.5
1,6-HDA	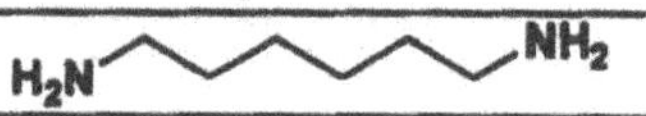 H_2N NH_2	1.0

[a] Reaction conditions: 121°C, 7.8 atm, formaldehyde/N-H=1.0.
   Catalyst: 5% Pd/carbon.
[b] 1,4-BDA=1,4-butanediamine.
[c] 2M-1,5-PDA=2-methyl-1,5-pentanediamine.

In attempting to explain the formation of formamides in N-methylation, one is tempted to invoke Cannizzaro chemistry or modifications thereof. For example, the N-methylol intermediate formed via condensation of formaldehyde with an amine could undergo reaction with additional formaldehyde to produce the N-formyl group, as follows:

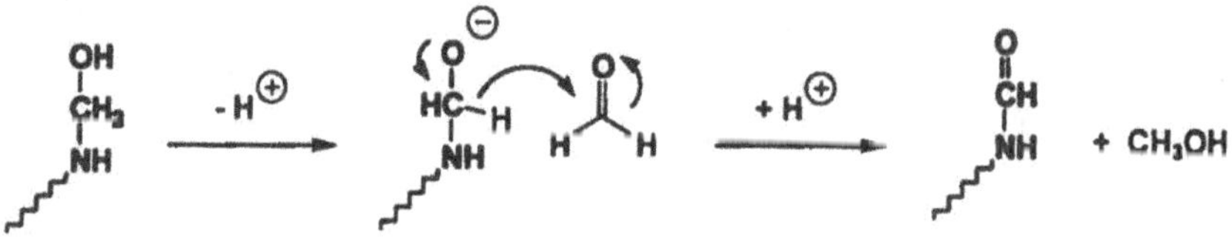

Another possibility is the classic Cannizzaro reaction, the formation of formic acid and methanol from formaldehyde and water, followed by direct amidation of the amine by the formate. Note that a key signature of the existence of Cannizzaro chemistry is the co-production of methanol in an amount equal to the formyl groups formed.

To investigate the possible occurrence of Cannizzaro reactions during N-methylation, a material balance on methanol was done for the permethylation of 1,3-PDA. To maximize the ability to detect relatively low quantities of methanol, Butyl Formcel® (40% formaldehyde/53% n-butanol/7% water) was used as the formaldehyde source, instead of Methyl Formcel®, because it contains no methanol. The selectivity to formamides for permethylation of 1,3-PDA using Butyl Formcel® was comparable to that observed using Methyl Formcel®. Interestingly, the moles of methanol produced was only 5% of the total moles of formyl groups formed. This result shows that a Cannizzaro-type reaction can be eliminated as a significant source of formamides for this system.

Additional insight into the 1,3-PDA system was obtained by sampling and analyzing the reactor contents during reductive N-methylation. Table 2 shows the major species detected in the reactor after 1, 2, and 4 molar equivalents of formaldehyde were added. The most significant observation is the presence of relatively high concentrations of the cyclic aminals (hexahydropyrimidines) after 1 and 2 equivalents of formaldehyde. The final product (4 equivalents) had very little of these cyclics, but formamides were present in high concentration.

Table 2 Major species observed during permethylation of 1,3-PDA^a.

^{a} Reaction conditions: 121°C, 7.8 atm. Catalyst: 5% Pd/carbon.
^{b} Moles of formaldehyde per mole of 1,3-PDA.

The observed hexahydropyrimidines are likely formed by the addition of formaldehyde to one of the amine nitrogens of diamine **5** to form the N-methylol-substituted **6**, followed by an intramolecular condensation reaction involving the other amine group to form the hexahydropyrimidine, **7**:

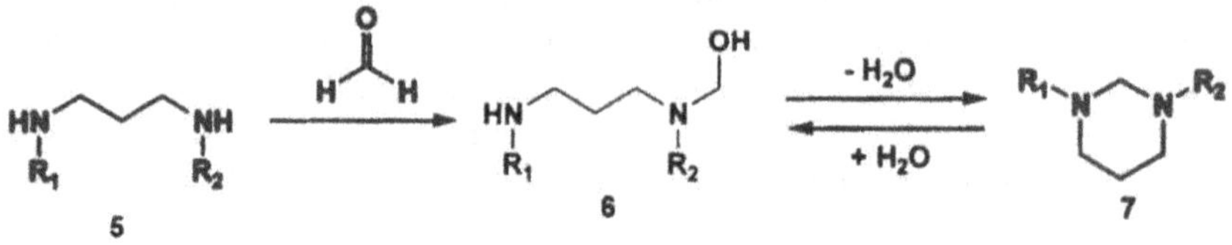

The formation of hexahydropyrimidines from formaldehyde and 1,3-PDA under mild conditions has been reported (10,11) and the reaction is known to be reversible. It is clear from the above scheme that, the formation of a hexahydropyrimidine requires not only a propylene bridge between the amine groups, but also at least one active H atom on each of the adjacent nitrogens.

While the above results suggest a link between cyclic intermediates and the production of formamides, additional support to this connection was provided by the results of permethylation of various derivatives of 1,3-PDA. Table 3 shows the formamide selectivity observed for the permethylation of several 1,3-PDA derivatives. For two of the substrates, **9** and **11**, the target amount of formaldehyde could not be added completely to the reactor because no uptake of hydrogen occurred. These experiments were terminated because the accumulated Methyl Formcel® in the reactor ultimately resulted in compression of the reactor headspace until the maximum pressure capability of the feed pump was exceeded. After termination of the experiment for both **9** and **11**, analysis of the reactor contents revealed high concentrations of hexahydropyrimidines and formamides. By contrast, selectivity to formamides was very low for substrates **10, 12**, and **13**. This is consistent with the fact that the formation of cyclic hexahydropyrimidines is precluded for those substrates because none has the required active H atom on each of two adjacent amine groups. Thus, these results establish a positive correlation between the formation of cyclic intermediates and the production of formamide by-products.

Basic molecular accounting in N-methylation of 1,3-PDA provides additional mechanistic information. Counting all of the N-methyl groups present in the product mixture and comparing that to the quantity of H_2 consumed in the reaction reveals that only 0.6 moles of H_2 were consumed per mole of N-methyl groups formed, while the expected stoichiometry is 1.0 mole of H_2 per mole of N-methyl. This indicates that roughly 40% of the N-methyl groups in the final product was produced without the consumption of molecular H_2. This intriguing result suggested the possibility that formaldehyde may react with the cyclic aminal and simultaneously forming a methyl group and a formyl group on adjacent N atoms.

Table 3 Permethylation of derivatives of 1,3-PDA[a]

Structure	Formamides Selectivity (GC area%)
8	54
9	(no H$_2$ uptake)
10	< 0.5
11	(no H$_2$ uptake)
12	< 1.0
13	< 1.0

[a] Reaction conditions: 121°C, 7.8 atm, formaldehyde/N-H=1.0. Catalyst: 5% Pd/carbon.

To investigate this possibility further, the reaction of N,N'-dimethylhexahydropyrimidine, **15**, with formaldehyde was selected as a suitable probe reaction:

Reactant **15** was prepared by the reacting N,N'-dimethyl-1,3-propanediamine, **9**, with formaldehyde, followed by recovery by vacuum distillation. The isolated **15** was then reacted further with formaldehyde, in the absence of H_2, and the product sampled and analyzed by FID/GC and GC/MS. In addition, various additives were also employed to investigate their influence on the reaction chemistry. Table 4 shows the results of these studies. Clearly, the postulated reaction chemistry does, in fact, occur, even at a temperature 40°C lower than that used for permethylation. Comparison of the first two lines of Table 4 indicates that the reaction is dramatically accelerated by the Pd/C catalyst used in the permethylation reactions. The results obtained for activated carbon and pure Pd powder (third and fourth lines) indicate that Pd is more effective than the carbon support in catalyzing the reaction. The results of Table 4 provide clear evidence of a pathway to formamides that proceeds through a cyclic aminal, which reacts with formaldehyde to simultaneously produce an N-methyl and N-formyl groups.

Table 4 Reaction of formaldehyde with N,N'-dimethylhexahydropyrimidine, **15**, in the presence of various additives[a].

Additive	Additive conc. (wt%)[b]	Reaction time (hr)	Conversion (GC area%)
none	----	72	1.5
5% Pd/C catalyst	1.1	17	84
carbon support	5.0	22	3.7
Pd powder	0.4	40	15.1

[a] Reaction conditions: 80°C, 1 atm, formaldehyde solution: Methyl Formcel®, initial charge (moles): formaldehyde/N,N'-dimethylhexahydropyrimidine ~0.6, no hydrogen.
[b] Additive concentration relative to compound **15**.

While the above results provide ample evidence of the direct involvement of the intermediate cyclic aminal (hexahydropyrimidine) species in the extensive formation of formamides for 1,3-PDA, the low selectivity to formamides observed for 1,2-EDA must be rationalized with this result, especially since the ethylene bridge in 1,2-EDA also enables the formation of cyclic aminals (imidazolidines) upon condensation with formaldehyde. In fact, the existence of imidazolidines has been observed during the permethylation of 1,2-EDA (12). As will be shown, imidazolidines, like hexahydropyrimidines, can react directly with formaldehyde to simultaneously form an N-methyl and an N'-formyl group. The reaction of N,N'-dimethylimidazolidine, **19**, with formaldehyde produces N,N,N'-trimethyl-N'-formyl-1,2-ethanediamine, **20**:

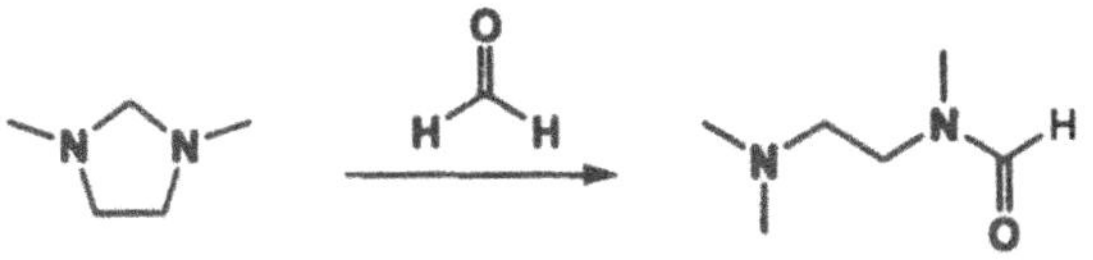

To compare the relative reactivity with formaldehyde of the imidazolidine ring to the hexahydropyrimidine ring, a competitive reaction experiment was done involving the reaction of formaldehyde with a mixture of N,N'-dimethylhexahydropyrimidine, **15**, and N,N'-dimethylimidazolidine, **19**. The results of this experiment are presented in Table 5, which shows the major species concentrations as a function of reaction time. Comparison of the rate of decrease of **19** vs. **15**, indicates that the imidazolidine reacts faster with formaldehyde than the hexahydropyrimidine.   Both **19** and **15** form the corresponding formyl substituted products, **20** and **16**, respectively. Interestingly, the permethylated products, **21** and **22**, are also formed:

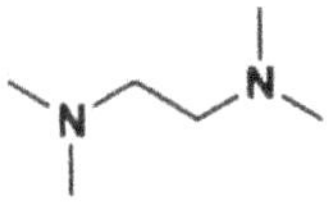

The exact mechanism that is responsible for the formation of the permethylated products is not clear.   One possibility is the base-catalyzed hydrolysis of the formyl group, followed by the N-methylation of the resulting amine using formic acid, formed from formaldehyde, as the reductant. This is the well-known Eschweiler-Clarke reaction (13).

The competitive experiment with formaldehyde does not help explain why formamide formation is much greater for 1,3-PDA than for 1,2-EDA. In fact, the results of that experiment alone suggest the opposite, since the imidazolidine reacts more readily with formaldehyde than the hexahydropyrimidine to form the formamide.   However, it is important to note that, during reductive permethylation, the intermediate cyclic aminals participate in competing chemistry, most notably the catalytic hydrogenolysis of the aminal to open the ring and form a methyl group on one of the N atoms.

Table 5 Competitive reaction of N,N'-dimethylimidazolidine, **19**, and N,N'-dimethylhexahydropyrimidine, **15**, with formaldehyde[a].

Time (hr)	Species Concentration (GC area%)					
	19	**20**	**21**	**15**	**16**	**22**
20	30	0.7	1.6	59	0.4	0.2
44	20	4.9	7.7	58	2.7	0.4
116	3.7	21	15	43	14	0.7
164	0.9	24	15	36	20	0.8

[a] Reaction conditions: 80°C, 1 atm, formaldehyde solution: Methyl Formcel®, initial charge (moles): formaldehyde/**15**/**19**= 4/1/1, no hydrogen.

Once again a competitive probe experiment was done to assess differences between the imidazolidine and the hexahydropyrimidine. In this experiment, equimolar amounts of **15** and **19** were dissolved in methanol and then reacted with hydrogen in the presence of the Pd/carbon catalyst. The results from the analysis of reactor samples for this experiment are shown in Table 6. Clearly, imidazolidine, **19** reacts much faster than hexahydropyrimidine, **15**. The main hydrogenolysis products from **15** and **19** are N,N,N'-trimethyl-1,3-propanediamine, **23**, and N,N,N'-trimethyl-1,2-ethanediamine, **24**, respectively. In addition to these products, others are formed, including the permethylated compounds, **21** and **22**. Evidently, some scrambling of methyl groups occurs in parallel, evidently via secondary reaction of the cyclic aminals with the open chain intermediates.

Table 6 Competitive hydrogenation of N,N'-dimethylimidazolidine, **19**, and N,N'-dimethylhexahydropyrimidine, **15**[a].

Time (min)	Species Concentration (GC area%)			
	19	**24**	**15**	**23**
0	42.6	0	56.3	0
45	1.6	26	60	0
240	0	27.3	33.8	15.5

[a] Reaction conditions: 80°C, 9 atm, initial charge (moles): **15**/**19**= 1/1, catalyst: 5% Pd/carbon.

Conclusions

This work has elucidated critical aspects of the reaction mechanism. The findings are summarized in Figures 1 and 2, which depict proposed reaction schemes for ethyleneamines and propyleneamines, respectively. A key feature of these schemes is the formation of intermediate cyclic aminals, when active H atoms are available on adjacent N atoms, and their intermediacy in the formation of formamide by-products.

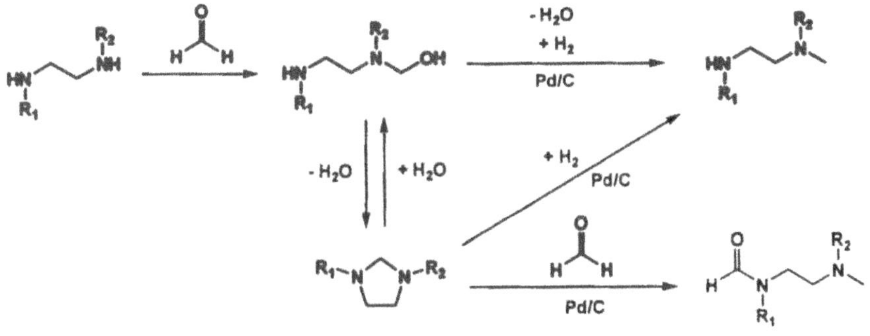

Figure 1   Reaction pathways in the N-methylation of ethyleneamines.

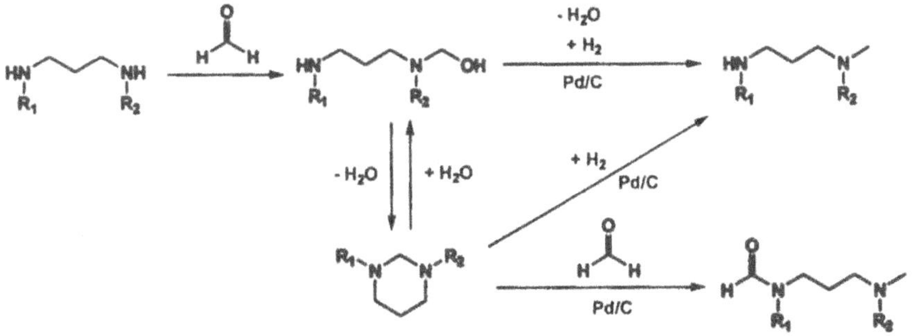

Figure 2   Reaction pathways in the N-methylation of propyleneamines.

Formamide formation for the propyleneamines is greater than that for the ethyleneamines for probably two reasons. Firstly, formation of the 6-membered hexahydropyrimidine ring is probably thermodynamically more favorable than formation of the 5-membered imidazolidine ring, which leads to a higher concentration of the former species during N-methylation. Secondly, catalytic hydrogenolysis of the hexahydropyrimidine is much slower than that for the imidazolidine, another factor which favors a high concentration during N-methylation. Details regarding the reaction of formaldehyde with the cyclic aminal to produce the formamide are not clear at this point; better understanding of that part of the mechanism would require further study.

Acknowledgments

The authors would like to thank Zane Barrall and Kristen Minnich, who performed most of the experiments, and Ann Kamzelski for the GC/MS analyses.

References

1.   GB Pat. 1,305,258, to BASF (1973).
2.   GB Pat. 1,403,569, to BASF (1975).
3.   T. Inagaki, T. Katagiri, and K. Maenoh, JP Pat. 61/152,643, to Lion Akuzo Co., Ltd. (1986).
4.   A. Gohdoh, T. Miyamoto, and S. Hashimoto, JP Pat. 62/281,846, to Toray K. K. (1987).
5.   K. Okabe, Y. Yokota, K. Matsutani, and T. Imanaka, US Pat. 4,757,144, to Kao Corp. (1988).
6.   M. Tanis and G. Rauniyar, US Pat. 5,105,013, to Dow Chemical Co. (1992).
7.   R. L. Zimmerman, W. C. Crawford, and R. F. Lloyd, US Pat. 5,646,235, to Huntsman Petrochemical Corp. (1997).
8.   A. P. Bonds and H. Greenfield, Chem. Ind. (Marcel Dekker), **52**, (*Catal. Org. React.*), 65 (1992).
9.   H. Greenfield, Chem. Ind. (Marcel Dekker), **53**, (*Catal. Org. React.*), 265 (1994) and references therein.
10.   R. F. Evans, *Aust. J. Chem.*, **20**, 1643 (1967).
11.   J. Dale and T. Sigvartsen, *Acta Chem. Scand.*, 45, 1064 (1991).
12.   R. V. C. Carr, unpublished results.
13.   J. March, *Advanced Organic Chemistry: Reactions, Mechanisms, and Structure*, 4th Ed., John Wiley & Sons, 1992, pp. 899.

Reductive Alkylation of 2-Methylglutaronitrile with Palladium Catalysts

Frank E. Herkes and Jay L. Snyder

DuPont Nylon Intermediates and Specialities, E. I. DuPont de Nemours & Co. Experimental Station, P. O. Box 80302, Wilmington, Delaware 19880-0302

Abstract

A study was made on the mono reductive alkylation of 2-methylglutaronitrile (MGN), a co-product in the synthesis of adiponitrile from butadiene and HCN, with dialkyl amines with the aim of understanding the chemistry and product selectivity associated with the use of Group VIII metals and Raney® catalysts for reductive alkylation of MGN to tertiary amines. This initiative was undertaken to find a selective catalyst for the synthesis of 5-diethylamino-2-methylvaleronitrile (4), a known precursor to novoldiamine, which is used in the synthesis of the anti-malarial drug, chloroquine. The isomer, 2-diethylamino-2-methylvaleronitrile (5) is also produced in the alkylation step.

The selectivity to 4 and 5 is strongly dependent upon the metal and catalyst support and is general for secondary amines. With carbon supports their yield is high and relative distribution is 74/26, but can be increased to 80-85/20-15 employing acidic metal oxide supports albeit with significant loss to higher molecular weight amines. Prolonged alkylations with excess secondary amines and Pd catalysts produce high yields of the bis-alkylated tertiary amines at mild conditions.

Introduction

In early 1992, we were approached by a consultant representing a large Indian Company in Bombay to make 5-diethylamino-2-methylvaleronitrile (4) by a reductive alkylation process employing 2-methylglutaronitrile (MGN) and diethylamine over a Pd/C catalyst. 4 is an intermediate in the synthesis of novoldiamine (1), a precursor for chloroquine (3). 1 is prepared (1) by hydrolysis of 4 with sulfuric acid producing the amide followed by treatment with NaOCl and base in a Hoffmann rearrangement. Coupling 1 and 4,7-dichloroquinoline (2) produces 3 in high yield. 3 is the oldest and most attractive workhorse drug in the anti-malarial market.

Several patents to National Distillers (1) indicated that **4** was the only product formed in the reductive alkylation of MGN with diethylamine using a Pd/C catalyst. A further search, however, uncovered a set of earlier patents (2) to Sterling Organics teaching the preparation of **1** from methyleneglutaronitrile. Under similar reaction conditions described in the National Distillers patent, both **4** and its expected isomer, 5-diethylamino-4-methylvaleronitrile (**5**) were reported in an 80/20 distribution from reductive alkylation of MGN. DuPont produces billions of pounds of adiponitrile along with significant amounts of MGN, making the latter an attractive low cost raw material for the synthesis of **1**. The focus of this study was to find a higher selectivity route to **4** based on MGN as a raw material building block.

Experimental

Batch hydrogenations of MGN in the presence diethylamine were performed in a 300 mL Stainless Steel Autoclave Engineers magnedrive packless autoclave equipped with a thermocouple, cooling coils, sample dip tube containing a stainless steel 5 micron Mott filter and Dispersimix turbine type draft tube agitator containing a rotating impeller. Hydrogen uptake kinetics were followed by the pressure drop in a 1-L hydrogen reservoir feeding the autoclave and transmitted to a Yokogawa HR1300 recorder. Hydrogen uptake data collected every minute throughout the run was monitored both graphically and electronically, and fed into a data file for analysis.

Product analysis of the reductive alkylation product was done using a Hewlett Packard 5890 gas chromatograph GC), equipped with a HP-5 (5% crosslinked phenyl-methyl-silicone) megabore column (30 m long, 0.33 ID, 0.25 um film thickness) and a flame ionization detector. The temperature program was 60°C for 2 min + 8°/min to 230°C for 15 min. The column flow rate was 1.5 cc/min helium and split vent flow rate of 60 cc/min helium. The injector and detector temperatures were 250°C and 265°C, respectively.

Product and by-product identification was made by GC/MS analysis. The assignment of **4** relative to **5** was based on the higher stereoselective reduction of the unhindered terminal nitrile group compared to the nitrile adjacent to the methyl substituent. The hydrogenation catalysts were used as received from the vendors. Diethylamine (Aldrich Chemical) was used without further purification.

In a typical batch run, 40g (0.37 mole, 99%) refined MGN and 81 g (1.11 moles, 3-fold excess) diethylamine were charged to the autoclave. A hydrogenation catalyst having the composition 5% Pd/C (2.5 g, dry basis, Engelhard) was next charged to the reactor. After closing, the reactor was purged 3x with hydrogen. The temperature was raised to 80°C under 50 psig hydrogen with very slow stirring. At reaction temperature, the pressure was raised to 500 psig with hydrogen and maximum (~ 1200 rpm's) stirring commenced. Under these conditions, reduction to the desired isomers **4** and **5** required 120 minutes. The molar hydrogen uptake was approximately 0.75-0.80 after the 120 run time. A theoretical uptake value of 0.73 mole hydrogen was calculated for reduction of one nitrile group. Total conversion of MGN was observed under these conditions resulting in a 67% yield to **4** (85% yield to **4**+ **5**).

Results and Discussion

To resolve the dilemma of the two-isomer possibility, MGN was hydrogenated in the presence of diethylamine (3 molar excess) employing a 5% Pd/C catalyst in a 300 mL stainless steel batch autoclave at 60-100°C and 500-700 psig. Hydrogen uptake was monitored with time and samples taken for kinetic analysis. In our hands, both **4** and **5** were produced in all cases in ratios ranging from 75/25 to 85/15. The high yield of **4** from MGN is due to the low reactivity of the CN group adjacent to the methyl group compared to the reactivity of the unhindered CN group. In addition to the expected bis (1,5-diethylamino)-2-methylpentamethylenediamine (**6**), formed by continued alkylation of **4** and **5** with diethylamine, we also observed the formation of condensed secondary aminonitriles (**7**, 3 isomer combinations) and its cyclized hydrogenation product, 1-(5-cyano-5-methylbutyl)-3-piperidine (**8**, 2 isomer combinations). Unlike reduction of aromatic nitriles (**3**) in the presence of secondary amines, where primary amines are exclusively produced with Pd/C in solvents, neat MGN produced the tertiary amine in high selectivity. The secondary amine **7** was the major side product with the Pd/C catalysts. It is produced in a parallel reaction along with **4** and **5** by competitive condensation of 5-amino-2-methylvaleronitrile (**9**) and 5-amino-4-methylvaleronitrile (**10**) with the intermediate primary imine formed from the initial addition of hydrogen to MGN.

In the initial stage of MGN reduction, two intermediate primary imines are formed with the less steric isomer predominating. Addition of a second mole of hydrogen produces the aminonitrile isomers, **9** and **10**. Previous reports (4) indicated only **9** was formed when Raney® Co was employed. However, we observe both isomers similar to that for the reduction of MGN with Raney Co in

MeOH containing base (5).   In the presence of excess (3X) diethylamine, competitive addition to the primary imines predominates to produce **4** and **5**.

Because primary amines are reactive (6), if not poisons, with nitriles in the presence of palladium, side reactions occur where the primary aminonitriles, **9** and **10** condense with the initial primary imine intermediates in a number of different combinations (7) to produce gem diamino compounds. These in turn can lose ammonia producing the respective aldimines that then add hydrogen to produce the three isomeric condensation products of **7**. Under the reaction conditions none of the intermediate aldimines are observed. Alternate direct hydrogenolysis of the gem diamino intermediates (7b) similar to that involved in the reductive alkylation process for tertiary amines, is another mode to producing **7**. Only low steady state concentrations of **9** and **10** are seen during the reduction. By-product ammonia build-up during secondary amine formation can also compete with diethylamine for the primary imine, reducing the selectivity to **4** and **5**.

When Raney® Ni is employed instead of 5% Pd/C with diethylamine under similar reaction conditions, less than 1% diethylamine alkylation is observed at 81% MGN conversion. A mixture of **9** and **10** are produced in 39% selectivity along with 2-methylpentamethylenediamine (MPMD) and 3-methylpiperidine in 1.6% and 3.4% selectivity, respectively.   By-product aldimines, produced by coupling of MPMD were observed in contrast to the formation of **7**. Four (two major) aldimines were observed by GC/MS at 81% MGN conversion. Their exclusive formation with Raney Ni resides in their facile desorption or displacement by diethylamine on the Raney Ni surface compared to that of Pd on carbon which shows complete reduction to three isomers of **7**.

A study was made to determine if the selectivity of **4** relative to **5** could be enhanced by different Group VIII metals and/or supports. A variety of catalysts were screened under similar reaction conditions. The results unequivocally show that Pd is by far the most selective metal vs. Pt, Ru, Rh and Raney Ni (Table 1) for producing high yields of **4+5**. Addition of a small

(0.5%) amount of Pt to a Pd/C catalyst showed a higher hydrogenation rate and 4+5 selectivity compared to Pd/C alone. Both displayed similar by-product selectivity to 7. The presence of Pt along with Pd also slightly increased the isomer distribution of 4 from 74/26 to 79/21.

Table 1. Effect of Group VIII catalysts on the reductive alkylation of MGN with diethylamine at 60°C and 500 psig.

Catalyst	% MGN Conv	4 + 5 % Sel	4/5 ratio	6 % Sel	7 % Sel	8 % Sel
5% Pd/C	99	86	74/26	1.7	7.7	3.3
4.5% Pd/0.5% Pt/C	99	85	79/21	1.4	8.7	3.8
5% Ru/C	< 1					
5% Pt/C	82	49	88/12	0.33	31	3.2
5% Rh/Al₂O₃	93	< 1		0	41	6.6
5% Pt/Al₂O₃	88	36	88/12	0.11	39	3.2

Pt on C or Al$_2$O$_3$ displayed surprisingly high distributions to 4 albeit in low selectivity and higher conversion to 7 and 8 because of the strong adsorption (8) of amines on the Pt catalyst surface. Ru on C or alumina was totally inactive for the reduction at these reaction conditions, even up to 100°C, and Raney Ni did not produce any of the reductive alkylated products as did Rh on alumina. Diethylamine appears to poison the Ru catalyst for nitrile reduction under these conditions. Rh, known (9) to be active for secondary and tertiary formation, produced mainly secondary amines such as 7 under these conditions

A carbon or basic support was found to be a requisite for reducing the amount of secondary amines by minimizing the re-adsorption of aminonitriles, 9 and 10 on the catalyst This effect was demonstrated when acidic oxide supports such as Al$_2$O$_3$, TiO$_2$, SnO$_2$ and BaSO$_4$ were employed. With these supports, the selectivity of 4 + 5 was much lower at high MGN conversions compared to carbon and basic supports such as calcium and barium carbonate (Table 2). Metal oxide supports are more apt to retain or enhance re-adsorption of the primary amines during aminonitrile formation leaving them predisposed for condensation with the primary imines produced from MGN. Since the surface area for carbon is higher vs. CaCO$_3$, it is preferred for maximum activity and product selectivity.

Table 2. Effect of supports with Pd on the selectivity of 4 + 5 for the reductive alkylation of MGN with diethylamine at 60° and 500 psig.

Catalyst	% MGN Conv (% cat loading)	4 + 5 % Sel	4/5 ratio	6 % Sel	7 % Sel	8 % Sel
5% Pd/C	99 (5)	86	74/26	1.7	7.7	3.3
4.5% Pd/0.5% Pt/C	99 (5)	85	79/21	1.4	8.7	3.8
5% Pd/BaSO$_4$	55 (5)	83	78/22	0.17	8.1	0.38
5% Pd/CaCO$_3$	90 (9.7)	75	79/21	0.34	19	3.2
5% Pd/BaCO$_3$	90 (9.7)	76	79/21	0.58	6.9	0.28
5% Pd/Al$_2$O$_3$	96 (12)	62	81/19	0.80	17	3.4
5% Pd/TiO$_2$	98 (4.8)	72	82/18	0.52	19	4.8
5% Pd/SnO$_2$	94 (12)	58	86/14	0.32	26	9.4

A more dramatic observation was the higher distribution of 4 vs. 5 with the metal oxide supports. With a carbon support, the ratio was 74/26 and increased to 86/14 employing a SnO$_2$ support. All the metal oxides display a significantly higher selectivity to 4 than the carbon supports. The order of increasing ratio to 4/5 by catalyst supports was SnO$_2$ > TiO$_2$ ~ Al$_2$O$_3$ > BaSO$_4$ ~ Pd/Pt/C >Pd/C. However, with the higher 4 selectivity there was observed higher condensation products, 7 and 8. Both impacted the overall yield of 4 + 5. Carbon supported catalysts displayed the highest (62-67%) yield to 4 because of lower secondary amine formation.

A closer examination of 7's three isomers (a:b:c)showed a distinctive distribution for several of the different supports indicating there may be a preferential addition of 9 and 10 with the primary imines intermediates on different supports. None of the 7 isomers were identified as to their structure. It would be anticipated that their formation would be based on their relative steric hinderence. With the carbon catalysts containing Pd and that with 0.5% Pt, the distribution of 7a:b:c was 7:41:52 and 5:40:55, respectively. The metal oxide supports such as alumina, titiana and tin showed different selectivities compared to carbon support catalysts. Their average selectivities were 13:52:35 for alumina and titiana, and 19:53:28 for the tin support. The latter selectivities were obtained from alkylations employing 12-16% catalyst loading (relative to MGN) compared to 5% loading for the Pd on carbon catalysts. However, little difference in distribution was seen with a 5% catalyst loading employing 5% Pd on alumina and titiana supports.

Carbon Supports

The type of carbon used as the support did show significant variation in activity, but not in 4 + 5 selectivity. An unreduced 5% Pd/C was the most active catalyst, even at a lower catalyst loading, in the 5% Pd/C series. Its reduced and stabilized form had slightly lower activity at all the same temperatures and a mixed form was even less active under similar reaction conditions (e.g., 180 min run time). Only a 4.5% Pd/0.5% Pt/C catalyst was similar in activity as the unreduced form (Table 3). All the 5% Pd/C catalysts displayed a 71-74% distribution of 4 in the 4+5 products. Selectivities of 4+5 were high (85-95%) using carbon supports, and only minimal selectivity (< 9%) to coupling products was observed. Some of the diamine, 6, was produced, but only in 2-3% selectivity.

Table 3. Effect of carbon supports containing Pd for the reductive alkylation of MGN with diethylamine at 60° and 500 psig.

Catalyst	% MGN Conv (% cat loading)	4 + 5 % Sel	4/5 ratio	6 % Sel	7 % Sel	8 % Sel
5% Pd/C (edge & unreduced)	99 (5)	86	74/26	1.7	7.7	3.3
4.5% Pd/0.5% Pt/C (edge & reduced)	99 (5)	85	79/21	1.4	8.7	3.8
4.5% Pd/0.5% Pt/C (edge & red, 50°)	99(5.5)	80	81/19	1.6	4.2	11
5% Pd/C (edge & reduced)	74 (5)	92	73/27	0.40	6.5	0.51
5% Pd/C (mixed & reduced)	65 (4.8)	92	72/28	0.38	6.0	0.55
5% Pd/C (80°C) (wood carbon)	26 (3.0)	95	70/30	0	2.8	0.14
5% Pd/C Sibunit	99 (4.8)	86	73/27	1.3	8.4	2.4
20% Pd(OH)$_2$/C	99 (4.8)	86	78/22	2.8	7.7	8.8

Addition of 0.5%Pt to a Pd/C catalyst not only enhanced its activity relative to 5%Pd/C, but also had a higher 4 yield. The MGN rate for the Pt doped catalyst was approximately 2.5X faster than the 5% Pd/C catalyst. Lowering the temperature to 50°C raised the distribution of 4 in 4+5 to 81% which is a 12% increase over that of the 5%Pd/C catalysts studied, but also resulted in an increase of 8.

A technique used to enhance alkylation rate is to purge ammonia during the reduction thereby maintaining a higher partial pressure of hydrogen throughout the run. This approach can minimize competitive side reactions to **9** and **10** by formed ammonia produced in the alkylation reaction employing diethylamine. Analysis of the product mixture after reduction at 104° and 350 psig for 120 minutes, with venting, using a 3% loading of a 5% Pd/C catalyst showed 93% MGN conversion and 91% selectivity to **4 + 5**. The 4/5 distribution was 73/27. Repeating the run without venting produced 93% selectivity to **4 + 5**, but the MGN conversion was only 84% after 150 minutes. A similar (73/27) distribution of **4** to **5** was also observed. Removing the ammonia by purging doubled the MGN rate (Figure 1).

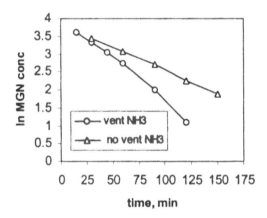

Figure 1. Effect of ammonia venting on the rate of MGN reductive alkylation with diethylamine at 104°C and 350 psig.

Effect of Solvents
Reductive alkylation of MGN with diethylamine in different solvents under the similar conditions, employing a Pt doped Pd/C catalyst, all showed high MGN conversion and comparable selectivity to **4** (Table 4). With the exception of methanol and NMP solvents, the selectivity to 4+5 was 80-81%. This was true of the control where no solvent was employed. Reductions in methanol produced higher **6** (11% vs. 2-3% selectivity) and less coupling products. High selectivities to **7** and **8** by-products were observed with NMP.

Table 4. Effect of solvents for the reductive alkylation of MGN with diethylamine at 65°C and 500 psig using a 4.5% Pd/0.5% Pt/C catalyst.

Solvent	% MGN Conv (5% cat loading)	4 + 5 % Sel	4/5 ratio	6 % Sel	7 % Sel	8 % Sel
Methanol	99	60	75/25	12	2.3	6.0
THF	99	80	78/22	2.1	6.6	7.0
NMP	99	68	81/19	2.4	11	9.4
Control	99	81	81/19	3.6	5.1	11
Diethylamine	99	80	80/20	3.2	5.0	8.7

Palladium on Acidic Supports

Nitriles can be reductively alkylated to tertiary amines with secondary amines over Pd catalysts, especially with metal oxide supports (10). Our scouting of metal oxide supports such as alumina, titiana, tin and miscellaneous acidic supports (Table 5) showed a much different product spectrum vs. carbon supported catalysts. Of the three metal oxides, titiana was the most active and least selective to **4** vs. carbon due to the higher formation of **7**. Distributions of 82/18 to 86/15 **4/5** were consistently observed with metal oxide catalysts with tin being the highest. Tin-oxide supported catalysts gave the highest **4** distribution, but their activity was quite low due to its relatively lower surface area vs. carbon. Other acidic supports such as Pd on HZSM, Pd on sulfated zirconia and BaSO$_4$ were much less active and only weakly selective. No shape selectivity enhancement was observed with Pd/HZSM. Reductions using a gamma alumina support were relatively fast at catalyst loadings > 9% showing lower **4+5** selectivities (e.g., 67-70%) than carbon, but higher **4/5** ratios. The major by-product was the secondary amine **7**, and at higher catalyst loadings (24%) it was converted to **8**. Little (< 1%) conversion to **6** was observed at most all the different catalyst loadings.

Table 5. Effect of metal oxide supports on the reductive alkylation of MGN with diethylamine at 60°C and 700 psig.

Catalyst	% MGN Conv (% cat loading)	4 + 5 % Sel	4/5 ratio	6 % Sel	7 % Sel	8 % Sel
10% Pd/SnO₂	26 (5.0)	78	83/17	0	4.8	0.25
10% Pd/SnO₂	94 (12)	58	86/14	0.32	26	9.4
5% Pd/TiO₂	97 (4.8)	72	82/18	0.52	19	4.8
5% Pd/TiO₂	99 (12)	54	85/15	5.8	1.0	26
5% Pd/Al₂O₃	73 (4.8)	70	82/18	0.15	22	1.7
5% Pd/Al₂O₃	96 (12)	62	81/19	0.80	17	3.4
5% Pd/BaSO4	55 (5.0)	88	78/22	0.17	8.1	0.38
5% Pd/HZSM-5	21 (9.7)	85	79/21	0	6.2	0
Pd/sulfated ZrO2	19 (12)	89	76/24	0	3.6	0

Use of a titiana support also showed good activity for the reductive alkylation. The distribution of 4 in 4+5 was 82-86% and was notably higher than that observed with alumina, being in the low 80's. This distribution appeared to increase with higher catalyst loading reaching a maximum of 86% at a 12% loading. Higher catalyst loading yielded high selectivities to 8. The yield of 4 was, nevertheless, 15% lower than that observed with a carbon support, but 4's distribution was higher (85% vs. 73%).

Reductive Alkylation with Dimethylamine and 3-Methylpiperidine

Mono reductive alkylation of MGN with a 3-fold excess of dimethylamine at 80° and 500 psig using a 2.5% loading of 5% Pd/C catalyst produced a mixture of 5-dimethylamino-2-methylvaleronitrile (11) and 5-dimethylamino-4-methylvaleronitrile (12) in 92% yield with a respective 70/30 distribution similar to that observed with diethylamine. A 6.2% yield of the N,N,N',N'-tetramethyl-2-methylpentamethylenediamine (13) was observed with only traces of 9 and 10 seen after a reduction time of 24 h (e.g., 99+% MGN conversion). Increasing the temperature to 110° after 24 hours at 80° increased the conversion of 11 + 12 to 13. After 24 hours at 110°, the selectivity to 13 (11a, b) increased to 82.2% with the major balance (15% sel.) being 11 and 12 in an 89/11 ratio. The change in distribution from 70/30 after 24 hours to 89/11 indicates a stronger preference for reductive alkylation of 12 vs. the sterically hindered isomer, 11.

Likewise, reduction (11c) of MGN at 80°C and 800 psig employing a 5% catalyst loading and 3-fold excess 3-methylpiperidine produced a mixture of

5-piperidino-2-methylvaleronitrile (14) and 5-piperidino-4-methylvaleronitrile (15) in 92% combined yield with an 82/18 ratio.

Conclusions

MGN can be mono reductively alkylated with diethylamine at moderate temperatures and pressure employing Pd on catalyst supports to produce 4 and 5 in high yield. Carbon supports produce yields of 75-85% 4 + 5 at 99% MGN conversion with coupling by-products of the primary aminonitriles as the major yield loss. A mixed metal catalyst, 0.5% Pt and 4.5% Pd on carbon produced the highest yield of the desired isomer, 4, (66%) and the highest productivity (2.5 X higher rate). Lower yields of 4 + 5 and higher coupling products are observed with Pd on metal oxide supports due to the strong adsorption of the primary amine intermediates on the acidic surfaces. Ratios of the less hindered product 4 to 5 are 75/25-79/21 and can be increased to 86/14 with the oxide supports albeit in lower yield compared to a carbon support. Other secondary amines such as dimethylamine and 3-methylpiperidine give similar selectivities and ratios. By the proper choice of metal, support and solvent, one can obtain high yields of the mono alkylated nitrile that can then be further hydrogenated to the primary amine if desired.

References

1. O. D. Frampton and J. B. Pedigo, US Pat. 3,673,251 to National Distillers (1972).
2. Brit. Pat. 1,157,637 (1966), Brit. Pat. 1,157,638 (1966) and Brit. Pat. 1,157,639 (1966) to Sterling Drugs.
3. P. N. Rylander, L. Hasbrouck and I. Karpenko, *Ann. N.Y. Acad. Sci*, 21 (1973), 100.
4. J. Feldman and M. Tomas, US Pat. 3,322,815 (1967) and US Pat. 3,350,439 (1967) to National Distillers.
5. S. Ziemecki, *Stud. Surf. Sci. Catal.*, 78 (1993), 283-90.

6.   J. L. Dallons, A. Van Gysel and G. Jannes, Chem. Ind. (Marcel Dekker), **47**, (Cat. Org. React.), 93, (1996).
7.   a) J. Braun, G. Blessing and F. Zobel, *Chem. Ber.* **56B**, 1988-2001 (1923); b) H. Greenfield, *Ind. Eng. Chem., Prod. Res. Dev.*, **6**, 1422 (1967).
8.   J. Volf and J. Pasek, *Stud. Surf. Sci. Catal.*, **27** (1986), 105-144.
9.   A. Galan, J. de Mendoza, P. Prados, J. Rojo and A. M. Echavarren, *J Org. Chem.*, **56** (1991), 452-454.
10. E. Fuchs, B. Breischeidel, R. Becker and H. Neuhauser, US Pat. 5,894,074 to BASF (1999).
11. a) T.Witzel, US Pat. 5,463,130 to BASF (1995); b) T. Witzel, US Pat. 5,557,011 to BASF (1996); c) J. Heveling, A Gerhard and U. Daum, WP 95/30666 to Lonza.

The Synthesis of Amines by Catalytic Hydrogenation of Nitro Compounds

E. Auer, M. Berweiler , M. Gross, J. Pietsch, Daniel Ostgard, and Peter Panster

Degussa-Hüls AG, Sivento Division
Research. Development. Technical Service. Chemical Catalysts and Zeolites
P. O. Box 1345 – 63403 Hanau / Germany

Abstract

This paper deals with the development of industrially applied precious metal powder catalysts (PMPC) and activated base metal catalysts (ABMC, otherwise known as sponge metal or skeletal catalysts) for the selective hydrogenation of substituted aliphatic and aromatic nitro compounds to the corresponding amines. Since 40 – 50 % of the hydrogenation reactions in fine, intermediate, and pharma-ceutical chemistry deal with the reduction of nitro groups, this paper reviews the efficient use of different catalysts in such processes. Test results of improved PMPC and ABMC are discussed for the hydrogenation of aliphatic nitro compounds and nitroarenes. The basic principles of nitro group reduction and concepts for developing tailored catalysts for each class of the aforementioned nitro compounds are also presented. This paper underscores the ongoing efforts in industry on catalyst improvements contributing to environmentally benign and highly efficient processes.

General Introduction

Aliphatic and aromatic amines are widely used as key intermediates or products in industrial organic chemistry (1). The manufacture of pharmaceutical and agrochemical products is an example that frequently includes the selective hydrogenation of a nitro group. Other applications are the synthesis of intermediates for azo dyes, pigments, and photographic chemicals. Aromatic amines represent the most important class of chemical compounds for the polymer industry (e.g., the isocyanate and polyurethane routes) and for the rubber industry (e.g., the N-alkylation route). The use of amines as antioxidants, corrosion inhibitors, and vulcanization accelerators are of minor importance. All these products require selective and efficient synthetic routes. In industry, the majority of aliphatic and aromatic amines are manufactured by the catalytic hydrogenation of the corresponding nitro compound using precious metal or activated base metal catalysts. Activated carbon, carbon black, alumina, or silica supported palladium, platinum, and iridium catalysts are commonly used as heterogeneous hydrogenation catalysts. The catalytic activity and selectivity of PMPC depends on the precious metal crystallite size, dispersion, and metal loading. Moreover, the use of catalyst modifiers like iron, copper, vanadium, manganese, bismuth, and other transition metals is becoming increasingly necessary to meet the specific requirements of catalytic hydrogenation reactions (2). Needless to say an excellent filterability is also required for powdered catalysts. Process conditions such as pressure, temperature, solvent, hydrogen supply, mixing dynamics, and catalyst concentration also play an important role in the efficiency of the hydrogenation.

Catalyst and process design is especially important when using Pt/C for the hydrogenation of aliphatic nitro compounds. Due to the higher activation energy for aliphatic nitro group hydrogenation, the transformation into the corresponding amine is much slower in comparison to nitroarenes (4). As a consequence, highly active catalysts must be used along with harsher reaction conditions when reducing nitroaliphatics. Most publications that describe aliphatic nitro reductions deal with special organic molecules used in the pharmaceutical sector (5) where the hydrogenation step is not emphasized. This topic has only been dealt with in general terms as far as catalyst optimization goes (6-8). In order to fill the gap between the industrial need to hydrogenate aliphatic nitro compounds and the general lack of such catalysts, one objective of this paper was to determine the activity of various PMPC along with their transition metal promoted varieties for the hydrogenation of 2-nitropropane.

The relatively low costs, high activity, long life, and durability of ABMC make them an important class of hydrogenation catalysts. New techniques for enlarging the active surface area of ABMC and improved activation processes for base metal-alumina alloys generate extremely porous and highly active systems. Similar to the aforementioned PMPC, the modification of ABMC with transition elements is a useful tool to prevent the formation of undesired products and to increase the activity in the hydrogenation of aromatic nitro compounds (3).

Results and Discussion

Hydrogenation of Aliphatic Nitro Compounds with modified Pt/C Catalysts
The test results for various activated carbon supported platinum catalysts given in Figure 1, displays an inverse relationship between the activity for 2-nitropropane hydrogenation and the induction time (listed as "start" in Figure 1) where the most active platinum catalyst started hydrogenating in less than a minute. However, most of the platinum on carbon catalysts did not give access to a major improvement of the hydrogenation activity, but were only slightly more active. A large number of catalysts were at best equal to the reference catalyst A.

Table 1: Properties of unmodified platinum catalysts for nitropropane reduction

Type	Pt-loading [wt.%]	activated carbon type	metal dispersion
A	5	steam activated	medium
B	5	chemically activated	medium
C	5	chemically activated	high
D	5	highly chemically activated	high
E	5	chemically activated, naturally promoted with iron traces	high

Only catalyst E showed a more than doubled hydrogenation activity. We attributed this result to an increased level of iron present in the activated carbon that had been used as support to make the catalyst. This observation prompted us to specifically promote platinum on carbon catalysts with transition metals and design a highly active catalytic system as discussed below. Moreover, relevant physico-chemical properties of the monometallic platinum catalysts A-E are listed in Table 1 and allow an unambiguous correlation between the different carbon supports, the metal dispersion, and the catalytic activity. It clearly turned out that the use of a chemically activated carbons accompanied by an increased metal dispersion lead to an enhanced nitropropane activity.

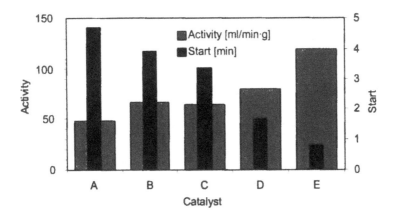

Figure 1: Nitropropane activity and induction time of Pt/C catalysts; A = reference

Table 2: Properties of modified platinum catalysts for nitropropane reduction

Type	Pt-loading [wt.%]	metal loading and type	activated carbon type	metal dispersion
A	5	none	steam activated	medium
A*	5	0.5 wt.% Fe	steam activated	medium
F	5	0.5 wt% Fe	steam activated	high
G	5	0.5 wt.% Fe	highly steam activated	high
H	5	0.5 wt.% Fe	highly chemically activated	high
I	5	0.5 wt.% V	steam activated	high
K	5	0.5 wt.% V	highly steam activated	high
L	5	0.5 wt% V	highly chemically activated	high

      As promoter elements to boost catalytic activity for the hydrogenation of 2-nitropropane we chose iron and vanadium. The idea to do so was also supported by publications on similarly promoted precious metal catalysts for the hydrogenation of nitroarenes (3,9). Here, the activity of palladium as well as platinum on carbon catalysts could be enhanced by the addition of vanadium or iron salts. The explanation for this effect was based on a "catalytic by-pass" mechanism, which accelerates the formation of the amine product from intermediate hydroxylamine species (Figure 2.).

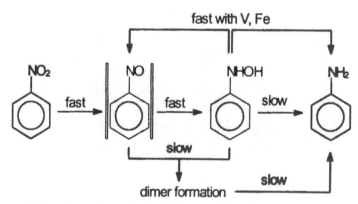

Figure 2: Reaction scheme for the hydrogenation of nitrobenzene in the presence of promoting elements such as iron and vanadium.

In our study 5 wt.% platinum on carbon catalysts were modified with 0.5 wt.% of the promoting element. As given in Table 2, the carbon supports were also varied to study the positive effect of a chemically activated carbon on the catalytic activity. Likewise we also obtained excellent results with both iron and vanadium promoted platinum on carbon catalysts (Figure 3). In contrast to Figure 1, there was no systematic relationship between the catalytic activity for 2-nitropropane hydrogenation and the length of the induction time. For the iron promoted catalysts instantaneous hydrogen consumption could be observed, whereas the more active vanadium promoted catalysts had somewhat longer induction periods. However, the latter still were found to be well below 60 seconds. All platinum on carbon catalysts promoted with either iron or vanadium showed a much higher activity for 2-nitropropane hydrogenation than the reference catalyst A. This was very clear from a) the extremely short induction periods (usually < 30 seconds), b) the low temperature at which the reaction started (usually < 35 °C) and c) the high hydrogen consumption observed for these catalysts (up to 350 ml/min·g). The most active catalysts were the vanadium promoted catalysts K and L, possessing a 6–7 times higher hydrogenation activity for 2-nitropropane compared to the activity level of the standard catalyst. A targeted level of iron promotion gave also very good results as shown with catalysts G and H. In this case the 2-nitropropane hydrogenation activity was approx. 4-5 times higher than the referenced level for catalyst A.

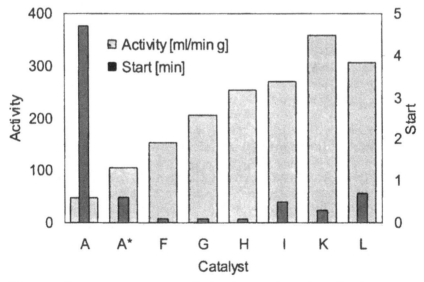

Figure 3: Nitropropane activity and induction time for activated carbon supported iron and vanadium promoted platinum catalysts

It is noteworthy that the synergistic effect observed for the iron promoted catalysts when going from catalyst G to catalyst H by simply changing the support material could not be discovered for the corresponding vanadium promoted catalysts K and L. The positive influence of the chemically activated carbon support is apparently confined to the unpromoted (cf. Figure 1, catalysts B and D) and iron promoted platinum catalysts (cf. Figure 3, catalysts G and H).

Hydrogenation of Aromatic Nitro Compounds with ABMC

Table 3 displays the nitrobenzene (NB) and dinitrotoluene hydrogenation (DNT) data of the ABMC studied here in the order of increasing average particle size (APS) where activated nickel catalyst-1 (ANC-1) has a smaller APS than ANC-5. The low pressure NB hydrogenation activity of the Cr and Fe doped catalysts was, as expected, inversely related to their APS. The only exception to this rule was catalyst ANC-4 that had the same APS as ANC-3, but by far the lowest measured pore volume of the group. Since Cr and Fe promotion increases the surface area of an ABMC (10-11), it is logical that these catalysts are more active than unpromoted ANC-2. Recent studies (12) have shown that the activity of ABMC for the hydrogenation of NB is directly related to the amount of activated metal. Thus, the comparison of NB activities to DNT reaction data suggests that it is the nature of the active site and not its quantity that determines the effectiveness of ABMC for the formation of toluenediamine (TDA). The TDA yields listed here are very impressive for the first three catalysts in table 1, however this drops off dramatically for the slower catalysts. In fact, one can see a strong inverse relationship between the catalyst's reaction time and the yield of TDA. This is not a case of secondary reactions via readsorption; otherwise the performance of ANC-4 with the lowest pore volume would have been better than that of ANC-3 which has the same particle size. The absence of ring hydrogenation and the presence of incompletely hydrogenated monomers strongly suggest that the concentration of hydrogen at the active site controls the yield of TDA. If the strongly adsorbed incompletely hydrogenated intermediates (e.g., hydroxylamines) are not provided hydrogen fast enough then they will be free to dimerize and further polymerize. Interestingly, the desirable performance of ANC-2 brings the need of promoters into question. In addition to increasing surface area, Cr and Fe are also electopositive species that coordinate electron-rich reactants and orient them to the active site (11). Typically this increases catalyst activity, however in the case of DNT and its incompletely hydrogenated species this may tip the balance of adsorption strength to hydrogen availability in the direction of dimerization as seen for ANC-4. If promoters are added in a way that also maintains hydrogen availability, then the performance of the catalyst can be improved as seen by ANC-3 with a TDA yield of 99.95%.

Table 3: Nitrobenzene (NB) and Dinitrotoluene (DNT) Reactions over ABMC

Type	Promoters	Nitrobenzene activity	Dinitrotoluene reaction data				
		$mlH_2/g\ min$	min	%TDA[a]	%RH[b]	%IH[c]	%D/O[d]
ANC1	Cr and Fe	82	45	99.67	0.00	0.11	0.22
ANC2	none	50	55	99.69	0.00	0.13	0.16
ANC3	Cr and Fe	71	33	99.95	0.00	0.00	0.05
ANC4	Cr and Fe	45	216	50.80	0.00	0.00	19.2
ANC5	Cr and Fe	45	85	97.16	0.00	0.22	2.59

[a] percentage of toluenediamine
[b] percentage of ring hydrogenated monomers.
[c] percentage of incompletely hydrogenated monomers.
[d] percentage of dimers and/or oligomers.

Conclusions

All Fe or V promoted Pt/C catalysts had higher 2-nitropropane activities (up to 350 ml/min·g) and lower induction times (usually < 30 seconds) than the reference catalyst. Promotion also reduced the initial temperature in which the reaction started (usually < 35 °C). The V promoted Pt/C catalysts were able to hydrogenate 2-nitropropane 6-7 times faster than the standard while Fe promotion increased activity 4-5 times above the reference. It is interesting to note that changing the support from steam to chemically activated carbon increased catalyst activity in both the unpromoted and Fe promoted cases while this was not observed for V promotion. The reason for this behavior is not yet understood, but it may have to do with the presence of other elements, such as phosphorus, in the carbon support.

While NB hydrogenation on ABMC is sensitive to the availability of activated nickel sites, the hydrogenation of DNT is more sensitive to the type of site. The addition of promoters does not always enhance TDA yield, and some catalysts perform very well without them. If used, then promoters must be added in a way that keeps the balance of adsorption strength to hydrogen availability away from the formation of dimers as was demonstrated by ANC-3 with a TDA yield of 99.95%.

Experimental

The low pressure hydrogenation of 2-nitropropane (8.9 g), as expressed in units of ml H_2/ min·g catalyst, was carried out under atmospheric pressure at 50 °C in a 4:1 MeOH:H_2O solution over 200 mg of Pt/C catalyst. The 5 wt.% Pt/C catalysts used here were either tested as is, or after modification with 0.5 wt.% of the promoting element. Similarly, the low pressure nitrobenzene hydrogenations were carried out over 1.5 grams of ABMC in 110 ml of a 9.1% nitrobenzene ethanolic solution at 25°C and atmospheric pressure. Both low pressure reactions were performed in a baffled glass reactor outfitted with a bubbling stirrer spinning at 2000 rpm. The hydrogenation of 50 grams of DNT in 200 grams of methanol stirred at 400 rpm over a 2% ABMC loading was carried out in a 500 ml autoclave at 50 bar H_2 and 50°C.

References

1. Ullmann, Enzyklopädie der technischen Chemie, Vol A2, 37 (1985).
2. E. Auer, A. Freund, J. Pietsch, and T. Tacke, Applied Catalysis A, 173, 259 (1998).
3. U. Sigrist, P. Baumeister, H.-U. Blaser, and M. Studer, (Marcel Dekker), 75, (Catal. Org. React.), 207 (1998).
4. P. Rylander, Catalytic Hydrogenation in Organic Synthesis, Academic Press, New York 1979.
5. D.E. Butler, J.D. Leonhard, B.W. Caprathe, Y.J. L'Italien, M.R. Pavia, F.M. Hershenson, P.H. Poschel, and J.G. Marriott, J. Med. Chem., 30, 498 (1987).
6. N.V. Andronova, L.D. Volkova, G.D. Zakumbaeva, and D.V. Sokolskii, Izv. Akad. Nauk Kaz. SSR, Ser. Khim., 27, 12 (1977).
7. V. Dubois, G. Jannes, J.L. Dallons, and A. Van Gysel, Chem. Ind., 53, 1 (1994).
8. V. Dubois, G. Jannes, and P. Verhasselt, Stud. Surf. Sci. Catal., 108, 263 (1997).
9. P. Baumeister, H.-U. Blaser, and M. Studer, Catal. Lett., 49, 219 (1997).
10. S.N. Thomas-Pryor, T.A. Manz, Z. Liu, T.A. Koch, S.K. Sengupta, and W.N. Delgass, (Marcel Dekker), 75, (Catal. Org. React.), 195 (1998).
11. P. Gallezot, P.J. Cerino, B. Blanc, G. Flèche, and P. Fuertes, J. Catal., 146, 93 (1994).
12. D.J. Ostgard, A. Freund, M. Berweiler, B. Bender, K. Möbus, and P.Panster, Chemie-Anlagen-Verfahren, 9, 118 (1999).

Hydrogenation of Cinnamaldehyde on Ru-MCM and Ru-beta Catalysts

V. I. Pârvulescu[1], V. Pârvulescu[1], S. Kaliaguine[2], U. Endruschat[3], B. Tesche[3] and H. Bönnemann[3]

[1]University of Bucharest, Department of Chemical Technology and Catalysis, B-dul Regina Elisabeta 4, Bucharest 70346, Romania, Fax.+4013320588
[2]Universite Laval, Departement de Genie Chimique, Ste-Foy, Quebec, Canada
[3]Max-Plank-Institut für Kohlenforschung, 45466 Mulheim/Ruhr, Germany

Abstract

Hydrogenation of cinnamaldehyde on Ru-MCM and Ru-beta catalysts provides different selectivities as a function of the metal dispersion, of the support characteristics and of the nature of the solvent.

Introduction

Catalytic hydrogenation of cinnamaldehyde received a keen interest because of the applications in the field of flavor and fragrances, and as building block in organic synthesis. The carbon-carbon double bond in the α, β- unsaturated aldehyde is more easily hydrogenated than the carbonyl double bond. Therefore, the chemoselectivity is a very important factor in this reaction. Several studies stressed that this reaction can be improved by modifying the surface properties of the catalyst (1). Most reported cases refer to the use of the various promoters such as metal chloride (2), or heteropolyacids (3). But the dispersion and the reduction state of the metal also are important factors in this reaction, both determining changes of the electronic state. Gallezot and Richard (1) underlined the necessity to have low dispersion. Arai and co-workers brought evidence for the fact that a metal surface is more selective for this reaction when it exhibits high Miller index planes (4, 5). In addition to the preparation conditions, the support is another key factor in controlling these properties. Titania (6), zirconia (7), and also microporous zeolites (8, 9) proved to have a good effectiveness. In all these cases, the improvement factor seems to be related with the presence of Lewis acidity, and not eventual shape selectivity effects induced by the microporous texture. The selective hydrogenation of the carbonyl bond can be achieved by the proper activation of dihydrogen into a hydride-proton pair (10). Strong Lewis sites would have a profound impact by the way of generating this.

The aim of this research was to investigate whether a controlled texture could have a positive influence on the selectivity in the hydrogenation of cinnamaldehyde. For such a purpose, it was checked if the deposition of Ru inside the pores of mesoporous MCM could indeed induce some steric hindrances. To achieve this, the deposition of Ru both on MCM-41 and MCM-48 using various precursors was considered. The activity of these catalysts was compared with that of Ru-beta prepared under similar conditions. Ru is one of the active species frequently considered for this reaction, and in such conditions, a comparison with other supports or catalysts can easily be made (7, 11-13).

Results and Discussion

Table 1 summarizes the properties of the investigated catalysts. Deposition of Ru on the various supports provoked a decrease of the surface area, particularly significant for the beta-zeolites. For the mesoporous catalysts, both the shape of the nitrogen adsorption-desorption isotherms (at 77 K) and the pore size distributions were only slightly affected. SAXS spectra were barely modified, confirming these data. The dispersion of Ru on these catalysts is rather small, and mainly depends on the nature of the precursor. However, the deposition of Ru on K-beta zeolites is consistent with a smaller dispersion than those measured for the MCM supports, except for the Ac catalysts which exhibit very close values. The measurement of both the dispersion and the pore size was confirmed by TEM and FTTEM analysis, which in addition showed an unmodified topography of the support after Ru deposition. It was inferred from TEM analysis that most of the Ru was inside the pores of MCM. The XPS Ru/Si atomic ratios also confirmed the results of the Ru dispersion. The same analysis carried out on Ru-beta catalysts showed that on this support, Ru seems to be mainly dispersed on the external surface. However, for the Ac catalysts, an external deposition of Ru occurred irrespective of the support. Information concerning the oxidation state was obtained both from Mössbauer (lower Ru^0/Ru^{IV} ratios) and XPS measurements (shift of the binding energy to higher values). Both methods confirmed that, except for the Ac catalysts, the deposition of Ru on beta-zeolite leads to a more oxidized Ru than on MCM ones. The oxidation state was also influenced by the nature of the precursor (Table 1).

Figure 1 presents the variation of the selectivity vs. conversion for the reaction performed in the presence of isopropyl alcohol as a solvent. Except for the cinnamyl and hydrocinnamyl alcohol, the analysis of the reaction products indicated the presence of hydrocinnamaldehyde diisopropyl acetal and hydrocinnamaldehyde, and small contents of benzoic aldehyde and of cinnamic acid. The same hemiacetal was also identified when the other alcohols were used as solvents, but in smaller amounts than in ethanol. The best conversions were obtained on Ru-beta catalysts. The data presented in Figure 1 show that also the selectivity was higher on the same catalysts. Concerning the nature of the precursor, its influence was found to be different for the two kinds of

Table 1. Characteristics of the catalysts

Property	Catalyst								
	NH-MCM-48	NH-MCM-41	NH-beta	HCl-MCM-48	HCl-MCM-41	HCl-beta	Ac-MCM-48	Ac-MCM-41	Ac-beta
BET area, m^2g^{-1}	621	702	168	665	865	148	887	940	132
Pore size, nm	3.8	2.5	0.72	3.8	2.5	0.69	3.8	2.5	0.64
Ru^o/Ru^{IV},[a]	0.25	0.26	0.21	0.51	0.35	0.25	0.21	0.20	0.14
H_2 uptake, $cm^3\,g^{-1}$	0.52	0.52	0.33	0.35	0.34	0.35	0.22	0.16	0.36
Dispersion, %	5.91	5.86	3.72	3.94	3.81	2.95	2.51	1.80	4.12
Metallic surface area m^2g^{-1}	0.86	0.86	0.54	0.58	0.56	0.58	0.37	0.26	0.31
Ru_{3p} energy, eV	462.0	461.9	462.6	462.0	462.1	462.7	462.9	463.0	463.0
Ru/Si XPS ratio x 10^2	1.11	1.41	1.12	1.51	2.56	1.21	8.97	10.3	3.21

[a] Determined from Mössbauer data

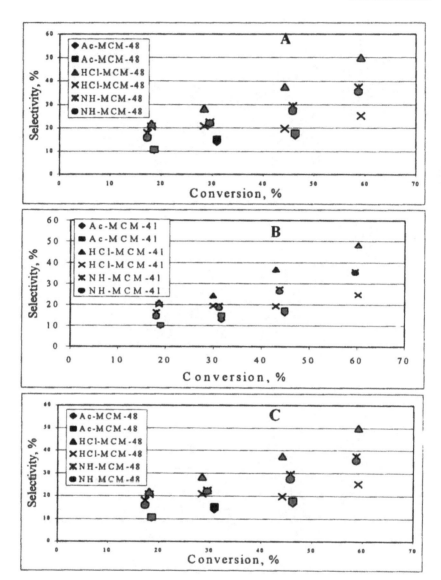

Figure 1. Variation of the selectivity vs. conversion: A: Ru-beta,
B: Ru-MCM-41, C: Ru-MCM-48 (♦,▲,✳- cinnamyl alcohol,
■,X,●-hydrocinnamyl alcohol (80 °C, 40 atm., isopropyl alcohol).

supports. For Ru-beta, the best results were obtained for Ac-beta, namely, the one with the poorer dispersion. On the contrary, for the MCM catalysts, the best results were observed for the HCl catalysts. However, it should be noted that for Ac catalysts, the dispersion was nearly equal. The type of MCM structure only had a small effect on the reaction. MCM-48 was only a little more efficient.

Based on these determinations, it is quite difficult to accept that the hydrogenation of cinnamladehyde is controlled by shape selectivity, as Gallezot et al. suggested (9). For Ru-beta catalysts, most of the pores are filled. Better conditions to exhibit a shape selectivity control have MCM catalysts for which the pores are still accessible. Therefore, we suppose that the behavior of beta catalysts is more probably related with the Lewis acid sites surrounding Ru, as suggested by other authors (2, 7, 10). The samples prepared from Ru(acac)$_3$ perhaps give the best opportunity to understand the activity data. A fine dispersion of Ru is produced on beta-zeolite that cannot be easily reduced to the metal. These ions are close to Al^{3+} ion since the most acidic protons reside in this site. The proximity of the strong Lewis sites in combination with few Ru metal atoms results in an environment that favors heterolytic scission of dihydrogen and thus favors selective hydrogenation of the carbonyl. When, Ru(acac)$_3$ is contacted with silaceous MCM, there are no strong acid sites, so fewer Ru ions are deposited on the surface, atomically. Multiple layers of Ru(acac)$_3$ form and collapse upon heating to make large crystallites of Ru^0 upon heating in hydrogen. Moreover, the absence of Al^{3+} ion, means that few Lewis sites will be present. Without these Lewis sites, the dihydrogen will be activated by an homolytic mechanism over metallic Ru so that saturation of the olefin becomes the predominant pathway.

In conclusion, we believe that on the Ru-beta catalysts, the reaction occurs mainly via activation of dihydrogen into a hydride-proton pair which will attack preferentially the CO group leading to better selectivities, whereas on the MCM catalysts, the reaction occurs inside the pores, and the selectivity could be considered, indeed, as the consequence of some steric hindrances.

Experimental

The MCM-41 support (1264 m^2 g^{-1}) was obtained by hydrothermal crystallization at 100 °C for 24 h of a gel having the composition: SiO$_2$ - 0.085 Na$_2$O - 0.16CTAB - 63H$_2$O. In the case of the MCM-48 (1204 m^2 g^{-1}), the gel had the composition: SiO$_2$ - 0.25 Na$_2$O - 0.65CTAC - 62H$_2$O, and the synthesis was carried out at 100°C for 96 h. The solid was then post-treated at 110 °C for 27 days, followed by a second post treatment for 21 days at 110 °C. The samples were dried in air at 100 °C, and then calcined for 6 h at 500 – 550 °C. Beta zeolite (442 m^2 g^{-1}) was from PQ Corporation (ZB25). Deposition of Ru on these supports was carried out from 0.4 M solutions of high purity $RuCl_3$, $[Ru(NH_3)_6]Cl_2$ or $Ru(acac)_3$. Samples with 5 wt % Ru were prepared. They were denoted as HCl, NH, and Ac according to the used precursor. The above catalysts were dried overnight in a vacuum stove at room temperature and then reduced in flowing hydrogen (50 ml min^{-1}) at 300 °C for 6 h, with a heating rate of 0.2 °C min^{-1}. All the catalysts were characterized using several techniques: adsorption-desorption of N_2, H_2-chemisorption, XRD, SAXS, XPS, Mossbauer spectroscopy, and TEM coupled with Fourier transform analysis (FTTEM). In this way, information on the texture, reduction state and metal dispersion was obtained. The details about the conditions in which these samples were analyzed have been previously described (14, 15). Hydrogenation of the substrate was carried out in a 50 ml stainless steel stirred autoclave (stirring speed 1600 rpm) under 20 - 40 bar hydrogen pressure at 80-100 °C, using a solution of 1.1 ml cinnamaldehyde in 5 ml anhydrous solvent (ethanol, isopropanol, butanol, acetonitrile or THF). Before the reaction, the catalyst (50 mg) was pretreated in hydrogen in the autoclave and then the vessel was purged three times with hydrogen. Analysis of the reaction products was carried out in a Hewlett-Packard GC/MS equipment using a 28 m Rtx-1 column.

References

1. P. Gallezot and D. Richard, *Catal. Rev.- Sci. Eng.*, **40**, 81 (1998).
2. S. Galvagno, A. Donato, G. Nerl, and R. Pietropaolo, *J. Mol. Catal.*, **49**, 223 (1989).
3. B. Liu, L. Lu, T. Cai, and Iwatani, K., *Appl. Catal. A: General*, **180**, 105 (1999).
4. M. Arai, K. Usui, and Y. Nishiyama, *Chem. Commun.*, 1853 (1993).
5. M. Arai, H. Takahashi, M. Shirai, Y. Nishiyama, and T. Ebina, *Appl. Catal. A: General*, 176, 229 (1999).
6. P. Gallezot, A. Giroir-Fendler, and D. Richard, *Catal. Lett.*, **5**, 169 (1990). B. Coq, P. S. Kumbhar, C. Moreau, P. Moreau, and F. Figueras, *J. Phys. Chem.*, **98**, 10180 (1994).
7. B. Coq, P. S. Kumbhar, C. Moreau, P. Moreau, and F. Figueras, J. Phys. Chem., 98, 10180 (1994).
8. D. G. Blackmond, R. Oukaci, B. Blanc, and P. Gallezot, *J. Catal.*, **131**, 401 (1991).
9. P. Gallezot, B. Blanc, D. Barthomeuf, and M. I. Pais da Silva, *Stud. Surf. Sci. Catal.*, **84**, 1433 (1994).
10. J. Halpern, T. Okamoto, and A. Zakhariev, *J. Mol. Catal.*, **3**, 65 (1977).
11. L. Mercadante, G. Neri, C. Milone, A. Donato, and S. Galvagno, *J. Mol. Catal. A: Chemical*, **105**, 93 (1996).
12. T. Braun, M. Wohlers, T. Belz, and R. Schlögl, *Catal. Lett.*, **43**, 175, 1997.
13. G. Neri, L. Bonaccorsi, L. Mercadante, and S. Galvagno, *Ind. Eng. Chem. Res.*, **36**, 3554, 1997.
14. V. I. Pârvulescu, S. Coman, P. Palade, D. Macovei, C. M. Teodorescu, G. Filoti, R. Molina, G. Poncelet, and F. E. Wagner, *Appl. Surf Sci.*, **141**, 164 (1999).
15. S. Coman, F. Cocu, V. I. Pârvulescu, B. Tesche, H. Bönnemann, J. F. Roux, S. Kaliaguine, and P. A. Jacobs, *J.Mol.Catal.*, **146**, 247 (1999).

Novel Preparation of 5α-Dihydroethisterone from Androst-4-ene-3,17-dione

Mike G. Scaros, Peter K. Yonan, Kalidas Paul, John Schulz and Jae C. Park

Department of Chemical Sciences, Searle/Monsanto Skokie, Illinois, USA

Abstract

Selective hydrogenations have been key reactions in the preparation of fine chemicals for many decades. The success in the development of an economically favorable route for the preparation of 5α-dihydroethisterone from androst-4-ene-3,17-dione hinged on selective hydrogenation of 3-ethoxyandrosta-3,5-dien-17-one. This paper addresses the importance of the solvent, catalyst, catalyst loading, hydrogen pressure and temperature in obtaining both the required selectivity and desired high yield.

Introduction

In response to a business opportunity, Searle initiated an investigation for an inexpensive process for the preparation of 17-hydroxy-5α,17α-pregn-20-yn-3-one (5α-dihydroethisterone). Three potential processes were investigated. The sequence involving the stereoselective hydrogenation of 3-ethoxyandrosta-3,5-dien-17-one at the A/B steroid junction was the most economically favorable route. This process consists of four steps involving three critical reactions:

1. High yield/high purity in preparation of 3-ethoxyandrosta-3,5-dien-17-one
2. Stereoselective hydrogenation to obtain 3-ethoxy-5α-androst-3-en-17-one
3. Ethynylation and isolation in high purity of the desired 17-hydroxy-5α-dihydroethisterone.

Results and Discussion

Three potential processes were explored in an attempt to develop an efficient process. The first process as outlined in **Scheme I** (enolacetate process) had the disadvantage of high cost and six steps. The second process as outlined in **Scheme II** (dehydroepiandrosterone process) involved an expensive starting material (dehydroepiandrosterone) that had a questionable availability. The

Scheme I (Enolacetate process)

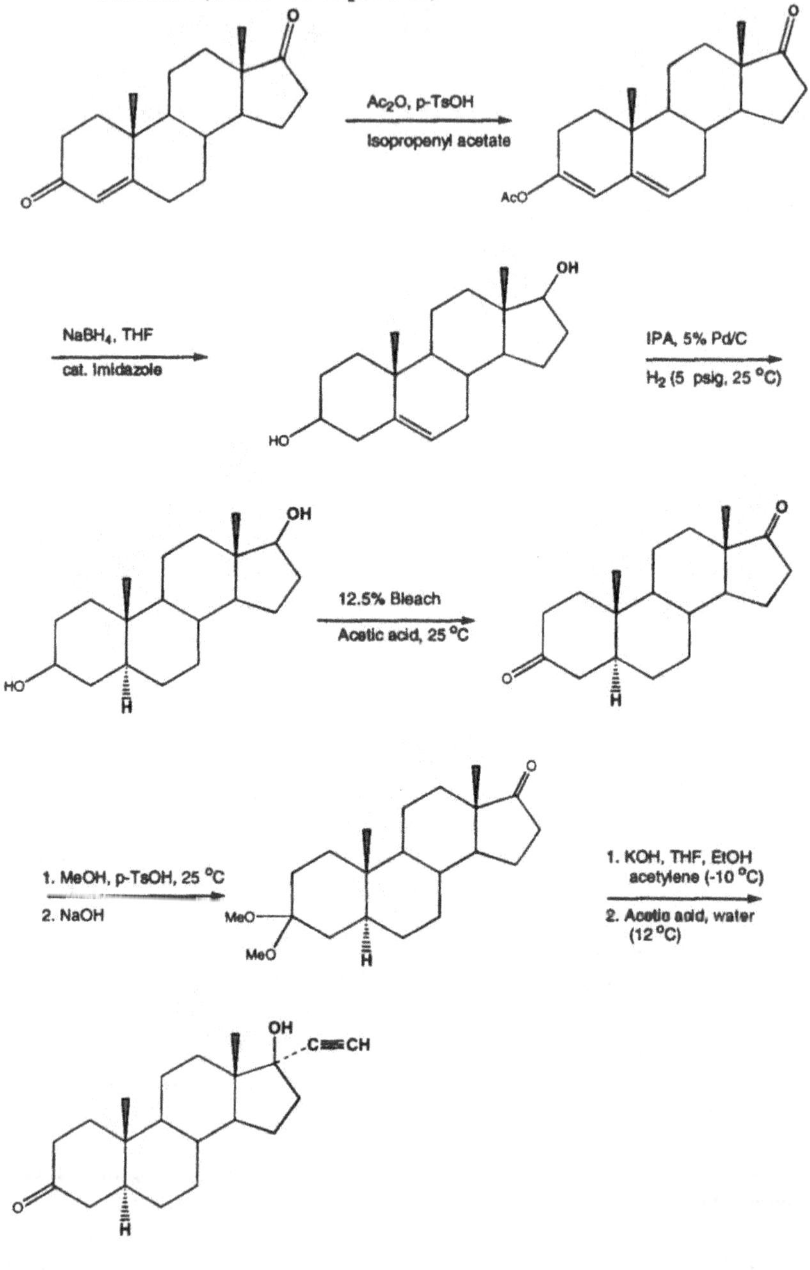
O
Ac2O, p-TsOH
Isopropenyl acetate
O
AcO
NaBH4, THF
cat. imidazole
OH
HO
IPA, 5% Pd/C
H2 (5 psig, 25 °C)
OH
HO
H
12.5% Bleach
Acetic acid, 25 °C
O
O
H
1. MeOH, p-TsOH, 25 °C
2. NaOH
O
MeO
MeO
H
1. KOH, THF, EtOH
acetylene (-10 °C)
2. Acetic acid, water
(12 °C)
OH
C≡CH
O
H

5α-Dihydroethisterone

process of choice for the preparation of 5α-dihydroethisterone was the one outlined in **Scheme III** (dienol ether process) which involved a secure inexpensive source of starting material (androstenedione) and had only 4 steps in an approximate 50% overall weight yield. For the purpose of this paper, only the "dienol ether process" will be discussed in detail.

The **first step** was preparation of the dienol ether of androstenedione using triethylorthoformate and p-toluene sulfonic acid monohydrate in ethanol. The procedure involved the reverse addition of the water scavenger triethylorthoformate to a slurry of androstenedione and p-toluene sulfonic acid monohydrate in ethanol (1,2). This minimized the formation of the 17-ketal and allowed for an 85-90% isolation of the desired product (see experimental).

The **second step,** the key to the success of the "dienol ether process", was the selective hydrogenation of the Δ-5 double bond of 3-ethoxyandrosta-3,5-dien-17-one to give predominately the 3-ethoxy-5α-androst-3-en-17-one (3). The best isomeric ratio was obtained when 2% Pd/SrCO₃ (reduced), 4 volumes of tetrahydrofuran and a 5% catalyst loading at 25 °C/5 psig was employed (see Table 1 and experimental). When a high catalyst loading was employed, the steroselectivity was maintained; however, a low yield was obtained. The low yield indicated overreduction to the 3-ethoxy compound where the 3,4 double bond is reduced and which is very soluble and nonhydrolizable. In order to ensure complete reduction, all hydrogenations were run for 1 h past the point where the hydrogen uptake ceased. Prolonged hydrogenation at the 5% catalyst loading did not result in overreduction compared to the 10 and 20% loading. In a few cases where a number of other Pd/C catalysts were investigated, the results indicated overreduction and a less than desired α/β ratio at the A/B steroid junction (see Table I).

Table I Catalyst Studies

Catalyst (catalyst loading, wt %)	Solvent	Conc. of Substrate (%)	Temp.°C/Press (psig)	Ratio 5α/5β	% Yield 5α-dihydro-ethisterone
2% Pd/SrCO₃[a] (20)	3A-EtOH[b]	3.3	25/60	5 spots seen on TLC	77.8
2% Pd/SrCO₃[a] (20)	THF	20	25/5	97/3	88.9
2% Pd/SrCO₃[a] (10)	THF	20	25/5	95/5	80.0
2% Pd/SrCO₃[a] (5)	THF	25	25/5	97/3	97.6
2% Pd/SrCO₃[a] (2)	THF	25	25/5	97/3	97.8

[a] The catalyst was obtained from Engelhard Industries (C-10024), [b] Denatured with ~5% methanol.

Scheme II (Dehydroepiandrosterone Process)

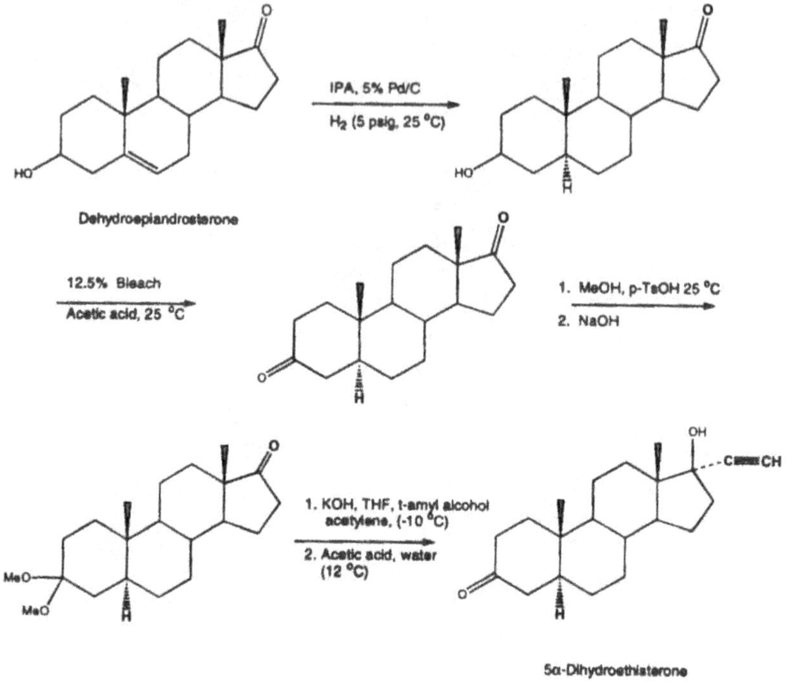

Dehydroepiandrosterone

5α-Dihydroethisterone

Scheme III (Dienol ether Process)

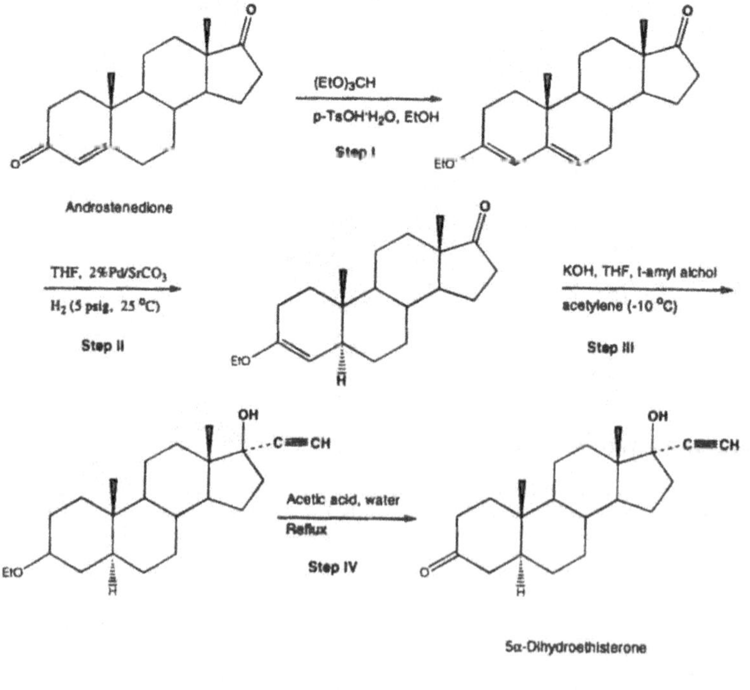

Androstenedione

5α-Dihydroethisterone

The **third and fourth steps** involved the standard procedure of generating the potassium salt in tetrahydrofuran and t-amyl alcohol. The salt was used to generate the potassium salt of acetylene which will react with the 17-carbonyl group of the steroid. The work-up involved hydrolysis of the enol ether using aqueous acetic acid followed by layer separation. This process gave consistently good quality material with high yields depending on the quality of starting materials (see experimental).

Experimental

Step I (Preparation of 3-ethoxyandrostra-3,5-dien-17-one)

To a three-necked, round-bottom flask equipped with a mechanical stirrer, thermometer and addition funnel was charged, under a nitrogen atmosphere, 28.6 g of 4-androsten-3,17-dione (0.10 mol), 0.47 g (0.0025 mol) of para-toluenesulfonic acid monohydrate and 171.6 mL of 2B ethanol. The resultant slurry was cooled to 5 °C and 22.2 g (0.15mol) of triethylorthoformate was added dropwise over a 15-20 minute period while maintaining the temperature between 0 -10 °C. Then 1.2 mL pyridine in 1.2 mL water was added and the mixture was warmed to 40 °C. The reaction mixture was maintained at 40 °C until the hydrolysis of the ketal was completed as indicated by TLC. Then 3 mL of triethylamine was added and the reaction mixture was heated to reflux. Additional 2B ethanol in 5 mL portions were added until a clear solution was obtained. The solution was cooled to 5 °C and then stirred for 30 minutes at a temperature between 0-5 °C. The solid was isolated by filtration and washed twice with 25 mL of 1% triethylamine in 2B ethanol solution at 5 °C. The solid was air dried to give 25.7 g (81.8%) of the title compound, 3-ethoxyandrosta-3,5-dien-17-one.

Step II (Preparation of 3-ethoxy-5α-androst-3-en-17-one)

A two-liter Parr Bottle under a nitrogen atmosphere was charged with 15.7 g of 2% palladium-on-strontium carbonate, 314.0 g (1.0 mol) of 3-ethoxyandrosta-3,5-dien-17-one (product from Step I), 1256 mL of tetrahydrofuran and 6.3 mL of triethylamine. The bottle was fitted to a Parr Shaker and hydrogenated at a constant pressure of 5 psig of hydrogen at room temperature until the reaction was complete as indicated by hydrogen uptake from an accumulator. The hydrogen was vented, the vessel was purged with nitrogen and the palladium catalyst was removed by filtration. The catalyst was washed and the wash was combined with the filtrate. The combined filtrate contained 3-ethoxy-5α-androst-3-en-17-one which was used in the procedure described in Step III.

Step III & IV (Preparation of 5α-dihydroethisterone)

To a Fisher/Porter bottle (reactor) equipped with a magnetic stirrer was charged under a nitrogen atmosphere 24.7 g of potassium hydroxide, 0.16 g of ethylenediaminotetraacetic acid, 150 mL of dry tetrahydrofuran and 9.78 g of t-amyl alcohol. The reaction mixture was heated to 40 °C with stirring and maintained at 40 °C for 35-40 minutes. The reaction mixture was then cooled to -10 °C and a solution of 25.0 g of 3-ethoxy-5α-androst 3-en-17-one in 100 mL of tetrahydrofuran was added dropwise at -10 °C over approximately 15 minutes. Acetylene was allowed to flow through the vented reactor containing the reaction mixture for approximately 15 minutes. The vent valve of the reactor was closed and the reaction was maintained at 2.5-3.5 psig and a temperature of -10 °C until the acetylene uptake had stopped. The reaction pressure was stabilized at 3-3.5 psig and the reaction was continued for an additional hour. The Fisher/Porter bottle was vented and sparged with nitrogen for 2-3 minutes. The cold bath was removed and 57.2 mL of water was added to the reaction at 10-20 °C. The reaction was heated to 50 °C with stirring and maintained at 50 °C for 30 minutes. Stirring was discontinued and the layers were allowed to separate. The aqueous layer was extracted with a solution of 15 mL tetrahydrofuran in 32 mL of water at 50 °C. The organic layers were combined and concentrated by distillation at reduced pressure to a final pot temperature of 35 °C. A solution of 200 mL of acetic acid in 200 mL of H_2O was added to the distillation residue. The slurry was then heated to 95 °C and held at 95-100 °C for 1 h. After the slurry was cooled to 25 °C, it was filtered and washed three times with 50 mL water and twice with 25 mL tetrahydrofuran. The product was dried in a forced air dryer at 60 °C.

Conclusion

In summary, we have demonstrated a cost effective synthesis of 5α-Dihydroethisterone employing the dienol ether process (Scheme III). This synthesis, based on the stereospecific hydrogenation, has the following advantages;

*   Secure source of starting material.
*   A four-step synthesis that involves proven chemistry.
*   A process that is cost effective.
*   5α-Dihydroethisterone which met the customer's specifications.

Acknowledgments

The authors would like to thank Srin Babu, Rosemary Bergeron, John Medich, E. Sy, Cara Weyker and Joe Wieczorek for their assistance and to Searle/Monsanto for permission to publish this work.

References

1. H. H. Inhoffen, *Ber.*, **84**, 361 (1951)
2. E. G. Meek, J. H. Turnbull, and W. Wilson, *J. Chem. Soc.*, 811 (1953)
3. R. L. Augustine in *"Heterogeneous Catalysis for the Synthetic Chemist"*, Marcel Dekker, New York, 1996, p 352.
4. H. A. Stansburg Jr. And W. R. Proops, *J. Org. Chem.*, **27**, 279 (1962)

Selective Epoxidation of Allylic Alcohols with Amine-Modified Titania-Silica Aerogels

Marco Dusi, Carsten Beck, Tamas Mallat, and Alfons Baiker

Laboratory of Technical Chemistry, Swiss Federal Institute of Technology,
ETH-Zentrum, CH-8092 Zürich, Switzerland

Abstract

Epoxidation of 2-cyclohexen-1-ol with a 20 wt% TiO_2 - 80 wt% SiO_2 aerogel and tert-butylhydroperoxide was studied. Rapid consumption of 2-cyclohexen-1-ol was due to acid-catalyzed, non-oxidative side reactions of the allylic alcohol, the epoxide representing a product in only low amount. Epoxide selectivities related to allylic alcohol, $S_{C=C}$, could be improved remarkably and acid-catalyzed side reactions suppressed by addition of amines. The best modifier was tritylamine (Ph_3CNH_2); 1 mol% amine (related to allylic alcohol) improved the epoxide selectivity $S_{C=C}$ from 22 % to 81 %, while maintaining the high rate of epoxide formation. It was found that amine additives have a much stronger positive effect on selectivities than application of lower reaction temperature and the use of a weakly basic solvent. The method of amine addition was also applied to the epoxidation of the bishomoallylic alcohol 3-cyclohexen-1-methanol.

Introduction

Microporous (1, 2) or mesoporous (3-5) sol-gel TiO_2-SiO_2 mixed oxides represent valuable alternatives to crystalline Ti-containing materials (6) and titania-on-silica type binary oxides (7, 8) as epoxidation catalysts. Highly dispersed titania in the silica matrix, mesoporous structure and high surface area are the properties of a 20 wt% TiO_2 containing aerogel obtained by this method (3). This catalyst showed outstanding activity and selectivity in the epoxidation of cyclic olefins with alkylhydroperoxides (5, 9) and was also applied in the epoxidation of allylic alcohols and unsaturated carbonyl compounds (10-12).

Generally, in the oxidation of functionalized olefins epoxide selectivities are often limited by various consecutive and parallel side reactions, such as the solvolysis of the epoxide (13), oxidation of a hydroxyl functional group (14) or various isomerization reactions (11). The difficulties in controlling the acidity of titania-silica mixed oxides are due to the presence of silanol groups (Brønsted sites) and Ti sites (Lewis acidic centers) (15). The latter are responsible for the epoxidation activity, but both can catalyze side reactions. Suppression of Brønsted acidity of the catalyst by neutralization by alkali metal exchange can

315

improve selectivity, but this treatment often leads to a loss of activity (11, 13, 16). The simple addition of inorganic bases to the reaction mixture represents a recent development in this direction. Even weak bases, such as NaHCO₃, which are poorly soluble in the apolar reaction medium, had a remarkable positive effect on selectivity in the epoxidation of β-isophorone (11) and some allylic alcohols (12).

N-bases (urea, isoquinoline, Bu₄NOH), when used as additives, have also been shown to suppress the undesired hydration of the epoxide during oxidation of cyclohexene with a TiO₂-SiO₂ mixed oxide and H₂O₂ (4). The enhanced selectivity and reduced activity in the presence of N-bases was explained by the effect on Lewis acidity of the catalyst.

We have previously shown that inorganic bases (zeolite 4Å, NaHCO₃) as additives provide high selectivities in the epoxidation of primary and secondary allylic alcohols (12). Using the same modifiers, epoxidation of cyclohexenols (and especially alkyl-substituted cyclohexenols) yielded the epoxides as minor products. Here we present a study on the influence of amines on the epoxidation of 2-cyclohexen-1-ol (1) and 3-cyclohexen-1-methanol (4) (Scheme 1).

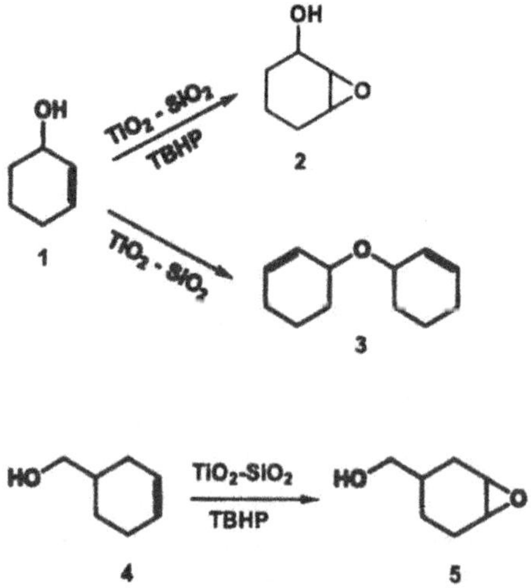

Scheme 1

Experimental Section

A titania-silica aerogel containing 20 wt% TiO_2 was prepared using a sol-gel method (3). The solvent was semicontinuously extracted with supercritical CO_2 at 313 K and 24 MPa. The raw aerogel was calcined in air at 673 K for 5 h. Details of the synthesis and characterization of the aerogel have been published earlier (3, 9, 17, 18).

In a standard epoxidation reaction, 0.1 g catalyst was dried by heating to 473 K in Ar for 1 h. To the *in situ* dried catalyst were then added solvent (toluene) and 0.5 ml internal standard (dodecane). This mixture was heated to 363 K. 10 mmol olefin was added and the reaction started by the addition of 2.5 mmol *tert*-butylhydroperoxide (TBHP, as a ca. 5.5 M solution in nonane) ([olefin]$_0$: [TBHP]$_0$ = 4 : 1). The total reaction volume was 10 ml. In most reactions some amine was also added (the amount used is given in mol% related to the olefin and is indicated in the Tables and Figures). The products were analyzed by GC, and an internal standard method was used for quantitative analysis. Hydroperoxide conversion was determined by iodometric titration. Two different types of selectivities are used: Epoxide selectivity (related to the reactant converted): $S_{C=C}$ = 100 • [epoxide] / ([reactant]$_0$ - [reactant]), and peroxide selectivity (epoxide related to the peroxide consumed): S_{TBHP} = 100 • [epoxide] / ([TBHP]$_0$ - [TBHP]. The olefin conversion is related to the initial TBHP concentration.

Table 1 Epoxidation of 2-cyclohexen-1-ol (**1**) with N,N-dimethylbutylamine as modifier.[a]

Amine concentration[b] (mol%)	$t_{50\%}$[c] (min)	$S_{C=C}$[d] (%)	S_{TBHP}[d] (%)	Diastereomeric ratio[d] cis : trans
—	2	22	n. d.	65 : 35
0.1	4	30	30	75 : 25
0.5	18	62	52	72 : 28
1	56	75	51	68 : 32
20	60[e]	8.3[f]	2.5[f]	n. d.

[a] Standard reaction conditions; T = 363 K. [b] Amine concentration is given in mol% related to the amount of **1**. [c] Time for 50 % conversion of **1**. [d] At $t_{50\%}$. [e] Time for 22 % conversion of **1**. [f] At 22 % conversion of **1**.

Results

The influence of amine concentration on 2-cyclohexen-1-ol (**1**) epoxidation is shown in Figure 1 and Table 1 using N,N-dimethylbutylamine as an example. Under standard reaction conditions, the allylic alcohol was readily converted non-oxidatively to the ether **3** and oligomers, yielding the epoxide in only low yield. Addition of N,N-dimethylbutylamine decreased the rate of cyclohexenol conversion and increased the selectivity $S_{C=C}$. With respect to $S_{C=C}$ and epoxide yield, 0.5 and 1 mol% amine were the optimum amounts. Interestingly, 1 mol% amine increased the rate of epoxide formation (Figure 1) and decreased dramatically the rate of olefin conversion ($t_{50\%}$, Table 1). Apparently, this amine can selectively block the silanol groups without reducing the activity of the Ti sites. Concerning the diastereoselectivity, only a minor shift towards the *cis*-diastereomer was observed in presence of amine.

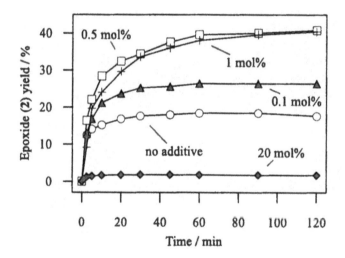

Figure 1   Epoxide (**2**) yield in dependence of the amine concentration. Standard reaction conditions and N,N-dimethylbutylamine as additive; T = 363 K.

The non-oxidative conversion of **1** to dicyclohex-2-enyl ether (**3**) and oligomers are the major side reactions in cyclohexenol (**1**) epoxidation. As an example, the formation of **3** is illustrated in Figs. 2/a-b. This reaction is catalyzed by the acidic sites on the catalyst ("no additive"). Note that the yield of **3** is related to **1** and not to TBHP, since this reaction does not require any oxidant. 0.5 or 1 mol% N,N-dimethylbutylamine neutralized some acidic sites and thus efficiently suppressed ether formation (Figure 2a), while a reasonably high catalyst activity was still maintained (see also Figure 1). It is evident from Figure 2b that the absence of TBHP even promoted the formation of 3, when compared to the standard reaction (with TBHP). Addition of tritylamine (Ph_3CNH_2) completely stopped this reaction, whereas the inorganic base $NaHCO_3$ did not affect ether formation.

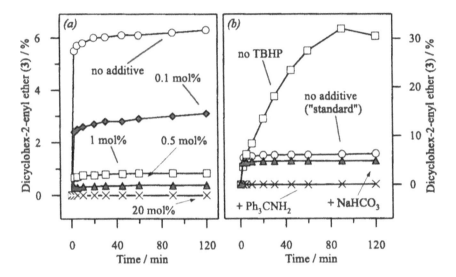

Figure 2 Influence of amine concentration *(a)* and of TBHP and organic and inorganic bases *(b)* on the formation of dicyclohex-2-enyl ether (3) by non-oxidative conversion of 2-cyclohexen-1-ol (1) (yield related to 1). Standard reaction conditions; T = 363 K. Additives: *(a)* N,N-dimethylbutylamine; *(b)* Ph_3CNH_2 (0.1 mmol), $NaHCO_3$ (2 mmol). Note that only a small fraction of $NaHCO_3$ dissolved in the medium.

Various other amines with different structure and basic strength were also tested for their potential as selective modifiers in the epoxidation of 1 (Table 2). Although the optimum conditions and the oxidation stability of the amine certainly varies for each amine, some conclusions can be drawn. At 50 % conversion of 1, most amines provided higher epoxide selectivities ($S_{C=C}$) and consequently higher epoxide yields (conversion $\cdot$ $S_{C=C}$) when compared to the standard reaction in the absence of any modifier. Considering the set of modifiers aniline, diphenyl- and triphenylamine, the decreasing selectivity $S_{C=C}$ and the increasing catalyst activity (characterized by $t_{50\%}$), which is indicative for side reactions, may be a consequence of decreasing basic strength and increasing steric demand. On one hand, the basicity of e.g. Ph_3N is too low to neutralize the acidic sites responsible for non-oxidative side reactions of 1, on the other hand, approach of the nitrogen atom to these sites is considerably hindered by the three bulky phenyl substituents. For benzylamine, diphenylmethyl- and triphenylmethylamine one would expect similar correlations since decreasing basicity comes along with increasing bulkiness. However, Ph_3CNH_2 provided the highest selec-

Table 2 2-Cyclohexen-1-ol (1) epoxidation with different amines as modifiers.[a]

Modifier	pK_a[b]	$t_{50\%}$[c] (min)	$S_{C=C}$[d] (%)	S_{TBHP}[d] (%)	Diastereo-meric ratio[d] cis : trans
—	—	2	22	n. d.	65 : 35
$BuNMe_2$	10.2	56	75	51	68 : 32
$PhNH_2$	4.6	9	32	31	66 : 34
Ph_2NH	0.9	2	25	35	68 : 32
Ph_3N	-5	1	15	43	67 : 33
$PhCH_2NH_2$	9.4	20	65	46	73 : 27
Ph_2CHNH_2	7.3[e]	4	53	51	73 : 27
Ph_3CNH_2	6.2[e]	8	81	59	71 : 29
$PhCH(CH_3)NH_2$	9.4	31	70	50	74 : 26
Quinuclidine	10.9	27[f]	56[g]	31[g]	65 : 35[g]
$(CH_3)_2N(CH_2)_2N(CH_3)_2$[h]	9.1 (5.6)	6	52	45	75 : 25
$CH_3NH(CH_2)_2NHCH_3$[h]	10.1 (6.8)	16	38	35	76 : 24

[a] Standard reaction conditions; 1 mol% amine (based on 1); T = 363 K. [b] In H_2O at 298 K. [c] Time for 50 % conversion of 1. [d] At $t_{50\%}$. [e] In dioxane : H_2O (6 : 4). [f] Time for 25 % conversion of 1. [g] At $t_{25\%}$. [h] 0.5 mol% amine was used.

tivity $S_{C=C}$ of all amines tested. In this set of amines, the differences in basic strength are smaller and the nitrogen is separated from the bulky phenyl groups by one carbon atom. Generally, more strongly basic amines ($pK_a > 9$) reduced the reaction rate considerably (as indicated by $t_{50\%}$), whereas weakly basic amines had a minor influence on the activity. A small change in the diastereomeric ratio in favor of the *cis*-isomer was observed in presence of amines, although no correlation with basicity and/or size of the amines could be established. During the reaction it was observed that the *trans*-epoxide was less stable than the *cis*-isomer.

Lower reaction temperature generally improves selectivity, but reduces also the reaction rate. An alternative for the suppression of acid-catalyzed side reactions is the application of a weakly basic solvent (16). The influence of this two parameters is shown in Table 3. It is evident that the positive influence of basic additive on the selectivity is much stronger than the lower temperature (333 K). For example, using dimethylbutylamine additive, the selectivity $S_{C=C}$ obtained at 363 K (75 %) was as high as $S_{C=C}$ at 333 K (77 %), but at much higher conversion of 1 (50 % at 363 K vs. 20 % at 333 K). Similarly, application of the weakly basic solvent butyronitrile was not sufficient for the suppression of side reactions, and again amine addition had a much stronger positive effect than the basic solvent alone.

Table 3 Influence of temperature and solvent on the epoxidation of 1.[a]

Solvent	Amine conc. (mol%)	T (K)	$t_{50\%}$[b] (min)	$S_{C=C}$[c] (%)	S_{TBHP}[c] (%)	Diastereomeric ratio[c] (*cis* : *trans*)
Toluene	—	333	8[d]	44[e]	41[e]	60 : 40[e]
Toluene	1	333	120[d]	77[e]	36[e]	59 : 41[e]
Toluene	—	363	2	22	n. d.	65 : 35
Toluene	1	363	56	75	51	68 : 32
Butyronitrile	—	363	2	28	36	79 : 21
Butyronitrile	1	363	114	63	40	72 : 28

[a] Standard reaction conditions; N,N-dimethylbutylamine as additive. [b] Time for 50 % conversion of 1. [c] At $t_{50\%}$. [d] Time for 20 % conversion of 1. [e] At $t_{20\%}$.

The results obtained in the epoxidation of 3-cyclohexen-1-methanol (4) are collected in Table 4. This bishomoallylic alcohol possesses a more electrophilic double bond than 2-cyclohexen-1-ol (1) and is thus expected to be more reactive towards epoxidation. On the other hand, assuming a coordination of the alcoholic functional group to the metal of the Ti-alkylhydroperoxy complex, the double bond may be too far away from the peroxidic oxygen. Under standard reaction conditions, the reactivity of 4 is lower than that of 1 and a higher $S_{C=C}$ is achieved. Apparently, 3-cyclohexen-1-methanol (4) is less prone to acid-catalyzed side reactions. Amine addition also had a positive influence on the selectivities, but the effect of modifiers is by far less pronounced than in the epoxidation of 2-cyclohexen-1-ol (1). For example, $S_{C=C}$ was increased from 36 % to 49 % in the epoxidation of 4, but from 22 % to 81 % in the epoxidation of 1. An experiment with recycled catalyst was performed under standard reaction conditions to test for catalyst deactivation (entry 2 in Table 4). The activity of the recycled catalyst was lower (indicated by $t_{50\%}$), but the selectivities were not strongly affected. Elementary analysis revealed that the used catalyst had organic impurities ($[C] = 7.50$ %, $[H] = 1.62$ %) even after recalcination at 673 K. The lowered activity was attributed to the formation of oligo- and polymers during reaction, which can block the active sites and the catalyst pores. Removal of these contaminants requires calcination at temperatures above 673 K.

Table 4 Influence of amine concentration on 3-cyclohexen-1-methanol (4) epoxidation.[a]

Amine concentration (mol%)	$t_{50\%}$[b] (min)	$S_{C=C}$[c] (%)	S_{TBHP}[c] (%)	Diastereomeric excess[c] (%)
—	7	36	36	24
—[d]	20[d]	34[d]	31[d]	44[d]
0.5	7	41	46	12
1	20	44	38	6
5	14	41	36	4
10	17	49	39	1

[a] Standard reaction conditions, Ph_3CNH_2 as additive. [b] Time for 50 % olefin conversion. [c] At $t_{50\%}$. [d] Second run with recycled catalyst.

Discussion

Selectivities in the epoxidation of 2-cyclohexen-1-ol (1) and 3-cyclohexen-1-methanol (4), catalyzed by a TiO_2-SiO_2 aerogel, could be improved by the addition of small amount of amine. Under standard reaction conditions (without amine) the reactant 1 was rapidly converted to mainly oligo- and polymers and dicyclohex-2-enyl ether (3). These side reactions were of non-oxidative nature, catalyzed by the acidic sites on the aerogel. Amine modifiers suppressed these acid-catalyzed side reactions and enhanced the apparent rate of epoxide formation (apparent rate = real rate - rate of decomposition reactions of the epoxide). In the best case, the epoxide selectivity $S_{C=C}$ in the epoxidation of 1 could be improved from 22 % to 81 % at 50 % conversion in the presence of only 1 mol% tritylamine (Ph_3CNH_2).

Three different types of acidic sites have been proposed to be present on the aerogel, which are probably involved in side products formation: (i) atomically dispersed Ti(IV), isolated by siloxygroups, represent Lewis acidic sites which are responsible for epoxidation activity (19); (ii) other Lewis acidic species are Ti(IV) atoms located in titania nanodomains (Ti connected via oxygen to Si and/or Ti atoms), which possess only moderate epoxidation activity (9); (iii) surface silanol groups which represent Brønsted acidic sites.

Suppression of side reactions is obviously due to neutralization of acidic sites on the catalyst by the amine additives. Upon addition of bases to TS-1 in epoxidations with H_2O_2, formation of a stable peroxo complex at the Ti sites by abstraction of the (acidic) proton of the Ti-hydroperoxy complex has been proposed (20). The low reactivity of this species was attributed to its reduced electrophilicity caused by the negative charge. However, this mechanism involves abstraction of a proton from the Ti-hydroperoxy complex formed with H_2O_2 and cannot be used to explain the effect of bases on the Ti-complex formed with TBHP. In our model, basic compounds, beside neutralization of surface silanol groups, can also coordinate to the titanium active site and reduce its acidity by electron donation. As a consequence of the suppression of acid-catalyzed side reactions, the apparent rate of epoxide formation increased considerably, as shown in Figure 1. The positive effect on activity is attributed to suppression of (acid-catalyzed) oligo- and polymerization. These non-volatile compounds can block the active site and the narrow pores, resulting in an apparent loss of epoxide formation.

As concerns the Brønsted sites in titania-silica, deactivation of the surface silanol groups does not require strong bases. It was shown that treatment of the catalyst with weakly basic NaN_3, and even neutral NaCl, suppresses the acid-catalyzed isomerization of β-isophorone (11). The inorganic salt, as the most polar compound of the reaction mixture, will adsorb preferentially on the polar surface silanol groups and prevent the interaction (acid-catalyzed side reaction) with the reactant or product.

References

1. S. Thorimbert, S. Klein and W.F. Maier, *Tetrahedron* **51**, 3787 (1995).
2. S. Klein, J.A. Martens, R. Parton, K. Vercruysse, P.A. Jacobs and W.F. Maier, *Catal. Lett.* **38**, 209 (1996).
3. D.C.M. Dutoit, M. Schneider and A. Baiker, *J. Catal.* **153**, 165 (1995).
4. H. Kochkar and F. Figueras, *J. Catal.* **171**, 420 (1997).
5. R. Hutter, T. Mallat, D. Dutoit and A. Baiker, *Top. Catal.* **3**, 421 (1996).
6. B. Notari, *Adv. Catal.* **41**, 253 (1996).
7. F. Wattimena and H.P. Wulff, British Pat. 1'249'079, to Shell (1971).
8. C. Cativiela, J.M. Fraile, J.I. García and J.A. Mayoral, *J. Mol. Catal. A* **112**, 259 (1996).
9. R. Hutter, T. Mallat and A. Baiker, *J. Catal.* **153**, 177 (1995).
10. R. Hutter, T. Mallat and A. Baiker, *J. Catal.* **157**, 665 (1995).
11. R. Hutter, T. Mallat, A. Peterhans and A. Baiker, *J. Catal.* **172**, 427 (1997).
12. M. Dusi, T. Mallat and A. Baiker, *J. Mol. Catal. A* **138**, 15 (1999).
13. G.J. Hutchings, D.F. Lee and A.R. Minihan, *Catal. Lett.* **39**, 83 (1996).
14. T. Tatsumi, M. Yako, M. Nakamura, Y. Yuhara and H. Tominaga, *J. Mol. Catal. A* **78**, L41 (1993).
15. M. Dusi, T. Mallat and A. Baiker, *J. Catal.* **187**, 191 (1999).
16. A. Sato, J. Dakka and R.A. Sheldon, *Stud. Surf. Sci. Catal.* **84**, 1853 (1994).
17. D.C.M. Dutoit, U. Göbel, M. Schneider and A. Baiker, *J. Catal.* **164**, 433 (1996).
18. D.C.M. Dutoit, M. Schneider, R. Hutter and A. Baiker, *J. Catal.* **161**, 651 (1996).
19. R.A. Sheldon, In *Aspects of Homogeneous Catalysis*, R. Ugo, Ed., D. Reidel, Dordrecht, 1981, p. 3.
20. M.G. Clerici and P. Ingallina, *J. Catal.* **140**, 71 (1993).

Mononuclear Heterogeneous Catalysts for the Epoxidation of Olefins by Aqueous H$_2$O$_2$

S. H. Holmes, F. Quignard and A. Choplin

Institut de Recherches sur la Catalyse-CNRS
2, Avenue A. Einstein, 69 626 Villeurbanne Cedex, France

Abstract

Titanium based catalysts were synthesized from tetraneopentyltitanium and different silicas, differing by their hydrophilic / hydrophobic character. The analysis of their catalytic properties for the epoxidation of cyclohexene by aqueous H$_2$O$_2$ highlights the importance of the isolation and number of anchoring bonds of the titanium centers, in accordance with the high sensitivity of the Ti-O bonds towards hydrolysis.

The same approach allowed the synthesis of zirconium based catalysts, which are unusually active for the same reaction.

Introduction

Isolated (or mononuclear) active sites are considered as responsible of the specific properties of a number of heterogeneous catalysts for various organic reactions. This is the case for the titanosilicate TS-1, a titanium substituted silicalite, for which the outstanding catalytic properties for various reactions of oxidation seem to be correlated to the specific environment of titanium (1-4). But, because of the steric limitations induced by the zeolitic framework, applications of this solid are restricted to reactions involving small molecules. Therefore, during the last fifteen years, many studies focused on the synthesis of large pore titanium substituted zeolites (5-8) and of non - microporous amorphous mixed titania-silica oxides (9-12).

We wish to describe here our approach of the synthesis of mononuclear titanium entities anchored on the surface of silica and the catalytic properties of the resulting solids for the epoxidation of cyclohexene with aqueous hydrogen peroxide. This is part of a long term research project on the synthesis of molecular heterogeneous catalysts for various applications in fine chemical synthesis (13-15). Schematically, our strategy comprises two main controlled steps. The first uses the reactivity of a precursor organometallic complex with an inorganic oxide to form a covalent bond between the metal center and some surface oxygen atoms. This step leads to a grafted mononuclear organometallic species. The second step consists in building around the metal center the ligand sphere which is adapted to the target reaction.

In the present case, the titanium site should lie in an environment as close as possible to that schematically drawn below, according to literature data (16) (scheme 1):

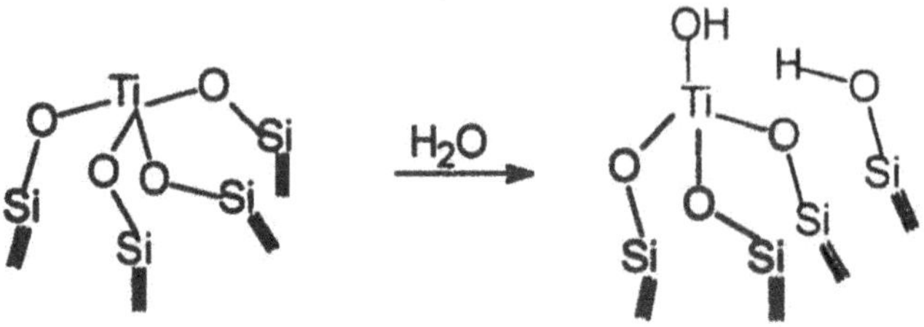

Scheme 1 Active titanium sites in TS-1 (16)

In order to reach this goal, tetraneopentyltitanium, $TiNp_4$, was chosen as the organometallic precursor complex for the following reasons:
- it is a mononuclear complex (indeed, association via alkyl ligands is not possible). It is a homoleptic complex: therefore, only one type of reaction can occur with the hydroxyl surface groups. Finally, it is also a stable complex, due to the steric hindrance of the neopentyl ligands; it can therefore be handled under easy to control conditions.
- the titanium-carbon bonds are hydrolyzed under very mild conditions. This is important because this type of reaction will be the anchoring reaction of titanium on silica. It is a clean reaction, which produces only neopentane, an inert compound.

Most of these results were published recently (17-18). We wish here to focus on the most pertinent features.

Experimental Section

All experiments were done under strict exclusion of air, using standard high vacuum line equipment and break and seal techniques. The complexes MNp_4 (M = Ti, Zr) were synthesized following a reported procedure (19). Four different silicas were used as support: Aerosil 200, R812 and R974, and Sipernat D17 all from Degussa. The first is hydrophilic, the three latter are hydrophobic. Only the latter is a precipitated silica. H_2O_2 (from ATOCHEM) was a 70% v/v aqueous solution.

The *in situ* IR spectroscopic studies were performed using a cell, designed to allow for thermal or vacuum treatments and for spectra recording (17). The complexes were either sublimed on a self supported disk of silica for the IR studies or introduced by impregnation of silica by a solution in hexane (larger samples for catalytic tests). The catalytic epoxidation of cyclohexene was performed under the conditions given under the tables (see results). These were not optimized, but allowed us to make easy comparisons with other catalysts synthesized in the laboratory.

Results

Synthesis and characterization of the catalysts

All four silicas were first evacuated at 500°C to remove the molecularly adsorbed water and to dehydroxylate the surface; only isolated silanol groups remain then on the surface (20). These reactions are of minor importance on the hydrophobic silicas, on which most hydroxy groups have been substituted by trimethylsiloxy or dimethylsiloxy groups (21). The reaction between $TiNp_4$ and silicas occurs at room temperature as evidenced by IR data: the disappearance of the $v(OH)$ vibrational band at 3747 cm^{-1}, characteristic of isolated surface silanol groups, is concomitant with the appearance of $v(C-H)$ bands between 2960 and 2880 cm^{-1}, attributed to the neopentyl ligands. This is shown on Figure 1 for silica Aerosil 200 and silica R812 respectively as the supports.

Interestingly, the same type of reaction occurs on both types of silica despite the small number of silanols available on R812 and their sterically more hindered environment. The resulting surface titanium complex has mainly three neopentyl ligands and one siloxy ligand (eq. 1). This was deduced from the elemental analysis of the solid (C/Ti = 12.6) and from the quantitative analysis of the neopentanol liberated by acidic hydrolysis of the complex obtained after reaction of 1 with O_2 (eq. 2) (3.0 ± 0.1 mole neopentanol /Ti).

$$TiNp_4 + \equiv Si\text{-}OH \rightarrow \equiv SiO\text{-}TiNp_3 + NpH \qquad (1)$$
$$\mathbf{1}$$

$$(\equiv SiO)_{4\text{-}x} TiNp_x + O_2 \rightarrow (\equiv SiO)_{4\text{-}x} Ti(ONp)_x \qquad (2)$$

The supported complex, $\equiv SiOTiNp_3$ (1), is as such not a catalyst; it also has not the target optimum environment for titanium (scheme 1). Two modifications must be made to reach this goal: – substitute the neopentyl ligands by hydroxy ligands, - increase the number of anchoring bonds, which may be an important parameter for the complex stability.

328   Holmes et al.

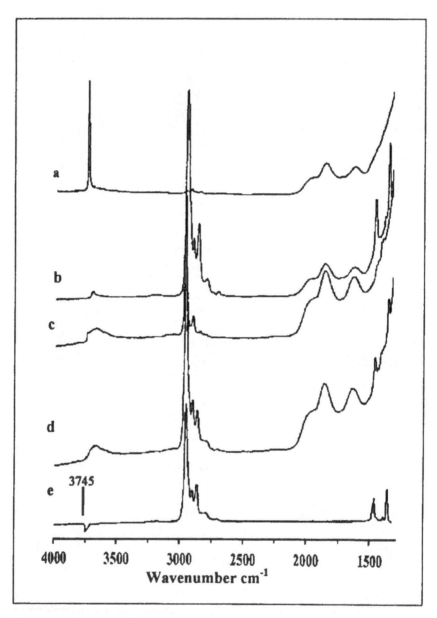

Figure 1: IR spectra (% absorbance, a.u.) of silica (a) Aerosil 200, (c): R812 evacuated at 500°C; then after the sublimation (b) on Aerosil 200, (d): on R812 of TiNp$_4$, followed by evacuation at 60°C. Spectrum (e) = (d) – (c).

The first type of modification can be achieved by reaction with water. The behavior of **1** towards hydrolysis is dramatically different whether it is supported on a hydrophilic or on a hydrophobic silica. Indeed, the removal of the neopentyl ligands from **1** (sharp intensity decrease of the v(CH) bands) occurs on both silicas (Fig. 2).

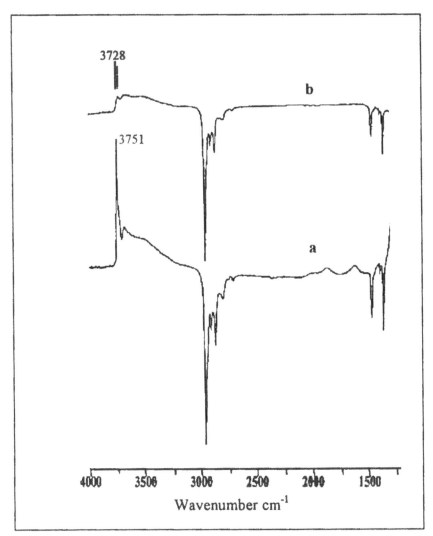

Figure 2 Hydrolysis of ≡SiOTiNp₃. The IR spectra (% absorbance, a.u.) are the differences between the spectrum of (**1**+H₂O, then evacuation) and the spectrum of **1** on (a): Aerosil 200 and (b): silica R812.

But hydrolysis restores simultaneously most of the silanol groups on Aerosil 200 but none on R 812. This strongly suggests that the titanium anchoring bond is hydrolyzed on the hydrophilic silica while it is essentially preserved on the hydrophobic silica. In the former case, an intermediate $Ti(OH)_4$ would be formed, which can then be transformed to TiO_2 particles. Given the small amount of titanium deposited (see table 1), these particles are probably small. This is in agreement with the UV spectrum which presents a broad band at 250 nm, attributed to small oxide particles and/or Ti(IV) in an octahedral environment (22, 23). In the case of the hydrophobic silica, $v(TiOH)$ bands are observed at 3728 cm^{-1}. This would suggest the formation of mononuclear entities such as $\equiv$SiOTi(OH)$_3$. The UV band of this solid is slightly shifted towards smaller wavelengths, when compared to the case of the hydrophilic silica supported species (λ= 240 nm).

In conclusion, the hydrolysis of $\equiv$SiOTiNp$_3$ leads to non anchored "$Ti(OH)_4$", precursors of small TiO_2 particles on hydrophilic silica and probably to surface anchored mononuclear $\equiv$SiO-Ti(OH)$_3$ complexes on hydrophobic silica. In the latter case, the hydrophobic trimethylsilyl groups may either hinder the approach of H_2O to the anchoring bond Si-O-Ti and /or isolate the Ti centers from each other.

The second modification which ought to be made on 1 is to increase the number of anchoring bonds between titanium and the surface of silica. On Aerosil 200, this can in principle be possible by increasing the initial concentration of hydroxyl groups on the surface i.e. by decreasing the temperature of dehydroxylation. Nevertheless, a careful study of the closely related system ZrNp$_4$/silica showed that, whatever the degree of dehydroxylation of the support, a mixture of mono, di and certainly tri-ligated zirconium entities is formed (24). This is not good for the catalytic properties (activity and selectivity) and the overall stability of the catalyst.

When 1 supported on the different silicas is calcined at 400°C (a classical mean used to remove all carbonaceous species and thus regenerate the catalyst), we observed *vide supra* that some of the solids based on hydrophobic silicas showed improved selectivity for epoxide formation. Their complete characterization is not straightforward. Nevertheless, the near IR spectrum of calcined 1/R812 shows that all alkyl groups, i.e. the neopentyl ligands and the trimethylsilyl groups are removed (no [$2v(CH)$] bands at 1710 and 1750 nm) and that simultaneously silanol groups (band at 1370 nm) are formed, in large excess when compared to the initial number. Finally, the UV spectrum, is characterized by a band at 235 nm, (UV spectrum recorded in air). These data, taken all together, are in favor of well dispersed titanium entities on the surface of these samples, a large portion of them being probably mononuclear.

Catalytic properties

We have tested the two families of solids obtained by hydrolysis or by calcination of ≡SiOTiNp₃ (1) and have determined the influence of the surface state of the silica support on their selectivity for epoxide formation (Table 1) The test reaction is the epoxidation of cyclohexene by aqueous H_2O_2 (eq. 3).

$$\text{cyclohexene} + H_2O_2 \longrightarrow \text{cyclohexene oxide} + H_2O$$

$$(3)$$

All solids are very selective for the epoxide formation when an organic peroxide such as ᵗBuOOH is used as the oxidant (95 %). This is common for titanium based solids if the reaction is performed under strict exclusion of water.

With aqueous H_2O_2, the performance of complex 1, when supported on the hydrophilic silica and then hydrolyzed, is only modest: the epoxide yield is medium, i.e. close to 40% (Table 1, entry 1)) The main by-products are the product of hydrolysis of the epoxide, cyclohexane-1,2-diol, and the products of allylic oxidation, 2-cyclohexen-1-ol and 2-cyclohexen-1-one. It is interesting to point out here the fact that epoxide ring opening hydrolysis is catalyzed by Ti sites; indeed this reaction is observed neither in the homogeneous phase nor in the presence of silica alone. When a calcination is made in place of hydrolysis, the solid shows similar properties (Table 1, entry 3). Both solids, i.e. hydrolyzed and calcined ≡SiOTiNp₃ decompose H_2O_2 into H_2O and O_2 to an extent of ca. 50%. This latter reaction is believed to occur on titanium oxide particles. The recycling of these solids (Table 1, entries 1-4) is possible without significant loss of activity or selectivity. The TiO_2 particles remain on the solid and do not significantly sinter under the reaction conditions. Titanium leaching into the solvent is very low (less than 0.6 % of total Ti introduced).

The catalysts supported on hydrophobic silicas (Table 1, 4-6) are much more selective, whatever the treatment which is applied to the precursor ≡SiOTiNp₃, except with D17 as the support. Thus, total epoxidation yield (epoxide plus diol) reaches values close to 80%, with R812 as the support. The efficiency of these solids decreases with nature of the support as follows:

R812 > R974> D17

which is not the order of decreasing hydrophobicity:

D17>> R812> R974

Table 1 Catalytic properties of ≡SiOTiNp₃ derived solids for the epoxidation of cyclohexene.

N°	silica	treatment	% Ti	% conv.[a]	% yield epoxide	diol
1	Aerosil 200	hydrolysis	1	95	37	21
2				94[b]	50	2
3	Aerosil 200	calcination	1.2	95	42	15
4	"			95[b]	41	16
5	R812	hydrolysis	0.27	96	64	10
6				94[b]	56	6
7	R812	calcination	0.30	93	63	15
8				95[b]	70	9
9	R974	hydrolysis	0.83	96	47	13
10	D17	hydrolysis	0.33	90	20	7
11	D17	calcination	0.19	93	36	0
12	Aerosil 200	none	0	35	10	0
13	R812	none	0	21	8	0
14	D17	none	0	92	5	0
15	none	none	0	15	6	0

Exp. cond.: T = 105°C, t_R = 2 h 30 min., 5.1 mmol H_2O_2, 50 mmol cyclohexene, 30 mL diglyme; [a] based on oxidant; [b]: second run.

Blank experiments revealed (Table 1, entries 12-14) that pure D17 is very active for the decomposition of H_2O_2, hampering its use as a support for any catalyst of epoxidation.

Synthesis of zirconium based catalysts

Given the lower sensitivity of the Zr-O bond towards hydrolysis, we applied the same strategy to prepare zirconium based catalysts, starting from $ZrNp_4$ as the organometallic precursor. Unexpectedly (25), these solids were as efficient as the titanium based solids; similarly, the use of hydrophobic silica is beneficial for the epoxide selectivity. Nevertheless, the behavior of these Zr based catalysts differs from that of the related Ti catalysts: the epoxide selectivity decreases with time (Fig. 3). It is decomposed by a reaction which is catalyzed by the same Zr centers and certainly involves the solvent (diglyme).

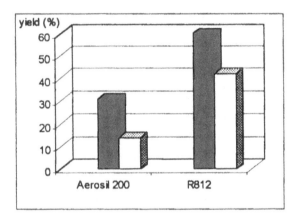

Figure 3 Catalytic properties of $ZrNp_4$/silica. Evolution with time of the epoxide yield. Exp. cond. as given below table 1. % Zr: 2% wt. (Aerosil 200), 0.8%wt. (R812). ■: t= 10 min; □: t = 100 min.

Discussion and Conclusions

Our strategy, based on the initial reaction of $TiNp_4$ with the surface hydroxyl groups of partially dehydroxylated silicas, allows for the synthesis of titanium sites anchored to the support through a covalent bond and characterized as being $\equiv SiOTiNp_3$ entities. These sites are mononuclear and clearly no Ti-O-Ti bonds are formed. Catalysts for the epoxidation of cyclohexene with aqueous H_2O_2 are obtained by mild hydrolysis or calcination of these grafted complexes.

The efficiency of these catalysts is correlated to a number of parameters, most of them involving directly or indirectly the silica support itself. First, the silica must be highly pure. Thus, silica D17, the precipitated silica, is active as such for the decomposition of H_2O_2 into H_2O and O_2, a property which may be attributed to iron present in non negligible amounts (0.03 % as Fe_2O_3). In sharp contrast, under the same reaction conditions, the silicas Aerosil 200 and R 812, given as highly pure, show properties very similar to that of the non catalyzed homogeneous system (Table 1, entries 12, 13, 15). It seems that the silanol groups, present in higher concentration on Aerosil 200, intervene essentially in the non productive decomposition of H_2O_2. Only small yields of epoxide are obtained with all the silicas tested, but no diol; this means that the ring opening hydrolysis of the epoxide is catalyzed by the Ti centers and not by the weakly acidic Bronsted sites of silicas.

Second, silylated silicas are preferred. Thus, the best results in terms of catalytic properties for the epoxidation of cyclohexene by aqueous hydrogen peroxide are obtained when using as the support a hydrophobic silica (R812), on which most surface hydroxy groups are substituted by trimethylsilyl entities, by reaction with hexamethylsilazane. These hydrophobic groups seem to protect the titanium anchoring bonds, $\equiv$SiO-TiR$_3$, from hydrolysis. This latter reaction is facile, occurs on hydrophilic supports as demonstrated by *in situ* IR spectroscopy and allows then for the formation of small particles of TiO_2, which are non selective towards epoxide formation. The importance of the extent of the substitution of the silanol groups by the SiMe$_3$ moieties is exemplified by the differences in the behavior of the titanium sites obtained by hydrolyzed $\equiv$SiOTiNp$_3$ on R812 and R974 respectively. One must assume that the less hydroxy groups are left on the surface, the more they are isolated. The optimum degree of substitution corresponds thus to the maximum number of hydroxy groups, necessary to chemically attach titanium on the surface, compatible with the isolated character. The lower degree of substitution for R974, evidenced by the larger amount of titanium which can be chemically grafted on the surface, may leave hydrophilic domains accessible to water and consequent rupture of the titanium anchoring bonds.

The fact that the hydrolysis and the calcination of $\equiv$SiOTiNp$_3$ supported on R812 lead to two solids with similar catalytic properties is intriguing: indeed, there are no more hydrophobic methyl groups on the calcined sample as revealed by near IR spectroscopy, while these entities are not sensitive to hydrolysis. This observation suggests that titanium has a very similar proximate environment on both solids, a statement which can be rationalized by the transformations shown on scheme 2. From IR data, one can reasonably assume that hydrolysis of $\equiv$SiOTiNp$_3$ leads to $\equiv$SiOTi(OH)$_3$; the grafting bond is protected against hydrolysis by the hydrophobic SiMe$_3$ groups. A calcination of the same entity, $\equiv$SiOTiNp$_3$, transforms the neopentyl ligands on titanium and the methyl ligands on silicon into CO_2 and H_2O and leads to the formation of hydroxy silicon and titanium complexes; these can easily condense at the temperature of calcination. The titanium atoms are then in an environment very similar to the target one (scheme 1).

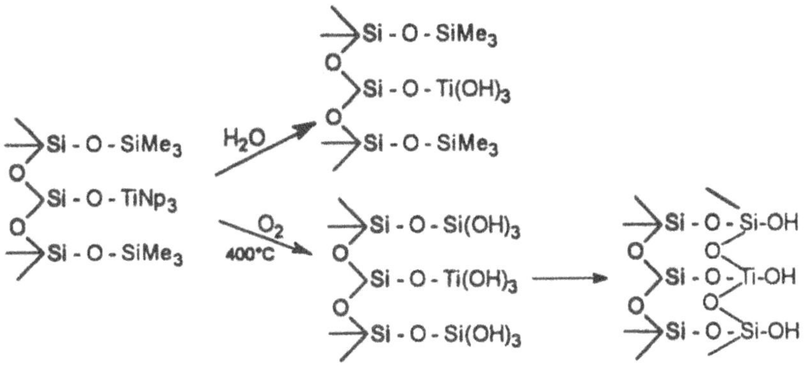

Scheme 2

Finally, it must be pointed out that the approach presented here is very different from that reported earlier (26); in the latter case, silylation was performed after the anchoring step in order to passivate the silanol groups, not consumed by the reaction between tetraethoxy- or tetrachloro-titanium and silica. Our strategy highlights the importance of site isolation for the obtainment of a catalyst selective for the epoxidation of olefins with aqueous hydrogen peroxide, a property which is achieved through preliminary isolation of the surface hydroxy groups on silica by their partial substitution with trimethylsilyl moieties. Such catalysts have no steric limitations and can easily be regenerated by simple calcination: studies are currently under investigation to extend their applications to a wide range of substrates.

Acknowledgements : ELF-ATOCHEM is acknowledged for financial support (S.H.) and many fruitful discussions with Drs. J. Kervennal and R. Teissier.

References

1. M. Taramasso, G. Perego and B. Notari, *US Patent* 4,410,501 (1983).
2. B. Notari, *Stud. Surf. Sci. Catal.*, **60**, 343 (1991).
3. B. Notari, *Catal. Today*, **18**, 163 (1993).
4. G. Bellussi and M.S. Rigutto, *Stud. Surf. Sci. Catal.*, **85**, 177 (1994).
5. M. Costantini, A. Corma, L. Gilbert, P. Esteve, A. Martinez and S. Valencia, *J. Chem. Soc. Chem. Commun.*, 1339 (1996).
6. A. Corma, M.T. Navarro and J. Perez-Pariente, *J. Chem. Soc. Chem. Commun.*, 147 (1994).
7. P T. Tanev, M Chibwe and T.J. Pinnavaia, *Nature*, **386**, 239 (1994).
8. A. Corma, Q. Kan and F. Rey, *Chem. Commun.*, 579 (1998)
9. D.C. Dutoit, M. Schneider and A. Baiker, *J. Catal.*, **153**, 165 (1995).
10. R. Hutter, T. Mallat and A. Baiker, *J. Catal.*, **153**, 177 (1995).

11.   Z. Liu, G.M. Crumbaugh and R.J. Davis, *J. Catal.*, **159**, 83 (1996).

12.   J.C. van der Waal and H. van Bekkum, *J. Mol. Catal.*, **124**, 137 (1997).

13.   A. Choplin, B. Coutant, C. Dubuisson, P. Leyrit, C. McGill, F. Quignard and R. Teissier, *J. Mol. Catal.*, **120**, L27 (1997).

14.   F. Quignard, O. Graziani and A. Choplin, *Appl. Catal. A: General*, **182**, 29 (1999).

15.   A. Choplin and F. Quignard, *Coord. Chem. Rev.*, **178-180**, 1679 (1998).

16.   S. Bordiga, S. Coluccia, C. Lamberti, L. Marchese, A. Zecchina, F. Boscherini, F. Buffa, F. Genoni, G. Leofanti, G. Petrini and G. Vlaic, *J. Phys. Chem.*, **98**, 4125 (1994).

17.   S.A. Holmes, F. Quignard, A. Choplin, R. Teissier and J. Kervennal, *J. Catal.*, **176**, 173 (1998).

18.   S.A. Holmes, F. Quignard, A. Choplin, R. Teissier and J. Kervennal, *J. Catal.*, **176**, 182 (1998).

19.   P.J. Davidson, M.F. Lappert and R. Pearce, *J. Organomet. Chem.*, **57**, 269 (1973).

20.   H.P. Boehm and H. Knözinger, *Catalysis, Science and Techology*, J.R. Anderson and M. Boudart (Eds.), vol. 4, ch. 2, Springer Verlag, Berlin, 1983.

21.   The hydrophobic silicas are obtained, according to Degussa, (Technical Note N° 11 (1995)), by reacting Aerosil silica with either hexamethyldisilazane (R812) or dichlorodimethylsilane (R974).

22.   Z. Liu and R.J. Davis, *J. Phys. Chem.*, **98,** 1253 (1994).

23.   A. Zecchina, G. Spoti, S. Bordega, A. Ferrero, G. Petrini, G. Leofanti and M. Padovan, *Stud. Surf. Sci. Catal.*, **69**, 251 (1991).

24.   F. Quignard, C. Lécuyer, A. Choplin and J.M. Basset, *J. Chem. Soc Dalton Trans.*, 1153 (1994).

25.   R.A Sheldon and J.A. van Doorn, *J. Catal.*, **31**, 427 (1973).

26.   R.A. Sheldon, *J. Mol. Catal;.* 7, 107 (1980).

Common Features and Differences in the Heterogeneously Catalysed Conversion of Substituted Methyl Aromatics to Aromatic Aldehydes and Nitriles

Andreas Martin and Bernhard Lücke

Institut für Angewandte Chemie Berlin-Adlershof e.V.
Richard-Willstätter-Str. 12, D-12489 Berlin, Germany

Abstract

A comparative study on the vapour phase oxidation and ammoxidation of toluene, p-methoxytoluene and p-chlorotoluene to the corresponding aldehydes and nitriles using V-containing catalysts is described. The effect of acid-base properties of catalyst surfaces in connection with electronic effects of the substituents on the catalytic performance is discussed. Common features and differences in mechanistic steps are demonstrated. The optimum catalyst composition for vapour phase partial oxidation of methyl aromatics to their aldehydes is quite different from that for vapour phase partial ammoxidation to nitriles.

Introduction

The selective oxidation of aromatic hydrocarbons both in liquid and vapour phase represents an important class of catalytic reactions and was developed into large scale operations for the production of commodities in the chemical industry (e.g. 1,2). First commercial oxidation of aromatics based on vapour phase processes started in the first decades of this century (e.g. phthalic anhydride production by oxidation of naphthalene over vanadium oxide catalysts in 1917). This exemplary reaction is attributed to electrophilic type oxidations whereas the more efficient conversion of o-xylene to phthalic anhydride in the vapour phase belongs to the nucleophilic reaction type (e.g. 2). Another aromatic bulk oxygenate, benzoic acid, is produced by a very selective liquid phase side-chain oxidation of toluene (e.g. 3).

The selective production of aromatic aldehydes in a vapour phase process is much more complicated because the aldehydes themselves are consecutively oxidised very fast to acids and deeper oxidised products (e.g. anhydrides and quinones). For example, benzaldehyde is produced in the vapour phase at low toluene conversions (10 - 20 %) per pass at short residence time (< 1 s); even then, it is only 40 - 60 % of the theoretical yield (4). This is the main reason why

337

benzaldehyde and various substituted aromatic aldehydes are rather manufactured in liquid phase oxidation processes or by the hydrolysis of benzal chlorides (4).

On the other hand, the vapour phase ammoxidation of aromatic side chains also belongs to the nucleophilic type oxidation and is the most simple and economically most profitable route for the production of aromatic nitriles. This reaction refers to the interaction of ammonia with a reducible organic material (alkanes, alkenes, alkyl aromatic and heteroaromatic compounds) in the presence of oxygen. Especially, the ammoxidation of alkyl aromatics to the corresponding nitriles is intensively studied and applied in industry in a wide range (e.g. 5). First communications on this reaction appeared in the 40s and 50s. Beside the production of benzonitrile from toluene and dinitriles from xylenes various substituted methyl aromatics as well as heteroaromatics were used as feedstock in such reactions (e.g. 5-7).

Substituted aromatic nitriles as well as aldehydes are of widespread use in fine chemical production. They can be converted to commercially interesting and valuable intermediates for the synthesis of several pharmaceuticals and dyestuffs as well as pesticides and final products of these lines.

From this point of view, this paper gives a survey on common features and differences in the oxidation and ammoxidation of toluene, p-methoxytoluene and p-chlorotoluene using vanadium-containing catalysts under comparable conditions (see Scheme 1) supported by results of in situ-ESR and FTIR spectroscopy.

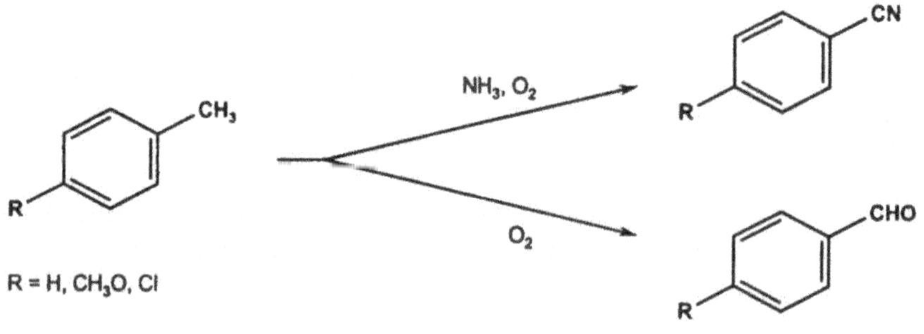

R = H, CH₃O, Cl

Scheme 1

Experimental

Catalysts

Different vanadium phosphate (VPO) catalysts were prepared using the well known VOHPO$_4$ · ½ H$_2$O solid as starting material. Precursor pretreatment under inert gas up to ca. 773 K leads to (VO)$_2$P$_2$O$_7$ (VPP) (e.g. 8) whereas heating under NH_3-air-water vapour leads to $(NH_4)_2(VO)_3(P_2O_7)_2$ solids that contain defined proportions of amorphous or poor crystalline VO_x (AVP_{gen}) (e.g. 9). Additionally, vanadium oxide catalysts that contain potassium were prepared by V_2O_5 impregnation with K_2SO_4 solutions (e.g. 10).

Catalytic measurements
The catalytic properties of the solids were determined during the ammoxidation and oxidation of toluene (TOL) to benzonitrile (BN) and benzaldehyde (BA), respectively, p-methoxytoluene (MTOL) to p-methoxybenzonitrile (MBN) and p-methoxybenzaldehyde (MBA), respectively, and p-chlorotoluene (CTOL) to p-chlorobenzonitrile (CBN) and p-chlorobenzaldehyde (CBA), respectively, using a fixed bed U-tube quartz-glass reactor. The catalysts (0.5 ml each, corresponding to ca. 0.5 g) were used as sieve fractions (1-1.25 mm) and mixed prior to oxidation runs with the equal portion of quartz glass (1 mm) to avoid local overheating. The oxidation runs were carried out using the following conditions: air : (substituted)toluene = 100 : 1, W/F = 1 g h mol^{-1} and reaction temperatures of ca. 623 – 673 K. The following reaction conditions were applied for the ammoxidation tests: (substituted)toluene : NH_3 : air : water vapour = 1 : 5 : 32 : 24, W/F = 10 g h mol^{-1} and reaction temperatures of ca. 653 – 703 K. The product stream was analysed on line or it was trapped in aqueous ethanol and determined by off line-capillary GC. The formation of carbon oxides was continuously followed by non-dispersive IR photometry.

In situ-catalyst characterisation
Various *in situ*-FTIR and ESR spectroscopic results obtained during the partial oxidation of toluene are described here to give some examples of a deeper insight into reactant-catalyst interaction; basics and detailed studies are reported elsewhere (e.g. 11,12).

The *in situ*-FTIR investigations were carried out on a Bruker IFS 66 FTIR spectrometer using self-supporting discs with a diameter of 20 mm and a weight of 50 mg mounted in an heated IR cell. The samples were pretreated by heating up to 723 K (β = 10 K/min) under vacuum for 1.5 h. The adsorption spectra were recorded, adopting the following procedure: the sample was heated up to the selected temperature under feed and then kept for 1 h, cooled down to room temperature followed by evacuation and recording the spectra.

In situ-ESR measurements were performed with the cw spectrometer ELEXSYS 500-10/12 (Bruker) equipped with a flow reactor and a gas-liquid supplying system, consisting of mass flow controllers and saturators. An amount of 0.14 g of the VPP solid (0.5-1 mm fraction) was applied as catalyst. The sample was tested under similar toluene oxidation conditions as described above.

Results and discussion

Recent studies (13) showed that during the ammoxidation of toluene (Fig. 1) on VPP benzaldehyde could be observed as a reaction intermediate during the reaction cycle.

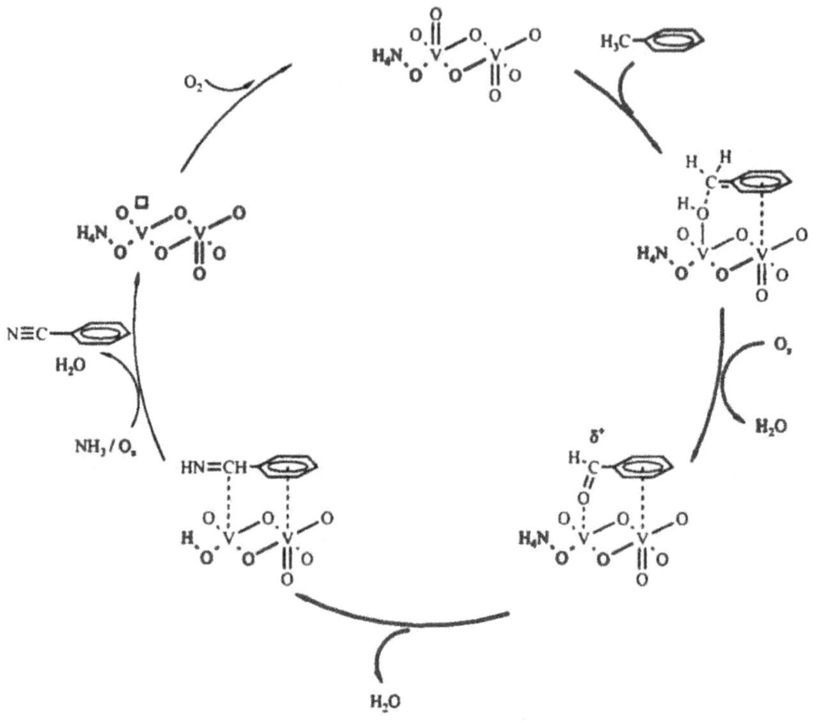

Figure 1 Schematic reaction cycle of the ammoxidation of toluene on a vanadyl dioctahedra unit of a $(VO)_2P_2O_7$ catalyst (13).

Therefore, it seemed entirely possible to use these catalysts for a conversion of toluene as well as substituted toluenes in the absence of ammonia, i.e. in their partial oxidation to synthesise aromatic aldehydes. The catalytic studies were carried out in the above mentioned manner with similar VPO catalysts; the ammoxidation runs were carried out using VPP or AVP_{gen} catalysts whereas VPP samples were used in the partial oxidation. The results are depicted in Fig. 2. The ammoxidation runs reveal the expected high conversion rates and high nitrile selectivities but the oxidation of the same reactant under similar reaction conditions on similar catalysts looks quite different with rather small conversion

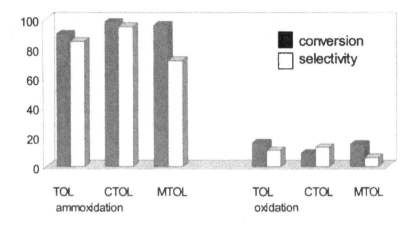

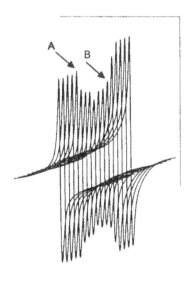

Figure 2 Conversion and nitrile as well as aldehyde selectivity in the ammoxidation (703 K) and oxidation (653 - 673 K) of toluene (TOL), p-chlorotoluene (CTOL) and p-methoxytoluene (MTOL) on VPO catalysts.

Figure 3 A sequence of *in situ*-ESR spectra taken during toluene oxidation on VPP

and aldehyde selectivities (mainly carbon oxides were found as balance); only the CTOL conversion seems to proceed in a little bit more selective way. This catalytic result was the reason to try to uncover activity and selectivity limiting factors during such vapour phase partial oxidation.

First *in situ*-investigations on the reactant-catalyst interaction in the case of the toluene partial oxidation revealed some of the reasons, leading during the catalytic reaction to such low aldehyde selectivities. Fig. 3 demonstrates a series of *in situ*-ESR V(IV) spectra on a VPP catalysts, showing a sequence taken at 673 K under N_2, toluene oxidation feed and again nitrogen. Switching from N_2 to an air-toluene feed (A) results in an immediately declined intensity of the ESR signal, reflecting a deterioration in exchange interaction of coupled V(IV) sites of the catalyst. Toluene, partial oxidised intermediates and/or coke-precursor-like compounds cover the active sites and push electron density into the catalyst bulk via strong chemisorption on Lewis-acid sites. This observation is proven by a re-switching to N_2 (B); the signal intensity returns only very slow, i.e. adsorbates desorb very slow (14).

A contribution to the identification of these strongly adsorbed intermediates is provided by FTIR spectroscopy as shown in Fig. 4. Beside the formation of benzaldehyde (1678 cm^{-1}) the generation of cyclic anhydrides (1852, 1783 cm^{-1}; mainly maleic anhydride as shown by GC/MS) is always present and comes to the fore at higher reaction temperatures (14).

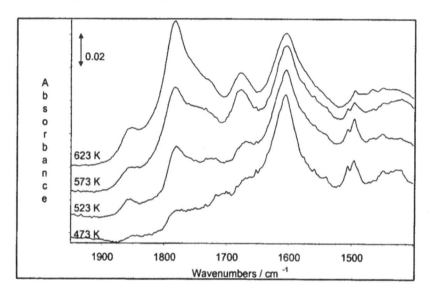

Figure 4 FTIR spectra of adsorbed species on a VPP catalyst after toluene oxidation (benzaldehyde 1678 cm^{-1}; cyclic anhydrides 1852, 1783 cm^{-1})

Reports on the oxidation of toluene and substituted toluenes to their aldehydes in the vapour phase are rather rare but appearing since ca. 25 years. Mainly tests for the oxidation of toluene to benzaldehyde in the vapour phase were carried out. For example, the use of V_2O_5-K_2SO_4 catalysts (10,15), the synthesis and application of novel ultrafine complex Mo-based oxide particles (16) and new V_2O_5-TiO_2 oxide phases doped with Te or K_2SO_4 (17) as well as the application of V-containing micro- and mesoporous materials (e.g. 18) and Fe-Mo substituted deboronated borosilicates (e.g. 19) were tested. Beside, also some results on the oxidation of substituted methyl aromatics became known. First results of intensive efforts on the vapour phase oxidation of MTOL to MBA were claimed in the 70s by Japanese researchers (20,21). They used V-P-

Cu-O catalysts doped with potassium sulphate and gave surprising high MBA yields up to ca. 70 wt.-%. The commercialisation of this reaction by Nippon Shokubai and Kagaku Kogyo using an alkali metal-containing vanadium catalysts was recently reported (22).

In between, mainly Japanese groups reported on the role of acid and base properties of vanadium oxide catalysts, containing base metal oxides (K, Rb, Cs, Tl, Ag) for partial oxidation of MTOL and other substituted toluenes (23-25). The most effective base metal oxide studied was the rather hazardous Tl_2O. In general, they stated that activity strongly depends on the amount and strength of acid sites whereas the reaction selectivity to substituted benzaldehydes is closely related to the basic properties of the catalyst. This statement is related to a strong enough chemisorption of the toluene molecule on coordinatively unsaturated (Lewis) sites coupled with a fast desorption of a rather basic aldehyde molecule from basic surface areas.

This finding matches the results of own studies on the above mentioned oxidation of toluene to benzaldehyde on the rather acidic VPP catalyst (14). VPP contains vanadyl dioctahedra units that possess Lewis-sites, being responsible for the chemisorption of toluene via the π-electron system of the nucleus. Actually, a freshly prepared VPP specimen shows no surface OH groups, but under oxidation working conditions water is formed and a large proportion of surface OH groups is erected by hydrolysis of V-O-V and/or V-O-P bonds. These OH groups (Brønsted sites) are responsible for the formation of ammonium ions during the ammoxidation, being involved in the reaction cycle but during the oxidation they cause a strengthening of the chemisorptive interaction of the intermediately formed aldehyde and the catalyst surface in addition to the molecule sorption by co-ordinatively unsaturated sites. The reason is the generation of hydrogen bonding between the formed carbonyl group of the aldehyde species and neighboured OH groups (26) as shown in Scheme 2 and as clearly demonstrated by FTIR results depicted in Fig. 5.

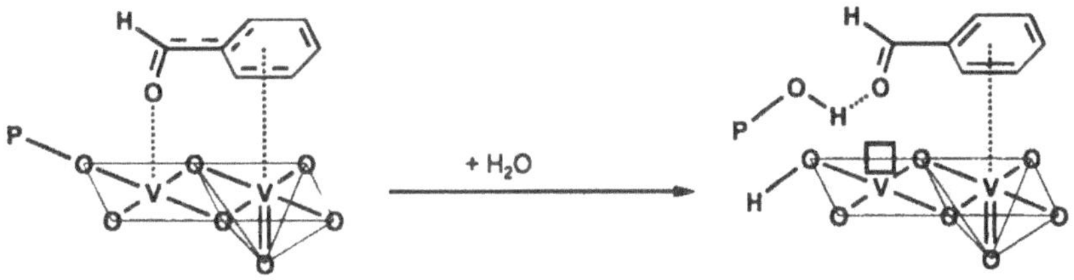

Scheme 2

Spectra *a)* (obtained after air-toluene exposure) and *b)* (additionally water vapour was dosed) reveal a very similar appearance, i.e. the admixture of water vapour does not significantly change product distribution as well as the acidic properties of the VPP surface. This means that the water formed during the reaction is able to generate a sufficient amount surface OH groups. Otherwise, spectrum *d)* points to a very strong adsorption of benzaldehyde on the water vapour pretreated VPP specimen (ca. 1680 cm^{-1}) by a 25 cm^{-1} shift to lower wavenumbers in contrast to benzaldehyde adsorbed on an untreated catalyst (ca. 1705 cm^{-1}, spectrum *c)*).

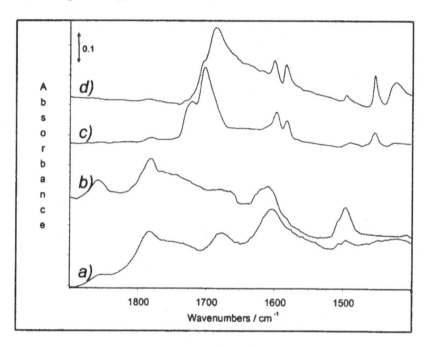

Figure 5 FTIR spectra of adsorbates on VPP after *a)* air-toluene flow, *b)* air-toluene flow after water vapour treatment, *c)* air-benzaldehyde flow and *d)* air-benzaldehyde flow after water vapour treatment

This action could be also proven by the addition of pyridine to the feed. The admixture of this base to the feed (note that pyridine does not react under these conditions) could cause a permanently blocking of the *in situ* generated Brønsted acid sites, probably, and would suppress the hydrogen bonding. Therefore, a speed up of the desorption of the aldehyde would be the result. This idea is depicted in Scheme 3. Figure 6 demonstrates FTIR spectra proving this assumption.

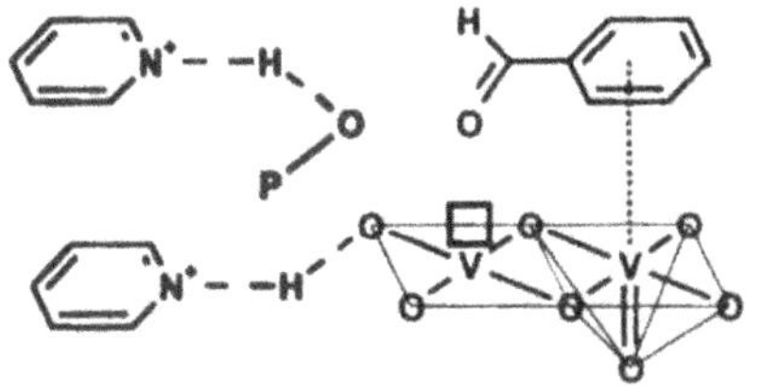

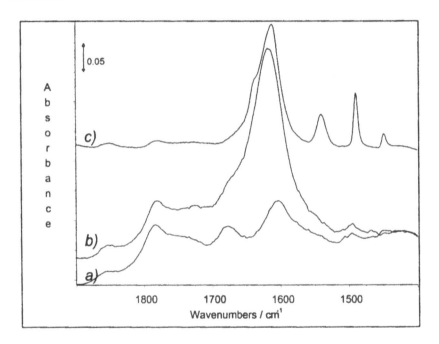

Scheme 3

Figure 6 FTIR spectra of adsorbates after *a)* air-toluene flow, *b)* air-toluene-water vapour flow and *c)* air-toluene-4% aqueous pyridine vapour flow

Spectra *a)* and *b)* reveal the known picture of adsorbed aldehyde and cyclic anhydrides but surprisingly spectrum *c)* depicts pyridine adsorbates, no trace of aldehyde and only low intense cyclic anhydride bands. The spectra demonstrate that no aldehyde as well as anhydride is still chemisorbed on the surface after the reaction. The catalytic test proved this result of the faster desorption of the aldehyde from the catalyst surface. A threefold aldehyde selectivity was obtained as shown below in Fig. 7. Pyridine seems to preferentially block the acid OH groups because nearly the same conversion rates were observed (26), i.e. the Lewis sites are not significantly affected.

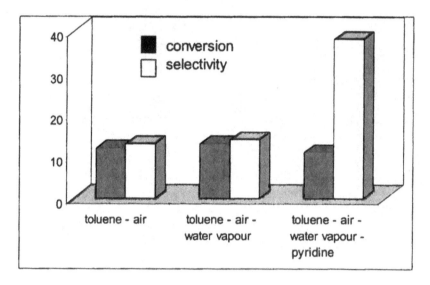

Figure 7 Comparison of benzaldehyde selectivity on VPP in the oxidation of toluene under comparable conditions, feeding toluene-air, toluene-air-water vapour and toluene-air-water vapour-pyridine (4 % aqueous solution)

In the same direction, the addition of alkali metal salts to vanadium-containing oxidation catalysts (e.g. 10,15) makes the desorption of the desired products easier by a surface basification and, additionally, alkali cations block Brønsted acid sites. Such catalysts (KVO) were also used for our studies, showing a significant increase in catalytic activity and aldehyde selectivity as demonstrated in Table 1. The development of new alkali metal-containing vanadium oxides and the investigation of the effect of the alkali metal on surface acid-base properties as well as catalytic behaviour are still under study.

Table 1
Oxidation of toluene and substituted toluenes and corresponding aldehyde selectivity obtained on KVO catalysts

Reactant	Temperature / K	Conversion / %	Selectivity / %
TOL	643	17	40
MTOL	661	24	17
CTOL	648	17	29

Conclusion

The formation of aromatic aldehydes by vapour phase oxidation of the corresponding methyl substituted reactants proceeds in its activation in equal pathways to the described ammoxidation to nitriles. However, the desorption of the intermediately formed aldehyde is restricted, e.g. by hydrogen bonding and, therefore, the aldehyde can be easily oxidised to anhydride structures, being precursor compounds of total oxidation. The admixture of inert base compounds can prevent this interaction. Similar effects can be reached using alkaline metal-containing vanadium oxides as catalysts.

In contrast to the usage of methoxytoluenes as reactant, the halogen substituent reveals a rather electron-withdrawing effect; the result is that the aldehyde selectivity can be increased by a faster desorption of the aldehyde molecule in this case. This behaviour is in line with results of the ammoxidation and the ideas suggested above on the interaction of reactants being different in their acid-base properties with more acid catalyst surface of an ammoxidation catalyst and with the more base surface of an oxidation catalyst.

The vapour phase oxidation and ammoxidation of substituted aromatic compounds will be continuously of interest. The reason for this is given by the growing importance of the application of heterogeneous catalysts for the production of fine and special chemicals on the one hand and of the future growth in the supply of aromatics by the expected change in the petroleum processing on the other hand. The efficiency of these processes will be improved with more selective catalysts; the amount of waste as well as the formation of undesired by-products will be significantly limited and complicated separation operations could be replaced by simpler ones.

Acknowledgement

The authors thank Dr. U. Bentrup and Dr. A. Brückner for carrying out the *in situ*-measurements and the Federal Ministry of Education and Research, Germany for financial support (project No. 03 C 0279 0).

References

1. J. Haber in: G. Ertl, H. Knözinger and J. Weitkamp (Eds.), Handbook of Heterogeneous Catalysis, Wiley-VCH, 1997, p. 2253 and references therein.
2. K. Weissermel and H.-J. Arpe, Industrielle Organische Chemie, VCH, Weinheim, 1994, S. 415.
3. K. Weissermel and H.-J. Arpe, Industrielle Organische Chemie, VCH, Weinheim, 1994, p. 381, 426.
4. F. Brühne and E. Wright in: Ullmann, 6th Edition 1998 Electronic Release (benzaldehyde entry).

5.   R.G. Rizayev, E.A. Mamedov, V.P. Vislovskii and V.E. Sheinin, Appl. Catal. A 83 (1992) 103.
6.   B.V. Suvorov, Ammoxidation of Organic Compounds, Nauka, Alma-Ata, 1971 (Russ.).
7.   R.K. Grasselli, J.D. Burrington, R. DiCosimo, M.S. Friedrich and D.D. Suresh, Stud. Surf. Sci. Catal. 41 (1988) 317.
8.   H.Berndt, K.Büker, A.Martin, A.Brückner and B.Lücke, J. Chem. Soc., Faraday Trans., 91(4) (1995) 725.
9.   Y. Zhang, M.Meisel, A.Martin, B.Lücke, K.Witke and K.-W. Brzezinka, Chem. Mater., 9 (1997) 1086.
10. A. Kaszonyi, M. Antol, M. Hronec, G. Delahay and D. Ballivet-Tkatchenko, Collect. Czech. Chem. Commun. 60 (1995) 505.
11. Y.Zhang, A.Martin, H.Berndt, B.Lücke and M.Meisel, J. Mol. Catal.: A, 118 (1997) 205.
12. A.Brückner, A.Martin, B.Lücke and F.K.Hannour, Stud. Surf. Sci. Catal., 110 (1997) 919.
13. A.Martin, H.Berndt, B.Lücke and M.Meisel, Top. Catal., 3 (1996) 377.
14. A. Martin, U. Bentrup, A. Brückner and B. Lücke, Catal. Lett. 59 (1999) 61.
15. M. Ponzi, C. Duschatzky, A. Carrascull and E. Ponzi, Appl. Catal. A 169 (1998) 373.
16. Y. Fan, W. Kuang, W. Zhang and Yi Chen, Stud. Surf. Sci. Catal. 110 (1997) 903.
17. A. Aguilar Elguézabal and V. Cortés Corberán, Catal. Today 32 (1996) 265.
18. G. Centi, F. Fazzini, L. Canesson and A. Tuel, Stud. Surf. Sci. Catal. 110 (1997) 893.
19. J.S. Yoo, Appl. Catal. A 143 (1996) 29.
20. US Patent 4,054,607 (1977), Tanabe Seiyaku Co.
21. H. Seko, Y. Tokuda and M. Matsuoka, Nippon Kagaku Kaishi 1979, 558.
22. B. Delmon, Stud. Surf. Sci. Catal. 110 (1997) 43.
23. N. Shimizu, N. Saito and M. Ueshima, Stud. Surf. Sci. Catal. 44 (1988) 131.
24. M. Ueshima, N. Saito and N. Shimizu, Stud. Surf. Sci. Catal. 90 (1994) 59.
25. M. Ueshima and N. Saito, Chem. Lett. 1992, 1341.
26. A. Martin, U. Bentrup, B. Lücke and A. Brückner, J. Chem. Soc., Chem. Commun. 1999, 1169.

Platinum-Catalyzed Oxidation of N-Alkyl Glyphosates: A New Glyphosate-Forming Reaction

David A. Morgenstern*, Yvette M. Fobian, and David S. Oburn

Monsanto Co.
800 N. Lindbergh Blvd.
St. Louis, MO 63167
314-694-8786
david.a.morgenstern@monsanto.com

Abstract

N-alkyl glyphosates in aqueous solution can be aerobically oxidized to glyphosate over platinum catalysts under mild conditions. Other noble metals and homogeneous catalysts are ineffective. Although supported platinum catalysts are active, most inorganic supports are unstable in the reaction medium while carbon supports lead to an undesired side reaction. Thus, platinum black is the most practical catalyst for the oxidation. N-alkyl glyphosate oxidation is a two-electron, nitrogen-centered oxidation and thus differs from the well-known oxidation of N-phosphonomethyl iminodiacetic acid ("glyphosate intermediate"). The latter, the most widely practiced method for preparing glyphosate, proceeds in two 1-electron steps via a Kolbe mechanism.

Most of our studies have focussed on NMG oxidation. The rate is first-order in NMG, but selectivity is improved if oxygen is made the limiting reagent by controlling flow. Under all conditions studied, the selectivity remains fairly constant until it reaches a critical conversion, typically about 80%. At this point, both rate and selectivity decline dramatically. The selectivity break can be pushed to higher conversion by increasing the catalyst loading or decreasing the oxygen flowrate, but cannot be eliminated. The break appears to result from over-oxidation of the catalyst.

Introduction

Glyphosate is already among the world's most widely used herbicides owing to its effective control of virtually all weeds and its lack of toxicity toward animals and even most bacteria (1). The continuing growth of demand for glyphosate-based herbicides, particularly Monsanto's Roundup® (a registered trademark of Monsanto Co.) has stimulated interest in identifying new ways to produce it.

$$H_2O_3P \diagdown \diagup \overset{\overset{H}{|}}{N} \diagdown \diagup CO_2H$$

glyphosate

Viewed retrosynthetically, the challenge of glyphosate synthesis resides in assembing the P-C-N bond ("phosphonomethylation"). Perhaps the fastest and easiest method of generating phosphonates, particularly on a large scale, is the condensation of phosphorous acid with formaldehyde and a secondary amine, first described by Moedritzer and Irani (2). Primary amines cannot be selectively mono-phosphonomethylated (3). In glyphosate synthesis, the logical secondary amine substrate is an N-substituted glycine (eqn 1). Traditional electron-withdrawing substituents such as amides and carbamates are ineffective protecting groups as they deactivate the amine toward phosphonomethylation and are unstable in aqueous acid. Thus there has been considerable interest in developing deprotection reactions (equation 2) which convert N-substituted glyphosates to glyphosate.

$$HO_2C \diagup\diagdown NH + CH_2O + H_3PO_3 \overset{\Delta}{\underset{H^+}{\longrightarrow}} HO_2C \diagup\diagdown N \diagup\diagdown PO_3H_2 \quad (1)$$
$$\underset{R}{|} \qquad\qquad\qquad\qquad\qquad\qquad \underset{R}{|}$$

N-alkyl glycine $+ H_2O$

$$HO_2C \diagup\diagdown N \diagup\diagdown PO_3H_2 \underset{?}{\longrightarrow} HO_2C \diagup\diagdown N \diagup\diagdown PO_3H_2 \quad (2)$$
$$\qquad\qquad \underset{R}{|} \qquad\qquad\qquad\qquad\qquad \underset{H}{|}$$

N-substituted glyphosate   **glyphosate**

A number of versions of reaction 2 have been described including hydrogenolysis of N-benzyl glyphosate (4) and dealkylation of N-t-butyl or N-benzyl glyphosate with strong acid (5) and N-isopropyl glyphosate in strong base. (6) In practice, however, by far the most successful deprotection reaction has proven to be oxidation of N-phosphonomethyl iminodiacetic acid (PMIDA), also commonly called "glyphosate intermediate" or "GI," (eqn 3). Reaction 3 can be conducted electrochemically, (7) or catalyzed by homogeneous (8) or heterogeneous (1,9)catalysts. The latter is practiced commercially by Monsanto.

$$HO_2C \frown N \frown PO_3H_2 \xrightarrow[\substack{\text{carbon} \\ \text{catalyst}}]{O_2} HO_2C \frown \underset{H}{N} \frown PO_3H_2 \quad (3)$$

$$HO_2C \diagup \qquad\qquad\qquad\qquad + 2CO_2$$

PMIDA

Equation 3 can be thought of electrochemically as a Kolbe reaction proceeding by decarboxylation of GI followed by oxidation of the resulting radical in a separate one-electron step. Carbon electrodes are well-known to be effective for Kolbe reactions (10), hence carbon catalysts are the usual catalysts for GI oxidation. GI oxidation is extremely selective and, unlike alternative reactions, produces CO_2 as its only byproduct. However, the reaction wastes two carbon atoms and more atom-efficient reactions have long been sought.

Experimental Section

Materials: Platinum black, sarcosine and other N-alkyl glycines and phosphorous acid were purchased from Aldrich and used as received. 20% Pt/Vulcan XC-72R was obtained from Alfa Aesar. N-alkyl glyphosates were synthesized by phosphonomethylation of the corresponding N-alkyl glycine. The following synthesis of N-methyl glyphosate is typical: 89.09 g sarcosine (1.00 mole), 82.0 g phosphorous acid (1.0 mole) and 110 g concentrated hydrochloric acid were mixed and refluxed in a 130°C oil bath. 89.3 g of 37% formalin (1.1 mole) was added dropwise over 20 minutes and the reaction was continued for an additional 85 minutes. After cooling to room temperature, 40 g (1 mole) sodium hydroxide was added followed by 250 g water leading to the formation of a white precipitate which was recovered by filtration and assayed by HPLC. Yield: 70.5%

Analytical: The results of N-methyl glyphosate (NMG) oxidations were assessed via HPLC analysis of the reaction mixture, suitably diluted. An ion exchange separation was used, and the analytes were detected using UV/visible detection following post-column reaction to form a phosphomolybdate complex. This method can distinguish between, NMG, glyphosate, and phosphoric acid, but AMPA and MAMPA coelute. Because AMPA and MAMPA have the same response factor, on a molar basis, the sum of the AMPA and MAMPA concentrations can be reliably determined. This value is reported as "(M)AMPA." Results of oxidation of N-alkyl glyphosates other than NMG were quantified by ^{31}P NMR.

Catalyst screening: Catalysts were generally screened by the following procedure. The data in the following table was obtained by heating to reflux a mixture containing 1 g. of NMG, 20 ml water, and sufficient catalyst to contain 5 mg of platinum metal in a magnetically-stirred roundbottom flask equipped with a reflux condenser. Oxygen was bubbled through for five hours using a needle. The catalyst was then removed by filtration and the filtrate analyzed by HPLC. The same method was used to measure the selectivity of the oxidation of N-alkyl glyphosates other than NMG.

Oxidation of NMG over palladium black: A solution consisting of 3.0 g. of N-methyl glyphosate, 0.3 g. of palladium black, and 57 g. of water were refluxed in air over a weekend under a water-cooled reflux condenser. NMR analysis indicated the following product distribution: N-methyl glyphosate (97.2%), glyphosate (2.8%), phosphoric acid (0.05%).

Oxidation of NMG in an autoclave: NMG oxidations were conducted in a 300 ml Autoclave Engineers reactor with a Hastelloy C body. Oxygen and nitrogen were introduced through a 0.5 μm cylindrical stainless steel frit and the flowrate controlled with a Brooks mass flow controller. Gases exiting the reactor passed through a condenser and then exited through a backpressure regulator. Another 0.5 μm frit for withdrawing liquids was located under the impeller.

The reactor was loaded with catalyst, NMG, and water and was then sealed and brought to the reaction temperature and pressure under nitrogen flow. When the temperature stabilized, oxygen flow was started and nitrogen flow was usually shut off. Samples were withdrawn periodically and immediately diluted to avoid glyphosate precipitation. At the conclusion of the reaction, oxygen flow ended and nitrogen flow resumed. The contents of the reactor were withdrawn through the liquid withdrawal frit.

Results and Discussion

We now report that a variety of N-alkyl glyphosates can be selectively oxidized to glyphosate over platinum catalysts, (eqn 4). Most of our results concern the oxidation of N-methyl glyphosate (NMG), because the formaldehyde byproduct of NMG oxidation is further oxidized to CO_2 (equation 5) providing environmental advantages and avoiding safety concerns. Oxidation of other N-alkyl glyphosates can lead to a mixture of oxygen and ketones in the reactor headspace, creating an explosion hazard.

$$\underset{\substack{R \quad R'}}{HO_2C\diagdown N \diagdown PO_3H_2} + 1/2\ O_2 \xrightarrow[Pt]{} HO_2C\diagdown \underset{H}{N} \diagdown PO_3H_2 \quad (4)$$

$$+ \underset{R \quad R'}{\overset{O}{\underset{\|}{C}}}$$

$$\underset{Me}{HO_2C\diagdown N \diagdown PO_3H_2} + 3/2\ O_2 \xrightarrow[Pt]{} HO_2C\diagdown \underset{H}{N} \diagdown PO_3H_2 \quad (5)$$

$$+\ CO_2$$

NMG

Reaction 4 can be conveniently conducted by bubbling oxygen through an aqueous solution of the N-alkyl glyphosate with platinum black at temperatures of 80-100C. This procedure was used to determine the selectivity of the oxidation for various N-alkyl glyphosates. The results are shown in Table 1.

Table 1 Selectivity of the oxidation of various N-alkyl glyphosates

R	Conversion	Selectivity
methyl	91%	95%
i-propyl	79%	98%
n-pentyl	62%	82%
benzyl	81%	89%
cyclohexyl	66%	11%

Of the N-alkyl glyphosates in Table 1, only the most sterically congested, N-cyclohexyl glyphosate, exhibits poor selectivity. This is consistent with a mechanism involving oxidation of the nitrogen center to an iminium ion which is stabilized by coordination to the platinum surface (Scheme 1). The mechanism of tertiary amine oxidations such as N-alkyl glyphosate oxidation is widely accepted as proceeding via a two-electron oxidation to the iminum ion (11). Three tautomers are possible, as shown Scheme 1, and it is of some interest to ask whether the distribution of tautomers is product-determining.

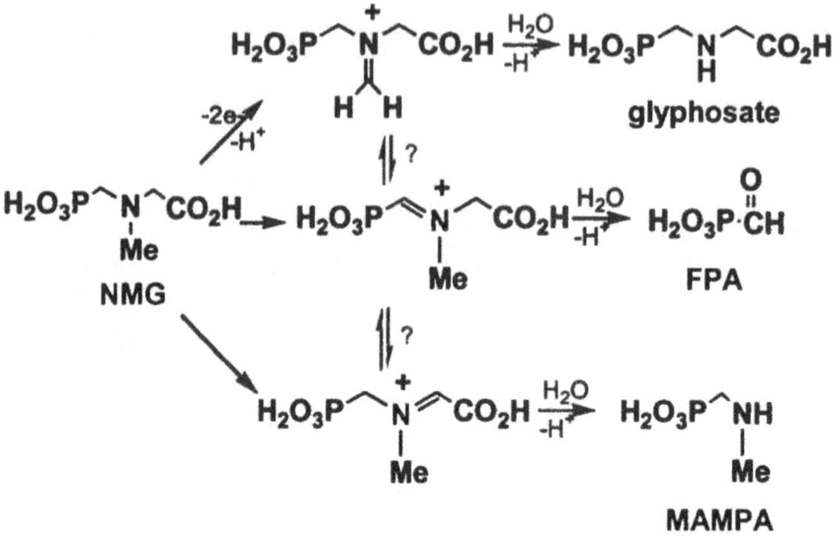

Scheme 1 Pathways for NMG oxidation assuming the formation of an iminium ion intermediate.

In the case of NMG, the predicted side products are N-methylaminophosphonic acid (MAMPA) and formyl phosphonic acid (FPA). MAMPA is indeed one of the major byproducts observed, but the other is phosphate, apparently derived from the hydrolysis of FPA to H_3PO_3 which is then oxidized. Under typical NMG oxidation conditions, H_3PO_3 (present as an impurity in the starting material) oxidizes to H_3PO_4 in about 10 minutes. The spectrum of products then is consistent with the expected mechanism, the phosphate arising from rapid oxidation of FPA. FPA was observed by ^{31}P NMR during slow (8-24 hour) oxidation of NMG over platinum black at 70C.

Oxidation of an even more hindered analog, bis-phonomethyl-1,4-piperazine (BPMP) over platinum black under the same conditions gave phosphate as the only phosphorus-containing product (eqn 6). This observation suggests that steric factors which inhibit the formation of a planar iminium ion, such as the puckered rings of BPMP and N-cyclohexyl glyphosate, deactivate tertiary amines toward oxidation on platinum.

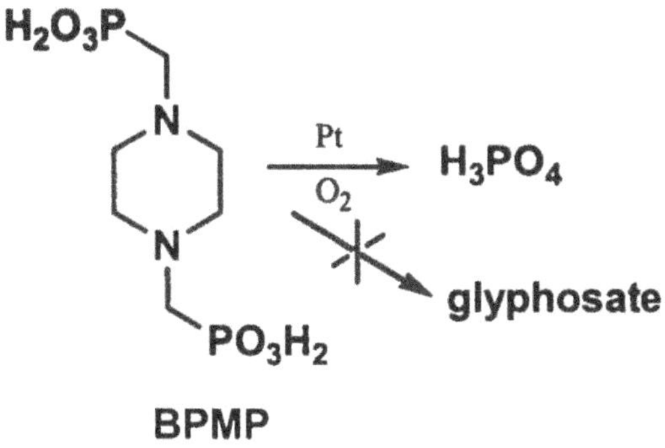

$$H_2O_3P \overset{Pt}{\underset{O_2}{\longrightarrow}} H_3PO_4 \quad\quad (6)$$

glyphosate

BPMP

Platinum is by far the most useful catalyst for the reaction. Palladium also selectively catalyzes the reaction at a low rate but requires very mild conditions and achieves low conversions. Supported platinum is initially as selective as platinum black, at least for non-carbon supports. Table 2 shows the selectivity of the various supported platinum catalysts for oxidation of 5% NMG at reflux in a standard screen (see Experimental Section for details). All supported platinum catalysts had the same selectivity pattern within experimental error.

Table 2 Oxidation of N-methyl glyphosate over various platinum catalysts

Catalyst	Conversion (%)	Glyphosate	Selectivity MAMPA	(%) H_3PO_4
Pt black	14.7	85.3	3.0	11.7
5% Pt/SnO$_2$	18.0	88.7	2.6	8.7
5% Pt/ZrO$_2$	13.9	89.5	7.3	3.2
5% Pt/BaSO$_4$	31.2	92.2	2.8	5.1
5% Pt/TiO$_2$	47.4	91.9	1.7	6.4
5% Pt/SiO$_2$	23.7	88.9	2.3	8.8

Other metal heterogeneous catalysts either leach metal under the reaction conditions (rhodium, silver, nickel) or are simply inert (gold). Oxidation of NMG using typical homogeneous oxidation catalysts including nickel and ruthenium salts and cobalt and vanadyl salen complexes leads to MAMPA and phosphate formation and no glyphosate. Stoichiometric oxidizing agents such as lead tetraacetate are equally ineffective.

Unfortunately, virtually all inorganic catalyst supports in common use lack the physical stability required for sustained use of the catalyst. Glyphosate is a good chelator and a moderately strong acid. Consequently, silica, alumina, zirconia, and other such supports lose their mechanical integrity over time in the reaction medium. Activated carbon supports are stable in the medium, but lead to high levels of MAMPA via a carbon-catalyzed Kolbe reaction similar to GI oxidation. Identification of a non-carbon support for platinum which is chemically and mechanically robust in the medium and which would allow for high platinum dispersion (particularly by comparison to platinum black) would be of great value.

The carbon support problem may be partially ameliorated by the use of graphitic supports. Table 3 shows the performance in our standard screen of three Pt/C catalysts using supports which vary from non-graphitic to highly graphitic. DeGussa's 5% Pt/F106 carbon catalyst is typical of the first while Johnson-Matthey's 20% Pt/Vulcan XC-72R catalyst is an example of an almost completely graphitic support. Intermediate between the two extremes is a class of carbons known as "Sibunit" manufactured by the Boreskov Institute of Catalysis in Russia. Table 3 shows that selectivity improves as the support becomes more graphitic. We have not yet determined whether the improvement in selectivity is simply a result of the fact that graphites have a lower surface area than conventional activated carbons or whether graphite is intrinsically inactive toward MAMPA formation. The former is more likely, since graphite electrodes are effective for Kolbe reactions (10).

Table 3 Oxidation of NMG over Pt/C catalysts

Catalyst	Conversion (%)	Glyphosate	Selectivity (%) MAMPA	H_3PO_4
5% Pt/F106 carbon	98.9	62.2	29.0	8.7
3% Pt/Sibunit carbon	53.7	73.7	18.1	8.2
20% Pt/Vulcan XC-72R	53.6	83.5	10.4	6.1

Higher rates can be achieved without loss of selectivity by conducting the reaction in an autoclave at 100-150C. The selectivity is improved by limiting the oxygen flowrate, making the reaction pseudo-first order. When excess oxygen is used, the reaction is first order in NMG.

Studies of NMG oxidation in this temperature range revealed a sharp break in selectivity at a critical conversion (Figure 1). The break is not due to poisoning by product. Addition of glyphosate to the reaction mixture does not affect the rate and selectivity of NMG oxidation.

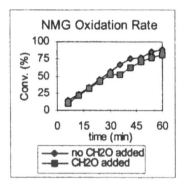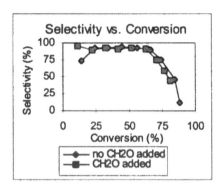

Figure 1 Oxidation of 150 g. of 25% aqueous NMG over 1.4 g. of platinum black at 150C using 200 sccm of pure oxygen at 90 psig. The effect of pumping in 0.7 ml/min of 7.4% formaldehyde beginning 30 minutes into the reaction is also shown.

It seemed possible that the break might be associated with the formaldehyde produced during NMG oxidation. As shown in Figure 1, continuous addition of formaldehyde beginning midway through the reaction stops NMG oxidation completely for a few minutes, but the reaction then resumes with the same rate and selectivity as before.

The selectivity break is most likely a result of the NMG concentration falling below a critical value, leading to over-oxidation of the platinum. Both NMG and formaldehyde help to keep the platinum reduced, counteracting the effect of oxygen, but the NMG evidently plays a more important role, based on the results of the formaldehyde addition experiment.

We tentatively view the selectivity break as the point at which oxidation of the catalyst by O_2, which is supplied at a constant rate, can no longer be balanced by reduction by NMG whose concentration declines as the reaction proceeds. Consistent with this hypothesis, we have observed that decreasing the oxygen flowrate moves the selectivity break to higher conversion, although the effect is small (Figure 2). We are continuing our investigation of the peculiar kinetics of this reaction.

358   Morgenstern et al.

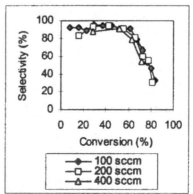

Figure 2 Oxidation of 150 g. of 25% aqueous NMG over 0.7 g. of platinum black at 150C using various flowrates of pure oxygen at 90 psig.

Acknowledgments

We thank Reed Herzig and Dan Gehrig for performing all of the HPLC analyses and Jerry Ebner and Mike Scaros for useful discussions.

References
1. (a) Franz J.E. In *The Herbicide Glyphosate*; Grossbard, E., Atkinson, D. Eds., Butterworths, Boston, 1985, pp 3-17.   (b) Franz, J.E.; Mao, M.; Sikorski, J.A. In *Glyphosate, A Unique Global Herbicide*; American Chemical Society Monograph; ACS; Washington DC, 1997 pp. 7-14.
2. (a) Moedritzer, K.; Irani, R.R. *J. Org. Chem.* **1966**, *31* 1603.
3. Reviews: (a) Redmore, D.   *Topics in Phosphorus Chemistry*; "The Chemistry of P-C-N Systems, *8*, 515-85 (1976); (b) Petrov, K.A.; Chauzov, V.A.; Erokhina, T.S. *Russ. Chem. Rev.* **1974**, *43*, 984-1006; (c) Mastalerz, P. in *Handbook of Organophosphorus Chemistry*, Engel, R. ed., Marcel Dekker, New York, 1992, 277-375.
4. (a) Maier, L., EP 0 055 695, 1981.   (b) Maier, L *Phosphorus, Sulfur, and Silicon* **1991** *61*, 65-7.
5. (a) Gaertner, V.R., U.S. Patent 3,927,080, 1975. (b) Parry, D.R.; Dudley, C.; Tomlin, S., U.S. Patent 3,956,370, 1976.
6. Balthazor, T.M., Miller, W.H., Eur. Pat. 187633, 1984.
7. Frazier, H.W.; Smith, L.R.; Wagenknecht, J.H., U.S. Patent 3,835,000, 1974.
8. (a) D.P Riley, D.L. Fields and W. Rivers, *J. Am. Chem. Soc.* **113**, 3371-8 (1991). (b) D.P Riley, D.L. Fields and W. Rivers, *Inorg. Chem..* **30**, 4191-78 (1991).

9.   Hershman, A.; Bauer, D.J. U.S. Patent 4,264,776, 1981.

10.   Lund, H.; Baizer, M.M. *Organic Electrochemistry: An Introduction and Guide*, 3rd ed., Dekker, New York, 1991, p 539.

11.   Note that *in free solution*, disproportionation of a 1-electron oxidized nitrogen-centered radical to form the enamine can be kinetically significant, but the second electron transfer should be faster than disproportionation on the catalyst surface. See P.J. Smith and C.K. Mann, *J. Org. Chem.* **34**, 1821-6 (1969).

Betaine Synthesis via the Aqueous Oxidation of Choline Salts

Dorai Ramprasad, W. Eamon Carroll, Francis J. Waller and Thomas Mebrahtu

Corporate Science and Technology Center and Analytical Technology Center, Air Products and Chemicals, Allentown PA 18195

Abstract

This paper describes a new synthetic method to produce betaine (trimethylglycine; $Me_3N^+CH_2CO_2^-$) via catalytic oxidation of a choline salt, $[Me_3NCH_2CH_2OH]^+X^-$, in the presence of a noble metal. The effect of the anion X^- on the oxidation, the characterization of co-products and the issue of catalyst stability are addressed. The best result obtained is a 95 % conversion of choline hydroxide with a 95 % selectivity to betaine using a 5 % Pd/C catalyst.

Introduction

The compound betaine (trimethylglycine; $Me_3N^+CH_2CO_2^-$) has a wide variety of applications (1). These include animal feed supplement, enzyme stabilizer, pharmaceutical intermediate and personal care ingredient. Currently natural source betaine is extracted as a product in the de-sugaring of beet molasses (1). The most common synthetic route for betaine is by the reaction of an excess of trimethylamine with monochloroacetic acid to give a mixture of betaine and trimethylamine hydrochloride (2). A variant of this method is to react trimethylamine with an alkali or alkaline earth metal salt of monochloroacetic acid. In either case there are disadvantages since the product is contaminated with salts such as sodium chloride, trimethylamine hydrochloride or trimethylammonium chloroacetate (3). Another route to betaine is the biochemical oxidation of choline salts $[Me_3NCH_2CH_2OH]^+$ X^- using the enzyme choline oxidase (4). This is not a synthetic method, but is rather used as an analytical method for choline by detecting the amount of hydrogen peroxide produced in the reaction.

Our approach was to selectively oxidize the alcohol functionality in choline to produce betaine. The oxidation of primary alcohols to the carboxylic acid salts using alkali metal hydroxide in the presence of precious metals such as Pd, Pt, Rh, Ru, Ir, Os, Re is well known (5). We have prepared aqueous solutions of betaine from the direct oxidation of choline using O_2 and a supported noble metal catalyst such as Pd/C. The reactions are shown in equations (1) or (2):

$$[Me_3NCH_2CH_2OH]^+Cl^- + O_2 + NaOH \longrightarrow Me_3N^+CH_2CO_2^- + NaCl + 2H_2O \quad (1)$$

$$[Me_3NCH_2CH_2OH]^+OH^- + O_2 \longrightarrow Me_3N^+CH_2CO_2^- + 2H_2O \quad (2)$$

The oxidation of choline chloride in the presence of base will produce NaCl as a contaminant. However, the method in equation (1) still offers advantages over the chloroacetate route in which trimethylamine is an additional contaminant and has to be removed. The oxidation of choline hydroxide as shown in equation (2) produces water as a co-product and has the added advantage that no salt is produced. The goal of this work was to study reactions represented by equations (1) and (2) in the presence of different noble metal catalysts.

Experimental Section

Choline hydroxide, 50 % in water, choline chloride, and sodium hydroxide were purchased from Aldrich. Catalysts were obtained from various sources including Aldrich, Engelhard and Degussa. The Degussa catalyst is designated as E101 and the Engelhard as HDC.

The apparatus consists of a four necked flask equipped with a sparger, mechanical stirrer, thermocouple and a water cooled condenser. The sparger is positioned underneath the blades of the mechanical stirrer to break up the gas bubbles and to ensure proper mixing. The O_2 flow rate was regulated using a mass flow controller. For this equipment 200 rpm was found to be an adequate stirring rate above which the reaction is not mass transfer limited.

Catalyst recycling procedure: Approximately 5g of catalyst (5%Pd/C) was added to 105ml of a 40 % choline hydroxide solution. O_2 was sparged at 57 cc/min and the temperature was set to 78°C with a stirrer speed of 200 rpm. After 5.5 h the catalyst was filtered and the filtrate was analyzed by ^{13}C NMR or by electrical conductivity measurement. The catalyst was washed with warm water, 5% acetic acid, warm water and used for the next run. A silicone antifoaming agent (10 ppm) was used to mitigate foaming problem which leads to a loss of reactants from the condensor. Only the liquid that remained in the flask was analyzed.

Conductivity was used to monitor the conversion of choline hydroxide to betaine. The conductivity of a choline hydroxide solution is in the milli-Siemens range whereas a betaine solution has conductivity in the micro-Siemens range.   Therefore we concluded that in a mixture of the two species the conductivity measured would   approximate the choline concentration. A calibration curve was plotted for choline hydroxide solution in the range of 0-0.1M. For the actual measurement 1g of reactor solution was diluted with 14 g de-ionized water after which the conductivity was measured. The calibration curve was used to determine the concentration of choline hydroxide in solution.

^{13}C NMR integrals were used to determine conversion and selectivity. The conversion was defined as (x) and calculated as (x)÷(1-x) = sum of integrals of carbonyl species ÷ integral of choline . Percent conversion was defined as 100(x). Betaine selectivity was determined by integral of betaine peak ÷ sum of all carbonyl species. The analytical technique was validated via other techniques such as ion chromatography.

Results and Discussion

The oxidation of an alcohol to a carboxylic acid in the presence of O_2 using a noble metal catalyst such as Pt or Pd supported on carbon is well known (5). The accepted mechanism is that the noble metal functions as a catalyst for the dehydrogenation of the alcohol to the aldehyde followed by oxidation of the hydrated aldehyde to the carboxylic acid. The O_2 serves to restore the catalyst to its original state with the formation of water. The addition of base such as NaOH or Na_2CO_3 catalyzes the hydration of the intermediate aldehyde and serves to drive the reaction to the right via neutralization of the carboxylic acid formed (5).

The initial experiments involved the oxidation of dilute (5 %) solutions of choline chloride in water using a Pt/C catalyst and the results are tabulated in Table 1.

Table 1 Oxidation of 5 % choline chloride in water at two different temperatures[a].

Temperature °C	Time (h)	Choline Conversion %	Betaine Selectivity %
25	18	7	100
75	4	35	85

[a] 300 ml of water was added to 23.5 g of a 70 % choline chloride solution, 8.6 g of NaOH , 3.3 g of a 5%Pt/C (Aldrich) with O_2 sparging.

The results show that room temperature oxidation is slow compared to the reaction at higher temperatures. The loss in selectivity at the higher temperature is indicative of side reactions and is discussed later in this paper. Since the halide ions present in the reaction may poison the catalyst, the oxidation was attempted with choline hydroxide which has the necessary base built into the structure and produces only water as byproduct. As a comparison, a 5% choline chloride solution was treated with one equivalent of NaOH and reacted with O_2 at 78°C in the presence of 5% Pt/C or 5% Pd/C for 3 hours. The analogous reaction was performed with choline hydroxide, and the amount of betaine formed was deduced from the ^{13}C NMR spectrum (integration of the betaine carbonyl peak as compared to the corresponding carbon on choline). The results are shown in Table 2 .

Table 2 Oxidation of dilute aqueous solutions of choline chloride and hydroxide.

Sample	Catalyst	Choline conversion %	Betaine selectivity %
choline chloride[a]	5% Pt/C Aldrich	17.5	60
choline hydroxide[b]	5% Pt/C Aldrich	14.2	70
choline chloride[a]	5% Pd/C Aldrich	33.7	60

[a] 275 ml of a 5 % choline chloride, 2.1g catalyst, 4.4 g NaOH, 78°C, 3 h O_2 sparging.
[b] 275 ml of a 5 % choline hydroxide, 2.1 g catalyst, 78°C, 3 h O_2 sparging.

Though there are small differences in conversion and selectivity, the results show that choline hydroxide is similar to choline chloride indicating that the presence of halide ion is not detrimental to the catalyst in dilute solution. Also a 5% Pd/C(Aldrich) was found to be ~2 times more active than a 5% Pt/C(Aldrich) on a per gram basis and equal on a per mole basis.

Literature reports on the oxidation of alcohols using noble metal catalysts indicate that the catalyst to alcohol ratio can play a major role in the conversion and selectivity to product (5). The goal therefore was to vary the amount of Pd/C catalyst for a fixed amount and concentration of choline chloride and the results are shown in Table 3.

Table 3 Oxidation of 6 % choline chloride using different amounts of Pd/C[a].

Catalyst weight (g)	Choline conversion %	Betaine selectivity %
2.1	33	60
4.2	28	63
6.3	47	83
10	55	73

[a] 275 ml of 6 % choline chloride, 4.4 g NaOH, catalyst (5% Pd/C, Aldrich), 78°C, 3 h O_2 sparging.

The results show that there is a general trend of increased conversion with additional amounts of catalyst.

Literature results describe that an excessive rate of O_2 supply can result in a drop in overall reaction rate, a phenomenon known as 'O_2 poisoning' (5). In the reactions described thus far, no efforts were made to control the O_2 flow rate. Since the choline solutions were all very dilute(5%) it was deduced that the catalyst surface was saturated with O_2 leading to low conversions. Learning from these results, an experiment was designed to maximize the conversion and selectivity. The amount of choline chloride was kept the same, however a more concentrated solution was used. It was felt that a higher concentration of alcohol would keep more sites on the Pd in the metallic state and prevent over oxidation. The results are shown in Table 4.

Table 4 Oxidation of concentrated (>14 %) choline solutions in water[a].

Substrate(weight %)	choline conversion %	betaine selectivity %
choline chloride(14 %)	88	85
choline chloride(40 %)	45	72
choline hydroxide(40 %)	89	87

[a] All reactions performed with 100 ml solutions, 10 g of 5 % Pd/C(Aldrich), 78°C, 5.5 h, one equivalent NaOH added to choline chloride.

The results show that by increasing the choline chloride concentration to 14% , a higher conversion (88%) can be obtained. However upon increasing the concentration to 40% , the conversion decreased to 45%. Interestingly enough, using a 40% choline hydroxide improves the conversion to 89%. This result implies that chloride ion is a catalyst poison at higher concentrations.

Once the chemistry of choline oxidation had been demonstrated with a 5 % Pd/C catalyst , we screened catalysts from several different vendors. The most promising were the Degussa E101 and an Engelhard HDC 5 % Pd/C catalysts. All catalysts screened had BET surface areas between 900-1000m^2/g and Pd crystallite size between 130-160 A$^\circ$. The reactions were monitored over time using conductivity measurements as described in the experimental section. Oxidation of a 40% choline hydroxide solution was attempted at 60°C using the E101 5% Pd/C catalyst. The results are shown in Figure 1:

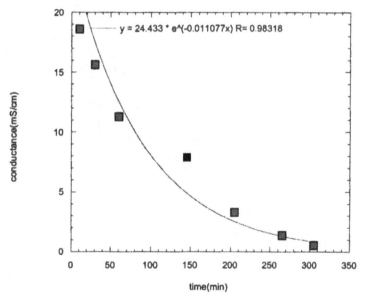

Figure 1 Monitoring the oxidation reaction via conductance versus time.

The conductivity of the solution decreases with time showing that this technique is an excellent way of monitoring the reaction. The shape of the plot suggests the reaction to be first order in choline. However, this result should be treated with caution, since there is an additional factor , adsorption of choline and betaine on the carbon support, which is also a function of concentration. The values of conductivity shown in the plot reflect only the concentration of choline in solution, not what is adsorbed though one would expect a proportionality with concentration of choline in solution. Both choline and betaine could compete for the sites on carbon and one may adsorb preferentially to the other.

An Engelhard HDC catalyst was recycled using a procedure described in the experimental section. The results are described in Table 5.

Table 5 Catalyst Recycle with 40 % Choline Hydroxide.

Run No.	Choline conversion %	Betaine selectivity %
1	92.6	92.8
2	89.9	92.2
3	95.0	95.5
4	94.3	95.5
5	95.4	96.0

The HDC catalyst appears to be stable over 5 runs indicated by consistently high conversions over 5.5 hr reaction time. It was observed that choline hydroxide foams excessively when sparged with O_2 whereas the product betaine does not. Therefore as the reaction progresses the foaming does subside. Increased foaming was noted in successive runs indicating change in the catalyst properties leading to slower kinetics. The filtrates were analyzed for Pd which was found to be ~0.2ppm corresponding to 0.01% metal loss suggesting that leaching of Pd is not a major issue.

The HDC catalyst was characterized using x-ray photoelectron spectroscopy (XPS) and x-ray diffraction (XRD). XPS results in the Pd 3d region of the fresh catalyst showed the surface of Pd particles to be mostly in an oxidized state (Pd(2+) at 337.4 eV), with a smaller contribution from metallic Pd at 335.7 eV. In contrast, the used catalyst had ~ 75% Pd(0) and 25% Pd(2+). In addition, the Pd/C surface atomic ratio decreased from 0.0181(fresh) to 0.0023(used). We believe this indicates the presence of organic species adsorbed on the surface, which may be the reason for the change in catalyst activity. All the results described above were also observed for the E101 catalyst. XRD patterns of used catalysts indicate a shift of the Pd(111) reflection to lower degrees 2 θ when compared to the virgin material. This may be explained by absorption of hydrogen and/or carbon into the bulk Pd lattice (6, 7). We found that we were able to partially form the metallic phase by heating the used HDC catalyst in N_2 flow at 200°C. Pd-H_x phases decompose under these conditions; however, Pd-C_x phases should remain stable up to 600°C in inert gas (7). As to why the Pd-H$_x$ phase is stable in air at room temperature, one possible reason is that the deposition of organic species on the surface slows the diffusion of hydrogen from the bulk. Further evidence of an organic species absorbed on the catalyst was obtained from TPD experiments and is shown in Table 6. The samples were heated and the gaseous products detected by IR.

Table 6 Temperature programmed desorption experiments with catalyst E101.

Catalyst	Temperature °C	Products
fresh in N$_2$	>120	H$_2$O, CO$_2$
used in N$_2$	>120	H$_2$O, CO$_2$, CH$_4$, TMA
fresh in air	>230	No thermal event
used in air	>185	rapid evolution of H$_2$O, CO$_2$, thermal event at 230°C
Betaine	>312	Me$_3$N, CO$_2$, Me$_2$NCH$_2$CO$_2$CH$_3$

The results in Table 6 show that the used catalyst appears to have some organic component strongly adsorbed which reacts vigorously with air above 185°C. It does not appear to be betaine since no Me$_3$N or NOx was detected.

^{13}C NMR was used to determine by-products in the oxidation reaction. Two species other than betaine were detected, [Me$_3$NCH$_2$CH$_2$OH]$^+$OAc$^-$ and Me$_3$N$^+$CH$_2$CH$_2$OCH$_2$CO$_2^-$. The formation of the latter is ascribed to the oxidation of [Me$_3$NCH$_2$CH$_2$OCH$_2$CH$_2$OH]$^+$OH$^-$ which is present as a 4% molar impurity in the starting material. The formation of choline acetate is possibly due to the Hoffman degradation of choline hydroxide with subsequent oxidation of acetaldehyde to acetic acid. The proposed mechanism is described in equations (3) and (4).

[Me$_3$NCH$_2$CH$_2$OH]$^+$OH$^-$ $\longrightarrow$ Me$_3$N + MeCHO + H$_2$O   (3)

MeCHO + 0.5O$_2$ $\longrightarrow$ MeCO$_2$H   (4)

Conclusions

The oxidation of choline hydroxide to betaine has been demonstrated. Analogous oxidations with choline chloride are problematic especially at higher concentrations due to halide ion deactivation of catalyst. The only significant by-product is choline acetate formed by thermal decomposition of choline hydroxide. Catalyst deactivation is due to the deposition of organic species on the surface.

Acknowledgements

The authors wish to acknowledge summer intern student, Tung Nguyen, for conductivity experiments, Diane Nelson for XRD and Terry Slager for TPD analysis, Ann Kotz for NMR experiments, and Air Products and Chemicals, Inc. for permission to publish.

References

1. H.O. Heikkila, J.A. Melaja, E.D. Millner and J.J. Virtanen, US Pat. 4,359,430, to Suomen Sokeri Osakeyhtio (1969)
2. G. Nagy, US Pat. 3,480,665 to Ugine Kuhlmann (1969)
3. H.E. Bellis, T.R. Jemison and O.B. Mathre, US Pat. 5,684,191 to DuCoa (1997)
4. S. Ikuta, Y. Horiuchi, H. Misaki, K. Matsuura, S. Imamura and N. Muto, US Pat. 4,135,980 to Toyo Jozo Kabushiki Kaisha (1979)
5. T. Mallat and A. Baiker, *Catal. Today.*, **19**, 247 (1994)
6. J.A. McCaulley, *J. Phys. Chem.*, **97**, 10372 (1993)
7. S.B. Ziemecki, G.A. Jones, D.G. Swartzfager, R.L. Harlow and J. Faber, Jr.. *J. Am. Chem. Soc.*, **107**, 4547 (1985)

Synthesis of Methylmaleic Anhydride (Citraconic Anhydride) by Heterogeneous Selective Oxidation of Isoprene with V/Ti/O Catalysts

G.L. Castiglioni[1], F. Cavani, C. Fumagalli[1], S. Ligi, and F. Trifirò

Dipartimento di Chimica Industriale e dei Materiali, V.le Risorgimento 4, 40136 Bologna, Italy
[1]Lonza Intermediates and Additives, Via E. Fermi 51, Scanzorosciate (BG)

Abstract

The synthesis of methylmaleic anhydride (citraconic anhydride, CA) by gas-phase selective oxidation of isoprene with a titania-supported vanadium oxide catalyst was studied. The effect of the main operating parameters on catalytic performance was checked. Both conversion and selectivity to CA were found to be controlled mainly by the availability of O-insertion sites at the catalyst surface.

Introduction

Citraconic anhydride (CA, methylmaleic anhydride) is an important intermediate for the preparation of synthetic resins, and is also employed for the synthesis of pharmaceuticals. Due to different functional groups which are present in this relatively small molecule, several synthetic routes can be envisaged, starting from different raw materials (1-4). CA can be synthesized either i) starting from condensation of acetone to yield mesityl oxide, which is then oxidized, or ii) from maleic anhydride, through methylation with formaldehyde or methanol, iii) via pyrolysis of citric acid, iv) from succinic anhydride, via condensation with formaldehyde, or v) by selective oxidation of iso-C_5 hydrocarbons, such as isoamylene or isoprene. The direct oxidation of iso-C_5 hydrocarbons is particularly interesting, since the C_5 fraction is one of the main component of pyrolytic gasoline from naphtha steam cracking. Isoprene in this fraction can be either separated or transformed to ter-amylmethylether; the residual C_5 raffinate mainly contains n- and isopentane, pentenes and cyclopentane. It is of industrial interest to find chemical uses for these hydrocarbons, as feedstocks for the synthesis of valuable intermediates for the petrochemical and polymer industry. In a previous paper, the reactivity of isopentane over various vanadium oxide-based systems was examined (5). Low selectivities to CA were in generally found, due to the preferential formation of maleic anhydride and of carbon oxides. More promising results were obtained starting from isoprene. The present study involved a thorough investigation of the main parameters which affect isoprene conversion and selectivity to CA over titania-supported vanadium oxide as the heterogeneous catalyst, in order to find conditions which maximize the yield to the desired product. In addition, this reaction represents an example of how the catalytic performance in the selective oxidation of hydrocarbons with vanadium oxide-based catalysts can be affected through control of the reaction parameters.

Experimental section

Titania-supported vanadium oxide-based catalysts were prepared by impregnation of TiO_2 anatase with ammonium vanadate. The weight ratio between titania and vanadia was 13. Catalytic tests were carried out in a tubular, fixed-bed laboratory reactor at atmospheric pressure. Three grams of catalyst were loaded, shaped in particles 0.3-0.5 mm in diameter. The condensable products were bubbled in acetone, and analyzed by gas-chromatography (FID detector, OV 101 semi-capillar column). CO and CO_2 were sampled on-line and analyzed by gas-chromatography (TCD detector, Carbosieve S packed column). Identification of the products was done by GC-MS.

Results and discussion

The oxidation of isoprene with molecular oxygen yields a number of products, amongst which the most important are i) products with 5 carbon atoms (besides CA, also 2-methyl-2-butenal and 3-furaldehyde, ii) products with less than 5 C atoms (acrylic acid, maleic anhydride and, mainly, carbon monoxide and carbon dioxide), and iii) products with more than 5 C atoms (methylphthalic anhydride). The formation of all these products occurs by direct, parallel reactions over the olefinic reactant (5). Many of these products clearly are formed through several steps and this means that all the compounds which are intermediate in the formation of these products remain adsorbed on the catalyst surface, and that the only products which desorb into the gas phase are either those that are so stable as not to undergo any further transformation at the adsorbed state, or that possess those features (i.e., in terms of acidity) which make them no longer compatible with the catalyst surface characteristics.

Figure 1 shows the effect of the reaction temperature on the conversion of isoprene and on the yields to the main products; details concerning the oxygenated by-products are reported in Figure 2. Total conversion of isoprene can be reached at around 340°C. At low temperature the formation of CA, as well as of the other products of partial oxidation, is practically negligible. The prevailing products are carbon oxides, the selectivity for which is very high in this low-temperature range. With increasing temperature, the formation of both CA and of other oxygenated by-products increases rather than the formation of carbon oxides, the selectivity for which thus decreases. This phenomenon is not common in the field of heterogeneous selective oxidation of hydrocarbons with molecular oxygen. It can be attributed to the fact that at low temperature the presence of a considerable amount of adsorbed compounds saturates the surface, and decreases the availability of selective O-insertion sites, necessary to perform the transformation up to oxygenated products. Under these conditions these compounds are easily attacked by non-selective oxygen species and burnt to carbon oxides. With increasing temperature, at steady state fewer compounds are adsorbed at the surface, and a greater number of selective oxidation sites becomes available to perform the transformation of isoprene up to CA and to be then reoxidized by gaseous O_2. Under these conditions the reaction becomes more selective.

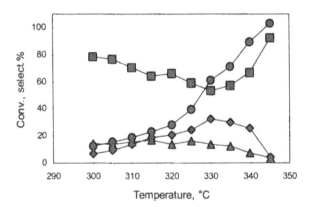

Figure 1 Conversion of isoprene (●) and selectivity to CA (♦), carbon oxides (■) and other by-products (▲) as functions of the temperature. Conditions: feed composition 0.5 mol.% isoprene, 20 % oxygen, rest He; W/F 4 g s / Nml.

At even higher temperature, above 330°C, the products of partial oxidation undergo combustion to carbon oxides, and the selectivity to the latter increases. Under these conditions the selectivity is thus controlled by the stability of the products towards consecutive combustion. Therefore, there exists an optimal temperature range, at around 330°C, where the selectivity to CA is the highest. Figure 1 and Figure 2 show that the maximum selectivity to the other by-products of partial oxidation occurs at a lower temperature (around 310°C)

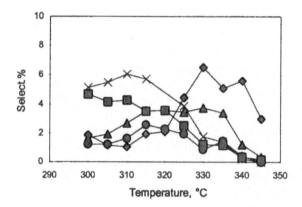

Figure 2 Selectivity to the "other" by-products as functions of the reaction temperature: acrylic acid (✖), 3-furaldehyde (▲), 2-methyl-2-butenal (■), MA (◆), methylphthalic anhydride (●). Reaction conditions as in Fig. 1.

than for CA, due to the different stability of the products towards combustion. The only exception is maleic anhydride, MA, the maximum for which corresponds to that of CA. This clearly comes from the similarity of the two products as regards the stability towards combustion.

In order to confirm the above mentioned hypothesis, tests were made by carrying out the reaction at temperatures lower than that for which the maximum in selectivity to CA is reached (i.e., at 310°C), and by studying the effect of O_2 partial pressure while keeping constant the isoprene partial pressure in the feed. Figures 3 and 4 show the effect of the O_2 partial pressure on catalytic performance.

The main effects obtained by increasing the O_2 partial pressure are i) an increase in isoprene conversion, and ii) an increase in the selectivity to CA with a corresponding decrease in the formation of COx. At very low oxygen-to-isoprene ratios in feed the catalyst surface is fully covered with adsorbed species, and the low number of available active sites leads to a low conversion of isoprene. Once again, under these conditions the selectivity to CA is very low. Increasing the O_2 partial pressure allows faster transformation of all adsorbed compounds and intermediates to CA and MA, due to restoration of selective oxidation sites at the surface. The same effect is obtained when the partial

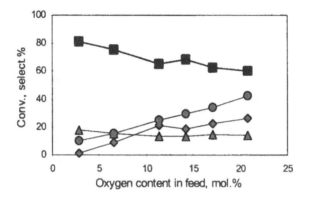

Figure 3 Conversion of isoprene and selectivity to the different classes of products as functions of the O_2 content in the feed. Symbols as in Fig. 1. Reaction conditions: T 310°C, isoprene 0.33 mol.%, W/F 4 g s / Nml.

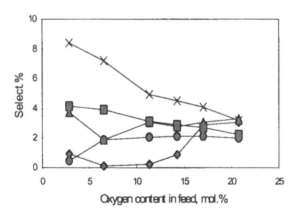

Figure 4 Selectivity to "other" by-products as a function of the O_2 partial pressure in the feed. Symbols as in Fig. 2. Reaction conditions as in Fig. 3.

pressure of isoprene is varied, at 310°C (Figures 5 and 6). Also in this case an increase in the isoprene-to-oxygen content in the feed leads to a decrease in the selectivity to CA, with a corresponding increase in the formation of carbon oxides. When instead the O_2 partial pressure is varied at high temperature (340°C), the selectivity to CA and that to the other by-products decreases as the oxygen-to-isoprene ratio in the feed increases. Therefore under these conditions the products of partial oxidation undergo a consecutive reaction of combustion, which is kinetically favoured at high O_2 partial pressure.

In conclusion, the catalytic performance of a V/Ti/O catalyst in the selective oxidation of isoprene is a function of the operating conditions. Temperature and feedstock composition affect the surface availability of selective oxidation sites, and thus direct the reaction pathway either towards the formation of CA (and of other oxygenated by-products) or towards the formation of carbon oxides. An analogous effect has been found in the oxidation of 1-pentene to phthalic anhydride over V/P/O-based catalysts, while in the oxidation of n-pentane no effect of surface saturation was observed, due to the weaker interaction of the paraffin with the catalyst with respect to the olefin (6).

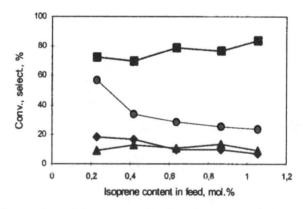

Figure 5 Conversion of isoprene and selectivity to the different classes of products as functions of the isoprene content in the feed. Symbols as in Fig. 1. Reaction conditions: T 310°C, W/F 4 g s / Nml, O_2 20 mol.%

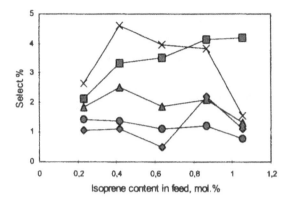

Figure 6 Selectivity to "other" by-products as a function of the isoprene partial pressure in the feed. Symbols as in Fig. 2. Conditions as in Fig. 5.

References
1. I.D. Huang, M.B. Sherwiu and A. Westner, US Pat. 3,987,064 (1976).
2. J.P.Harrison, US Pat. 3,968,127 (1976).
3. R.G. Berg, US Pat. 3,701,805 (1971).
4. J.E. Strojny, Can. Pat. 1,066,712 (1979).
5. C.I. Cabello, G.L. Castiglioni, F. Cavani, C. Fumagalli, L. Grasso, S. Ligi and F. Trifirò, *Science and Technology in Catalysis 1998*, Kodansha, Tokyo, 1999, p. 359
6. F. Cavani and F. Trifirò, *Appl. Catal., A: General*, **157**, 195 (1997).

Homogeneous Catalysis for Fine Chemicals: The Heck Reaction as a Clean Alternative for Friedel-Crafts Chemistry

Massoud S. Stephan and Johannes G. de Vries*

DSM Research, Dept. Fine Chemicals, P.O. box 18, 6160 MD Geleen, The Netherlands
e-mail: Hans-JG.Vries-de@dsm-group.com

Abstract

Functionalization of aromatic compounds using Friedel-Crafts chemistry is accompanied by a large amount of waste. The Heck reaction is an industrially viable alternative for carbon carbon bond formation to arenes. However, all existing methods make use of halide containing arylating agents such as aryl iodides and bromides, which also leads to salt waste. This dilemma was solved by the invention of aromatic anhydrides as the aryl donor. This new decarbonylative Heck reaction is catalyzed by $PdCl_2$, which is promoted, ironically, by a small amount of halide salt such as chloride or bromide. Benzoic acid, which is recycled to the anhydride and CO, which is burned to CO_2 are the only side products. The reaction needs neither bases nor phosphine ligands. Olefins with electron-withdrawing groups are excellent substrates, but electron-donating substituents gave rise to poor yields of the α-arylated olefins. Unsubstituted olefins are rapidly isomerized leading to isomeric mixtures of arylation products. The mechanism is still under investigation.

Introduction

The production of fine chemicals, in particular those made on a scale of 1-100 tons/year is largely based on the use of stoichiometric chemistry. This has the advantage of easy and rapid scale-up and results in reliable processes, which is dictated by market demand. Unfortunately, it also leads to the formation of a relatively large amount of waste (1).

The production of intermediates containing an aromatic ring is a case in point. These compounds are currently largely produced by use of the Friedel-Crafts reaction. Particularly with deactivated aromatic compounds this requires more than stoichiometric amounts of catalysts such as $AlCl_3$. Consequently, the amount of NaCl and Al_2O_3 waste grossly exceeds the amount of product.

In recent years, the fine chemical industry has discovered the Heck reaction (2) as a method to form carbon carbon bonds between aromatic rings and olefins (Scheme 1). Examples are the agrochemical Prosulfuron™ (3), the

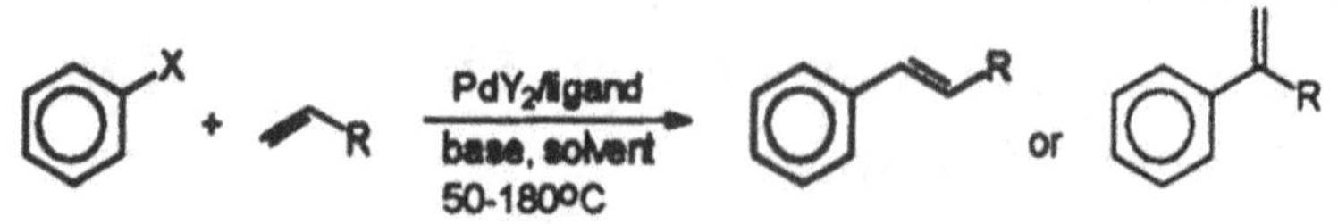
PdY2/ligand
base, solvent
50-180°C
or

X = I, Br, Cl, COCl, OSO $_2$R, $N_2^+X^-$
Y = Cl, OAc
Ligand = None, phosphine, phosphite, phosphoramidate
Base = Et_3N, $NaHCO_3$, K_2CO_3, KOAc, KH_2PO_4

Scheme 1 The Heck reaction.

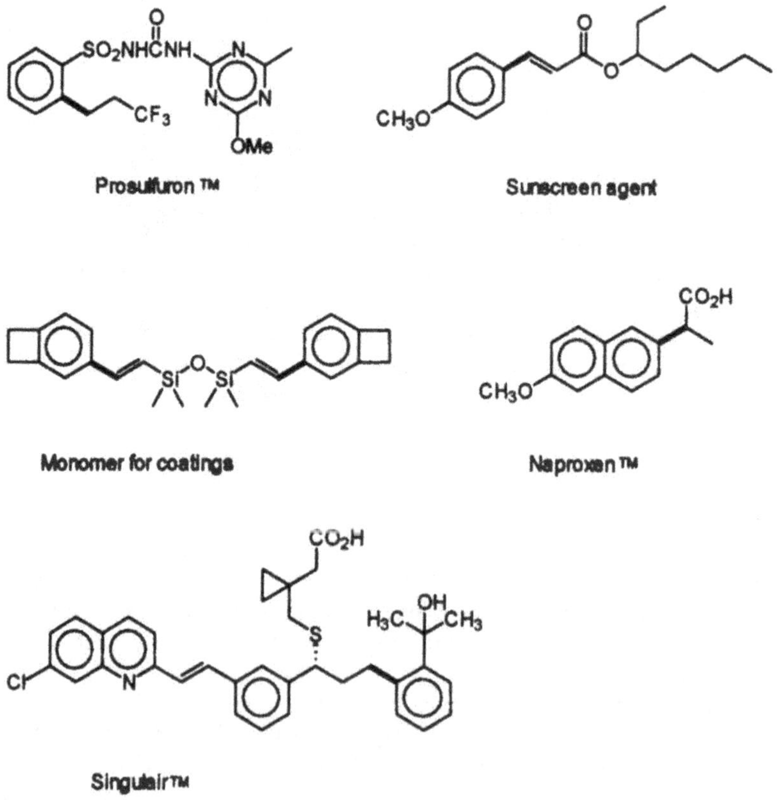
Prosulfuron ™
Sunscreen agent
Monomer for coatings
Naproxen ™
Singulair™

Scheme 2 Use of the Heck reaction for commercial products and intermediates (the **bold** bond was created using the Heck reaction).

anti-inflammatory Naproxen™ (4), the sunscreen agent octyl *p*-methoxycinnamate (5), the anti-asthma agent Singulair™ (6) and monomers for coatings (7) (Scheme 2).

The regiochemistry of the Heck reaction is usually determined by the nature of the group R on the olefin: Electron-withdrawing groups lead almost exclusively to the β-arylated products, whereas electron donating substituents usually lead to a large proportion of the α-arylated product. The Heck reaction can accommodate various substituents on the aromatic or hetero-aromatic ring. Consequently, the method has a large scope. Though it is obvious that this method is much cleaner than Friedel-Crafts chemistry, the arylating agents always contain a halide (8), leading to the formation of one equivalent of sodium halide as a waste product. As part of our basic research program aimed at the use of homogeneous catalysis for the production of fine chemicals, we have developed a new variant of the Heck reaction that is free of salt waste.

A waste free Heck reaction

As the halide waste in the Heck reaction stems from the haloarene we set out to find a halide free aryl equivalent. After testing several candidates, we finally found that aromatic anhydrides can be used for this purpose. Thus, reacting benzoic anhydride with *n*-butyl acrylate in the presence of $PdCl_2$ (0.25 mol%) and a small amount of NaBr (1 mol%) as promotor at 160°C in NMP as solvent, led to complete conversion of benzoic anhydride after 90 min. and formation of *n*-butyl cinnamate in 77% isolated yield (Scheme 3) (9).

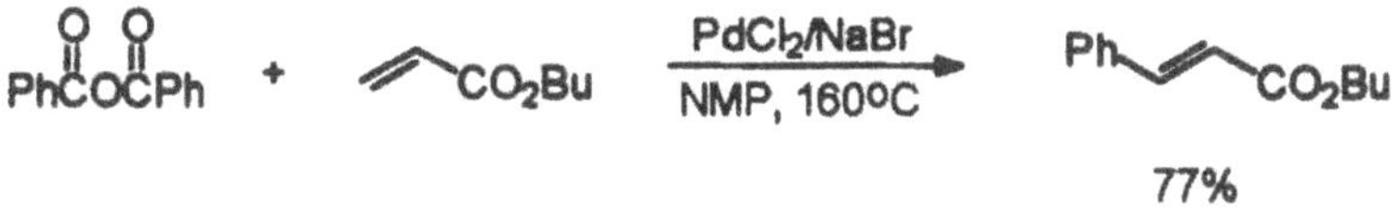

$$
PhCOCPh \ + \ \diagup\!\!\!\diagdown CO_2Bu \ \xrightarrow[\text{NMP, 160°C}]{PdCl_2/NaBr} \ Ph\diagup\!\!\!\diagdown CO_2Bu
$$

77%

Scheme 3 The DSM Heck reaction.

Side products in this decarbonylative Heck reaction are benzoic acid, which can be recycled and CO, which is burned off to CO_2. This new variant has many advantages particularly in relation to an industrial application:
1. No salt waste.
2. No bases are needed. In addition to the obvious cost reduction, this will also simplify the isolation procedure.
3. No phosphines are needed.
4. Fast reaction. It is also possible to lower the catalyst/substrate ratio to 1:1000.

It is known from an older patent that benzoic anhydride can be produced by simply heating benzoic acid till slightly below its boiling point (10). Water is collected over the top of the distillation unit and benzoic anhydride is collected from the bottom. This establishes the waste free production of the starting material. The next aspect to be addressed was the waste free isolation. We have developed two different methods. In case the boiling point of the product is sufficiently different from benzoic acid simple distillation can be used. In case this is not possible, the following procedure is recommended: The solvent NMP is distilled off, after which benzoic acid is isolated by extraction of the residue with hot water. The product, which remains in the organic phase can now be purified by distillation or crystallization.

Scope of the DSM Heck reaction

In general, the reaction works well for olefins containing electron-withdrawing groups. In practically all cases, good isolated yields of the β- arylated products (3) were obtained (Scheme 4, Table 1). Particularly gratifying was formation of

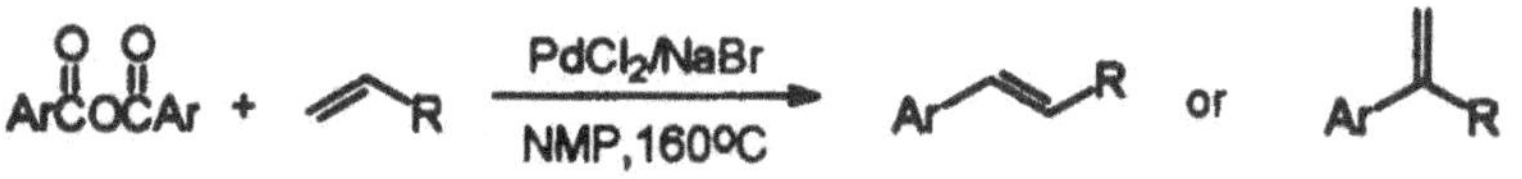

Scheme 4 Regiochemistry in the DSM-Heck reaction depends on the nature of the substituent R.

Table 1 Palladium catalyzed arylation of olefins (2) with aromatic carboxylic anhydrides (1).[a]

Run	Ar	Olefin	Time (min)	Yield of 3 (%)
1.	Ph	$CH_2=CHCO_2Bu$	90	77
2.	Ph	$C_8H_{17}CH=CH_2$	90	60
3.	Ph	$PhCH=CH_2$	120	76 (α:β=13:87)
4.	Ph	Cyclooctene	120	54
5.	Ph	$CH_2=CHOC_6H_{11}$	120	25[b] α
6.[c]	Ph	$CH_2=CHCN$	180	60 (E:Z=75:20)
7.	Ph	$CH_2=C(Me)CO_2Bu$	90	62
8.[d]	Ph	$Z\text{-}BuO_2CCH=CHCO_2Bu$	90	72 (E:Z=7:2)
9.	Ph	N-vinyl-pyrrolidone	90	25 (α:β-E= 20:5)
10.[d]	p-MeO-C_6H_4	$CH_2=CHCO_2Bu$	60	75
11.	2-Furanyl	$CH_2=CHCO_2Bu$	90	77

[a] Conditions: Aromatic anhydride : olefin : $PdCl_2$: NaBr = 100 : 120 : 0.25 : 1 mmol in 100 ml NMP, T = 160°C unless indicated otherwise. [b] GC Yield [c] T= 140°C [d] T = 190°C

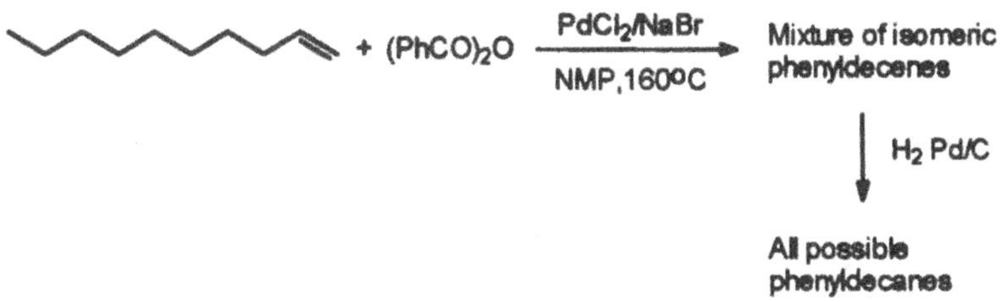

Scheme 5 Double bond isomerization and lack of regioselectivity in arylation of unsubstituted olefins.

the sunscreen agent n-butyl p-methoxycinnamate in 75% yield from p-methoxybenzoic anhydride (entry 10), in spite of the higher reaction temperature of 190°C, which was needed for this reaction. Apparently, electron-donating substituents on the aryl ring have a negative impact on the rate of reaction. Even butyl methacrylate (entry 7) was arylated in good yield, though double bond isomerization of the product resulted in a mixture of isomers. The Heck reaction of styrene gave 76% of stilbene and 11% of the α-arylated product 1,1-diphenyl-ethene as a side product.

   The arylation of unsubstituted olefins was also investigated. The reaction of 1-decene gave a mixture of arylated decenes in 60% yield (Scheme 5). Obviously, under the conditions of this reaction (Pd/acid) double bond isomerization will readily occur. What remained to be investigated was the selectivity of the arylation reaction. Is there a preference for terminal over internal double bonds? To simplify the analysis we hydrogenated the product mixture over Pd/C. The hydrogenation product was analyzed by GC and showed a mixture of all 5 possible decylbenzenes. The arylation of cyclooctene also led to a mixture of regioisomers. Thus, we must conclude that arylation of unfunctionalized olefins takes place without any useful selectivity. Addition of phosphine ligands did not reduce the isomerization substantially.

   Reaction of olefins containing electron-donating substituents leads to mainly α-arylated products with the classical Heck reaction. We found that vinyl esters were poor substrates as insertion of palladium in the vinyl-oxygen bond takes place, leading to the formation of styrene and stilbene in moderate yields. A more stable enol ether such as cyclohexyl vinyl ether (entry 5) could be arylated in a modest 25% yield, but gave exclusively the α-arylated product. Likewise, reaction of benzoic anhydride with N-vinyl-pyrrolidone gave a 4:1 mixture of α– and β-N-styryl-pryrrolidone in 25% yield with 58% of N-benzoyl-pyrrolidone as a major side product.

The arylating agent

In addition to benzoic anhydride and p-methoxy-benzoic anhydride heterocyclic anhydrides also were excellent arylating agents. Reaction of furanoic anhydride with butyl acrylate gave butyl 2-(2-furyl)-acrylate in 77% yield. Looking from the industrial perspective, the number of readily available aromatic anhydrides is probably limited to benzoic anhydride. Obviously, the anhydride could be produced by the distillation method referred to before, but this would entail an extra unit operation and thus have a negative impact on the cost of production. For this reason, a method that induces formation of the anhydride from the acid during the Heck reaction would tremendously enhance the scope of the method. As a first attempt, we have explored the use of mixed anhydrides prepared from acetic anhydride and benzoic acid (Scheme 6). However, both formation *in situ* as well as pre-formation resulted in very low

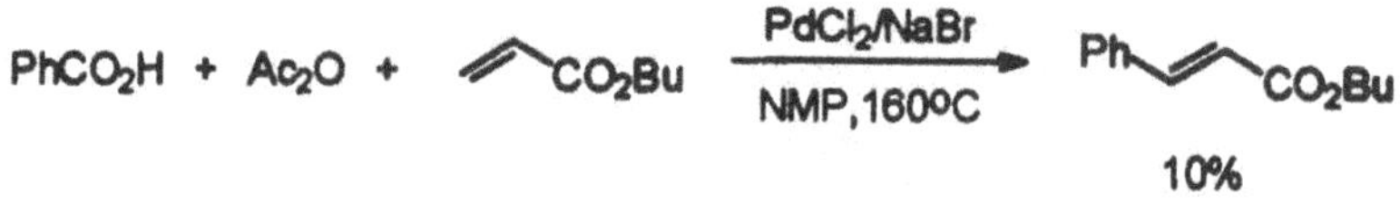

PhCO$_2$H + Ac$_2$O + ⟍⟍CO$_2$Bu $\xrightarrow[\text{NMP,160}^\circ\text{C}]{\text{PdCl}_2/\text{NaBr}}$ Ph⟍⟍CO$_2$Bu

10%

Scheme 6 Attempted use of mixed anhydrides in the DSM-Heck reaction.

yields of cinnamate esters. The mixed anhydrides are not very stable under the conditions of the DSM-Heck reaction and disproportionate to a mixture of benzoic and acetic anhydride (11). The low yield might be related to the negative effect that acetate exerts as counter ion to palladium (See Table 3). Not surprisingly, Ac$_2$O itself failed as methylating agent, presumably for the same reason.

An even better proposition would be the catalytic formation of aromatic anhydrides from the carboxylic acids during the Heck reaction. Consequently, a number of acids, such as *p*-toluenesulfonic acid, acidic ion exchange resins and Lewis acids were screened, but to no avail. It is possible, however, to use benzoic acid with stoichiometric P$_2$O$_5$ (12). This leads to the formation of cinnamates in synthetically useful yields, but the reaction is much slower.

Catalysts

A wide range of catalysts has been screened for this reaction, but only palladium compounds were catalytically active (Table 2). Phosphine ligands can be used but retard the reaction and have no benefits. Most useful catalyst precursors were PdCl$_2$ or PdCl$_2$(PhCN)$_2$. From screening the anions its also became apparent that either chloride or bromide is necessary to activate the catalyst; iodide and other anions have no effect or even retard (Table 3). The results also indicate that this is not a cation effect as sodium, lithium or tetra-alkylammonium had comparable effects. The differences that are observed probably are related to the solubility of these compounds. Figure 1 clearly

Table 2 Performance of Pd-catalysts in the DSM Heck reaction.[a]

Run	Catalyst (Mol%)	Additive (Mol%)	Conv. of anhydride (%)	Yield of cinnamate (%)
1	$[PdCl_2(PhCN)_2]$ (1.0)	None	36	27
2	$[PdCl_2(PhCN)_2]$ (0.5)	Bu_4Br (2.0)	100	89
3	$PdCl_2$ (0.25)	None	12	10
4	$PdCl_2$ (0.25)	NaBr (0.5)	67	57
5	Na_2PdCl_4 (0.1)	None	61	50
6	Na_2PdCl_4 (0.1)	NaBr (0.5)	77	63
7	Na_2PdCl_6 (0.1)	None	29	15
8	Na_2PdCl_6 (0.25)	Bu_4NBr (1.0)	100	81
9	$[Pd_2(dba)_3]$ (0.25)	None	0	1
10	$[Pd_2(dba)_3]$ (1.0)	Bu_4NBr (5.0)	72	59
11	$[Pd(Ph_3P)_4]$ (0.25)	None	4	4
12	$[Pd(Ph_3P)_4]$ (0.25)	Bu_4NBr (1.0)	29	11
13	$NiCl_2.6H_2O$	None	0	0
14	$CuCl_2$	None	0	0

[a] Reaction in NMP at 155°C during 1 h. Conversion and yield determined by GC

Table 3 Effect of additives on conversion of 1 and yield of 3 in the DSM Heck reaction.[a]

Run	Catalyst	Additive	Conv. of anhydride (%)	Yield of cinnamate (%)
1	$PdCl_2$	Bu_4NCl	68	57
2	$PdCl_2$	Bu_4NBr	75	53
3	$PdCl_2$	Bu_4NI	24	22
4	$PdCl_2$	Bu_4NOAc	24	16
5	$PdCl_2$	Bu_4NHSO_4	3	3
6	$PdCl_2$	Bu_4NOTf	26	22
7	$PdCl_2$	Bu_4NBF_4	38	29
8	$PdCl_2$	Me_4NCl	55	47
9	$PdCl_2$	Bu_4PCl	65	51
10	$PdCl_2$	LiCl	63	50
11	$PdCl_2$	NaCl	44	34
12	$PdCl_2$	CsCl	38	32
13	$PdCl_2$	LiBr	63	55
14	$PdCl_2$	NaBr	67	57
15	$PdBr_2$	Bu_4NBr	52	39
16	PdI_2	Bu_4NI	4	3
17	$Pd(OAc)_2$	Bu_4NOAc	1	1
18	$PdSO_4$	Bu_4NHSO_4	1	1

[a] Reaction in NMP at 155°C with anhydride:olefin:Pd:additive=100:110:0.25:0.5

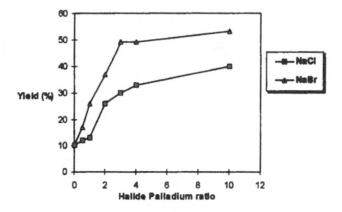

Fig. 1 Dependence of yield of cinnamate on halide palladium ratio.

shows the relation between the amount of halide ion and the yield of the reaction[*]. The fact that the effect levels off with 4 equivalents of halide suggests that it is related to the coordination chemistry of the catalyst. An alternative explanation would entail the reaction of benzoic anhydride with NaCl or NaBr to form the corresponding acyl halide, which is known to be an effective arylating agent in the Heck reaction (13). However, in separate experiments we could not find any evidence for this reaction.

Mechanism

The oxidative addition of benzoic anhydride to a Pd(0) phosphine complex has been observed by Yamamoto, who characterized the intermediate Pd(II) benzoyl benzoate complex, and also showed it could be hydrogenated to benzaldehyde and benzoic acid at 80°C (14). In view of this fact it seems likely that in our reaction also the first step is oxidative addition of benzoic anhydride to a Pd(0) species. Addition of an excess of Hg entirely stops the reaction, which confirms the presence of Pd(0) (15). The dependence of the reaction on halide ion suggests that anionic halide ligated Pd(0) species may be involved (16). A catalytic cycle can be written for this reaction as depicted in Scheme 7. Some other ligand will be necessary to create a stable and monomeric Pd(0) species, or it could be dimeric or even polymeric. This species than undergoes oxidative addition with benzoic anhydride, followed by decarbonylation. Exchange of CO for olefin is followed by olefin insertion to give the Pd-alkyl species, which will

---

[*] The reaction was performed at somewhat lower temperature and for a shorter period of time to ensure that the yield actually reflects differences in rate.

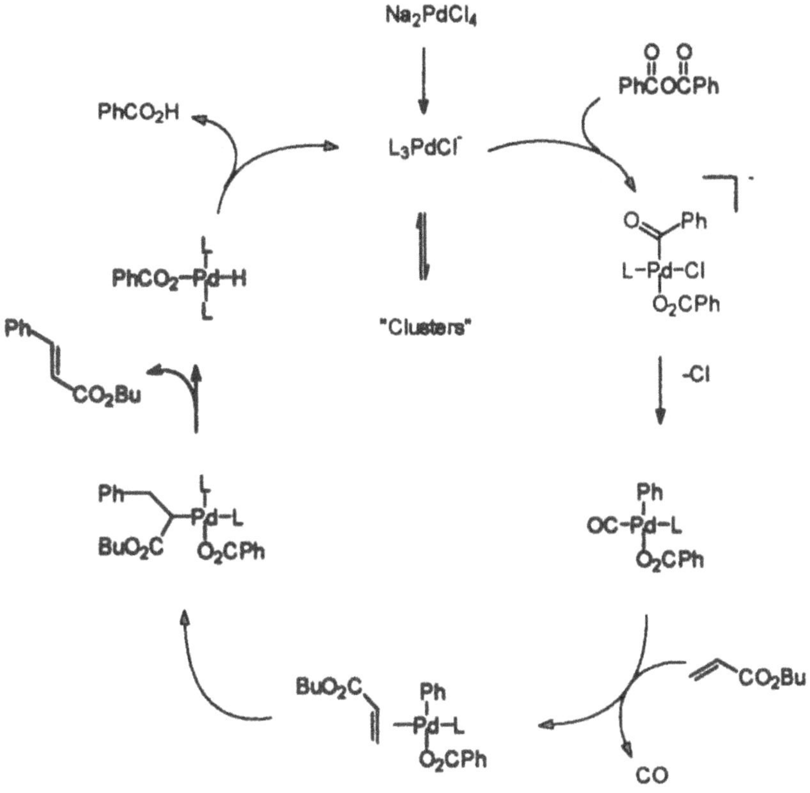

Scheme 7 Proposed mechanism for the DSM Heck reaction.

immediately undergo β-hydride elimination to give the cinnamate and a Pd(II)-hydrido benzoate species. It has always been assumed in the classical Heck reaction that base is necessary to reconvert this species back to Pd(0). Apparently, there is no need for this with the weaker acid benzoic acid. We assume that the oxidative addition of benzoic anhydride is the rate determining step. An assumption, which seems to be supported by the fact that electron-donating substituents on the aryl ring retard the reaction. To lend credibility to this notion we next performed electrospray MS on the crude reaction mixture at several intervals during the reaction to determine the exact nature of the L_3PdCl^- species. However, the only Pd-species we were able to detect were $[PdCl_3]^-$ or more likely the bisanionic dimer, $[PdCl_3(CO)]^-$ and $[PdCl_3(C_6H_5CO_2H)]^-$. The latter presumably is a complex with a hydrogen bond between chloride and benzoic acid. As these are all Pd(II) species we reconsidered the possibility of a Pd(II)/Pd(IV) catalytic cycle. To gain more information we next performed EXAFS on the reaction mixtures at several intervals (17). Initially, very poor results were obtained as the time for one measurement is about 30 minutes and

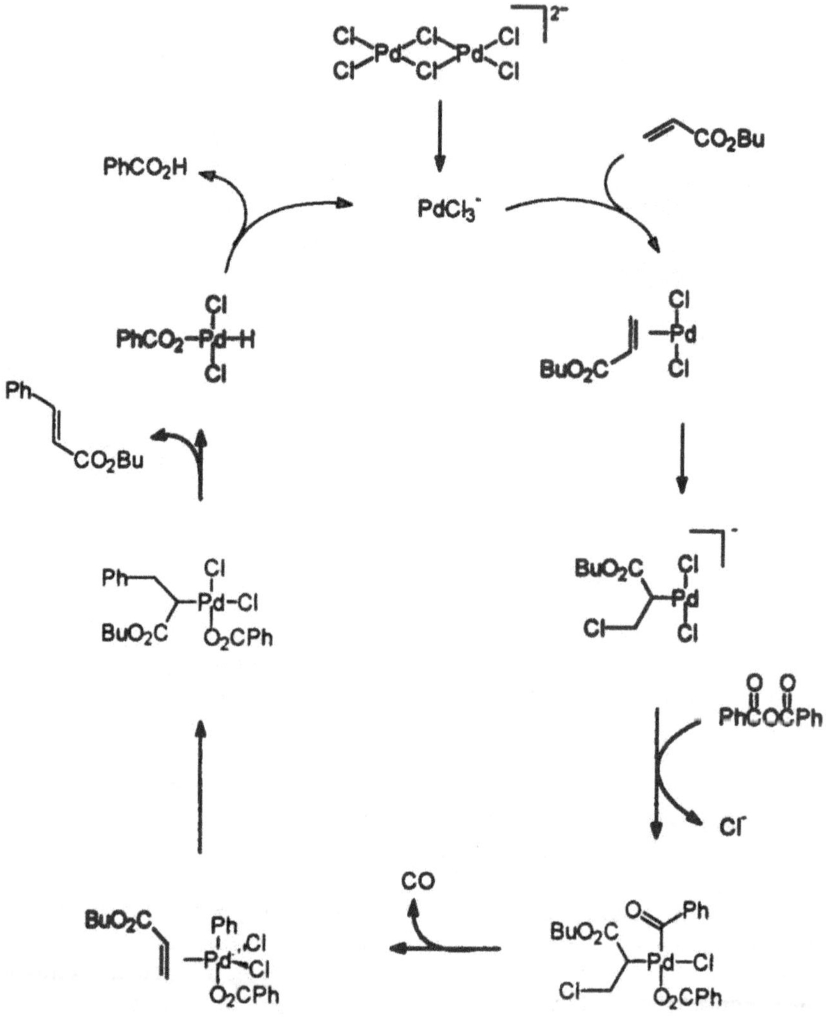

Scheme 8 Pd(II)/Pd(IV) mechanism for the DSM Heck reaction based on Shaw.

many changes take place in the reaction mixture during that period. Much better results were obtained by freezing the reaction mixture in liquid nitrogen. From these experiments it became clear that $PdCl_3^-$ is present in the initial stages, but as the reaction progresses more and more Pd-Pd bonds start to form, indicative of the formation of Pd-clusters. The picture that emerges form this is catalysis by an unsaturated Pd(0) species, which has a tendency to oligomerize into clusters. At the same time the reverse reaction also takes place as Reetz has shown that Pd-clusters are viable albeit slower catalysts in the Heck reaction (18).

An alternative mechanism, which we cannot exclude at this stage is a mechanism resembling the one proposed by Shaw for the Herrmann-Beller palladacycle catalyzed Heck reaction (19), which is based on a Pd(II)/Pd(IV) cycle (20). In this mechanism the role of the halide anions is to promote the first step, the formation of the β-haloalkyl-palladium intermediate (Scheme 8). This mechanism takes into account the fact that all reagents need to be present before the catalytic cyle can start, i.e. no reaction occurred when $PdCl_2$/NaBr was heated to 160°C with benzoic anhydride in a stochiometric reaction. A similar mechanism has been proposed for the rhodium catalyzed benzoylation of styrene with benzoic anhydride (21).

References

1.   R.A. Sheldon, *J. Mol. Catal. A Chem.*, **107**, 75 (1996).
2.   a) R.F. Heck, *Org. React.*, **27**, 345 (1982); b) R.F. Heck in *Comprehensive Organic Synthesis* Vol.4, (Eds.: B.M. Trost and I. Fleming), Pergamon, Oxford, 833 (1991); c) A. de Meijere and F.E. Meyer, *Angew. Chem. Int. Ed. Engl.* **33**, 2379 (1994); d) W. Cabri and I. Candiani, *Acc. Chem. Res.* **28**, 2 (1995); e) J. Tsuji, Palladium Reagents and Catalysts-Innovations in Organic Synthesis, Wiley, Chichester, UK, 1995. f) M. Beller, T.H. Riermeier and G. Stark in Transition Metals for Organic Synthesis; Building Blocks and Fine Chemicals, ed. M. Beller and C. Bolm, Wiley-VCH, Vol 1, 1998, 208. g) S. Bräse and A. de Meijere in Metal-catalyzed Cross-coupling Reactions, eds. F. Diederich and P.J. Stang, Wiley-VCH, Weinheim, 1998, 99.
3.   a) R.P. Bader, P. Baumeister and H.-U. Blaser, *Chimia*, **50**, 99 (1996). b) P. Baumeister, G. Seifert and H. Steiner, EP 381,622, to Ciba-Geigy AG (1990).
4.   a) J. McChesney, *Spec. Chem.*, **19**, 98 (1999). b) T.-C. Wu, US 5,315,026, to Albemarle Corporation (1994). c) T.-C. Wu, US 5,536,870, to Albemarle Corporation (1996).
5.   a) A. Eisenstadt, Chem. Ind. (Marcel Dekker), **75** (*Catal. Org. React.*), 415 (1998).

6.   I. Shinkai, A.O. King and R.D. Larsen, *Pure & Appl. Chem.*, **66**, 1551-1556 (1994).

7.   R.A. DeVries, P.C. Vosejpka and M.L. Ash, Chem. Ind. (Marcel Dekker), **75**, (*Catal. Org. React.*), 467 (1998).

8.   The only exception is the triflates. However, these are relatively expensive and require corrosive chemicals for their production.

9.   M.S. Stephan, A.J.J.M. Teunissen, G.K.M. Verzijl and J.G. de Vries, *Angew. Chem. Int. Ed. Engl.*, **37**, 662 (1998).

10.   H. Schäfer, W. Riemenschneider and E. Pászthory, German Patent 1,141,282, to Farbwerke Hoechst AG, 1962.

11.   In fact this is the basis of a published procedure for the preparation of benzoic anhydride: T.H. Clark and E.J. Rahrs, *Org. Synth. Coll. Vol. I*, 91 (1941).

12.   F.J. Parlevliet and J.G. de Vries, unpublished results.

13.   H.-U. Blaser and A. Spencer, *J. Organomet. Chem.*, **233**, 267 (1982).

14.   K. Nagayama, F. Kawataka, M. Sakamoto, I. Shimizu and A. Yamamoto, *Chem. Lett.*, 367 (1995).

15.   This is not sufficient to prove that Pd(0) is actually involved in the catalytic cycle.

16.   For a discussion on the role of chloride ligated palladium phosphine complexes in oxidative addition reactions see: a) C. Amatore, M. Azzabi and A. Jutand, *J. Am. Chem. Soc.*, **113**, 8375 (1991); b) C. Amatore, A. Jutand and A. Suarez, *J. Am. Chem. Soc.*, **115**, 9531 (1993).

17.   J.H. Bitter, D.C. Koningsberger and J.G. de Vries, to be published.

18.   M.T. Reetz, R, Breinbauer and K. Wanninger, *Tetrahedron Lett.*, **37**, 4499 (1996).

19.   W.A. Herrmann, C. Brossmer, C.-P. Reisinger, T.H. Riermeier, K. Öfele and M. Beller, *Chem. Eur. J.*, **3**, 1357 (1997).

20.   a) B.L. Shaw, *New J. Chem.*, 77 (1998). b) B.L. Shaw, S.D. Perera and E.A. Staley, *Chem. Commun.*, 1361 (1998).

21.   K. Kokubo, M. Miura and M. Nomura, *Organometallics*, **14**, 4521 (1995).

In Situ Generation of Formaldehyde for Environmentally Benign Chemical Synthesis

Eric H. Shreiber[a] and George W. Roberts

Department of Chemical Engineering, North Carolina State University, Box 7905, Raleigh, North Carolina 27695-7905

[a] *Current address: Praxair, Inc., 175 E. Park Ave., Tonawanda, NY 14150*

Abstract

Formaldehyde is a hazardous, self-reactive, and costly chemical that traditionally has been avoided in chemical synthesis. However, a number of important chemicals, e.g., methyl methacrylate and styrene, might be produced more safely and with much less pollution from formaldehyde than by their current commercial processes. This research introduces the concept of in situ generation of formaldehyde, i.e., the generation and consumption of formaldehyde in the same reactor, so that there is no net production.

Methanol dehydrogenation to formaldehyde was studied in a slurry reactor at 598K over three different catalysts: Raney copper, copper chromite, and 3% Mn on copper chromite. This reaction was rapid, and approached equilibrium at some operating conditons for the three catalysts. Under this circumstance, side reactions such as formaldehyde dimerization to methyl formate, and formaldehyde and/or methyl formate decomposition to carbon monoxide, were favored. The formation of methyl formate is not negative per se, because methyl formate appears to serve as a "reservoir" of formaldehyde.

Introduction

The cost and risk of producing, handling, shipping, and storing hazardous chemicals has motivated the chemical industry to avoid the use of such compounds as raw materials for chemical synthesis. Although this strategy can be valid, some of the production alternatives also have serious safety and environmental concerns. A novel approach to the problem of using hazardous and reactive chemicals is in situ generation.

In situ generation involves carrying out reactions that produce and consume the hazardous chemical in a single reactor, with no net production of the chemical, and therefore no need to purify, store, or transport it. Some of the best examples of the potential impact of in situ generation involve

391

formaldehyde, a versatile chemical intermediate that is a basic building block for a myriad of chemicals with a wide variety of end uses (1). Formaldehyde is classified as a suspect and probable human carcinogen by several organizations. Furthermore, formaldehyde is too self-reactive to be isolated, shipped, and stored in its pure form.

Commercial forms of formaldehyde include trioxane, methylal, and formalin. The expense of trioxane and methylal warrants against their use as commodity raw materials. Although formalin is cheaper, the large quantity of water it contains can: a) inhibit the reaction of formaldehyde with other molecules, both thermodynamically (2) and kinetically (e.g., 3); b) complicate the design of a product separation system; and c) increase energy consumption.

If a safe and inexpensive source of formaldehyde were available, several hazardous and polluting processes might be replaced with environmentally benign alternatives. Two potential examples are the production of methyl methacrylate and styrene. All of the methacrylate monomers produced in the United States and about 90 percent of worldwide production are made via the Acetone Cyanohydrin process (4). One of the raw materials, hydrogen cyanide, is toxic, making it difficult and costly to handle and introducing the possibility of a serious environmental incident. Moreover, reaction stoichiometry requires the production of one mole of ammonium bisulfate per mole of methacrylate. This by-product has little or no economic value. The most common disposal techniques have been to neutralize and sewer the stream or to pump it to deep wells.

In the principal route for styrene production, benzene is alkylated with ethylene, typically in the liquid phase with AlCl3 as the catalyst, to produce ethylbenzene, which then is dehydrogenated to styrene (5). Benzene is toxic and carcinogenic, thereby introducing the same handling issues previously discussed with hydrogen cyanide. The disposal of "spent" AlCl3 also creates environmental hazards. About 1 pound of catalyst must be discarded per 100 pounds of styrene produced (6).

Both methacrylate and styrene monomers can be produced via in situ generation of formaldehyde. Formaldehyde is formed via methanol dehydrogenation (Reaction 1 in Figure 1). Formaldehyde then can react with methyl propionate to form methyl methacrylate (Reaction 2 in Figure 1) or with toluene to form styrene (Reaction 3 in Figure 1). Combining Reaction 1 with either Reaction 2 or 3 leads to the overall reactions for methyl methacrylate and styrene production, Reactions 4 and 5 in Figure 1.

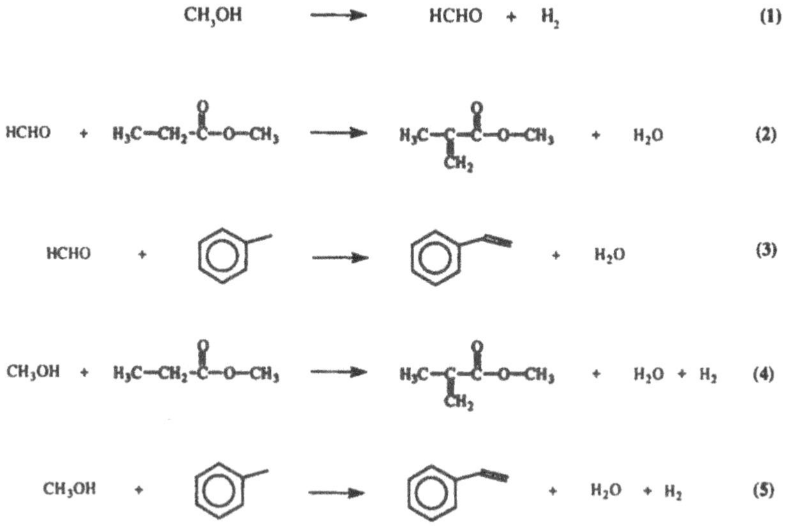

$$CH_3OH \longrightarrow HCHO + H_2 \tag{1}$$

$$HCHO + H_3C-CH_2-\overset{O}{\underset{}{C}}-O-CH_3 \longrightarrow H_3C-\overset{}{\underset{CH_2}{C}}-\overset{O}{\underset{}{C}}-O-CH_3 + H_2O \tag{2}$$

$$HCHO + \text{(styrene precursor)} \longrightarrow \text{(styrene)} + H_2O \tag{3}$$

$$CH_3OH + H_3C-CH_2-\overset{O}{\underset{}{C}}-O-CH_3 \longrightarrow H_3C-\overset{}{\underset{CH_2}{C}}-\overset{O}{\underset{}{C}}-O-CH_3 + H_2O + H_2 \tag{4}$$

$$CH_3OH + \text{(ethylbenzene)} \longrightarrow \text{(styrene)} + H_2O + H_2 \tag{5}$$

Figure 1 - Component and overall reactions for methyl methacrylate and styrene production.

The above routes to methyl methacrylate and styrene are much "greener" than the current commercial processes. Water and hydrogen are the only by-products required by the stoichiometry of Reactions 4 and 5, and the raw materials for these reactions are much less hazardous than those used in the current syntheses. If formaldehyde generation is properly balanced with formaldehyde consumption, there will be no net formaldehyde production. Through in situ generation, formaldehyde becomes a reaction intermediate rather than a raw material.

Reactions 2 and 3 proceed with high rates and selectivities in the temperature range of 573 and 673K over various heterogeneous catalysts (e.g. 7-10). However, studies of methanol dehydrogenation to formaldehyde are limited in this temperature range. Such studies typically have involved copper-based catalysts (11-14). The reaction rates in the studies with these catalysts ranged from 1×10^{-3} to 1×10^{-2} moles HCHO/g cat-hr, with selectivities as high as 100%. The major by-products were methyl formate, carbon monoxide, and carbon dioxide.

Use of a continuous slurry reactor is an ideal approach to carrying out two reactions in series, each of which requires a different catalyst. The catalysts are suspended in an inert liquid medium. The reactants are fed as a gaseous stream and the products are removed in the gas that leaves the reactor.

A more detailed understanding of formaldehyde generation from methanol at temperatures between 573 and 673 K is essential to the successful coupling of Reaction 1 with a formaldehyde-consuming reaction. The objective of this research was to compare three different copper-based catalysts: Raney copper, copper chromite, and 3% Mn on copper chromite, for methanol dehydrogenation at 598K in a slurry reactor.

Experimental

Raney copper catalyst (100 wt% Cu) was obtained from W.R. Grace Co. Both copper chromite (75 wt% CuO, 25 wt% CuCr2O4) and 3% Mn on copper chromite (73 wt% CuCr2O4, 21 wt% CuO, 5 wt% MnO) were supplied by Engelhard Corporation. The liquid in the slurry reactor was 1,2,3,4-tetrahydroquinoline (THQ). This material is thermally stable at 648 K (15). Reagent-grade methanol and THQ were obtained from Fisher Scientific. Nitrogen (99.99% purity) was obtained from a local distributor. Methanol and THQ were sparged with nitrogen prior to use to remove any dissolved oxygen.

For each experiment, 10 g of catalyst and 100 mL of THQ were charged to a 300 mL stirred autoclave reactor (16). Nitrogen was delivered to the reactor, below the impeller, through an activated charcoal trap to remove any impurities. The reactor was brought to the final operating pressure under nitrogen at ambient temperature, then heated to 598K over a 2.5 hour period in flowing nitrogen. Once the reactor had reached the final temperature, methanol feed was begun using a syringe pump. The reactor then operated continuously.

An overhead system was used to condense and return the THQ that vaporized from the reactor. Liquid samples were taken from the overhead system periodically and were analyzed using gas chromatography/mass spectrometry (GC/MS). The liquid in the reactor was also analyzed via GC/MS after each experiment.

The BET surface area and pore volume of fresh and spent catalyst samples were measured using a Micromeritics Flowsorb II 2300 Surface Analyzer. The "spent" catalysts from each experiment were washed thoroughly in acetone prior to analysis to remove any trace liquids that may have accumulated on the catalyst pores. The measurement procedure is outlined in Reference 17. The reported BET surface areas and pore volumes were derived from desorption peaks (18).

The reaction products and unconverted reactants were analyzed by on-line gas chromatography. Injections were made simultaneously into two columns. A 15 ft. Carboxen 1000 column was used to analyze hydrogen, nitrogen, carbon monoxide, and carbon dioxide via a thermal conductivity detector. A 6 ft. 8 in. HayeSep QS column was used to analyze combustible compounds via a flame ionization detector.

Results

Figure 2 shows the rate of methanol disappearance for the three catalysts at 598K, 2.75 MPa, a feed ratio of 10 mol nitrogen/1 mol methanol, and a gas hourly space velocity (GHSV) of 12000 sL/kg cat-hr. For each catalyst, the rate of methanol disappearance increased during the first 50 to 75 hours, while the catalyst was stabilizing and the liquid in the reactor was reaching a steady-state composition. Each catalyst eventually reached a rate that remained approximately constant during the latter 70 hours of each experiment. The Mn-promoted copper chromite catalyst had the highest steady state rate of methanol disappearance, approximately 5×10^{-1} moles/g cat-hr. The unpromoted copper chromite and Raney copper catalysts gave a steady state rate of approximately 4×10^{-1} moles/g cat-hr. There was no visible evidence of catalyst deactivation at 598K for any of the catalysts, once operation had stabilized.

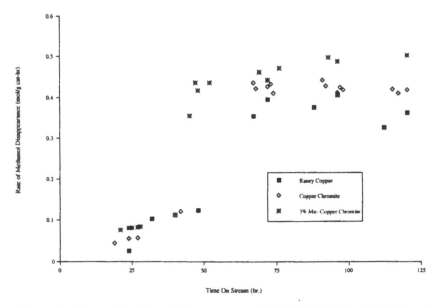

Figure 2 - Rate of methanol disappearance (598K, 2.75 MPa, N2/Methanol Feed Ratio = 10/1, GHSV = 12000 sL/kg cat-hr).

Figure 3 shows the selectivity to formaldehyde for each of the catalysts, based on reacted methanol. For both Raney copper and the Mn-promoted copper chromite, the initial formaldehyde selectivity was as high as approximately 40%. This selectivity declined to a value of about 10% for both catalysts, and remained approximately constant during the final 70 hours of each experiment. For unpromoted copper chromite, the formaldehyde selectivity remained essentially constant at around 10% during the entire experiment.

Since the steady state selectivity to formaldehyde, based on reacted methanol, was about 10%, the steady state rate of formaldehyde production was about 5×10^{-2} mol/g cat-hr. These experiments show that methanol dehydrogenation to formaldehyde can occur at rates faster than shown under similar conditions in earlier research by others (11-14).

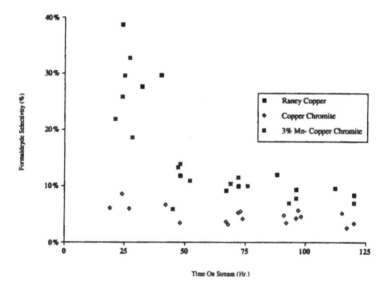

Figure 3 - Formaldehyde selectivity for the experiments of Figure 2.

Figure 4 shows that methyl formate was the primary product formed by each of the three catalysts. Methyl formate probably resulted from formaldehyde dimerization, Reaction 6.

$$2\,HCHO \rightleftarrows HCOOCH_3 \tag{6}$$

The steady state selectivity to methyl formate was as high as 80% with the unpromoted copper chromite catalyst, and as low as 60% with 3% Mn on copper chromite. Figure 5 shows the carbon monoxide selectivity.

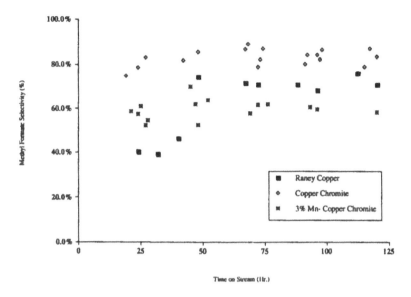

Figure 4 - Methyl formate selectivity for the experiments of Figure 2.

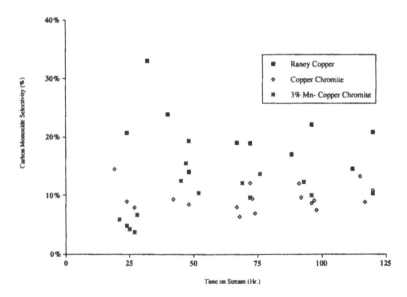

Figure 5 - Carbon monoxide selectivity for the experiments of Figure 2.

Raney copper gave the highest steady state CO selectivity, approximately 20%. Both of the copper chromite catalysts had CO selectivities of approximately 10%. The Mn-promoted copper chromite catalyst also showed about 10% selectivity to each of dimethyl ether and carbon dioxide. Dimethyl ether probably is formed from methanol dehydration.

$$2 \ CH_3OH \ \rightleftarrows \ H_3COCH_3 + H_2O \qquad (7)$$

The water produced in Reaction 7 may have reacted with CO to produce CO2 via the water gas shift. No dimethyl ether was detected at 598K using any of the other catalysts. Methane and carbon dioxide were detected in trace amounts with both Raney copper and unpromoted copper chromite.

Figure 6 is a comparison of the value of (PHCHO x PH2)/PMeOH, which was calculated from the experimental data, with the equilibrium constant for Reaction 1, which was calculated from thermochemical data (2). This figure shows that Reaction 1 was close to equilibrium during most of the experiment for both Raney copper and 3% Mn on copper chromite. For the unpromoted copper chromite, Reaction 1 was close to equilibrium initially, but was about an order of magnitude below equilibrium during the latter 70 hours of operation. Therefore, the rate of formaldehyde formation mentioned in connection with Figure 2 reflects a substantial equilibrium influence.

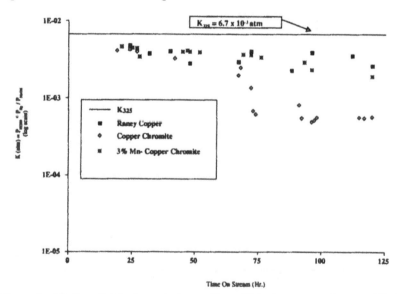

Figure 6 - Methanol dehydrogenation to formaldehyde: approach to equilibrium for the experiments of Figure 2.

Table 1 shows the BET surface area and pore volume for fresh and "spent" samples of the catalysts from the experiments of Figure 2. The measurements for the fresh Raney copper are consistent with values previously reported (19). The measurements for both fresh copper chromite catalysts are comparable to the values reported in product literature provided by Engelhard Corporation. The "spent" Raney copper and 3% Mn on copper chromite experienced minor losses in surface area and total pore volume. However, the BET surface area and total pore volume for the "spent" unpromoted copper chromite catalyst were only about 60% of those for the fresh catalyst. This may have resulted from inadequate washing prior to measurement.

Table 1 - BET Surface Area and Total Pore Volume Measurements

Catalyst		BET Surface Area (m^2/g)	Total Pore Volume (cm^3/g)
Raney Copper	fresh	28.2 ± 0.4	0.053 ± 0.005
	spent	25.4 ± 0.8	0.049 ± 0.001
Copper Chromite	fresh	13.1 ± 0.4	0.511 ± 0.04
	spent	7.7 ± 0.3	0.297 ± 0.05
3% Mn-Copper Chromite	fresh	35.3 ± 0.7	0.124 ± 0.02
	spent	27.6± 0.4	0.093 ± 0.002

The GC/MS analyses of liquid samples from each experiment showed evidence of both THQ dehydrogenation to quinoline and alkylation of either or both THQ and quinoline. For all three catalysts, the THQ concentration was very low in the final liquid, which was comprised of mostly C10 and C11 compounds.

Discussion

Figure 6 shows that methanol dehydrogenation to formaldehyde can be close to equilibrium at 598K. Therefore, the rates of methanol disappearance measured in this study can reflect a substantial equilibrium influence. These rates should be substantially higher at conditions where the approach to equilibrium is lower. The high selectivity to methyl formate that was observed in this research may be a consequence of the close approach to equilibrium.

Methyl formate is a dimer of formaldehyde, and formaldehyde can be formed by the reverse of Reaction 6. To the extent that methyl formate serves as a "reservoir" of formaldehyde, the high selectivity to methyl formate is not a serious negative. To test this hypothesis, methyl formate was fed into the stirred autoclave over Raney copper at 598K. Figure 7 shows that methyl formate "dedimerization" to formaldehyde, and methanol dehydrogenation to formaldehyde both were at equilibrium.

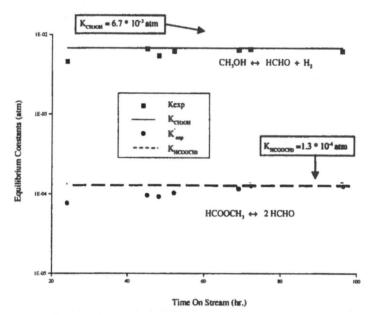

Time On Stream (hr.)

Figure 7 - Methanol dehydrogenation to formaldehyde and methyl formate "dedimerization" to formaldehyde: approach to equilibrium at 598K (2.75 MPa, N2/Methyl Formate Feed Ratio = 10/1, GHSV = 12000 sL/kg cat-hr).

Ideally, the reaction that is intended to consume formaldehyde, e.g. Reaction 2 or 3, should be fast enough to intercept formaldehyde before it is converted to methyl formate. The selectivity to carbon monoxide also may be reduced if the formaldehyde-consuming reaction is sufficiently fast, since carbon monoxide appears to be formed from methyl formate (16,17).

There was no evidence of catalyst deactivation at 598K. The data in Figure 2 and Table 1 suggest that the Raney copper and 3% Mn on copper chromite catalysts can experience modest losses in BET surface area and total pore volume without affecting their activities. Significant deactivation of the Raney copper catalyst was observed above 598K (16,17). The unpromoted copper chromite was stable at 623 and 648K. At 673K, this catalyst showed some initial deactivation, but eventually appeared to reach a stable level of activity.

Although the slurry liquid, THQ, alkylated during the reaction, it is not known specifically where on the ring structure the alkylation occurred. Furthermore, it is not known, for example, if a compound with two alkyl carbons was a dimethyl- or an ethyl- molecule. Nuclear Magnetic Resonance (NMR) and Liquid Chromatography (LC) studies are being performed on the "spent" liquids to determine their precise compositions.

Conclusions

Methanol dehydrogenation to formaldehyde can be so rapid with Raney copper, copper chromite, and 3% Mn on copper chromite at 598K, 2.75 MPa, and 12,000 sL/kg. cat-hr that this reaction approaches equilibrium. Side reactions, primarily formaldehyde dimerization to methyl formate, predominate at these conditions. Methyl formate can "dedimerize" to formaldehyde over these catalysts. Therefore, carbon monoxide formation is the major source of yield loss. No catalyst deactivation was apparent at 598K. The slurry liquid, THQ, was alkylated during operation. This reaction is another potential source of yield loss. However, all of these yield losses should be reduced significantly if formaldehyde is consumed selectively to form desired products such as methyl methacrylate and styrene.

References

1.   H.R. Gerberich and G.C. Seaman, in Kirk-Othmer Encyclopedia of Chemical Technology, J. I. Kroschwitz and M. Howe-Grant, Eds., Vol. 11, Wiley-Interscience, New York 1992, p. 929.
2.   E.H. Shreiber, J.R. Mullen, M.R. Gogate, J.J. Spivey, and G.W. Roberts, Ind. Eng. Chem. Res., 35, 2444 (1996).
3.   M. Ai, J. Catal., 107, 201 (1987).
4.   J. Periera, in Chemical Economics Handbook, SRI International, 1994, p. 674.4500D.
5.   K. Ring, in Chemical Economics Handbook, SRI International, 1995, p. 694.30000A
6.   K. Weissermel and H.J. Arpe, Industrial Organic Chemistry, VCH, Weinheim, 1993, p. 339.
7.   P.M. Bruylants and R.L. Garten, U.S. Patent 4,479,024 (1984).
8.   W.S. Weiland, R.J. Davis, and J.M. Garces, J. Cat., 173, 490 (1998).
9.   M. Ai, Bull. Chem. Soc. Jpn., 63, 199 (1990).
10.   J.S. Yoo, Appl. Cat. A, 102, 215 (1993).
11.   F. Merger and G. Fouquet, U.S. Patent 4,354,045 (1982).
12.   K. Takagi, Y. Morikawa, and T. Ikawa, Chem. Lett., 527 (1995).
13.   S.K. Ghosh and D.N. Ghosh, Chem. Ind. Dev. CP&E, 15 (1973).
14.   E. Miyazaki and I. Yasumori, Bull. Chem. Soc. Jpn., 40, 2012 (1967).
15.   G.W. Roberts, M.A. Márquez, M.S. McCutchen, C.A. Haney, and I.D. Shin, Ind. Eng. Chem. Res., 36, 4143 (1997).

16.   E.H. Shreiber, M.D. Rhodes, and G.W. Roberts, Appl Cat B, 23, 9
       (1999).
17.   E.H. Shreiber, Ph.D. Thesis, North Carolina State University, 1999.
18.   J.M. Thomas and R.M. Lambert, Characterization Of Catalysts, Wiley
       and Sons, Chichester, England, 1980.
19.   J.R. Mellor, N.J. Coville, A.C. Sofianos, and R.G. Copperthwaite,
       Appl. Catal. A, 164, 185 (1997).

Copper-Based Chromium-Free Hydrogenation Catalysts

Jürgen Ladebeck and Tiberius Regula

Süd-Chemie AG, Catalyst Division, Waldheimerstr. 13, 83052 Bruckmühl-Heufeld, Germany

Abstract

Copper-based chromium-free hydrogenation catalysts were prepared by standard methods and investigated in the hydrogenation of lauric acid and lauric acid ester. Catalyst stability in terms of metal soap formation and crush strength retention in presence of fatty acids was compared. The results obtained with Cr-free experimental catalysts were compared with standard copper chromite. The lowest copper solubility and best mechanical stability in fatty acids was observed with the classical copper chromite catalyst. The following order of stability was found: copper chromite > Cu on Al_2O_3 > Cu on Al_2O_3/SiO_2 > Cu on Al_2O_3/ZnO > Cu on ZnO. In fixed-bed methyl ester hydrogenation, other than copper chromite, only Al_2O_3-supported copper exhibits low metal leaching and reaches an acceptable mechanical stability.

Introduction

The higher monohydric aliphatic alcohols up to a chain length of more than 12 C atoms are widely used as raw materials in plasticisers and detergents. The industrially preferred main routes for the production of higher alcohols are (a) the hydrogenolysis of fatty acids and fatty acid esters, (b) the ZIEGLER-ALFOL process starting with ethylene, (c) the hydroformylation of olefins and the subsequent hydrogenation of the aldehydes to oxo-alcohols and (d) to a lesser extent the partial oxidation of paraffinic hydrocarbons. Except for the latter mentioned process, copper-based mixed oxide catalysts play a major role in the industrial, large scale production of fatty alcohols (1,2). The so-called fatty alcohols today are mostly based on natural sources like natural fats, oils, and waxes. Depending on the desired hydrocarbon chain length a limited range of natural raw materials can be applied for the production of fatty alcohols: coconut and palm kernel oil (C_{12} - C_{14}), palm oil, soybean oil, tallow or lard (C_{16} - C_{18}), and rapeseed oil (C_{20} - C_{22}). Main factors governing the selection of a fatty acid-based or a methyl ester-based process are the quality of the available feedstock, the ultimate use of the product and whether there is a necessity of unsaturated alcohols. After raffi-

403

nation, the triglycerides are pressure split to yield fatty acids, the methyl ester for fatty alcohol applications is obtained by transesterification.

Hydrogenation of the carboxylic group at high pressure and in the presence of a copper catalyst was discovered almost simultaneously by various groups, published and patented in the early 1930's (3-5). Analogous to the known Ni catalysts for fat hardening, kieselguhr as a support for the copper catalyst was applied. In the following years the copper chromite catalysts gained importance in technical applications. Meanwhile the copper chromites represent versatile heterogeneous catalysts, active in many reactions besides hydrogenation, including dehydrogenation, methanol steam reforming, water gas shift reaction and oxidation reactions in environmental catalysis. However, in the last decade chromium containing catalysts have become increasingly a subject of discussion because of the toxic nature of the Cr^{6+} content in the catalyst precursor or in improperly re-oxidized catalysts after use.

Numerous articles and patents were published in the recent years describing Cr-free copper-based catalysts for the production of alcohol by hydrogenation of fatty acids or fatty acid esters. The publications describe the catalyst application in the classical slurry hydrogenation process, in trickle-bed hydrogenation processes and in the gas phase (1, 2, 6).

In the present paper, the stability of various copper-based hydrogenation catalysts are investigated and compared with a classical copper chromite. For the investigation, four typical types of mixed oxide catalysts were selected as representative for the literature mentioned substitutes of copper chromites: ZnO-supported copper, Al_2O_3-supported copper, copper on Al_2O_3/ZnO and copper on alumina silicate. All catalysts were applied in a laboratory scale slurry phase hydrogenation of fatty acids and in fixed-bed trickle phase hydrogenation of fatty acid esters. As criteria for the catalysts stability, the metal solubility in the feed and product, and the grain and tablet mechanical strength were investigated. For a commercial application, these properties are of major importance and influence the decision of loading a given catalyst or not, even more than differences in activity or selectivity between catalysts as obtained in lab scale investigations (7).

Experimental Section

A series of copper-based catalysts were prepared by a standard precipitation procedure utilizing the metal nitrates. The precipitation method is described in the open literature as well as in numerous patents (8-12). A solution containing the metal nitrates in the appropriate concentrations and a second solution of sodium carbonate and sodium aluminate were combined slowly and simultaneously under agitation into a vessel containing water and the carrier in case of the silica containing catalyst. The precipitation pH was maintained at about 6.5 by varying the addition rate of the nitrate solution. The resulting precipitate was recovered by filtration, washed to remove the residual salts and spray dried at 110 to 120°C. The dried material was calcined at 600°C for 1 hour. For the application in fixed bed hydrogenation, binderless tablets (3x3 mm) were prepared by adding 2 wt.% graphite to the powder. A commercial copper chromite was used as a reference catalyst for the hydrogenation tests and for stability investigations.

Complete chemical and physical characterization of the catalyst precursor in the form of the powder and tablets was performed as far as necessary for industrial application (Table 1).

Table 1   Chemical and physical data of the prepared catalysts

		Cu/Cr	Cat A	Cat B	Cat C	Cat D
CuO	[%]	46	48	44	31	52
Cr_2O_3	[%]	48	-	-	-	-
MnO_2	[%]	6	-	-	-	-
ZnO	[%]	-	-	56	32	-
Al_2O_3	[%]	-	52	-	37	38
SiO_2	[%]	-	-	-	-	10
Specific surface area	[m²/g]	30	60	40	50	110
Powder particle size D_{50}	[μm]	19	25	21	18	22
Bulk density Powder	[g/L]	780	740	630	550	850
Side crush strength	[N]	80	90	70	90	105
Bulk density Tablet	[g/L]	1500	1455	1490	1450	1420
Specific pore volume	[mL/g]	0,16	0,18	0,16	0,18	0,14

Hydrogenation Reactions
Although it was not the main purpose of the work to compare catalyst conversion and selectivity all prepared samples were employed in comparative hydrogenation tests to demonstrate their principal capabilities as hydrogenation catalysts. The catalysts in powder form were used in the slurry phase hydrogenolysis of lauric acid, the palletized catalysts were subjected to a fixed-bed hydrogenation test of lauric acid methyl ester. Commercial copper chromite was used as reference.

Lauric acid hydrogenolysis
The hydrogenolytic splitting of fatty acids in the presence of a fatty alcohol can be considered as a two step reaction. In the first step, the esterification takes

place forming the wax ester and water according to the following reaction scheme **(1)**.

$$R\text{-COOH} + R^1CH_2\text{-OH} \rightleftharpoons R\text{-COO-CH}_2R^1 + H_2O \qquad (1)$$

In the second stage of the reaction, the hydrogenation of the ester leads to the alcohols **(2)**.

$$R\text{-COO-CH}_2R^1 + 2H_2 \rightarrow R\text{-CH}_2\text{-OH} + R^1CH_2\text{-OH} \qquad (2)$$

Both reactions occur simultaneously when carried out in a high pressure batch reactor. Commercially this process is carried out in the slurry phase and represents the very important industrial production of fatty alcohols via the acid route.

The laboratory slurry phase hydrogenation tests were carried out in a high pressure 2 liter autoclave reactor system with an electronically controlled stirrer (Fig. 1). The catalyst powder was suspended in alcohol and filled into the autoclave. After reduction of the catalyst the acid was injected with compressed hydrogen into the autoclave and the hydrogenation was performed.

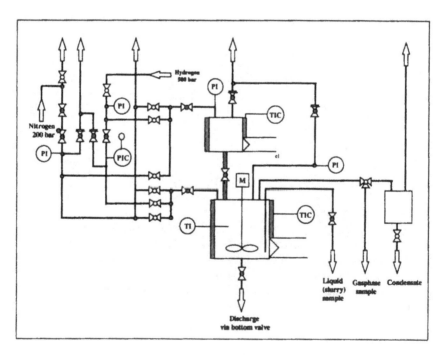

Figure 1   Reactor System for Slurry Phase Fatty Acid Hydrogenolysis

The following standard conditions were applied for investigation of the catalysts:

Lauryl alcohol (Dodecanol)	900 g
Lauric acid	100 g
H_2 pressure	300 bar
Temperature	300°C

In addition, the liquid samples taken from the reactor were analyzed for dissolved metals and the spent catalysts were investigated for filtering properties, and for crystalline phases and copper sintering by XRD.

Hydrogenolysis of fatty acid methyl ester
The technically important process starts with the transesterification of triglycerides in the presence of an alkaline catalyst. The equilibrium reaction is shifted towards the desired ester by an excess of methanol and by the removal of glycerol. The hydrogenolytic splitting of fatty acid methyl ester proceeds according to the following equation (3) and is in principle identical with the previously described reaction step (2):

$$R\text{-}CH_2COO\text{-}CH_3 + 2H_2 \rightarrow R\text{-}CH_2\text{-}CH_2\text{-}OH + CH_3\text{-}OH \qquad (3)$$

The hydrogenation of the fatty acid methyl ester in technical scale is widely applied in the trickle bed technique, only to a limited extent in the gas phase or in suspension. As often mentioned, one advantage of the ester hydrogenation process in comparison to the acid hydrogenation in terms of catalyst stability is the absence of free acid in the feed and consequently only limited attack on the catalyst. Important industrial representatives of the process are those of Henkel, Kao Soaps and others.

Figure 2 gives the flow scheme of the fixed-bed lab reactor system used in the described catalyst stability tests. After reduction at 200°C with H_2 under atmospheric pressure the following standard conditions were applied for investigation of the catalysts:

Substrate	Lauric Acid Methyl Ester
Pressure	280 bar (g)
GHSV	20,000 (H_2)
LHSV	0.7 (Ester)
H_2/Ester mol ratio	300
Temperature range	160 to 240°C

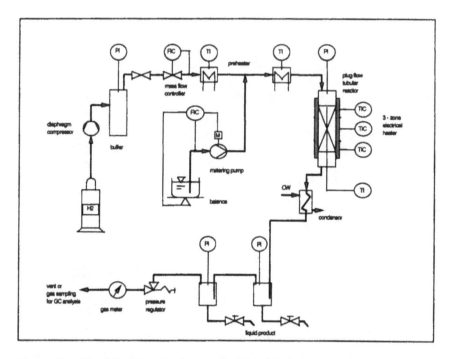

Figure 2   Fixed Bed Reactor System for Fatty Methyl Ester Hydrogenation

Liquid and gas phase product analysis in order to determine conversion and by-product formation were performed by capillary gas chromatography. In addition, the liquid samples from the reactor collected over 100 hours were analyzed for dissolved metals. The spent catalysts were investigated for lateral tablet crush strength, and for crystalline phases and copper sintering by XRD.

Results and Discussion

Catalyst Performance Tests
Before comparing the stability properties all catalyst samples were subjected to comparative hydrogenation tests to prove their suitability. The results are summarized in Table 2 and Figure 3. In the slurry hydrogenation of fatty acids the half life and the total conversion after 60 minutes are compared. The conversion is defined as $X (\%) = 100 (SV_0 - SV_{60})/SV_0$ were SV_0 is the saponification value at the beginning of the run and SV_{60} the value after 60 minutes reaction time. The half life $t_{1/2}$ is derived from the Arrhenius plot by $r^1 = k_0 * e^{-EA/RT}$ and $t_{1/2} = \ln 2/r^1$.

Table 2   Slurry Phase Lauric Acid Hydrogenation Results

		Cu/Cr	Cat A	Cat B	Cat C	Cat D
Activity (Half Life) $t_{1/2}$	[min]	7	6	9	6	7
Conversion after 60 min	[%]	88	88	76	89	89

Fixed-bed methyl ester hydrogenation tests in the temperature range between 160 and 240°C were performed. Liquid and gaseous samples were analyzed by gas chromatography and the ester conversions and dodecanol yields were compared. As can be seen from the results, under the chosen test conditions all catalysts exhibit sufficient conversions so that a technical application could be considered, but in both applications copper on ZnO (Cat B), showed disadvantages because of lower conversions (Figure 3).

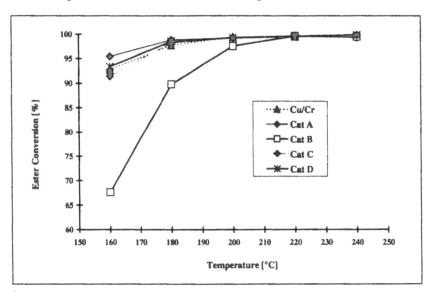

Figure 3   Fixed Bed Lauric Acid Methyl Ester Hydrogenation Results

Catalyst Stability
In slurry phase fatty acid hydrogenation the amount of dissolved copper in the product alcohol is 10 to 20 times higher for Cr-free catalysts than in the case of copper chromite. Besides a higher copper solubility the zinc containing catalysts suffer from a remarkable zinc leaching, which exceeds by far a tolerable amount. The high metal leaching is accompanied by a catalyst particle degradation and deteriorated filtration properties. By far the best results were obtained with classical copper chromite (Table 3).

Table 3   Catalyst Stability in Slurry Phase Hydrogenation of Lauric Acid, $T_{react} = 300°C$

Metal dissolved		Cu/Cr	Cat A	Cat B	Cat C	Cat D
Cu	[ppm]	1	18	22	11	1
Cr	[ppm]	0.5	-	-	-	-
Zn	[ppm]	-	-	170	190	-
Al	[ppm]	-	2	-	1.5	5
Si	[ppm]	-	-	-	-	12
Relative filtration time		1	2.2	1.4	1.3	1.1
Cu crystallite size	[Å]	85	97	187	95	105

In the methyl ester hydrogenation tests the effect on the copper seems to be less severe and the amount of dissolved metal in the product alcohol is lower than in the static slurry phase hydrogenation experiments. Comparing the high lateral side crush strength of the spent copper chromite with that of the experimental catalysts, especially the Al_2O_3 supported copper catalysts, seems to meet as well technical requirements (Table 4). The main reason for the apparently better stability of the tablets in contrast to the powder in terms of dissolved metals, is the absence of free acid in the ester feed. Even in an industrial feed the acid number is usually below the detection limits, due to a deacidification step before transesterification.

Table 4   Catalyst Stability in Fixed Bed Hydrogenation of Lauric Acid Methyl Ester, $T_{react} = 200°C$

Metal dissolved		Cu/Cr	Cat A	Cat B	Cat C	Cat D
Cu	[ppm]	0.7	2.5	9	6	3
Cr	[ppm]	0.3	-	-	-	-
Zn	[ppm]	-	-	47	38	-
Al	[ppm]	-	1	-	1	1
Si	[ppm]	-	-	-	-	55
Crush retention	[%]	96	80	65	70	50
Cu crystallite size	[Å]	105	120	210	135	130

The addition of water to the ester should result in an increasing concentration of free acid, according to the saponification equilibrium as given in the reaction scheme (1). This can be easily demonstrated by adding water to the feed ester in presence of reduced catalyst and analyzing the acid number, as shown below in Table 5. To avoid any acid formation by oxidation the water was added to the ester under inert conditions and kept for 10 hrs at 170°C.

Table 5   Water Addition to Methyl Ester Feed (170°C, N_2 atmosphere)

Water added [%]	Acid number (mg KOH/g sample)	
	Titrated:	Calculated:
0.15	not detectable	3.3
0.5	0.54	10.3
1	3.3	20.5
2	5.0	40.6

The addition of 1 and 2 % water to the ester and the subsequent treatment of the catalyst pellets at 170 °C caused a severe damage in particular to the non chrome catalyst samples (Table 6).

Table 6   Catalyst Stability in Water Containing Lauric Acid Methyl Ester, Water Content 2 %, T_{treat} = 170 °C

Metal dissolved		Cu/Cr	Cat A	Cat B	Cat C	Cat D
Cu	[ppm]	2.5	14	35	18	15
Cr	[ppm]	1.2	-	-	-	-
Zn	[ppm]	-	-	123	145	-
Al	[ppm]	-	1	-	1	1
Si	[ppm]	-	-	-	-	85
Crush retention	[%]	85	35	30	30	<10

As a consequence the plant operator should avoid any water content in the ester feed exceeding the limit of solubility by careful drying, or allowing the water to settle in an appropriate period of time before introducing the ester into the reactor.

Introduction of water into the reactor which cannot be avoided takes place prior to the hydrogenation during the catalyst reduction process. Usually the reduction water is removed together with the purge gas, collected, and measured for calculation of the achieved degree of catalyst reduction. In certain cases it is unavoidable for economic reasons or because a given plant or process is designed to perform the reduction process at high pressure.

Table 7 summarizes the results of a fixed-bed fatty ester hydrogenation at 200 °C after high pressure catalyst reduction at 40 bar. After the reduction the hydrogenation was started at 280 bar without an intermediate pressure relief. As can be seen from the data, the amount of dissolved metals in the product alcohol increased and the copper crystallite sintering was favored.

Table 7   Catalyst Stability in Fixed Bed Hydrogenation of Lauric Acid Methyl
Ester, T_{react} = 200°C, (reduction with 40 bar H_2)

Metal dissolved		Cu/Cr	Cat A	Cat B	Cat C	Cat D
Cu	[ppm]	1.1	12	23	13	6
Cr	[ppm]	0.8	-	-	-	-
Zn	[ppm]	-	-	63	55	-
Al	[ppm]	-	1	-	1	1
Si	[ppm]	-	-	-	-	35
Crush retention	[%]	90	40	30	35	20
Cu crystallite size	[Å]	110	160	240	145	140

The crush retention in particular for the non-chrome catalysts was further decreased. A possible reason for the degradation of the chromium-free catalysts could be the saponification of the feed ester by the reduction water, trapped inside the catalyst pores. The fatty acid formed inside the pore system can directly attack the texture and damage the pellets of the non-chrome catalysts. In the case of the copper chromite the stable spinel structure can resist the acid attack.

Conclusion

Although all tested non-chrome catalysts are active enough in lauric acid and lauric acid ester hydrogenation, besides the copper chromite only the alumina supported copper catalyst shows minimum metal soap formation and has a sufficient stability to resist acid attack. The following order in terms of catalyst stability was found: copper chromite > copper on alumina > copper on aluminum silicate > copper on Al_2O_3/ZnO > ZnO supported copper. Mainly the catalysts suffer from ZnO or SiO_2 leaching. By applying chromium free catalysts in fatty methyl ester hydrogenation the deacidification of the feed and the water removal becomes more important than it is in a process utilizing a copper chromite catalyst. The fatty acids formed by saponification of the ester in presence of water is leading to severe degradation of the chromium free catalysts and has to be avoided. Any excess of water carried into the reactor together with the feed or trapped inside the catalyst pore system during the reduction can cause catalyst damage and has to be removed carefully before starting the hydrogenation reaction. Highly recommended is the use of pre-reduced and stabilized catalyst. The free water in the reactor, and in consequence the acid number and catalyst damage by metal leaching will be minimized.

References

1. Ullmann´s Encyclopaedia of Industrial Chemistry, W. Gerhartz, VHC Verlagsgesellschaft, Weinheim, Vol. 10, 4th ed., 1985, pp. 245-296.
2. R. A. Peters, Alcohols Higher Aliphatic, in Kirk-Othmer Encyclopedia of Chemical Technology, John Wiley & Sons, Vol. 1, 3rd ed., 1978, pp.716-739.
3. W. Normann, Z. Angew. Chem., **44**, 714 (1931).
4. W. Schrauth, O. Schenk, K. Stickdorn, Ber. Dtsch. Chem. Ges., **64**, 1314 (1931).
5. H. Adkins, K. Folkers, J. Am. Chem. Soc., **53**, 1095 (1931).
6. Fatty Alcohols: Raw Materials, Methods, Uses, Henkel KGaA, Düsseldorf ZR-Fettchemie, 1st ed., 1982.
7. J. Ladebeck and T. Regula, Studies in Surface Science and Catalysis, (H. Hattori and K. Otsuka), **121**, (Science and Technology in Catalysis), 215, (1998)
8. T. Fleckenstein, J. Pohl, F. Carduck,, EP 0300 347, to Henkel (1992).
9. D. S. Thakur, T. J. Sullivan, B. D. Roberts, A. L. Vichek , EP 0 424 069, to Engelhard (1991)
10. M. Schneider, K. Kochloefl, G. Maletz, US Pat. 5,386,066, to Süd-Chemie (1994)
11. D. S. Thakur, T. J. Sullivan, B. D. Roberts, A. L. Vichek , US Pat. 5,345,005 to Engelhard (1994)
12 M. Schneider, K. Kochloefl, G. Maletz, EP 0 552 463 (1996)

Highly Effective Catalytic Enantioselective Deprotonation of *meso*-Epoxides with Isopinocampheyl Based Chiral Lithium Amides[*]

Sanjay V. Malhotra

Department of Chemical Engineering, Chemistry and Environmental Science
New Jersey Institute of Technology, University Heights, Newark, NJ-07102
Fax: 973-596-8436, E-mail: malhotra@adm.njit.edu

ABSTRACT

A highly enantioselective deprotonation of *meso*-epoxides resulting in the formation of corresponding optically active allylic alcohol in moderate to high enantiomeric excess (ee) has been achieved. Several new chiral secondary amines based on α-pinene were prepared, and the lithium salts of these chiral amines were used in a catalytic amount in combination with excess lithium diisopropylamide to obtain a catalytic cycle. An enantiomeric excess of up to 95 % was achieved in case of cyclohexene oxide, while a modest ee was realized in the opening of cyclopentene oxide and cyclooctene oxide. A systematic study shows that the isopinocampheyl moiety plays an important role in orienting epoxide and the lithium salt to achieve enantioselective deprotonation and thereby achieving the product allylic alcohol in high enantiomeric excess.

---

[*] Dedicated to Prof. Herbert C. Brown.

INTRODUCTION

Enantioselective desymmetrization of achiral materials is an attractive and extremely powerful concept in asymmetric synthesis. A number of strategies are known which demonstrate its viability and application in targeted synthesis to provide compounds with high enantiomeric excess (ee). Asymmetric synthesis by the use of chiral lithium amides is emerging as a useful method for the preparation of non-racemic compounds (1).

The deprotonation of an epoxide with a lithium amide to obtain an allylic alcohol was first reported in 1970 by Thummel and Rickborn in a deuterium labeling study (2). The reaction is thought to proceed via a cyclic six membered transition state, formed by a 1:1 epoxide-base complex, where the base coordinates to the lone pair of electrons on oxygen thereby facilitating the β-hydrogen removal (Scheme 1). If the epoxide is prochiral then the deprotonation using a chiral lithium amide would result in an enantioselective deprotonation to give an optically active product. The chiral lithium amides have been exploited by a variety of efficient enantioselective reactions such as deprotonation of prochiral cyclic ketones (3), kinetic resolution of racemic ketones (4), enantioselective dehydohalogenation (5), alkylation of achiral ketones (6), deracemization of chiral ketones by protonation (7), etc.

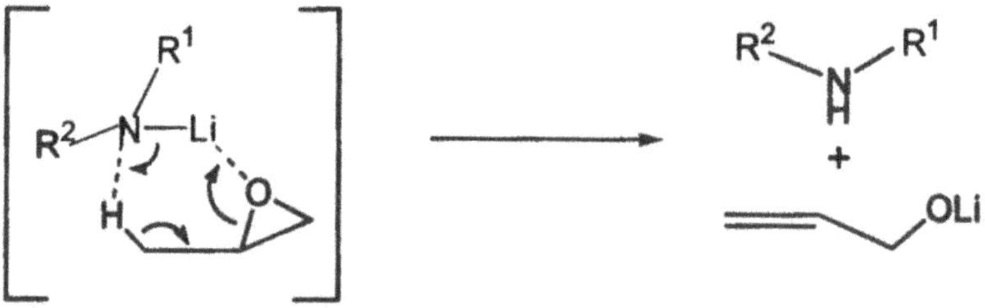

Scheme 1

The rearrangement of epoxides to allylic alcohols induced by chiral lithium amides is one such application, which is emerging as an important tool in asymmetric synthesis (8,9). The first non-enzymatic asymmetric version of this rearrangement was reported by Whitsell and Felman (8), who treated cyclohexene oxide (1) with a variety of mono- and dialkyl chiral lithium amides. However, the resulting 2-cyclohexen-1-ol (2) was obtained in no greater than 31 % ee (eq. 1).

$$ee = 31\%$$

The enantioselective deprotonation reaction has been applied to the synthesis of a number of key intermediates, which can be elaborated to important biologically active compounds such as prostaglandin (10). In most cases, the procedure requires more than a stoichiometric amount of chiral bases. Thus, the development of an effective catalytic system using a chiral lithium amide is currently a significant challenge in synthetic organic chemistry. There have been only a few reports where a chiral lithium amide has been employed in catalytic amount (11-14), and only three reports where the application has been used for the rearrangement of an epoxide to allylic alcohol (11-13). The first such study by Asami et al. (11), using catalytic amount of a series of chiral lithium amides derived from (*S*)-proline reported a maximum of 79% ee for the conversion. In a subsequent study, however, they improved the ee to 94% (13). While another study with bis-lithium amide base from chiral diamines reported the ee of up to 77% (12).

In the current study a number of chiral secondary amines based on α-pinene were synthesized. These amines were employed in catalytic amounts in combination with an excess amount of an achiral lithium amide to achieve the deprotonation of prochiral epoxides to obtain the corresponding optically active allylic alcohol.

EXPERIMENTAL SECTION

Materials: *B*-Chlorodiisopinocampheylborane (DIP-Cl$^{\circledR}$), n-butyllithium, diisoprylamine, Ethyl ether, THF, cyclopentene oxide, cyclohexene oxide and cyclooctene oxide, were obtained from Aldrich Chemical Co., and used as such. (*R*)-(+)-α-Methoxy-α-(trifluoromethyl) phenylacetic acid (MTPA) was obtained from Aldrich Chemical Co. and converted to the acid chloride using Mosher's procedure (15). All operations were carried out under nitrogen atmosphere (16). The chiral amines were prepared by amination of DIP-Cl (18).

Representative procedure: n-Butyllithium (2.9 mmol) was added slowly to a mixture of N-cyclohexyl-N-isopinocampheylamine (**A2**, 0.4 mmol) and diisopropylamine (2.5 mmol) in dry THF (8 ml) at 0 °C under a nitrogen atmosphere. The mixture was stirred for 30 min and cyclohexene oxide (2.0 mmol) was added slowly through a micro syringe. The mixture was stirred for 30 min and allowed to warm up to room temperature (22 °C) and the reaction continued for 15 hr. The solution was acidified with dil. HCl (3N, 8 ml). The organic layer was separated, washed with water and dried over anhydrous MgSO₄. The solvents were removed to obtain a pale brown liquid product, 2-cyclohexen-1-ol. The MTPA ester of this alcohol was prepared and analyzed using a gas chromatograph fitted with SPB-5 capillary column to establish the ee.

RESULTS AND DISCUSSION

The enantioselectiove deprotonation of an epoxide to an allylic alcohol with a chiral lithium amide base has received much attention in the recent years (1). With the exception of only few reports (11-14), the reactions that use chiral lithium amides, all require more than a stoichiometric amount of chiral auxiliary. Much effort has been directed towards improving the level of asymmetric induction in this reaction by designing a variety of chiral lithium amides. The development of a highly effective catalytic system with such chiral auxiliaries that can be employed for complex organic syntheses would be of much interest. Therefore, the next challenge in this field is to come up with catalytic system(s), which give very high levels of asymmetric induction in all the applications where chiral lithium amides have been studied thus far. The study by Asami et al (11), to develop such a system for the rearrangement of cyclohexene oxide (1) to 2-cyclohexene-1-ol (2) using (S)-2-(pyrrolidin-1-ylmethyl) pyrrolidine gave the product in 79% ee (11). A modification of this chiral auxiliary improved the ee to 94% (13). In the study with the lithium salts of chiral diamines only 77% ee of the product was realized (12).

(A1)

(A2)

(A3)

(A4)

A remarkable success has been achieved with a large number of pinane-based reagents, for various organic transformations resulting in the formation of optically active compounds (17). Considering the results obtained by all these studies, the obvious motivation was to design and study the chiral lithium bases having the α-pinene moiety in their structure. The primary goal of this investigation was to see the possibility of developing a catalytic process with lithium salt of diisopinocampheylamine (DIP-AM, A1).

Also, N-cyclohexyl-N-isopinocampheylamine (A2), N-benzyl-N-isopinocampheylamine (A3) and N-isopropyl-N-isopinocampheylamine (A4) were chosen to see the steric effects of isopinocampheyl moiety in opening of a *meso*-epoxide to allylic alcohol. Earlier study with stoichiometric amount of the reagent produced remarkable results. We achieved >99% ee of the product 2 on deprotonation of cyclohexene oxide, using the lithium salt of A1 (18).

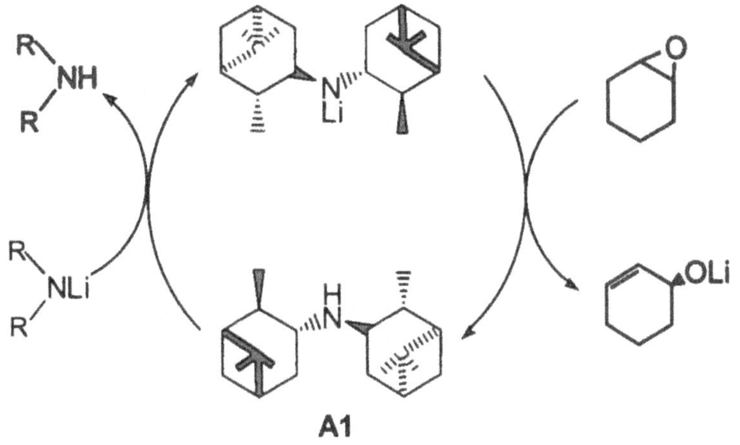

A1

Scheme 2

Though the reagents based on α-pinene have shown tremendous success in various organic transformations, these have been used only in stoichiometric amounts. Therefore, the obvious challenge has been to develop a catalytic cycle through which the chiral auxiliary could be regenerated. At the outset, a number of achiral lithium amides from bases such as diethylamine, diisopropylamine and pyrrolidine were tested to evaluate their reactivity with cyclohexene oxide, in comparison with chiral lithium amide from A1. It was found that the chiral lithium amide from diisopinocampheylamine was more reactive toward epoxide than the achiral lithium amides. This clearly indicated that it would be possible to regenerate (A1) and other chiral amines in the reaction mixture, with the help on an appropriate achiral lithium amide, *via* a cycle shown in scheme 2. The most satisfactory results were obtained with lithium diisopropyl amide (LDA), therefore, it was used in the entire study for the regeneration of chiral amines. The catalytic enantioselective deprotonation of cyclohexene oxide (eq. 2) was studied both at room temperature and at 0 °C with varying amounts of **A1** as shown in Table 1. As the entries 1-9 show, better results were obtained at 0 °C than at room temperature. Therefore, study with other chiral amines (**A2-A4**) were carried out at 0 °C.

$$\qquad\qquad(2)$$

As Table 1 shows, the best results were obtained in each case when the chiral auxiliary was used in 0.2 molar equivalent. A maximum ee of 95 % for the product was realized with lithium salt of DIP-AM, **A1**. However, on substituting one isopinocampheyl group in **A1** with cyclohexyl, the ee for the product **2** dropped to 78 %. On replacing isopinocampheyl with benzyl or isopropyl moiety the ee dropped even further. This clearly indicates that the isopinocampheyl moiety plays a significant role in orienting the epoxide and the lithium salt in such way that the deprotonation takes place enantioselectively. In other words, when two isopinocampheyl groups are present 97.5% of the deprotonation occurs from one phase resulting in the product allylic alcohol **2** with 95 % ee.

When the similar study was examined with cyclopentene oxide (eq. 3), a maximum ee of 80 % for the product **4** was obtained with DIP-AM, as shown in Table 2.

$$\qquad\qquad(3)$$

Table 1. Catalytic Enantioselective Deprotonation of Cyclohexene Oxide 1.

Entry	Amine	Mol (%)	Temp.	Yield (%)[a]	Ee (%)[b]
1	A1	5	rt	62	77
2	A1		0 °C	59	80
3	A1	5	rt	68	74
4	A1	10	0 °C	66	81
5	A1	10	rt	73	79
6	A1	15	0 °C	69	83
7	A1	15	rt	82	87
8	A1	20	0 °C	77	95
9	A1	20	rt	79	83
10	A1	25	0 °C	78	89
11	A2	25	0 °C	55	65
12	A2	5	0 °C	57	74
13	A2	10	0 °C	66	76
14	A2	15	0 °C	70	78
15	A3	20	0 °C	51	34
16	A3	5	0 °C	53	39
17	A3	10	0 °C	58	46
18	A3	15	0 °C	57	48
19	A4	20	0 °C	50	43
20	A4	5	0 °C	51	47
21	A4	10	0 °C	55	51
22	A4	15 20	0 °C	58	52

[a] Isolated yield. [b] Ee of the MTPA derivative.

Table 2. Catalytic Enantioselective Deprotonation of Cyclopentene Oxide **3**.

Amine[a]	Temp.	Yield (%)[b]	Ee (%)[c]
A1	rt	70	73
A1	0 °C	67	80
A2	rt	68	68
A2	0 °C	68	71
A3	rt	55	39
A3	0 °C	61	45
A4	rt	62	53
A4	0 °C	57	58

[a] 20 mol %. [b] Isolated yield. [c] Ee of the MTPA derivative.

Further extension of this study to the deprotonation of cyclooctene oxide (eq. 4), resulted in optically active product 2-cyclooctene-1-ol, **6**. The selectivity of this reaction using THF as solvent, as obtained in this study (78 % ee), is significantly higher than 53 % reported previously (13). The results are given in Table 3.

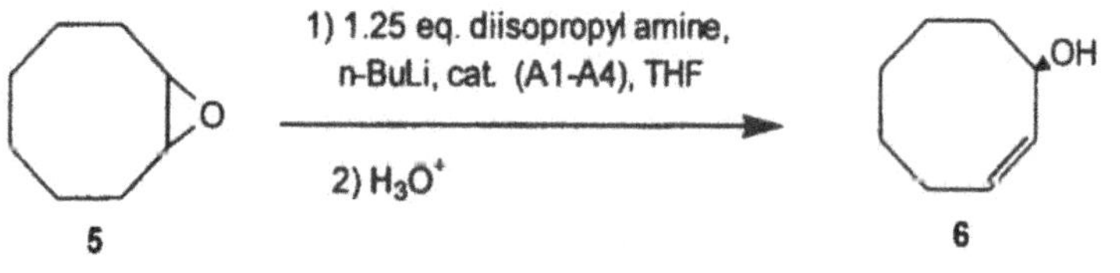

1) 1.25 eq. diisopropyl amine,
   n-BuLi, cat. (A1-A4), THF

2) H_3O^+

(4)

5   **6**

In all the studies reported here, the chiral secondary amines could be recovered in fairly good yields (74-83%).

CONCLUSION

This study has led to the discovery of a catalytic system in the enantioselective deprotonation of *meso*-epoxides, using various novel chiral secondary amines based on α-pinene. It is noteworthy that the highest recorded ee through a catalytic system for the conversion of cyclohexene oxide **1**, to 2-cyclohexen-1-ol **2**,

Table 3. Catalytic Enantioselective Deprotonation of Cyclooctene Oxide **5**.

Amine[a]	Temp.	Yield (%)[b]	Ee (%)[c]
A1	rt	57	73
A1	0 °C	51	78
A2	rt	52	66
A2	0 °C	46	72
A3	rt	43	20
A3	0 °C	39	24
A4	rt	42	31
A4	0 °C	39	33

[a] 20 mol %. [b] Isolated yield. [c] Ee of the MTPA derivative.

has been achieved with diisopinocampheylamine. This chiral auxiliary DIP-AM also gives the best results for other epoxides studied. Both isomers of α-pinene are readily available in very high ee; and optical upgradation of the commercial material is easily attainable. As a result, both isomers of the chrial amines can be easily prepared and utilized to obtain the desired isomer of the product alcohol. These results with diisopincampheylamine certainly promise the application of such a catalytic system in the complex synthesis of compounds such as natural products.

REFERENCES AND NOTES

1. P.J. Cox and N.S. Simpkins, *Tetrahedron: Asymmetry* 2, 1 (1991).
2. R.P Thummel and B. Rickborn, *J. Am. Chem. Soc.* 92, 2064 (1970).
3. E.J. Corey and A.W. Gross, *Tetrahedron Lett.* 25, 755 (1980).
4. H.D. Kim, H. Kawasaki, M. Nakajima and K. Koga, *Tetrahedron Lett.* 30, 6537 (1989).
5. L. Duhamel, A. Ravard, J.-C. Plaquevent and D. Davoust, *Tetrahedron Lett.* 28, 5517 (1987).
5. M. Murkata, M. Nakajima and K. Koga, *J. Chem. Soc., Chem. Commun.* 1657 (1990).
7. T. Yasukata and K. Koga, *Tetrahedron: Asymmetry* 4, 35 (1993).
8. J. K Whitsell and S.W. Felman, *J. Org. Chem.* 25, 755 (1980).
9. H. S. Simpkins, *Pure & Appl. Chem.* 68, 691 (1996).
10. M. Asami, *Bull. Chem. Soc. Jpn.* 63, 1402 (1990).
11. M. Asami, T. Ishizaki and S. Inoue, *Tetrahedron: Asymmetry* 5, 793 (1994).
12. J. P. Tierney, A. Alexakis and P. Mangeney, *Tetrahedron: Asymmetry* 7, 1019 (1997).
13. M. Asami, T. Suga, K. Honda and S. Inoue, *Tetrahedron Lett.* 38, 6425

(1997).

14. T. Yamashita, D. Sato, T. Kiyoto, A. Kumar and K. Koga, *Tetrahedron* **53**, 16987 (1997).

15. J.A. Dale, D.L. Dull and H.S. Mosher, *J. Org. Chem.* **34**, 1316 (1969).

16. H.C. Brown, G.W. Kramer, A.B. Levy and M.M. Midland, In *Organic Synthesis via Boranes*, Wiley- Interscience: New York, 1975, chapter 9.

17. H.C. Brown and P.V. Ramachandran, in *Reductions in Organic Synthesis*, A.F. Abdel-Magid, ed. American Chemical Society, Washington DC, (1996).

18. S.V. Malhotra and H. C. Brown, Unpublished results.

Diastereoselective Heterogeneous Catalytic Hydrogenation of Chiral Aromatic N-heterocyclic Compounds

Antal Tungler,[a] László Hegedûs,[a] Viktor Háda,[a] Tibor Máthé,[b] and László Szepesy[a]

[a]Department of Chemical Technology, Technical University of Budapest, H-1521 Budapest, Hungary
[b]Research Group for Organic Chemical Technology, Hungarian Academy of Sciences, Technical University of Budapest, H-1521 Budapest, Hungary

Abstract

High diastereoselectivities were obtained in the heterogeneous catalytic hydrogenation of some chiral pyridine and pyrrole derivatives with complete conversion, in non-acidic medium. The effects of catalytic metals, solvents, temperature and pressure on the conversion and the diastereomeric excess (d.e.) were investigated. The hydrogenation of N-picolinoyl-(S)-proline methyl ester resulted in 79% d.e. over 10% Pd/C, in methanol, at moderate pressure and temperature (50 bar, 50°C). Very high d.e. (94%) was achieved in the saturation of N-nicotinoyl-(S)-proline methyl ester under similar conditions. The hydrogenation of quaternized N-picolinoyl-(S)-proline methyl ester took place already at RT and 20 bar with almost complete diastereoselectivity (98%), over palladium on carbon, in methanol. Similar high d.e. (90%) was obtained in the saturation of N-(1'-methylpyrrole-2'-acethyl)-(S)-proline methyl ester over 5% Rh/C, in methanol, at 20 bar and RT. These are successful examples of diastereoselections in the hydrogenation of prochiral N-heterocyclic compounds.

Introduction

Stereoselective synthesis of chiral saturated N-heterocycles is of interest due to their biological activity. Chiral piperidine and pyrrolidine compounds are important and valuable pharmaceutical intermediates. However, their syntheses via diastereoselective heterogeneous catalytic hydrogenation have never been described.

In this work the diastereoselective hydrogenation of picolinic and nicotinic acid, as well as 1-methylpyrrole-2-acetic acid derivatives over precious metal catalysts (Pd/C, Rh/C, Rh/Al$_2$O$_3$, Ru/C and Pt/C), in non-acidic medium was investigated. Since the hydrogenation of the above mentioned compounds cannot be found in the literature, the saturation methods of the pyridine or pyrrole ring are shown through the examples of other pyridine and pyrrole derivatives, respectively.

Palladium, platinum, rhodium, ruthenium and nickel catalysts can be used for the hydrogenation of pyridines and pyrroles. These nitrogen containing heterocycles cannot be easily saturated, due to poisoning the metal catalysts by secondary or tertiary N. The poisoning can be eliminated by adding protic acid to the reaction mixture (1-3).

Among the platinum metals, rhodium is the most effective catalyst under mild conditions. For example, nicotinamide was converted to nipecotinamide over 5% Rh/C (40% by weight of reactant), at 60°C and 2.7 bar, in water with 86.6% yield (4). The use of ruthenium as RuO_2, gave also good results in the reduction of pyridines, but at relatively high pressure (100 bar) and temperature (70-100°C) (5). Platinum oxide has been widely used for the saturation of pyridine ring but only in the presence of acids, in order to prevent poisoning of the catalyst (6). Nevertheless, the inhibiting effect of basic nitrogen could be avoided by internal neutralization; picolinic and isonicotinic acids were completely reduced over PtO_2, at 2.5 bar and room temperature, in water, i.e., in a neutral medium (7). Palladium was used seldom and it acted only in acidic media (e.g. acetic acid) and above 70°C (8). However, the palladium mediated hydrogenation of 2-[(2',6'-dimethylanilino)carbonyl]-1-methylpiridinium methosulfate was carried out already at 40-60°C and 3-6 bar, in non-acidic medium (water). The reduction gave the corresponding piperidine derivative, which is a local anaestheticum, with 90% yield (9). Using Raney nickel, the temperature and pressure were much higher than in the reactions where platinum metals were used (10).

Previously we reported the heterogeneous catalytic hydrogenations of some pyrrole derivatives (1-methyl-2-pyrroleethanol, methyl 1-methyl-2-pyrroleacetate, 1-methyl-pyrrole and 1H-pyrrole) in non-acidic medium, over different supported noble metal catalysts (Rh/C, Rh/Al$_2$O$_3$, Ru/C and Pd/C), under mild conditions (6 bar, 25-80°C) with complete conversion and selectivity (11-14). According to our hydrogenation results, the saturation of pyrrole ring proceeded relatively easily, besides it was unnecessary to use acids to avoid the deactivation of the catalysts used.

There are two possibilities to hydrogenate aromatic compounds in stereoselective way. One is the diastereoselective hydrogenation of a chiral precursor, and other is the enantioselective hydrogenation of a prochiral substrate with a chiral catalyst. Asymmetric synthesis of a chiral piperidine derivative was described by Blaser and coworkers (15). Ethyl nicotinate was hydrogenated to ethyl nipecotinate in two steps. In the first step, the starting material was converted to the 1,4,5,6-tetrahydro derivative over Pd/C, at 50 bar and RT with 76% yield. Hydrogenation of ethyl 1,4,5,6-tetrahydronicotinate was investigated with 10,11-dihydrocinchonidine modified noble metal catalysts. The highest enantiomeric excesses (*e.e.*) were obtained at relatively low conversions over Pd/C, in DMF (19% *e.e.*, 12% conversion) and Pd/TiO$_2$, in a DMF/H$_2$O/AcOH solvent mixture (24% *e.e.*, 10% conversion), at 130 bar and 50°C.

In order to obtain optically active piperidine and pyrrolidine compounds, we have chosen the diastereoselective route, i.e., the hydrogenation of chiral pyridine and pyrrole derivatives, respectively. High diastereoselectivities were obtained in the asymmetric hydrogenation of substrates containing (*S*)-proline part. For example, in the hydrogenation of *N*-(2-methylbenzoyl)-(*S*)-proline methyl ester 50% diastereomeric excess (*d.e.*) was achieved over rhodium on carbon modified with ethyldicyclohexylamine (16). The hydrogenation of *N*-acetyl-dehydrophenylalanyl-(*S*)-prolinanilide over 10% Pd/C, in toluene resulted in 68% *d.e.* (17). In the enantioselective hydrogenation of isophorone, in the presence of stochiometric amount of (*S*)-proline, a C=C double bond was reduced over palladium on carbon, with 60% *e.e.*, through an intermediate containing an oxazolidine ring, which was formed in a condensation reaction between isophorone and (*S*)-proline (18). On the basis of these experiences we have also used this chiral auxiliary in order to obtain high asymmetric induction during the saturation of pyridine or pyrrole ring. In this study, the effects of catalytic metals, solvents, temperature and pressure on the conversion and the *d.e.* are discussed in the hydrogenation of some pyridine and pyrrole carboxylic acid derivatives.

Experimental

Materials. The reagents: (*S*)-proline (98%), nicotinic acid (99%) and *N,N'*-dicyclohexyl-carbodiimide (99%) were supplied by Merck-Schuchardt (Hohenbrunn, Germany), whilst picolinic acid (99%) and 1-methylpyrrole (98%) were purchased from Aldrich (Steinheim, Germany). The solvents: methanol, chloroform, ethyl acetate, toluene, *N,N'*-dimethylformamide and dichloro-methane were supplied by Reanal (Fine Chemicals, Budapest, Hungary), in *pro analysi* grade.

The catalysts were also commercial products: 5% Pt/C (Heraeus, Karlsruhe, Germany), 5% Ru/C (Aldrich), 5% Rh/C (Aldrich), 5% Rh/Al$_2$O$_3$ (Degussa, Hanau, Germany) and 10% Pd/C Selcat (19) (Finomvegyszer Fine

Chemicals, Budapest, Hungary). The dispersion of the catalysts, determined by H_2-, O_2- and CO-chemisorption measurements, are the following: $D_{5\% \text{ Pt/C}} = 0.36$, $D_{5\% \text{ Ru/C}} = 0.38$, $D_{5\% \text{ Rh/C}} = 0.42$, $D_{5\% \text{ Rh/Al}_2\text{O}_3} = 0.49$ and $D_{10\% \text{ Pd/C}} = 0.50$.

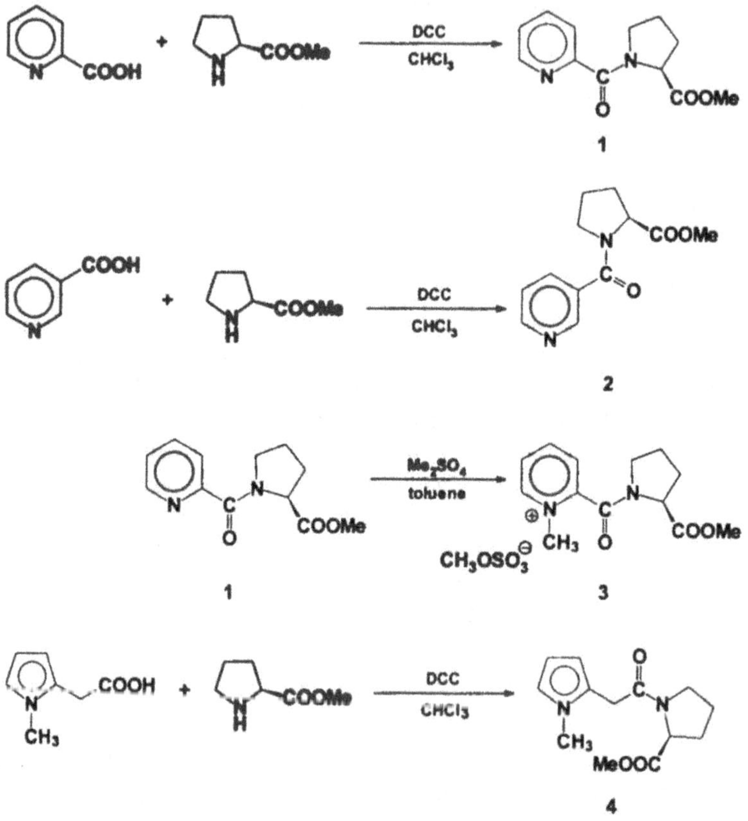

Scheme 1 Preparation of substrates for the diastereoselective hydrogenations.

    (S)-Proline methyl ester hydrochloride was prepared from (S)-proline, in methanol with $SOCl_2$, according to literature procedure (20). The free base was discharged with equivalent amount of triethylamine in the synthesis of starting materials, *in situ*. The starting materials for the diastereoselective hydrogenations, such as N-picolinoyl-(S)-proline methyl ester (1), N-nicotinoyl-

(*S*)-proline methyl ester (**2**), (*2'S*)-2-[2'-(methoxycarbonyl)-pyrrolidino-carbonyl]-1-methylpiridinium methosulfate (**3**) and *N*-(1'-methylpyrrole-2'-acethyl)-(*S*)-proline methyl ester (**4**), were prepared from picolinic and nicotinic acids, as well as 1-methylpyrrole-2-acetic acid (Scheme 1), on the basis of the procedure described in (21).

Synthesis of compound **1**: 13.8 g (66.9 mmol) *N,N'*-dicyclohexyl-carbodiimide was dissolved in 50 ml CHCl₃, and to this solution 8.2 g (66.9 mmol) picolinic acid in 70 ml CHCl₃ was added gradually. A solution of 11.1 g (66.9 mmol) *(S)*-proline methyl ester hydrochloride and 6.8 g (66.9 mmol) triethylamine in 70 ml CHCl₃ was added gradually to the former reaction mixture, which was stirred for 24 h. The chloroform was distilled off in vacuum. The afforded triethylamine hydrochloride and *N,N'*-dicyclohexylurea were crystallized from ethyl acetate and filtered off. After removal of the solvent, 10.0 g (64%) **1** was obtained as yellowish oil. Synthesis of compound **2**: it was prepared by using the same procedure as for the preparation of **1**. From 8.2 g (66.9 mmol) nicotinic acid 13.7 g (88%) **2** was obtained as yellowish oil. Synthesis of compound **3**: 4 g (17.1 mmol) **1** was dissolved in 80 ml toluene and 2.15 g (17.1 mmol) Me₂SO₄ was added. The reaction mixture was stirred for 5 h under reflux. After removal of the solvent, 4.7 g (76%) **3** was obtained in the form of a red oil. Synthesis of compound **4**: it was prepared according to the same procedure used for the preparation of **1**. From 5.0 g (36 mmol) 1-methylpyrrole-2-acetic acid, which was synthesized in a three-step synthesis developed in our laboratory, 8.54 g (95%) **4** was obtained as yellowish oil. The purity of the starting materials **1**, **2**, **3** and **4**, according to HPLC analyses, was 96-98%.

Hydrogenations. The catalytic hydrogenations were carried out in 100 or 250 cm³ stainless steel autoclaves (Technoclave, Budapest, Hungary) equipped with a magnetic stirrer (stirring speed: 1100 rpm) and a digital manometer, at 10-50 bar and 50-100°C, as well as in a 0.5 dm³ ZipperClave™ (Autoclave Engineers, Erie, USA) autoclave equipped with a magnetically driven turbine stirrer (MagneDrive® II, stirring speed: 1800 rpm) and an automatic gas pressure-controlling unit, at 100 bar and 25-50°C. The reactor containing the starting material, catalyst and solvent, was flushed with nitrogen and hydrogen, then charged with hydrogen to the specified pressure and heated up to the given temperature. After the hydrogenation was completed, the catalyst was filtered and the solvent was distilled off in vacuum. The retained products were analyzed by HPLC and NMR. All products, when the conversion was complete, were prepared with 97-99% yields.

Analysis. The HPLC analyses were carried out on a Purosphere® RP 18-e column (125×4 mm), using gradient elution methods. In order to determine the ratio of the diastereomers we have developed two separation methods, which are the following. Method A: the starting eluent was 5% methanol and 95% water, the gradient was 0→20% acetonitrile and 95→75% water in 15 min; method B: the starting eluent was 10% acetonitrile and 90% water, the gradient was 10→100% acetonitrile in 15 min. The UV-absorbance was measured at 220 nm. From the chromatograms the conversion and the d.e. values were determined. Diastereomeric excesses were calculated according to the following equation:

$$d.e. \ (\%) = \frac{[A]-[B]}{[A]+[B]} \cdot 100$$

where

[A] ... concentration of major diastereomer,

[B] ... concentration of minor diastereomer.

The NMR spectra were recorded on a Bruker DRX500 spectrometer, in CDCl₃. Optical rotation data were measured with a Perkin-Elmer 241 automatic polarimeter (c=1, MeOH). Circular dichroism (CD) spectra were recorded on a Jobin Yvon Dichograph Mark VI, in ethanol.

Results and Discussion

Hydrogenation of N-picolinoyl-(S)-proline methyl ester

The effect of the catalytic metals on the conversion and the d.e. values in the hydrogenation of 1 (Scheme 2), in methanol is summarized in Table 1.

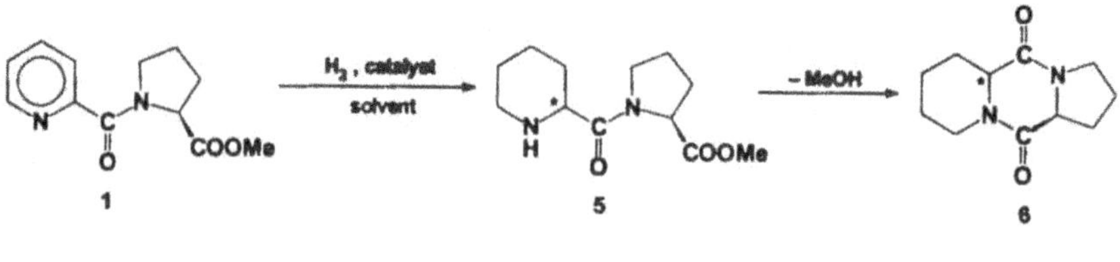

Scheme 2 Hydrogenation of N-picolinoyl-(S)-proline methyl ester.

The highest diastereoselectivity (64%) was achieved over palladium on carbon with 100% conversion after 3 h, at 98°C. The pyridine ring of 1 was also saturated completely with rhodium on carbon, but it required longer reaction time (5 h) and the d.e. was much lower (43%). Using Pt/C and Ru/C catalysts both the conversion and the d.e. values were low. In the hydrogenation of 1 the chemoselectivity to 6 was 100% over each catalytic metal.

During identification of the hydrogenated products, the ¹H- and ¹³C-NMR measurements showed that the expected *N*-pipecolinoyl-(*S*)-proline methyl ester (5) was completely converted to a tricyclic diketopiperazine derivative. The singlet of OCH₃ group (δ 3.74) was lacked in the ¹H-NMR spectrum and the peak of the COO group was shifted from (δ 173.6) towards (δ 164.7) in the ¹³C-NMR spectrum, which signed that is, in fact, a carbonyl group. On the basis of these data we came to the conclusion that an intramolecular *N*-acylation, which is a well-known reaction of the aminoacid esters, took place during the hydrogenation of 1. Due to this "side-reaction" mainly the (5a*S*,11a*S*)-perhydropyrido[1,2-*a*]pyrrolo[1,2-*d*]pyrazine-5,11-dione (6) diastereomer was formed. The absolute configuration of the product was determined by means of NMR and CD spectra, as well as optical rotation and m.p. data. According to the literature (22), the specific rotation of the optically pure 6 is $[\alpha]_D^{25} = -8.8°$ (*c*=0.3, MeOH), whilst the m.p. is 150-154°C.

Table 1 Hydrogenation of 1 over carbon supported precious metal catalysts.[a]

No.	Catalyst type	Catalyst/reactant ratio (gg⁻¹)	Reaction time (h)	Conversion (%)	d.e. (%)
1	5% Rh/C	0.2	5	100	43
2	10% Pd/C	0.3	3	100	64
3	5% Pt/C	0.3	3	21.3	23
4	5% Ru/C	0.3	5	52.1	38

[a] 1 g reactant, 30 ml methanol, 10 bar, 98 °C. Method A was used in the HPLC analysis.

10 bar pressure reductions had to be carried out at near 100°C, because at room temperature or 50°C no conversion was observed. The applied catalyst/reactant ratio (0.3) seems relatively high, but it was necessary to use this amount of catalyst in order to complete the hydrogenations.

It is well-known that in the catalytic hydrogenations the selectivity of the reaction, as well as the activity of the catalyst can be influenced by using appropriate solvents (23). The results of the hydrogenation of 1 in different solvents, over palladium on carbon are summarized in Table 2. There were no significant differences between the diastereoselectivities (64-67%) in methanol (MeOH), in ethyl acetate (EtOAc) and in dichloromethane. In methanol the hydrogenation of 1 was the fastest, the reduction was complete after 3 h. In ethyl acetate, after the same time, the conversion of 1 was low (34.8%) and the *d.e.* value was only 54%, whereas after 8 h the hydrogenation of 1 was complete and the diastereoselectivity increased to 65%. Both the conversion and the *d.e.* were significantly lower in *N,N'*-dimethylformamide (DMF) than in methanol.

Table 2 Hydrogenation of 1 in different solvents, over palladium.[a]

No.	Solvent	Reaction time (h)	Conversion (%)	*d.e.* (%)
1	MeOH	3	100	64
2	EtOAc	3	34.8	54
		8	100	65
3	DMF	3	79.3	52
4	Dichloromethane	3	83.5	67

[a] 1 g reactant, 10% Pd/C catalyst, catalyst/reactant ratio: 0.3, 30 ml solvent, 10 bar, 98 °C.

The lower temperature is favourable for the higher diastereoselectivity, therefore the effect of temperature and pressure on the conversion and the *d.e.* was also investigated in the hydrogenation of 1. At lower temperature, it was necessary to apply higher pressure to keep the activity of the catalyst. Decreasing the reaction temperature from 98°C to 50°C, the pressure was increased to 50 bar to complete the saturation of the pyridine ring of 1. According to the data of Table 3, at 50 bar and 50°C, in methanol, over palladium on carbon higher *d.e.* (79%) was achieved than at 10 bar and 98°C (64%), but the reaction time became much longer. Similarly to the hydrogenation carried out at higher temperature, the reaction was also slower in EtOAc than in MeOH. There are further similarities in these hydrogenations: in ethyl acetate the low conversion (41.2%) was also accompanied with low diastereoselectivity (53%). At room temperature no conversion was observed even at 100 bar pressure, therefore the temperature had to be raised to 50°C. Under these conditions the reduction was fast, but resulted in lower *d.e.* (62%) than at 50 bar. Applying high pressure increased the activity of the catalyst, whereas the diastereoselectivity decreased. It seems, the long reaction time, i.e., the low reaction rate is more favourable for the high stereoselectivity.

Table 3 Effect of temperature and pressure in the hydrogenation of **1**.[a]

No.	Solvent	Temperature (°C)	Pressure (bar)	Reaction time (h)	Conversion (%)	d.e. (%)
1	MeOH	50	50	12	97.4	79
2	EtOAc	50	50	9.5	41.2	53
3[b]	MeOH	25	100	4.0	0	—
		50	100	3.5	92.4	62

[a] 1 g reactant, 10% Pd/C catalyst, catalyst/reactant ratio: 0.3, 30 ml solvent.
[b] 5.6 g reactant, 160 ml solvent.

According to these results, the pyridine derivative (**1**) can be completely hydrogenated to **6** with almost 80% diastereoselectivity, over palladium on carbon, in methanol, at 50 bar and 50°C.

Hydrogenation of N-nicotinoyl-(S)-proline methyl ester

In the hydrogenation of **2** (Scheme 3) the most active catalytic system (10% Pd/C and MeOH) was used at 0.3 catalyst/reactant ratio. The hydrogenation results are summarized in Table 4.

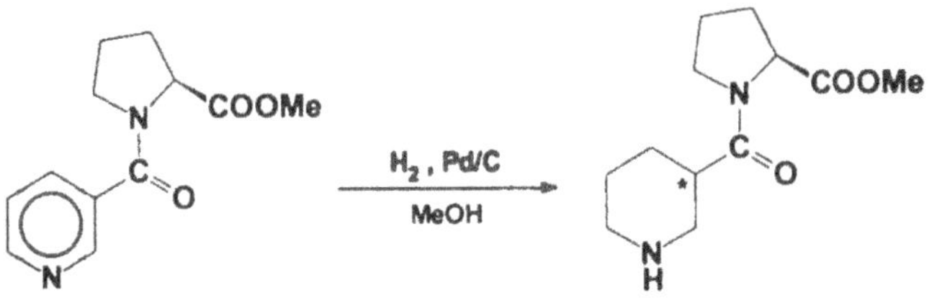

Scheme 3 Hydrogenation of *N*-nicotinoyl-*(S)*-proline methyl ester.

Comparing the *d.e.* values in the hydrogenation of **1** and **2** at 10 bar and 98°C, much lower diastereoselectivity (49%) was obtained in the reduction of **2** than in that of **1** (64%). On the contrary, at 50 bar and 50°C the *d.e.* was over 90%, which is much higher than in the hydrogenation of **1** (79%).

According to the NMR spectra, this hydrogenation resulted in the *N*-nipecotinoyl-*(S)*-proline methyl ester (**7**) diastereomers, i.e., intramolecular *N*-acylation did not take place. On the basis of NMR and CD spectra the major diastereomer is the (3*R*,2'*S*)-*N*-nipecotinoylproline methyl ester. Its absolute configuration was determined on the basis of the analogy with **6**, namely on the basis of their CD spectra.

Table 4 Hydrogenation of **2** over palladium on carbon.[a]

No.	Temperature (°C)	Pressure (bar)	Reaction time (h)	Conversion (%)	d.e. (%)
1	98	10	3.0	100	49
2	50	50	10.5	100	94

[a] 1 g reactant, 10% Pd/C catalyst, catalyst/reactant ratio: 0.3, 30 ml methanol. Method B was used in the HPLC analysis.

From NMR measurements we have got some hints for the explanation of the different temperature dependence of *d.e.* in the hydrogenation of **1** and **2**. The approximate ratio of peaks (9÷10), which can be assigned to the two conformers of **1** (e.g. δ 8.43 and 8.57 of Ar-H), did not change upon heating it up to 140 °C in the spectrometer, i.e., this pyridine derivative should have a very stable, rigid structure even at near 100°C. Spectra of **2** also contain the sign of conformers, which have hindered rotation (δ 3.42 and 3.75 of OCH_3 group), but the ratio of them is about 1÷2. This can serve for the explanation of the higher *d.e.* in the hydrogenation of **2** at lower temperature.

Hydrogenation of quaternized N-picolinoyl-(S)-proline methyl ester

In order to avoid the intramolecular *N*-acylation and decrease poisoning effect of nitrogen in the reduction of **1**, the hydrogenation of the quaternary pyridinium salt **3** was also tested (Scheme 4). In the hydrogenation of **3** the most active catalytic system was used, i.e., the reduction was carried out over 10% Pd/C, in methanol. The product free base (**8**) was discharged by adding 20% NaOH-solution.

Contrary to the hydrogenation of **1** and **2**, this *N*-alkylated picolinic acid derivative could be reduced already at room temperature and 20 bar. At 0.3 catalyst/reactant ratio the hydrogenation of **3** was complete after 4 hours with almost complete diastereoselectivity (98%). Since the spectroscopic data of **8**, similarly to compound **7**, cannot be found in the literature, its absolute configuration was also determined by means of its NMR and CD spectra, on the analogy of **6**. According to these measurements, the major diastereomer is the (2*S*,2'*S*)-*N*-1-methylpipecolinoylproline methyl ester.

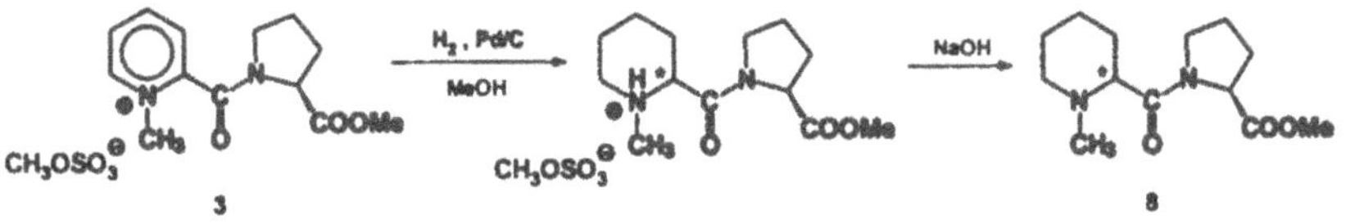

Scheme 4 Hydrogenation of quaternized *N*-picolinoyl-*(S)*-proline methyl ester.

Hydrogenation of N-(1'-methylpyrrole-2'-acethyl)-(S)-proline methyl ester

The effect of the catalytic metals on the conversion and the *d.e.* values in the hydrogenation of **4** (Scheme 5), in methanol is shown in Table 5.

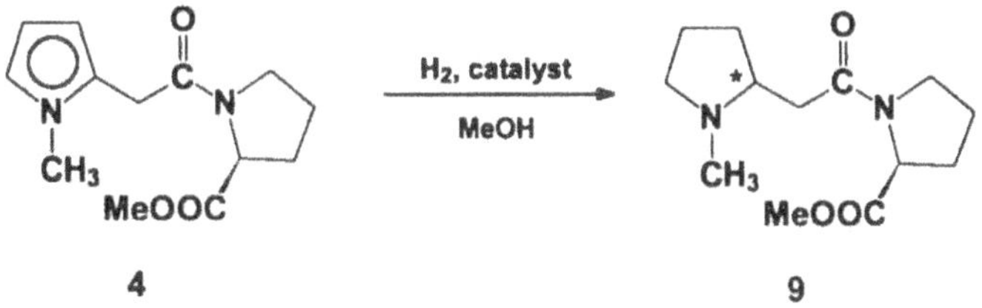

Scheme 5 Hydrogenation of *N*-(1'-methylpyrrole-2'-acethyl)-(*S*)-proline methyl ester.

The highest diastereoselectivity (90%) was achieved over rhodium on carbon with almost complete conversion, at RT and 20 bar. The pyrrole ring of **4** was also saturated completely with alumina supported rhodium, but it required higher temperature (80°C) and the *d.e.* was much lower (50%). Using Pd/C and Ru/C catalysts both the conversion, even at 80°C, and the *d.e.* values were low.

Due to the less rigidity of the molecule of **4**, compared with the structure of **1**, high diastereoselectivity could be obtained only at low temperature in the hydrogenation this pyrrole derivative. On the basis of the NMR and CD spectra, the major diastereomer is the (2'*S*,2*S*)-*N*-(1'-methylpyrrolidine-2'-

acethyl)proline methyl ester. Its absolute configuration was also determined on the analogy of 6.

Table 5 Hydrogenation of 4 over supported precious metal catalysts.[a]

No.	Catalyst type	Catalyst/reactant ratio (gg^{-1})	Temp. (°C)	Reaction time (h)	Conv. (%)	d.e. (%)
1	10% Pd/C	0.3	80	12	60	22
2	5% Ru/C	0.2	80	6	69	48
3	5% Rh/Al$_2$O$_3$	0.25	80	6	100	50
4	5% Rh/C	0.2	24	10	84	90
5[b]	5% Rh/C	0.3	24	5	99	90

[a] 2 g reactant, 50 ml methanol, 10 bar. [b] 1 g reactant, 40 ml methanol, 20 bar. Method A was used in the HPLC analysis.

Conclusions

The asymmetric heterogeneous catalytic hydrogenation of various chiral pyridine and pyrrole carboxylic acid amide derivatives was carried out in diastereoselective reactions, in non-acidic medium. In the saturation of pyridine ring the highest diastereoselectivities (79-98%) were obtained over 10% Pd/C catalyst and in methanol solvent, at moderate pressure (50 bar) and temperature (50°C). The quaternization of pyridine nitrogen made possible the almost complete diastereoselectivity of the hydrogenation. In the hydrogenation of pyrrole ring the highest d.e. (90%) was achieved over a rhodium on carbon catalyst, in methanol, at relatively low pressure (20 bar) and RT. According to these results, palladium is the best catalyst for the saturation of pyridine ring, whilst rhodium is the best one for that of pyrrole ring. The diastereoselective heterogeneous catalytic hydrogenation could be a useful method for the preparation of various chiral piperidine and pyrrolidine derivatives, which are important and valuable pharmaceutical intermediates. The proline based substituent proved to be again an effective synthon inducing good diastereomeric excess. These hydrogenations are successful examples of diastereoselections in the saturation of pyridine or pyrrole ring with complete conversion.

Acknowledgments

The authors acknowledge the financial support of the Hungarian OTKA Foundation under the contract number T 029557 and T 025041, as well as that of the Ministry of Education, FKFP 0017/1999. L. Hegedûs gratefully acknowledges the financial support of the Foundation for the Hungarian Higher Education and Research.

References

1. P.N. Rylander, *Catalytic Hydrogenation over Platinum Metals*, Academic Press, New York, 1976, p. 375.
2. P.N. Rylander, *Catalytic Hydrogenation in Organic Syntheses*, Academic Press, New York, 1979, p. 213.
3. M. Freifelder, *Practical Catalytic Hydrogenation*, John Wiley, New York, 1971, pp. 577-582.
4. M. Freifelder, R.M. Robinson and G.R. Stone, *J. Org. Chem.* **27**, 284 (1962).
5. M. Freifelder and G.R. Stone, *J. Org. Chem.* **26**, 3805 (1961).
6. T.S. Hamilton and R. Adams, *J. Am. Chem. Soc.* **50**, 2260 (1928).
7. M. Freifelder, *J. Org. Chem.* **27**, 4046 (1962).
8. G.N. Walker, *J. Org. Chem.* **27**, 2966 (1962).
9. I. Beck, E. Jákfalvi, I. Simonyi, J. Halmos, A. Dietz, A. Tungler and T. Máthé, *HU Patent* 198 017 (1986).
10. M. Freifelder, *Advances in Catalysis*, Academic Press, New York, 1963, Vol. 14., p. 203.
11. L. Hegedûs, T. Máthé and A. Tungler, *Appl. Catal. A* **143**, 309 (1996).
12. L. Hegedûs, T. Máthé and A. Tungler, *Appl. Catal. A* **147**, 407 (1996).
13. L. Hegedûs, T. Máthé and A. Tungler, *Appl. Catal. A* **152**, 143 (1997).
14. L. Hegedûs, T. Máthé and A. Tungler, *Appl. Catal. A* **153**, 133 (1997).
15. H.-U. Blaser, H. Hönig, M. Studer and C. Wedemeyer-Exl, *J. Mol. Catal. A* **139**, 253 (1999).
16. M. Besson, B. Blanc, M. Champelet, P. Gallezot, K. Nasar and C. Pinel, *J. Catal.* **170**, 254 (1997).
17. A. Tungler, Á. Fürcht, Zs.P. Karancsi, G. Tóth, T. Máthé, L. Hegedûs and Á. Sándi, *J. Mol. Catal. A* **139**, 239 (1999).
18. A. Tungler, T. Máthé, J. Petró and T. Tarnai, *J. Mol. Catal. A* **61**, 259 (1990).(1990).
19. T. Máthé, A. Tungler and J. Petró, *U.S. Patent* 4 361 500 (1982).
20. S. Guttmann, *Helv. Chim. Acta* **44**, 721 (1961).
21. K. Soai and H. Hasegawa, *J. Chem. Soc., Perkin Trans. I*, 769 (1985).
22. J. Vicar, J. Smolíková and K. Bláha, *Collect. Czech. Chem. Commun.* **37**, 4060 (1972).
23. P.N. Rylander, In *Catalysis in Organic Syntheses*, W.H. Jones, ed., Academic Press, New York, 1980, p. 155.

Immobilized Chiral Rhodium Complexes on Silicas Show Enhanced Enantioselectivity in Hydrogenation of Enamide Esters

Francis M. de Rege[a], David K. Morita[a], Kevin C. Ott[a], William Tumas[a] and Richard D. Broene[a,b]

[a]Division of Chemical Science and Technology
CST-18, MS J514 , Los Alamos National Laboratory, Los Alamos, NM 87532
[b]Department of Chemistry, Bowdoin College, 6600 College Station, Brunswick
ME 04011

Abstract

Heterogenization of the asymmetric hydrogenation catalyst [(R,R)-Me-(DuPHOS)Rh(COD)]OTf on the mesoporous silica MCM-41 is reported. Immobilization is accomplished by slurrying the organometallic reagent with MCM-41 in methylene chloride. Weak interactions immobilize the catalyst on the mesoporous surface which allows the catalyst to be recovered and recycled by simple filtration. As evidenced by XRD, pore volume, surface area, elemental analysis, TGA and NMR, hydrogen bonding between the triflate anion and SiO-H groups on the silica surface fixes [(R,R)-Me-(DuPHOS)Rh(COD)]$^+$ inside the pores of MCM-41. Deactivation due to mesopore fouling by the product of hydrogenation occurs after several cycles, and is avoided by decanting the solvent product mixture from the catalyst followed re-suspension in fresh solvent. Enantioselectivities and relative rates for the reaction of several enamide esters in hexane with the heterogeneous [(R,R)-Me-(DuPHOS)Rh(COD)]OTf/MCM-41 material show it to be a rare example of a heterogenized catalyst which is as, and in some cases more, active and selective as the homogenous catalyst under optimized conditions.

Introduction

Heterogeneous catalysts have several process advantages over their homogeneous counterparts such as stability, ease of handling, and are often reusable. Importantly, the ease of separation of the catalytic material from the product is greatly enhanced, as they are often phase separated from the reactants in solution by simple filtration. While attaching a transition metal complex to a support is, in principle, an attractive way to couple the benefits of a homogenous catalysts with those presented by a heterogeneous catalyst, in practice heterogenization often diminishes the catalytic activity and selectivity (1,2). Solid supports for heterogenizing enantioselective catalysts have been the objects of considerable research over many years (3), especially with regard to the method of immobilizing the catalyst on the surface. The simplest method in principle, is a sorption method where a homogeneous enantioselective catalyst is held to the support via interactions strong enough to prevent dissociation under the reaction conditions (3). Electrostatic forces, hydrogen bonds and collective van der Waals forces, are examples of catalyst-support interactions that may be suitable for immobilizing a homogeneous catalyst on or in the support under reaction conditions. Adsorption methods preclude the need to chemically modify the ligands or treat the resulting complex further. MCM-41 materials are convenient solid supports for heterogenizing homogenous catalysts, because of their large controllable pore size and regular porous topology. The pore size is effectively controlled by using cationic surfactants as templates, where longer chain length correlates with larger pore size (4). The present work demonstrates the ability of the mesoporous silica MCM-41 to heterogenize a chiral catalyst by adsorption. This paper discusses the synthesis, characterization, and hydrogenation reactions of MCM-41-supported (R,R)-Me-(DuPHOS)(COD)RhOTf, a chiral, cationic, rhodium complex. The immobilized complex is a heterogeneous and recyclable catalyst that shows high enantioselectivities in the hydrogenation of a series of prochiral enamide esters. It shows, in some cases, higher selectivities and activities than seen with the homogeneous analog.

Experimental Section

Synthesis of MCM-41

MCM-41 was synthesized using a previously published procedure (5). Cetyltrimethylammonium chloride (29g, 22 mmol) and ammonium hydroxide (0.83g, 6.9 mmol) were combined with stirring and dissolved in tetramethylammonium hydroxide hydrate (4.6g, 25 mmol). To this was added trimethylammonium silicate (Sachem, 30g, 25 mmol) resulting in precipitate formation. Cab-O-Sil M5 (7.5g, 125 mmol) was stirred into the mixture with a Teflon spatula until the mix was homogenized and the mixture heated statically at 140 °C for 48 h. The resulting material was washed on a fritted funnel with de-ionized water (1L) and subsequently dried in air at 100 °C overnight, resulting in a solid white product. This material was calcined at 1 °C/min to 550 °C and then 20 h at 550 °C under a steady flow of purified dry air. Mass loss upon calcination was 48%. The calcined material was crushed in a mortar with a pestle and calcined for 3 h at 550 °C. Powder X-ray diffraction indicated a hexagonal phase with a d spacing of 40.6Å. A BET analysis of nitrogen adsorption into the material indicated the material has a surface area of 953 m^2/g and a mesopore volume (BJH) of 1.003 cm^3/g.

[(R,R)-Me-(DuPHOS)Rh(COD)]OTf/MCM-41 (**1**)

[(R,R)-Me-(DuPHOS)Rh(COD)]OTf (200 mg, 3.00 x 10^{-4} mol) was dissolved in CH_2Cl_2 and combined with 3 g of MCM-41 in CH_2Cl_2. The slurry was stirred overnight during which time the solution became colorless and the heterogeneous material took on an orange color. The solids were isolated by filtration, washed with CH_2Cl_2 (3x 10 mL) and placed under vacuum for several hours to provide a free flowing light yellow powder. Powder X-ray diffraction is unchanged from the unfunctionalized MCM-41 with hexagonal phase having a d spacing of 40.7 Å. A BET analysis of nitrogen adsorption into the material indicates a surface area of 854 m^2/g and a mesopore volume (BJH) of 0.854 cm^3/g. Thermal gravimetric analysis (TGA) gave a 5.26 wt% loss (calculated: 5.29 wt%). Elemental analysis gave 1.03 wt% rhodium (calculated: 0.96 wt%). ^{31}P NMR (CH_2Cl_2, δ) 76.6 (bd, J_{Rh-P} = 147 Hz, $\upsilon_{1/2}$ = 130 Hz). ^{19}F NMR -76 (bs, $\upsilon_{1/2}$ = 575 Hz).

[(R,R)-Me-(DuPHOS)Rh(COD)]OTf-Davisil

[(R,R)-Me-(DuPHOS)Rh(COD)]OTf (0.200 g) was dissolved in CH_2Cl_2 (10 mL) and to this solution was added Davisil (grade 636, 35-60 mesh, 60 Å, surface area 480 m^2/g, pore vol. 0.75 cc^3/g, 3g). The heterogeneous mixture was stirred overnight with slow uptake of the solution's color by the solid. The solid was collected on a fritted funnel, washed with CH_2Cl_2 and placed under vacuum for several hours. This gives a free flowing yellow/orange powder. Elemental analysis for Rh gave 0.23 wt % which corresponds to a catalyst loading of 1.5%.

General Hydrogenation Procedure. A typical hydrogenation reaction was carried out as follows. To an oven dried 0.5 L Fischer-Porter bottle equipped with a sampling port and a pressure gauge in the drybox was added the substrate (25 mg), the catalyst (20 mg), and 20 mL hexane. The flask was sealed and moved to a Schlenk line. The vessel was degassed by exposure to vacuum (2 x ~10 sec) followed each time by exposure to hydrogen and finally brought to the desired pressure. Vigorous stirring was maintained by a 3 cm magnetic stirring bar with a constant setting on the magnetic stirrer. In the vessel used, the volume of solvent was nearly the depth of the stirring bar thickness, which allowed for very good agitation. Aliquots were removed through a valve attached to a septum port with a pressure syringe. The course of the reaction was followed by GC-MS but, in general, was allowed to proceed 16 hours. For control reactions in hexane, the more hexane soluble [(R,R)-Me-(DuPHOS)Rh(COD)]BArF was used as a catalyst. The reactions were carried out analogously, with the exception that 2 mg of [(R,R)-Me-(DuPHOS)Rh(COD)]BArF were used. Enantiomeric excesses were calculated in the normal fashion from the integrated chromatograms from the chiral Val-L column and are reported as the average of two or more reactions. For the chiral separations, typical column conditions were: He carrier gas flow rate was 0.8 mL/min. with the isothermal conditions depending on the substrate-160 °C for the product of **A** (ret. times: 0.79 and 0.85 min. (minor enantiomer)), 160 °C for the products of **B** (ret. times 1.36 and 1.47 (minor enantiomer)), and 170 °C for the products of **C** (ret. times 2.21 and 2.36 min. (minor enantiomer)).

Analogous reactions were performed for the reduction of the enamides with [(R,R)-Me-(DuPHOS)Rh(COD)]BArF, except that 2 mg of the catalyst were used.

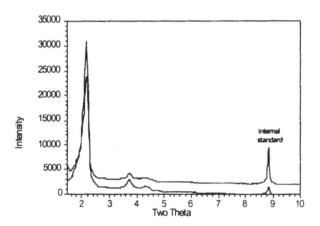

Figure 1 Powder XRD of MCM-41 (bottom) and 1 (top).

Results and Discussion

Synthesis and Characterization of Materials. The MCM-41 was synthesized by a previously published procedure (5) using cetyltrimethylammonium chloride as the templating agent. Analysis of the powder X-ray diffraction (XRD) (Figure 1) indicates the material is hexagonal with a d spacing (a_0) of 40.6 Å. Analysis of the nitrogen adsorption isotherms provided a surface area of 953 m^2/g, a mesopore volume of 1.00 cc/g and a narrow pore size distribution, typical of MCM-41 material prepared in this way (4). The catalyst chosen for the currentstudies was (R,R)MeDuPHOS-Rh(COD)OTf (COD= cyclooctadiene, OTf= trifluoromethylsulfonate) (6).

For the preparation of the heterogeneous catalyst, [(R,R)-Me-(DuPHOS)-Rh(COD)]OTf immobilized on MCM-41 (1, [(R,R)-Me-(DuPHOS)-Rh(COD)]OTf/MCM-41), orange-colored [(R,R)-Me-(DuPHOS)Rh(COD)]OTf was dissolved in CH_2Cl_2 and combined with MCM-41 slurried in CH_2Cl_2. Over time, the solution becomes colorless and the powder takes on the orange color. The powder was isolated by filtration as a free flowing light yellow powder. The powder XRD is unchanged from the unfunctionalized MCM-41 (Figure 1) showing that the hexagonal structure is intact. The loading of the organometallic complex was near quantitative as evidenced by several techniques. Thermal gravimetric analysis (TGA) conducted under dry air gave a weight loss of 5.26 %. From a theoretical 6.1 wt% loading of (R,R)-Me-(DuPHOS)Rh(COD) on MCM-41, the 5.26 wt% loss observed in the TGA is consistent with pyrolysis of the organic portions of (R,R,)-Me-(DuPHOS)Rh(COD)]OTf/MCM-41; i. e. only Rh metal remains on the MCM-

41 after the TGA experiment. Elemental analysis for Rh by ICP gave a rhodium content of 1.03 wt% (theory is 0.96 wt%). The expected organometallic loading based on the Rh elemental analysis is 6.67 wt%, and based on the TGA data is 6.22 wt%, both of which are close to the value of 6.1 wt% expected for quantitative loading. This indicates very efficient, near quantitative loading of [(R,R)-Me-(DuPHOS)Rh(COD)]OTf onto MCM-41. The surface area shows a decrease from 953 m^2/g in the unfunctionalized MCM-41 to 854 m^2/g in the [(R,R)-Me-DuPHOSRh(COD)]OTf on MCM-41 and a decrease in mesopore volume from 1.003 cm^3/g to 0.840 cm^3/g. These results are consistent with mesopore filling by the organometallic catalyst (7-12).

To compare the very high loading of (R,R,)-Me-(DuPHOS)Rh(COD)]OTf onto MCM-41 with other silicas, the commercially available silica (Davisil, 35-60 mesh) was slurried with [(R,R)-Me-(DuPHOS)Rh(COD)]OTf in CH_2Cl_2 under the same conditions to give a dark orange, free-flowing powder after drying. Elemental analysis on the resulting material showed it to be 0.23 wt% Rh, which corresponds to a catalyst loading of 1.5% [(R,R)-Me-(DuPHOS)-Rh(COD)]OTf. This loading is about 25 percent of the loading found for MCM-41 and correlates with the higher surface area and higher Si-OH content of MCM-41 relative to conventional silica (13). Pugin has pointed out the importance of high catalyst loading (and high activity) when heterogenizing homogeneous catalysts due to the need to minimize the amount of support material required in a reactor; however, high loadings are not always beneficial (1, 14).

The counter ion is very important for the successful immobilization of the catalyst onto MCM-41. Whereas the triflate anion effective for loading the complex onto the silicas, the lipophilic, poor hydrogen bonding (15) [(R,R)-Me-(DuPHOS)Rh(COD)]BArF (BArF = B[3,5-(CF$_3$)$_2$Ph]$_4$) salt failed to load under otherwise identical conditions. No loss of the orange color within the CH_2Cl_2 solution was observed over the course of the reaction. In addition, the solid

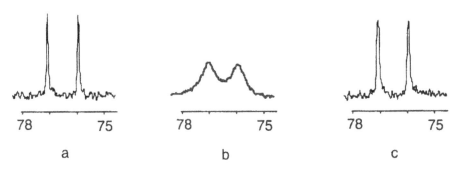

Frequency (ppm, δ vs H₃PO₄)

Figure 2 ³¹P NMR (a) [(R,R)-Me-(DuPHOS)Rh(COD)]OTf in CH₂Cl₂ (b) after addition of MCM-41 (c) **1** after addition of NaBArF.

material isolated lacked the characteristic orange color seen in MCM-41 after filtration. In fact, addition of NaBArF to a slurry of 1 causes orange coloration of the solvent, characteristic of dissolved cationic [(R,R)-Me-(DuPHOS)Rh(COD)].

The ^{31}P NMR of homogeneous [(R,R)-Me-(DuPHOS)Rh(COD)]OTf in CH_2Cl_2 shows a doublet centered at -76.6 ppm where the half width, $\upsilon_{1/2}$, is about 30 Hz (Figure 2a). The solution NMR of a slurry of [(R,R)-Me-(DuPHOS)Rh(COD)]OTf immobilized on MCM-41 in the same solvent shows a doublet with the same chemical shift which is considerably broader ($\upsilon_{1/2}$ = 130 Hz, Figure 2b) than the homogeneous resonance. The corresponding ^{19}F NMR spectra are shown in Figure 3. The sharp singlet ($\upsilon_{1/2}$ = 50 Hz) for the homogenous catalyst (Figure 3a) is broadened to $\upsilon_{1/2}$ = 575 Hz in the spectrum of a slurry of 1. The $\upsilon_{1/2}$ broadening seen in both the ^{19}F and the ^{31}P NMR spectra upon addition of MCM-41 to a solution containing [(R,R)-Me-(DuPHOS)Rh(COD)]OTf is consistent with restricted mobility of the organometallic complex within MCM-41, as would be expected for an immobilized molecule. Blümel has shown similar $\upsilon_{1/2}$ broadening in the slurry ^{31}P NMR spectra of phosphines tethered to silica (16).

Addition of NaBArF to a suspension of 1 in CD_2Cl_2 leads to a ^{31}P NMR spectrum where the peaks are nearly identical to the homogeneous catalysts in both chemical shift and line width ($\upsilon_{1/2}$ = 35 Hz, Figure 2c). Addition of NaBArF, however, does not lead to significant sharpening of the ^{19}F NMR peak corresponding to the free triflate ion (Figure 3c). Visual examination of the NMR sample treated with NaBArF shows the solution becomes yellow/orange— the color of the homogeneous catalyst in solution. Similar behavior was observed when polar solvents (THF, MeOH or diethyl ether) were added to solid 1.

The narrowing of the ^{31}P line width but not the ^{19}F line width upon addition of NaBArF is consistent with dissolution of the [(R,R)-Me-(DuPHOS)Rh(COD)]$^+$ fragment from the surface, while the triflate anion remains immobilized on the surface. This implies that in the absence of polar solvents and other competing counterions such as BArF⁻, triflate interacts with and binds the [(R,R)-Me-(DuPHOS)Rh(COD)]$^+$ fragment to the MCM-41 surface. This is in accord with the observation that attempted immobilization of

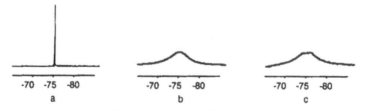

Frequency (ppm,   δ vs CCL $_3$F)

Figure 3 ^{19}F NMR (a) [(R,R)-Me-(DuPHOS)Rh(COD)]OTf in CH_2Cl_2 (b) after addition of MCM-41 (c) 1 after addition of NaBArF.

[(R,R)-Me-(DuPHOS)Rh(COD)]BArF in CH$_2$Cl$_2$ on MCM-41 does not lead to sorption of this complex onto the support. The experiments indicate that a weak interaction between the MCM-41 with the triflate anion immobilizes the [(R,R)-Me-(DuPHOS)Rh(COD)] cation on the surface. Triflate is known to be a hydrogen bond acceptor, while BArF is a very poor hydrogen bond acceptor (17,18). Using a hydrogen bonding acceptor anion to attach a cationic organometallic species to silica surfaces has also been implicated in the host-guest interaction of a manganese salen complex with aluminum-substituted MCM-41 (19). Polar solvents such as methanol, diethyl ether, and THF presumably provide a better solvation environment for the cationic rhodium complex, allowing a more favorable partitioning of the cation into solution and away from the surface-bound triflate anion as a solvent separated ion pair. Polar solvents may also compete with the triflate anion for hydrogen bonding sites.

To investigate the hypothesis of a hydrogen bonding attachment of the cationic rhodium complex to the support via the immobilized triflate ion, MCM-41 was first treated with trimethylsilyl chloride. Since this treatment silylates the surface hydroxyl groups, hydrogen bonding should be diminished in the silylated material (12) and the catalyst loading should decrease. In the experiment, the silylated MCM-41 was slurried with a CH$_2$Cl$_2$ solution of [(R,R)-Me-(DuPHOS)Rh(COD)]OTf. Under these conditions, the loading of (R,R)-Me-(DuPHOS)Rh(COD)]OTf onto silylated MCM-41 was 1.9 wt % (vs. 6.67 wt % for the untreated MCM-41), consistent with the proposition that surface silanol H-bonding to the triflate contributes to the sorption of the catalyst. It is possible that IR might give some indication for the interaction of the metal complex with surface silanol groups, however, at 6.1 wt% loading only about 3% of the surface hydroxyl groups on MCM-41 would be affected by triflate hydrogen bonding. Experimentally, no decrease in the IR intensity of the SiO-H at 3743 cm^{-1} was observed. A summary of our proposals for the immobilization of [(R,R)-Me-(DuPHOS)Rh]$^+$ to silica surfaces is illustrated in Figure 4.

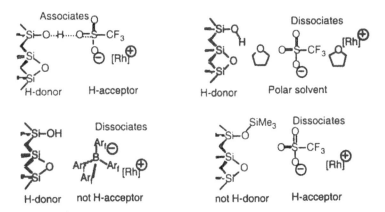

Figure 4 Schematic of surface anion interactions

These experiments put constraints on the conditions acceptable for the enamide hydrogenations we chose as our test reactions: the substrates for the hydrogenations are shown in Figure 5. Because leaching of an active species from a surface can lead to the false assignment of "heterogeneous" to an actually homogeneous catalyst (20), this problem must be carefully addressed in our experiments due to the high hydrogenation activity of [(R,R)-Me-(DuPHOS)-Rh(COD)]OTf in solution. As discussed above, the hydrogenations must be carried out in non-polar, non-hydrogen bonding solvents. Further experimentation showed that the combination of 1 and substrate in methylene chloride led to slow leaching of the catalyst, as evidenced by the slow coloration of the solvent. Suspension of 1 in hexane did not lead to any visible color change of the liquid, which suggested hexane as the reaction solvent (21).

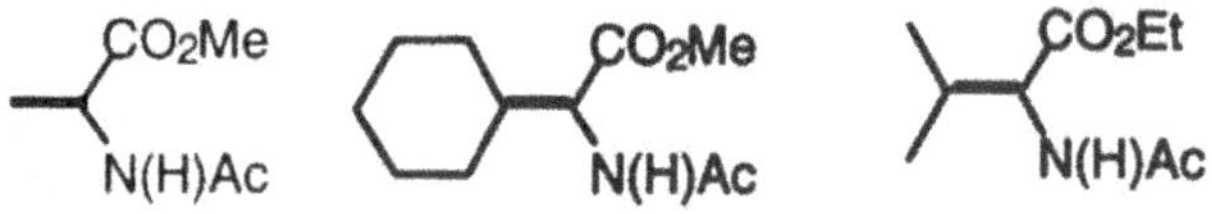

Figure 5 Hydrogenation substrates

The reactions generally were carried out in a 1 L Fisher-Porter bottle in 20 mL hexane at 8 psig H_2 with a 2.5 cm stir bar and magnetic stirring. These conditions maximize the surface of the gas liquid interface and increase diffusion of H_2 into the solution. Since there was some effect of stirring upon the reaction rate, as evidenced by incomplete reaction in the standard amount of time when stirring was interrupted, all reactions were performed on a single stir plate with a constant stirrer setting. The results of the reactions are found in Table 1.

The parent enamide methyl-N-Acyl-2-aminopropenoate (A) was hydrogenated with both high yield and high enantioselectivity within 30 min. by the MCM-41-supported catalyst. The homogeneous control reaction with [(R,R)-Me-(DuPHOS)Rh]BArF in hexane (22) was very sluggish in comparison to both the MCM-41-supported catalyst and the homogeneous reaction performed in MeOH (23). This is likely due to the poor solubility of the cationic rhodium complex in hexane, as evidenced by solid orange material observed in the flask during the hydrogenation. The enantioselectivity for this substrate was about the same as that seen for the reaction in MeOH, but higher than that seen when the reduction was performed in hexane.

More interesting are the results seen for the "notoriously difficult" β,β–disubstituted substrates **B** and **C** (23), where the MCM-supported catalyst not only appears to be more active and selective than the "homogeneous" catalyst in hexane, but is also more selective than the homogenous catalyst under optimized conditions in MeOH (23). For instance the cyclohexylidenamide ester **B** is hydrogenated with 98% ee with the MCM-Rh catalyst while its reaction

Table 1 Hydrogenation Results.   Conditions: 20 mL Hexane, 8 psig H_2.
Substrate **A**, 30 min., **B**, **C** 16 hours.

Substrate,Support		Solvent	Temp	Anion	Conv.	ee
A,	MCM-41	hexane	R.T.	OTf	>99%	99%
A,	none	hexane	R.T.	OTf	>99%	87%
A,	none	MeOH	R.T.	OTf	>99%	99%
B,	MCM-41	hexane	R.T.	OTf	>99%	97%
B,	MCM-41	hexane	R.T. 50 psig	OTf	>99%	97%
B,	MCM-41	hexane	50 °C	OTf	>99%	99%
B,	none	hexane	R.T.50 psig	BArF	92%	93%
B(23),	none	MeOH	25 °C	OTf	>99%	96%
C,	MCM-41	hexane	R.T.	OTf	>99%	98%
C,	MCM-41	hexane	50 °C	OTf	>99%	98%
C,	MCM-41	hexane	R.T.50 psig	OTf	>99%	98%
C,	SiO_2	hexane	R.T.16 hr	OTf	65%	97%
C,	none	hexane	R.T. 40 psig	BArF	29%	85%
C(23),	none	MeOH	R.T.	OTf	99	96%

with unsupported [(R,R)-Me-(DuPHOS)Rh]BArF in hexane achieves only 93 % ee at 92% conversion. The reaction in MeOH gives 96% ee at complete conversion (90 psig, 25 °C). The reaction has some degree of temperature dependence with a 99% ee seen at 50 °C and 8 psig; this is consistent with the accepted mechanism for hydrogenations using cationic Rh diphosphines (24). There does not seem to be a pressure dependence for the enantiomeric excess observed at pressures up to 50 psig, although it is possible that the relative amount of H_2 in hexane does not change significantly between 8 and 50 psig (25). The difference in rate is more apparent with the dimethyl compound **C**, where conversion is only 26 % and 85% ee after 22 hours at 40 psig while the MCM-supported catalyst is complete after 16 h at 8 psig with 98% ee. We note that the Davisil-supported material also shows high selectivity, but that the activity drops to 77% conversion in 16 hours: this is a consequence of lower rates corresponding with the four fold lower loading of the cationic Rh complex on silica compared to that seen for MCM-41. The hydrogenation of **B** with **1** compares favorably with the literature enantioselectivity for the homogeneous reaction of 96% (MeOH, 90 psig, 25 °C) (23). Few reports of such a positive influence on activity and selectivity for heterogenized catalysts exist, in fact the opposite is often true (1).

While the sluggishness of the "homogeneous" reactions in hexane above suggested that the reaction is not occurring by leaching of the catalyst into hexane from the support, further leaching experiments were carried out. In the first experiment, hydrogenation was conducted as previously outlined with substrate **A** in Table 1. The catalyst was filtered in a helium glove box and the filtrate tested for activity with additional **A**; no further conversion of **A** was observed. This suggests the reaction is occurring on the heterogeneous material rather than by a small amount of highly active material leaching into solution. A second experiment was conducted by hydrogenation of **B** under the conditions outlined in Table 1. The used catalyst was then filtered from the solution and washed with toluene. The solvents from the collected filtrate and washings were removed under vacuum and fresh hexane was then added along with substrate

A. Hydrogen was added and the solution vigorously stirred for 24 hours. No conversion of **A** by GC analysis was observed. This also suggests the reaction occurs on the support rather than in solution.

The leaching experiments do not address the possibility of catalyst deactivation and may lead to erroneous conclusions, therefore several recycling and stability experiments were performed. In the first experiment, substrate **A** (1 mmol) and **1** (100 mg, ~1 x10^{-5} mol Rh based on loading) were reacted under a hydrogen atmosphere for 30 min. An aliquot was removed, the mixture was stirred for 2 hours, after which time fresh substrate was added, keeping the reaction flask under a hydrogen atmosphere. This cycle was repeated 3 times and then the solution was left to stir. After 16 hours, fresh substrate was added and the reaction continued. As can be seen in Table 2, the catalyst remains active over 6.5 hours and the addition of new aliquots of **A**, while over longer periods of time the catalyst deactivates. Since the catalyst does not leach and the enantioselectivity is unchanged throughout this recycling experiment, compound **1** is sill the active catalytic material. Isolation of **1** by filtration after the experiment showed it to be coated with a viscous material identified as the hydrogenation product by ^{1}H NMR in CD$_3$OD.

TGA of the deactivated, coated catalyst showed a 70 wt% drop. Accounting for the 5 wt% associated with the organometallic in the catalyst, 65 wt% of product is associated with the MCM-41. It is likely that the deactivation of the catalyst is due to filling the mesopores of MCM-41 with product, impeding access of fresh substrate to the catalytic sites. Other deactivation processes would likely result in a lowering of the ee as well as a lowering of the activity. A second possibility however, is that the active catalyst decomposes over time to a species which has no hydrogenation activity.

Because some active catalysts decompose in the absence of substrate or solvent, we attempted a similar experiment where the solid material was isolated by decanting the supernatant rather than filtering the solids under vacuum. In this experiment, the materials were reacted under normal conditions for 30 min., the vessel vented, and the catalyst allowed to settle for 5 minutes. The supernatant was decanted leaving **1** as a solid in the vessel. The bottle was recharged with hexane, **A**, and hydrogen and the cycle was repeated 4 times. For the fifth cycle, the solvent was decanted and the vessel recharged with hexane only, allowed to stand covered for 16 hours, and then charged with **A** and hydrogen. This last reaction was continued for 30 min. and the vessel exposed to air. The results for each cycle is shown in Table 3. Under these product isolation conditions, the catalyst retains its activity over 24 hours, and

Table 2 Recycling Experiment 1.

Cycle #	Time to aliquot removal (h)	% Conv.	% ee
1	0.5	>99	99
2	2.5	>99	99
3	4.5	>99	99
4	6.5	>99	97
5	22.5	40	99

Table 3 Reaction of 99 mg 1 and A in hexane with reuse preceded by decanting the mixture. Reaction time: 31 +/- 1 min., time between runs - 10 min.

Run	A (mg)	Hexane (mL)	Product (mg)	Solvent (mL)	Conv. (%)	ee (%)
1	101	95	56	93	>99	99
2	100	90	77	88	>99	>99
3	102	100	83	96	99	>99
4	102	100	95	96	99	99
5	101	102	88	92	>99	>99
6	100	100	85	100	>99	>99

the conversion and enantioselectivity do not suffer. Importantly, the mass balance of the product isolated levels off after the second experiment, suggesting that the ratio of product partitioned between the solid catalyst and hexane solution has reached equilibrium. This experiment implies that the decrease in catalytic activity in the previous experiments due to both a fouling mechanism (Table 2) and a decomposition pathway, both of which can be avoided by careful design of the reaction conditions. We note that isolation of the catalyst by vacuum filtration in a He glove box with either hexane or toluene washing of the solid results in slow deactivation after 3 cycles.

Conclusion

This work clearly shows a chiral cationic rhodium catalyst can be simply and efficiently immobilized onto silicas without modification of the ligands. The loading of the mesoporous material MCM-41 is about 4 times higher than that of the amorphous silica which has positive implications for process chemistry. The surface-bound triflate counterion immobilizes the cationic Rh complex onto the surface of the MCM-41 and this interaction is stable in non-polar solvents. Sorption of a Rh cation correlates strongly with the hydrogen bond acceptor properties of the counterion. Importantly, the results show that binding [(R,R)-Me-(DuPHOS)Rh]OTf to a MCM-41 surface has a positive effect on enantioselectivity and activity in the hydrogenation of prochiral enamides when compared to the homogeneous catalyst. Heterogenization of the catalyst allows for its recovery and reuse.

Acknowledgments

Support for this work was generously provided by the US Department of Energy, Los Alamos National Laboratory Directed Research and Development. RDB thanks the National Science Foundation (CAREER Award). We thank J. Rau, N. Clark, G. Brown, and, C. Hijar (Los Alamos National Lab) for MCM characterization and material synthesis.

References

1.   B. Pugin, *J. Mol. Cat.* **107**, 273 (1996).
2.   Exceptions to this can be found in: A. Corma, M. Iglesias, F. Mohino, F. Sanchez, *J. Organomet. Chem.*, **544**, 147 (1997); A. Carmona, A.

Corma, M. Iglesias, A. San Jose, F. Sanchez, *J. Organomet. Chem.* **492**, 11 (1995).

3. H.-U. Blaser, B. Pugin, in *Suported Reagents and Catalysts in Chemistry*, B.K. Hodnett, A.P. Kybett; J. H. Clark, K. Smith, Eds. Spec. Pub., R. Soc. Chem, London, 1998, v. 216, p. 101.

4. J. Y Ying, C.P. Mehnert, M.S. Wong, *Angew. Chem., Int. Ed. Engl.* **38**, 56 (1999).

5. C.-Y.Chen, H.-X. Li, M.E. Davis, *Microporous Mater.* 2, 17(1993)

6   M.J. Burk, J.E. Feaster, W.A. Nugent, R.L. Harlow, *J. Am. Chem. Soc.* **115**, 10125 (1993).

7. M. Kruk, M. Jaroniec, A. Sayari, *J. Phys. Chem. B.* **101**, 583 (1997).

8   S.-S. Kim, W. Zhang, T.J.. Pinnavaia, *Cat. Lett.* , **43**, 149 (1997).

9. C. Liu, Y. Shan, X.Yang, X. Ye, Y. Wu, *J. Cat.* , **168**, 35 (1997).

10. C. P. Mehnert, J. Y. Ying, *Chem. Commun.* 22 (1997).

11. R. Anwander, R. Roesky, *J. Chem. Soc., Dalton Trans.* 137 (1997).

12. R. Burch, N. Cruise, D. Gleeson, S.C. Tsang, *Chem. Commun.* 951 (1996).

13. J.S. Beck, J.C. Vartuli, W.J. Roth, M.E. Leonowicz, C.T. Kresge, K.D. Schmitt, C.T.-W. Chu, D.H. Olson, E.W. Sheppard, S.B. McCullen, J.B. Higgins, J.L. Schlenker, *J. Am. Chem. Soc.* **114**, 10834 (1992).

14. B. Pugin, *J. Mol. Cat. A* **107**, 273 (1996).

15. M. Brookhart, B. Grant, A.F. Volpe, *Organometallics* **11**, 3920 (1992).

16. K.D. Behringer,J. Blümel, *Z. Naturforsch.* **50b**, 1723 (1995).

17. S. Spange, A. Reuter, W. Linert, *Langmuir* **14**, 3479 (1998).

18. O. Kristiansson, M. Schuisky, *Acta. Chem. Scand.* **51**, 270 (1997).

19. L. Frunza, H. Kosslick, H. Landmesser, E. Höft, R. Fricke, *J. Mol. Cat. A* **123**, 179 (1997).

20. R.A. Sheldon, M. Wallau, I. W. C. E. Arends, U. Schuchardt, *Acc. Chem. Res.* **31**, 485 (1998).

21. Benzene forms stable intermediates with the Rhodium cation and has been found to be a poor solvent for RhDuPHOS cationic hydrogenations. This suggests non-polar aromatic solvents such as toluene are inappropriate solvents. M.J. Burk, C.S. Kalberg, A. Pizzano, *J. Am. Chem. Soc.* **120**, 4345 (1998).

22. The lipophilic BArF anion is more soluble than OTf and provided more consistent results than the OTf salt. See also ref. 15.

23. a) M.J. Burk, M.F. Gross, J.P. Martinez, *J. Am. Chem. Soc.* **117**, 9375 (1995). b) M.J. Burk, S. Feng, M.F. Gross,W. Tumas, *J. Am. Chem. Soc.* **117**, 8277 (1995).

24. C.R. Landis, J. Halpern, *J. Am. Chem. Soc.* **109**, 1746 (1987).

25. Y. Sun, R.N. Landau, J. Wang, C. LeBlond, D.G. Blackmond, *J. Am. Chem. Soc.* **118**, 1348 (1996).

Heterogeneous Synthesis of Flavanone over MgO

Michele T. Drexler and Michael D. Amiridis

Department of Chemical Engineering, University of South Carolina, Swearingen Engineering Center, Columbia, SC 29208

Abstract

The heterogeneous catalytic synthesis of flavanone from benzaldehyde and 2-hydroxyacetophenone over a solid magnesium oxide catalyst was studied in a batch slurry reactor system. This synthesis involves a Claisen-Schmidt condensation reaction to form a 2'hydroxychalcone intermediate, followed by the isomerization of this intermediate to form the flavanone product. The results indicate that the MgO catalyst is active for these reactions. Kinetic data were collected at different temperatures and rate constants were extracted for the two steps of the flavanone synthesis process. In addition, mass transfer effects were evaluated by varying the stirring rate, the catalyst particle size, and the catalyst weight percentage in the reactor.

Introduction

Fine chemicals and pharmaceuticals are largely synthesized today via homogeneous catalytic processes. This type of chemical synthesis requires the use of significant amounts of organic solvents and creates large quantities of hazardous waste. Recently, there is a concerted effort to reduce the amount of waste that is inherent in these synthetic procedures. Preferably such a reduction would not only be the result of "end-of-pipe" treatments, but it would also result from the onset of new synthetic techniques that can prevent waste formation completely (1,2). The use of heterogeneous catalysts is a step in this direction, since it eliminates the need for large amounts of organic solvents and provides ease of separation of the solid catalyst which can then be recycled and reused (3,4).

Flavanone and its isomer 2'hydroxychalcone are members of a family of two-ringed aromatic compounds known as flavanoids. These compounds can be found naturally in higher plant species and protect these plants from bacteria, parasites, and harmful radiation (5,6). The synthesis of flavanone and 2'hydroxychalcone has been examined in several studies, focusing on homogeneous catalytic synthesis (7-16). The importance of flavanone and 2'hydroxychalcone stems from the fact that they are key intermediates in the

451

synthesis of more complex flavanoids, which are desired for their medicinal properties (7,8).

The overall synthesis of flavanone consists of two steps, as seen in Figure 1. First, benzaldehyde and 2-hydroxyacetophenone are condensed to form 2'hydroxychalcone and water in an acid- or base-catalyzed Claisen-Schmidt condensation reaction. The 2'hydroxychalcone formed is then isomerized to flavanone. This second step is also catalyzed by acids or bases. There is very limited information in the literature regarding the heterogeneous synthesis of flavanone. Corma et al. (6,18) have demonstrated that the reaction can indeed take place over several acid and basic solid catalysts, with relatively high yields. Nevertheless, little has been reported regarding the kinetic parameters of this system and the effect of the operating conditions on these parameters. This current study focuses on the synthesis of flavanone over MgO and, in particular, on the kinetic parameters of this system. This represents the first step in our effort to understand the fundamental chemistry of this important heterogeneous synthesis.

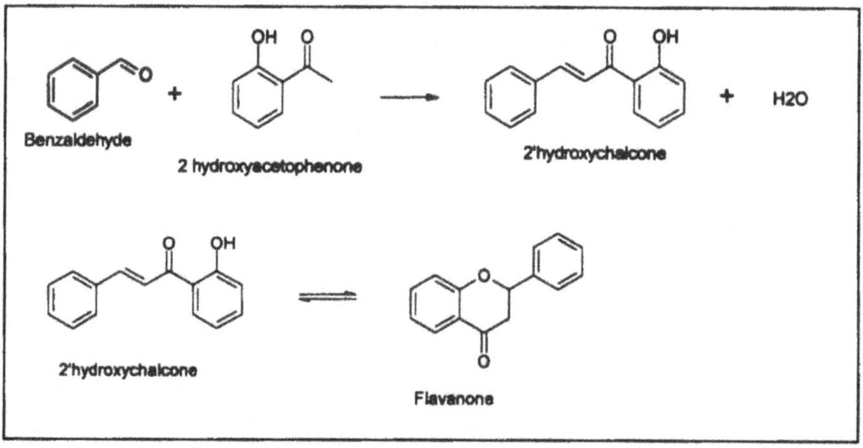

Figure 1. Synthesis of flavanone from benzaldehyde and 2 hydroxyacetophenone through the flavanone isomer, 2'hydroxychalcone

Experimental

Activity measurements were conducted in a 400-ml Pyrex batch reactor (8.0 cm diameter) operating under total reflux conditions. The reactor was further equipped with a mixer (IKA EUROSTAR, 5.0 cm diameter impeller) and a K-type thermocouple (Omega). The reactor was initially charged with equal volume quantities (75 ml each) of benzaldehyde and 2-hydroxyacetophenone (Aldrich). The system was then heated from room temperature to the reaction temperature at a rate of approximately 1°C/min. After reaching the reaction temperature, a sample was taken and the catalyst was added to the reaction mixture (time zero). Subsequently, small samples (approximately 0.5 ml) were taken periodically during the course of the reaction. Each sample was centrifuged at 10,000 rpm for 10 minutes in a VWR Scientific Model V Microcentrifuge to separate reactants and products from the solid catalyst particles that were also obtained during sampling. Once separated, a 5μl quantity of the liquid sample was removed and diluted in 1 ml of CH_2Cl_2 (Fischer, ACS grade). This diluted sample was then analyzed by gas chromatography (SRI Instruments 8610C GC; 5% phenyl methyl siloxane capillary column; FID detector).

Pure MgO (Aldrich) was used as the catalyst in these studies. Prior to its use, the catalyst was calcined in air at 475°C for 5 hours and sieved at the 60/80-mesh particle size.

Results and Discussion

Several runs were conducted initially to determine the effect of mass transfer limitations on the observed kinetic parameters. In particular, the parameters that can affect the mass transfer processes (i.e., stirring rate, catalyst loading, and catalyst particle size) were varied systematically during these runs. For example, four different particle sizes of MgO were used (in the range of 77-354 μm) and the reaction was carried out at 150°C. Similar rate constants were obtained in all four cases indicating that the system is not limited by internal mass transfer limitations. Similarly, experiments were carried out with different catalyst loadings (0.1 to 1wt%) and stirring rates (100-1000 rpm) to assess the importance of external mass transfer limitations. The results indicate that external mass transfer limitations are also not significant in the system containing less than 0.5wt% MgO, as long as the stirring rate exceeds 300 rpm. These results were further compared with estimates obtained from published literature correlations (3), which appear, however, to overestimate the importance of mass transfer limitations (i.e., a requirement for a higher stirring rate is predicted as compared to what was experimentally observed). Based on the experimental results, subsequent kinetic studies were conducted with 177-250 μm (60/80 mesh) catalyst particles and 0.1wt% catalyst loading at a stirring rate of 500 rpm. Under these conditions mass transfer limitations are not significant, and, thus, our results are believed to represent intrinsic kinetic parameters.

The reaction temperature was varied next in the 120°C to 160°C temperature range. The observed 2'hydroxychalcone and flavanone concentrations are shown as functions of the reaction time in Figures 2 and 3, respectively. During the course of these experiments, it is not expected that the reverse isomerization of flavanone to 2'hydroxychalcone will play a significant role due to the low concentrations of flavanone. As a result, this reaction was not considered during the kinetic analysis of the system. Rate constants for the two steps of the process were extracted from the experimental data shown in Figures 2 and 3 by the use of a simple reaction model. This model assumes that the Claisen-Schmidt condensation reaction is second order overall and first order in each one of the benzaldehyde and 2-hydroxyacetophenone reactants (an assumption that appears to be supported by the kinetic data collected thus far), and that the isomerization reaction is first order in 2'hydroxychalcone. Based on these two assumptions, the following set of differential equations can be obtained for the concentrations of reactants and products in the reactor.

$$\frac{dC_{benzaldehyde}}{dt} = -k_1 C_{benzaldehyde} C_{2-hydroxyacetophenone} \tag{1}$$

$$\frac{dC_{2-hydroxyacetophenone}}{dt} = -k_1 C_{benzaldehyde} C_{2-hydroxyacetophenone} \tag{2}$$

$$\frac{dC_{2'hydroxychalcone}}{dt} = k_1 C_{benzaldehyde} C_{2-hydroxyacetophenone} - k_2 C_{2'hydroxychalcone} \tag{3}$$

$$\frac{dC_{flavanone}}{dt} = k_2 C_{2'hydroxychalcone} \tag{4}$$

This model can be optimized (in terms of the two rate constants, k_1 and k_2) to give a good fit to the experimentally observed rates of formation of 2'hydroxychalcone and flavanone, as shown by the curves in Figures 2 and 3. The values for the constants obtained through the optimization process at different temperatures are listed in Table 1. As shown in Figure 4, these values appear to follow an Arrhenius dependence on temperature with activation energies of 98 and 69 kJ/mol for the Claisen-Schmidt condensation and the isomerization reactions, respectively.

The kinetic studies of flavanone synthesis over MgO are on going, focusing specifically on experiments with variable reactant concentrations to verify the assumed kinetic orders. Future plans also include the measurement of the rate of the isomerization reaction independently using 2'hydroxychalcone as the reactant.

Temperature (°C)	k_1 x 10^6 (L/mol-sec)	k_2 x 10^5 (sec^{-1})
120	0.6	1.3
140	1.7	2.8
150	5.0	6.9
160	9.2	8.1

Table 1. Observed rate constant for the Claisen-Schmidt synthesis of 2'hydroxychalcone (k_1) and the subsequent isomerization of 2'hydroxychalcone to flavanone (k_2).

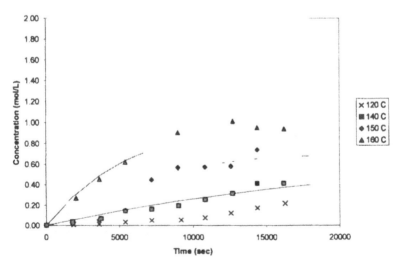

Figure 2. 2'hydroxychalcone concentration versus time at different temperatures (points represent experimental data and lines represent model predictions).

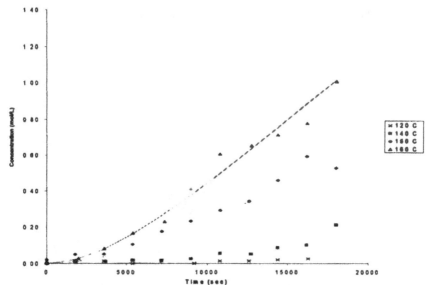

Figure 3. Flavanone concentration versus time at different temperatures (points represent experimental data and lines represent model predictions).

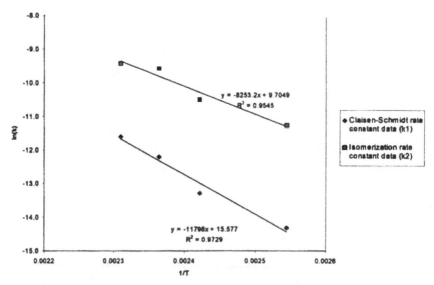

Figure 4. Arrhenius plot of observed rate constants vs. temperature for determination of the activation energies of the Claisen-Schmidt condensation and the 2'hydroxychalcone isomerization reaction.

References

1. S.C. Stinson, *C&EN* **73**, 10 (1995)
2. R. A. Sheldon, *J. Mol. Catal. A* **107**, 75 (1996)
3. P. L. Mills and R. V. Chaudhari, *Catal. Today* **37**, 367 (1997)
4. B. Gates, Catalytic Chemistry, John Wiley & Sons, New York, 1992
5. G. Britton, The Biochemistry of Natural Pigments, Cambridge Press, London, 1983
6. M.J. Climent, A. Corma, S. Iborra, and J. Primo, *J.Catal.* **151**, 60 (1995)
7. J.J.P. Furlong and N.S. Nudelman, *J. Chem. Soc.: Perkin Trans. II*, 633 (1985)
8. N.S. Nudelman and J.J.P. Furlong, *J. Phys. Org. Chem.* **4**, 263 (1991)
9. V.L. Arcus, C.D. Simpson, and L. Main, *J. Chem. Res. (S)*, 80 (1992)
10. L.J. Yamin, S.E. Blanco, J.M. Luco, and F.H. Ferretti, *J. Mol. Struct. (Theochem)* **390**, 209 (1997)
11. A. Cisak and C. Mielczarek, *J. Chem. Soc.: Perkin Trans. II*, 1603 (1992)
12. N.S. Nudelman and J.J.P. Furlong, *Can. J. Chem.* **69**(5), 865 (1991)
13. S.E. Blanco, N.B. Debattista, J.M. Luco, and F.H. Ferretti, *Tet. Lett.* **34**(29), 4615 (1993)
14. C.O. Miles and L. Main, *J. Chem. Soc.: Perkin Trans. II*, 1639 (1985)
15. C.O. Miles and L. Main, *J. Chem. Soc.: Perkin Trans. II*, 1623 (1989)
16. C.O. Miles and L. Main, *J. Chem. Soc.: Perkin Trans. II*, 195 (1988)
17. M.J. Climent, J. Primo, H. Garcia, and A. Corma, *Catal. Lett.* **4**, 85 (1990)

Pd-catalyzed Coupling Reactions in Supercritical Carbon Dioxide

N. Shezad*, R.S. Oakes, A.A. Clifford and C.M. Rayner

School of Chemistry, University of Leeds, Leeds LS2 9JT, UK,
email: c.m.rayner@chem.leeds.ac.uk.

Abstract

Pd(OCOCF$_3$)$_2$ and Pd(F$_6$-acac)$_2$ are excellent sources of palladium for carrying out a variety of coupling reactions (Heck, Suzuki, Stille, biaryl coupling) in supercritical carbon dioxide when used in conjunction with a range of phosphine ligands.

Introduction

The use of supercritical carbon dioxide (scCO$_2$) as an environmentally benign solvent for carrying out synthetic organic transformations is currently receiving much attention. In our own laboratories we have shown that using scCO$_2$ as a reaction medium also offers other very significant advantages, such as dramatic enhancements in diastereoselectivity simply by switching from conventional solvents to scCO$_2$.(1,2) New methods of carbon-carbon bond formation such as those catalyzed by transition metals, are also of great importance to synthetic chemists owing to their versatility and power in assembling complex and often synthetically challenging carbon frameworks. It is with this in mind that we embarked on a program to develop such reactions which can be carried out in scCO$_2$.

Palladium mediated coupling reactions in scCO$_2$

Recent reports by Holmes (3) and Tumas (4) have described the design and synthesis of fluorinated phosphine ligands **1** and **2** respectively, which when used in conjunction with Pd(OCOCH$_3$)$_2$ and Pd$_2$(dba)$_3$, allow a variety of

459

coupling reactions to be carried out in $scCO_2$. However, such ligands are generally not commercially available, which limits their application, particularly if simple alternatives are possible. For efficient coupling reactions in $scCO_2$ it is necessary to be able to form active catalyst species, and this is usually carried out *in situ*. For this to be possible in this case, all the components must have at least some solubility in $scCO_2$.(5,6,7) Simple phosphines would be expected to have good to moderate solubility (PPh₃ is borderline), however the palladium source might be expected to be more of a problem.

Hence we chose not to prepare any new fluorinated phosphine ligands and instead decided to investigate commercially available ligands used with a variety of Pd-sources which may be expected to show enhanced solubility in $scCO_2$. We believed this to be important if these procedures were to be adopted widely. We initially chose to investigate the classic Heck coupling reaction of iodobenzene and methyl acrylate in $scCO_2$ (Scheme 1, Table 1). Both triphenylphosphine and tris(2-furyl)phosphine (also reported by Tumas (4)) gave promising results with Pd(OCOCH₃)₂ and Pd₂(dba)₃, although at the temperatures we were using, high catalysts loadings and prolonged reaction times were required for complete conversion (entries 3 and 4). Tricyclohexylphosphine (PCy₃) was a poor ligand requiring even longer reaction times, with incomplete reaction (entry 5). Changing base from triethylamine to diisopropylethylamine (DIPEA) gave a modest improvement in conversion (*cf.* entries 2 and 6), which could be further increased using Pd₂(dba)₃ as the palladium source (entry 7). However, the catalyst loadings and reaction times were still somewhat unsatisfactory. We reasoned that Pd(OCOCH₃)₂ and Pd₂(dba)₃ were not ideal sources of palladium for reactions in $scCO_2$ because of their own insolubility and that of their ligands. Alternative commercially

available sources of palladium, which already contain fluorinated ligands are
$Pd(OCOCF_3)_2$ **3** and $Pd(F_6\text{-acac})_2$ **4**. We believed these would show enhanced
initial solubility in $scCO_2$ due to their fluorinated nature and hence investigated
them as catalysts.

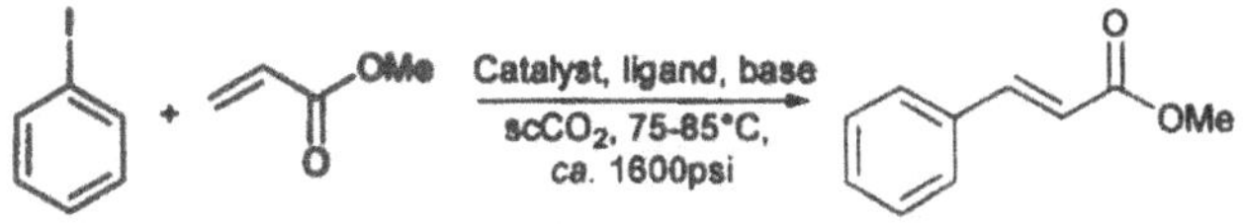

Scheme 1.

	Cat. (%)[a]	Ligand[a,b]	Temp.,°C	Time, h	Conv.	% Yield[c]
1	A (7)	PPh₃	75	18	25	-
2	A (7)	P(2-furyl)₃	75	18	26	-
3	A (7)	PPh₃	75	40	>95	84
4	A (14)	P(2-furyl)₃	75	40	>95	-
5	A (14)	PCy₃	75	90	66	-
6	A (7)	P(2-furyl)₃	75	18	35	-
7	B (7)	P(2-furyl)₃	75	18	55	-
8	C (2)	P(2-furyl)₃	75	15	>95	92
9	C (2)	PPh₃	80	15	71	-
10	C (2)	P(o-tolyl)₃	80	15	91	-
11	C (2)	PCy₃	80	15	81	-
12	C (2)	PBu₃	80	15	>95	-
13	C (6)	None	85	18	81	80
14	D (6)	None	85	18	88	88
15	D (2)	P(2-furyl)₃	80	17	>95	96
16	5 (4)	-	80	15	43	-
17	6 (2)	-	85	24	>95	94
18	7 (2)	-	80	15	20	-

[a]A, $Pd(OCOCH_3)_2$; B, $Pd_2(dba)_3$; C, $Pd(OCOCF_3)_2$; D, $Pd(F_6\text{-acac})_2$. [b]2 equiv. phosphine relative
to palladium; base also added, entries 1-5, NEt₃ (2.5 eq.); entries 6-18, DIPEA (1.5 eq.); [c]isolated,
after chromatography.

Table 1: Results of Heck Couplings

It rapidly became clear that they had very significant advantages over
those reported previously (Table 1). Reduced catalyst loadings were required for
complete conversion (2%), and while tris(2-furyl)phosphine was still the best

ligand, others which had previously been of little use and are usually regarded as poor ligands for Heck reactions, (12) notably PCy$_3$ and tri-n-butylphosphine, now gave acceptable conversions at low temperatures (80°C) with 2% catalyst (entries 11 and 12), and both were superior to triphenylphosphine (entry 9), presumably due to their greater solubility in scCO$_2$. Interestingly, tri(o-tolyl)phosphine was also significantly better than triphenylphosphine (entries 9 and 10). In the absence of any phosphine, a 6% catalyst loading was required for complete conversion using either Pd(OCOCF$_3$)$_2$ or Pd(F$_6$-acac)$_2$ indicating that while not absolutely essential, the phosphines do play an important role in the efficiency of the reaction. Pd(F$_6$-acac)$_2$ in the presence of tris(2-furyl)phosphine also gave impressive results (entry 15).

There is much current interest in catalysts for the Heck reaction based on palladium(II) metallacycles.(8-11) We chose to investigate 5 and 7,(8) and the trifluoroacetate 6.(2) These were also active in scCO$_2$ at a similar catalyst loading to those required above, although the trifluoracetate 6 (entry 17) was significantly better than the acetate 5 (entry 16) and the chloride bridged complex 7 (entry 18). Importantly these reactions showed no sign of catalyst decomposition, unlike those using 3, 4, Pd(OCOCH$_3$)$_2$ (4) or Pd$_2$(dba)$_3$ as palladium sources, which gave dark precipitates in the crude product, presumably due to metallic palladium residues. Attempted generation of 6 *in situ* using Pd(OCOCF$_3$)$_2$ and tri(o-tolyl)phosphine gave an efficient reaction, but the dark precipitation characteristic of catalyst decomposition was also observed (entry 10). Interestingly, using PBu$_3$ as ligand (entry 12), no significant catalyst decomposition was observed. Importantly, these catalyst systems could also be used in other palladium-mediated coupling reactions such as other Heck reactions, Stille (4) and Suzuki (3,4) couplings, again with good efficiency (Scheme 2).

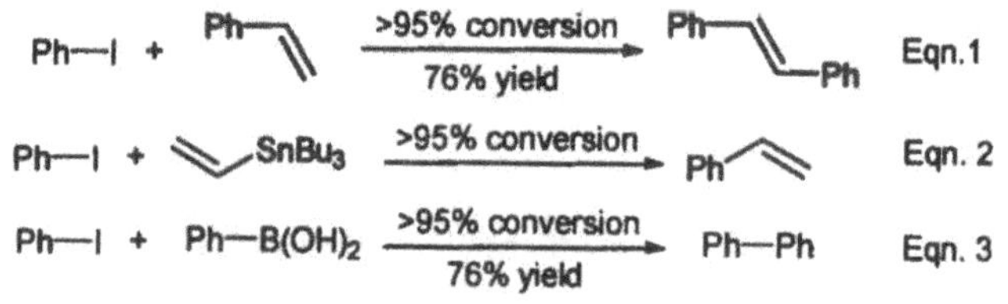

Reagents:Pd(OCOCF$_3$)$_2$ (2mol%), tris(2-furyl)phosphine (4mol%),
DIPEA (1.5 eq.) except eqn. 2, 1600psi CO$_2$, 80°C, 15-24h

Scheme 2.

Heck couplings using acrolein as one component are generally unsuccessful. We investigated the reaction between acrolein and iodobenzene in scCO$_2$ to see whether the unusual reaction medium offered any advantages; however, and as expected, none of the desired coupled product was observed. Interestingly, all the iodobenzene had been consumed and converted into biphenyl, along with oxidation of DIPEA. Repeating the reaction in the absence of acrolein gave the same result, and indeed a variety of aryl iodides could be converted into biaryls using this methodology (Scheme 3).

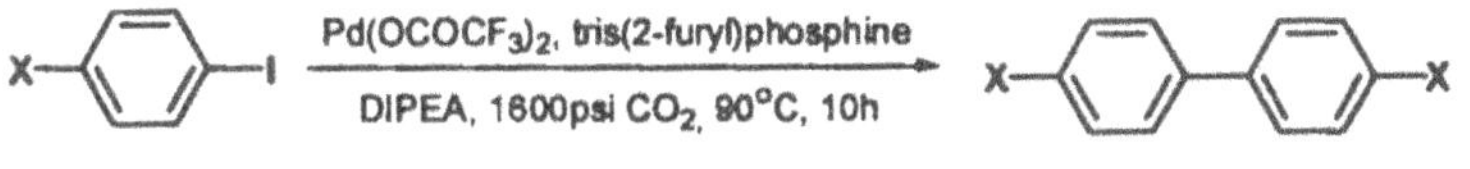

X	Yield
H	90%
MeO	86%
CO$_2$Me[a]	50%
NO$_2$[a]	50%

[a] up to 50% protiodeiodinated product also formed

Scheme 3

Good yields are generally obtained, however with electron-deficient aryl iodides some reduction is also observed. Interestingly, attempting the reaction in toluene under otherwise identical conditions gave no reaction indicating the unusual reaction medium played a crucial role, although the precise reason for this as yet is unclear. In the presence of suitable alkenes the expected Heck product is still observed exclusively. Related reactions have been reported although these usually require an additive such as stoichiomeric Bu$_4$NBr,(13,14) or involve use of a palladacycle at 110°C in DMF.(15) Further work on the mechanism of this reaction is currently in progress.

In summary, we have shown that the use of fluorinated palladium sources for a variety of coupling reactions in scCO$_2$ can give superior results to those previously reported, including low catalyst loading at moderate temperatures, and allows the use of ligands usually considered to be poor in conventional Heck reactions. We are currently investigating such reactions in more detail, the results of which will be published in due course.

Acknowledgements

   We are very grateful to the following members of the Leeds Cleaner Synthesis Group and their respective companies for funding and useful discussions: Dr. Andrew Bridge, Rhône-Poulenc Rorer; Dr. Mike Loft, GlaxoWellcome; Julie MacRae, Pfizer Central Research; Dr. William Sanderson, Solvay Interox; and Dr. Ken Veal, SmithKline Beecham Pharmaceuticals. CMR also wishes to thank Professor B.L. Shaw (University of Leeds) for useful discussions.

References and notes

1. R.S. Oakes, A.A. Clifford, K.D. Bartle, M. Thornton-Pett, and C.M. Rayner, *Chem. Commun.*, 247 (1999). This paper also contains details of the reactor, typical experimental procedures, and citations of other recent publications in this area.
2. N. Shezad, R.S. Oakes, A.A. Clifford and C.M. Rayner, *Tetrahedron Lett.*, **40**, 2221 (1999).
3. M.A. Carroll and A.B. Holmes, *Chem. Commun.*, 1395 (1998) and refs. cited therein.
4. D.K. Morita, D.R. Pesiri, S.A. David, W.H. Glaze and W. Tumas, *Chem. Commun.*, 1397 (1998) and refs. cited therein.
5. A. Furstner, D. Koch, K. Langemann, W. Leitner and C. Six, *Angew. Chem., Int. Ed. Engl.*, **36**, 2466 (1997).
6. P.G. Jessop, T. Ikaraya and R. Noyori, *Science*, **269**, 1065 (1995).
7. S. Hadida, M.S. Super, E.J. Beckman and D.P. Curran, *J. Am. Chem. Soc.*, **119**, 7406 (1997).
8. a)W.A. Herrmann, C. Brossmer, C.-P. Reisinger, T.H. Riermeier, K. Ofele and M. Beller, *Chem. Eur. J.*, **3**, 1357 (1997).
9. B.L. Shaw, S.D. Perera and E. Staley, *Chem. Commun.*, 1361 (1998).
10. B.L. Shaw and S.D. Perera, *Chem. Commun.*, 1863 (1998).
11. B.L.Shaw, *New J. Chem.*, **22**, 649 (1998).
12. R.F. Heck, *Comprehensive Organic Synthesis*, B.M. Trost, ed., Pergamon Press, Oxford, vol. 4, chapter 4.3, p.844.
13. M.D.Cliff and S.G. Pyne, *Synthesis*, 681 (1994).
14. M. Brenda, A. Knebelkamp, A. Greiner, W. Heitz, *Synlett*, 809 (1991).
15. F.-T. Luo, A. Jeevanandam and M.K. Basu, *Tetrahedron Lett.*, **39**, 7939 (1998).

Chiral Modifiers for Supported Metals

Anthony C. Testa and Robert L. Augustine

Chemistry Department, Seton Hall University, South Orange, NJ 07079, USA

Abstract
The extensive work reported concerning the cinchona alkaloid modified platinum catalyzed hydrogenation of α ketoesters and acids has provided us with the information needed to define what would be required for an effective chiral modifier other than the cinchona alkaloids and the chiral synthetic materials related to them. Simply put, these criteria are:

The presence of at least a binuclear aromatic ring for adsorption of the modifier onto the metal surface.

An appropriate 'chiral pocket' in the modifier located a sufficient distance from the binuclear anchoring 'foot' so the pocket surrounds the active site on the metal surface.

Some chemical entity which would interact with the substrate in order to hold it inside the 'chiral pocket' in a prochiral manner for hydrogenation.

Finally, the best modifiers should interact with the substrate in such a way as to accelerate the hydrogenation reaction.

We have used these principles to design some potential catalyst modifiers based on the chiral 1,1'-binaphthol system but we were unable to incorporate into them any substrate activating groups. Even with this limitation, one of these materials, (R) 2-amino-2'-(1-methoxynaphthyl)-1,1'-binaphthalene, was shown to be as effective as cinchonidine in the palladium catalyzed hydrogenation of 2-phenylcinnamic acid.

It is hoped that the data presented will lead the way in the development of other, more effective chiral catalyst modifiers.

Introduction
Enantioselective syntheses using chiral catalysts have been the object of extensive research for a number of years.(1) Most of this work has used soluble chiral organometallic complexes to promote a variety of synthetically useful reactions. Since these catalysts are molecular species they have a single 'active site' surrounded by an appropriate chiral pocket defined by the ligands present in the complex. With the appropriate chiral ligand such species are usually very selective, frequently giving products having enantiomeric excesses (ee's) in the 95%-99% range. These catalysts being in solution with the reaction products can be difficult to separate from the reaction products and once separated are not

easily reconstituted for potential re-use. From an economic standpoint, this loss of the catalyst can be an appreciable economic concern, not so much because of the loss of the metal but, more importantly, the chiral ligand which can be many times more expensive than the metal.

Heterogeneous catalysts, on the other hand, are insoluble in the reaction medium and are, therefore, easily removed from the product. There are, however, only two viable enantioselective heterogeneous catalyst systems. One, nickel modified with tartaric acid and sodium bromide, is useful only for the hydrogenation of β ketoesters and β diketones to the corresponding chiral β hydroxy esters, β hydroxy ketones or β dihydroxy compounds.(2,3) With the addition of pivalic acid the enantioselective hydrogenation of methyl ketones can also be accomplished.(4) The second system, platinum modified with a cinchona alkaloid, is useful only for the enantioselective hydrogenation of α keto esters and acids.(3,5,6)

This latter system has been the subject of extensive research aimed at developing an understanding of the nature of the substrate:modifier:catalyst interactions so that other modifiers might be synthesized for the enantioselective hydrogenation of different functional groups in a prochiral molecule. While there is disagreement on the details of these interactions, there are some experimental findings which are important in the development of a more general understanding of the types of modifiers which may be effective for different heterogeneously catalyzed enantioselective hydrogenations.

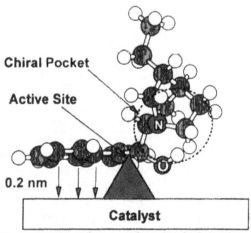

Fig. 1. Representaion of proposed mode of adsorption of cinchonidine on a Pt catalyst near an active site.(6)

An illustration of such a modifier:catalyst orientation is given in Fig. 1 which depicts our hypothesis for the orientation of a cinchonidine molecule near an adatom active site on the Pt catalyst surface.(5) An interaction between the quinuclidine nitrogen and the carbonyl group of the pyruvate not only holds this substrate within the 'chiral pocket' but also increases the rate at which the carbonyl group is hydrogenated.

The first thing that has to be recognized is that in contrast to the chiral homogeneous catalysts, the active sites on a supported metal catalyst do not have sufficient surface unsaturation for the direct interaction of a chiral entity with the metal atom of the active site and still be able to adsorb the substrate and a hydrogen molecule. This means that the chiral modifier must be adsorbed on the catalyst surface near the active site. Since there is no reason to expect that the adsorption of a modifier molecule would take place at exactly the proper orientation near an active site, the adsorption of the modifier cannot be so strong that it cannot migrate across the catalyst surface nor so weak that it is easily displaced. It appears that for a modifier to be effective it must have at least a binuclear aromatic ring by which it is adsorbed on the catalyst.(7)

This aromatic system also must be attached to an appropriate chiral entity to provide a 'chiral pocket' for interaction with the substrate. Further, this 'chiral pocket' must be attached to the binuclear aromatic system through sufficient bonds to place the 'pocket' in the proper position to provide chirality around the active site of the catalyst. This 'chiral pocket' should also contain appropriate functionality to provide a specific interaction with the substrate molecule and to hold it in the proper position for the chiral reaction at the active site to occur. Since there will always be non-modified active sites present on the catalyst surface, the modifer:substrate interaction should also increase the hydrogenation rate of the substrate so that the products formed at the modified sites will predominate in the reaction.

Since some of the more general chiral ligands in use in homogeneous catalysis are based on the chiral binaphthyl moiety, we decided to use these concepts in making some chiral modifiers based on 1,1'-binaphthol (1) and to test their applicability in the hydrogenation of some prochiral substrates over supported Pt and Pd catalysts.

(1)

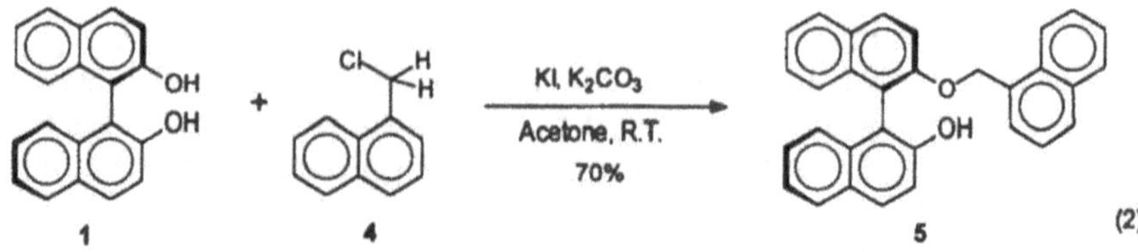

(2)

Modifier Synthesis

Since a bicyclic aromatic ring is apparently required for adsorption of the modifier on the metal catalyst surface, we examined various methods of incorporating a naphthalene 'foot' onto the binaphthol moiety. The first approach was the preparation of the acetal, **3**, (Eqn. 1) by reaction of dichloromethylmaphthalene, **2**, with **1**.(8) While this reaction did not proceed very well, some material was obtained for use as a potential modifier. This material, however, did not incorporate any functionality for interaction with potential substrates. Further work aimed at introducing such functionality into **3** was unsuccessful so other, more simplified, approaches were then used. In one of these the reaction of **1** with chloromethylnaphthalene, **4**, gave the monoether, **5** (Eqn. 2) which had the binuclear 'foot' and an aromatic hydroxy group for substrate interaction. A final synthesis started with the optically active amino-

(3)

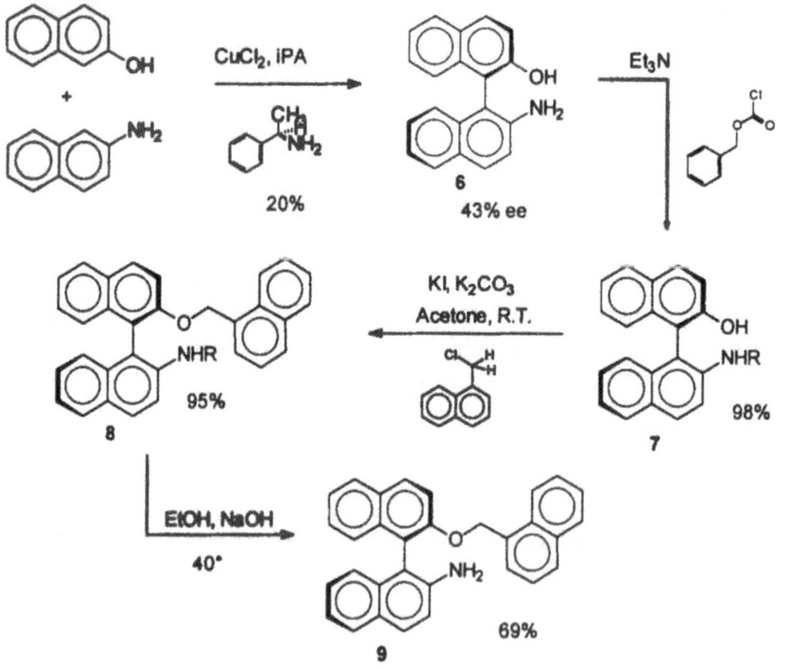

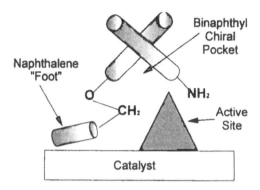

Fig. 2. Representaion of proposed mode of adsorption of **9** on a Pt
catalyst near an active site.

hydroxy binaphthalene, **6**, which was prepared in low yield with 43% ee by
coupling β naphthol with β-naphthylamine in the presence of an optically active
amine.(9) The desired amino-monoether, **9**, was produced by first blocking the
amine as the amide, **7**, amido-ether, **8** ,formation, and amide hydrolysis to give **9**
(Eqn. 3) which is similar to **5** but with an amine group for substrate interaction.
Fig. 2 shows how we believe **9** would adsorb on a metal catalyst surface to
provide a 'chiral pocket' near an active site.(10)

Results

The modifiers, **3**, **5** and **9** along with cinchonidine as a standard, were used in the
hydrogenations of ethyl pyruvate, **10**, methyl 2-acetamidoacrylate, **11**, and a
phenylcinnamic acid, **12**, over either 5% Pt/Al$_2$O$_3$ or 5% Pd/Al$_2$O$_3$ at 25°C and
1 atm pressure. Several concentrations of each modifier were used to provide a
range of modifier/catalyst ratios for each reaction. In the hydrogenations of **10**
and **11** over Pt/Al$_2$O$_3$ with **3**, **5**, or **9**, essentially no chirality was observed in the
products. However, as the data in Table 1 show, hydrogenation of **12** over
Pd/Al$_2$O$_3$ modified by **9** gave products with ee's higher than those observed
using cinchonidine under our reaction conditions. No attempt was made at
optimizing these results.

| 10 | 11 | 12 |

Table 1. Product ee data for the hydrogenation of **12** over Pd/Al$_2$O$_3$ using various chiral modifiers.[a]

	Product ee % (S)[b]		
Modifier	0.53 μmoles	1.05 μmoles	1.58 μmoles
3	5	7	6
5	5	6	5
9[c]	14	21	14
Cinchonidine	10	13	17

[a] 15 mg 5% Pd/Al$_2$O$_3$ stirred with modifier solution in 10 mL THF, 2.3 mmole **12** added in 5 mL THF and hydrogenation run at 25°C and 1 atm H$_2$.
[b] ± 2%.
[c] Data adjusted to 100% ee for **9**.

Conclusions

The data reported here shows that chiral modifiers based on the binaphthol moiety can be somewhat successful in promoting some enantioselective hydrogenations. However, much more work in this area is needed to expand on this concept and to optimize the nature of the modifier used for specific substrates.

Acknowledgements

This work was supported by grant CTS-9708227 from the US National Science Foundation.

References

1. I. Ojima "Catalytic Asymmetric Synthesis", VCH, New York, 1993.
2. Y. Izumi, Adv. Catal., 1983, 32, 215.
3. G. Webb and P. Wells, Catal. Today, 1992, 12, 319.
4. T. Osawa, T. Harada and A. Tai, *Catal. Today*, 1997, 37, 465.
5. R.L. Augustine, S.K. Tanielyan and L.K. Doyle, Tetrahedron:Asymmetry, 1993, 4, 1803;J. Mol. Catal., 1996, 112, 93; 1997, 118, 79.
6. H.-U. Blaser, H.-P. Jalett, M. Muller and M Studer, Catal. Today, 1997, 37, 441.
7. K.E. Simmons, G. Wang, T. Heinz, T. Giger, T. Mallat, A. Praltz and A. Baiker, Tetrahedron: Asymmetry, 1995, 6, 505.
8. A. Jazzaa and J. Clark, Chem. Lett., 1983, 89.
9. M. Smrcina, M. Lorenc, V. Hanus, P. Sedmera and P. Kocovsky, J. Org. Chem., 1992, 57, 1917.
10. Experimental details can be found in A. C. Testa, Ph.D. Dissertation, Seton Hall University, 1998.

Kinetic Study on Stereoselectivity of Catalyzed Cycloadditions of Cyclopentadiene to Pentenoate

Zhiyu Chen and Rosa M. Ortuno
Dept. De Quimica, Universitat Autonoma de Barcelona, Barcelona, Spain

Hong Yang
College of Environmental and Energetic Engineering, Beijing Polytechnic

University, Beijing, P. R. China

ABSTRACT

Cycloadditions to pentenoate are key processes for preparing synthetic precursors of carbocyclic nucleosides. The stereoselectivity of the reaction is highly enhanced under Lewis acid catalysis. The kinetic study explicates the stereoselectivity (*endo/exo)* and facial diastereoselectivity (*syn/anti*) of the catalytic reaction by means of rate constants, activation parameters and comparisons under catalyzed and thermal conditions.

INTRODUCTION

High stereoselectivities (*endo/exo* and *syn/anti*) have been achieved in asymmetric Diels-Alder reactions of cyclopentadiene and either (Z)-pentenoate **1**, or butenolide **2** (Scheme 1, Table 1) (1, 2) catalyzed by Lewis acids. The adducts **3** and **4** in the scheme are important intermediates for the family of carbocyclic nucleosides (3), and (-)-**5** is a key precursor in the synthesis of the antineoplasic antibiotic (-)-Aristeromycin ((-)-**6**), (-)-Neplanocin A ((-)-**7**) and antiviral agent (-)-Carbovir ((-)-**8**).

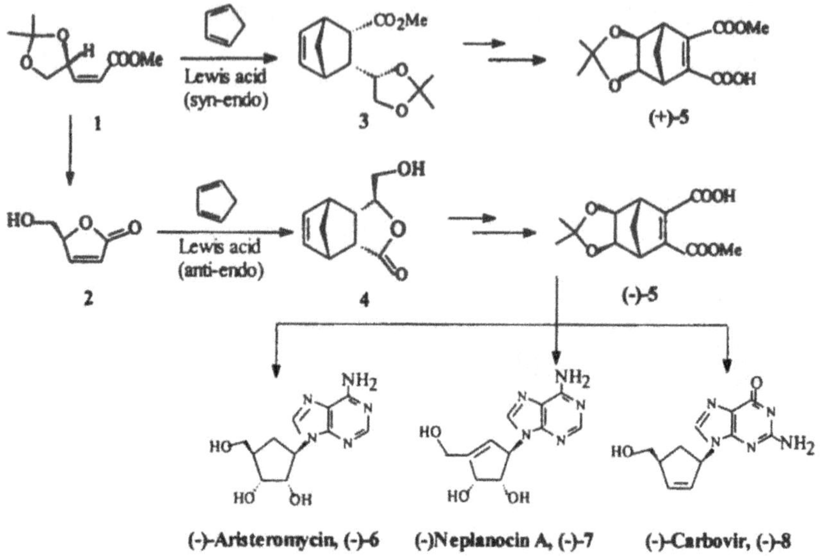

Scheme 1

Table 1 Lewis acid catalyzed cycloadditions of cyclopentadiene to pentenoate 1 and butenolide 2

Entry	Dieno-phile	Catalyst	Adduct	Temp. °C	Time hr.	Ratio of endo/exo	Ratio of syn/anti	Yield %
1	1	--	3	80	8	6	14	94
2	1	Et$_2$AlCl	3	-23	1.5	38	19	91
3	2	--	4	100	22	4	--	60
4	2	ZnI$_2$	4	15	21	10	--	95

Since the catalytic asymmetric Diels-Alder reactions are key processes for the synthesis of stereogenic centers in the following molecules, our investigation begin with a kinetic study of the stereoselectivity of the cycloaddition (reaction: 1 → 3) catalyzed by Lewis acids. Among the Lewis acid catalysts, titanium chloride, zinc chloride and alkyl aluminum complex (2), the diethyl aluminum chloride was chosen as a catalyst for its mild catalytic activity which yielded good kinetic behavior for the cycloaddition reaction. Table 1 shows that reactivity of the dienophile 1 in the uncatalyzed reaction is much slower than that in the catalyzed reaction. Therefore, it is reasonable to assume that the contribution of the parallel thermal reaction to the total rate of catalytic reaction is negligible. In the kinetic study, the molar equivalent of the catalyst is kept less than that of the dienophile because excess diethyl aluminum chloride will catalyze undesirable side reactions. Moreover, the method of initial rate is employed in determining rate constants and activation parameters, minimizing the effect of these parameters due to the formation of by-products.

RESULTS AND DISCUSSION

The rate equation of *endo* product 3 (Scheme 2) is expressed as: $r_0 = dC/dt = k_3[1]^\alpha[CPD]^\beta[cat]^\gamma$. A plot of natural logarithm of initial reaction rate, $\ln r_0$, vs that of catalyst concentration shows that the index of the catalyst concentration, γ, equals one, which means that reaction order of the catalyst is of the first order. The concentration of dienophile 1 does not affect the rate constant (k_3). For example, the dienophile 1 is increased from 0.07M to 0.2M which is two times the concentration of the catalyst and the free or uncomplexed dienophile do not result in a meaningful change in the rate constants. The reagent in the catalyzed reaction is the Z-enoate--Et$_2$AlCl complex rather than free dienophile. The reaction order of the cycloaddition (Scheme 2) was obtained by plotting the natural logarithm of product yield (y) vs time (t), based on a variety of reaction order assumptions. The plot is linear with time based on the second-order assumption, therefore, the cycloadditions catalyzed by Lewis acid are considered to be pseudo second-order reaction, while the thermal reaction is first-order.

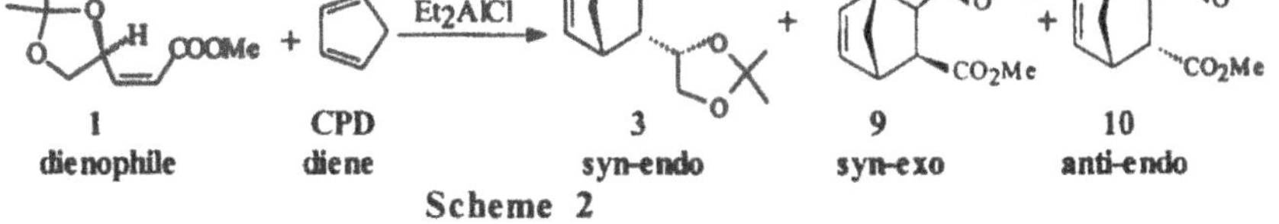

| 1 | CPD | 3 | 9 | 10 |
| dienophile | diene | syn-endo | syn-exo | anti-endo |

Scheme 2

| **Table 2** Stereoselectivities of the catalytic reaction at different temperatures |

Temp. °C	-29	-23	-9
Ratio (endo/exo)	50	39	36
Ratio (syn/anti)	23	19.2	17.8

| **Table 3** Natural logarithm of ratios of rate constant at different temperatures |

Temp. °C	-29	25[a]	Δ
$\ln(k_{exo}/k_{endo})$	-3.9	-3.2	0.7
$\ln(k_{anti}/k_{syn})$	-3.0	-2.5	0.5

a: extrapolated value; $\Delta = \ln(k_{xxx}/k_{yyy})_{25} - \ln(k_{xxx}/k_{yyy})_{-29}$

Table 2 shows that the stereoselectivity (*endo/exo*) and facial diastereoselectivity (*syn/anti*) are dependent on the temperature. Typical reaction curves (yield (y) vs time (t)) under the catalyzed conditions show that the formation of adducts is under kinetic control at the early stage and the ratios of adducts (*endo/exo*, *syn/anti*) are considered to be equal to that of the specific rate constants, k_{endo}/k_{exo}, k_{syn}/k_{anti} (Table 3 and Table 4).

Table 3 shows that the lower temperature ranging from 25 to −29 °C is favor to the increase of ratios of *endo/exo* and *syn/anti*. Owing to $\Delta[\ln(k_{exo}/k_{endo})]_{25-(-29)}=$ 0.7 > $\Delta [\ln (k_{anti} /k_{syn})]_{25-(-29)}= 0.5$, the *endo/exo* selectivity is more sensitive to the change of temperature than facial selectivity (*syn/anti*). And the increase of rate of the *exo*-isomer is greater than that of the anti-isomer when increasing temperature (Table 4). Stereoselective differences between the catalyzed cycloaddition and the thermal counterpart (Table 1, Entry 1 and 2) can be seen clearly in Table 5 by their activation energies, free energies of activation, rate constants, and stereoselectivities.

We examine the differences of activation energies between the two reactions, there is a greater reduction in activation energy for the formation of the *endo* isomer via the catalytic reaction (12.33 kJ/mol) than the exo isomer (9.53 kJ/mol) (Table 5).

Table 4 Rate constants of cycloadducts in the catalytic reaction at different temperatures

Temp. °C	$10^3 \times k_3$ (M s)$^{-1}$	$10^3 \times k_{exo (9)}$ (M s)$^{-1}$	$10^3 \times k_{anti(10)}$ (M s)$^{-1}$	$k_{endo(3+10)}$ (M s)$^{-1}$	$k_{syn(3+9)}$ (M s)$^{-1}$
-29	30.56	0.639	1.366	0.03192	0.03119
-23	44.67	1.20	2.386	0.04705	0.04587
-9	84.59	2.46	4.890	0.08948	0.08705
0	161.4				

Table 5 Activation parameters, rate constants and stereoselectivties under catalyzed and thermal conditions at 25 °C

	$Ea_{,th}-Ea_{,cat}$ (kJ/mol)	$\Delta G_{th}^{\circ *}-\Delta G_{cat}^{\circ *}$ (kJ/mol)	k_{cat} $(M\ s)^{-1}$	$10^6 \times k_{th}$ $(s)^{-1}$	Stereoselectivity
endo	12.33	32.01	0.3776	0.914	endo/exo:
exo	9.53	29.25	0.0165	0.122	$(k_n/k_x)_{cat}/(k_n/k_x)_{th}=$ 3.1
syn	12.77	31.86	0.3647	0.954	facial:
anti	10.54	31.59	0.0294	0.082	$(k_s/k_{an})_{cat}/(k_s/k_{an})_{th}=$ 1.1

n = *endo*; x = *exo*; s = *syn*; an = *anti*; cat = catalyzed; th = thermal

However, this reduction does not fully explain the increase in *endo/exo* selectivity since the activation energy of *syn* products also decreases almost the same as that of the endo products, $(Ea_{,th}-Ea_{,cat})_{syn}$ = 12.77 (kJ/mol), with a little increase of facial selectivity. The ratio of catalyzed *endo/exo* and thermal *endo/exo*, i.e. $(k_n/k_x)_{cat}/(k_n/k_x)_{th}$= 3.1 (Table 5), describes quantitatively that the *endo/exo* selectivity in the catalyzed reactions have increased about 3 times that of the thermal reactions at 25 °C. The differences of free energy of activation between the two reactions show a good correlation with the fact that catalysts generally enhance *endo* selectivity (Table 5), $(\Delta G_{th}^{\circ *}-\Delta G_{cat}^{\circ *})_{endo}$ = 32.01 (kJ/mol); $(\Delta G_{th}^{\circ *}-\Delta G_{cat}^{\circ *})_{exo}$ = 29.25 (kJ/mol). The catalytically active complex is more favorably attacked by the *endo* stereochemisry than the *exo*. Similarly, the difference of *syn* or *anti* free energy of activation between the two reactions also corresponds to an increase in facial selectivity.

CONCLUSION

The kinetic study on the stereoselectivity of the cycloaddition of cyclo-pentadiene to (Z)-pentenoate **1** indicates that the catalyzed reaction is a pseudo second-order reaction. The stereoselectivity of the catalyzed cycloaddition is increased by decreasing temperature. A quantitative comparison of the catalyzed reaction with the uncatalyzed was made by measuring their activation parameters.

EXPERIMENTAL SECTION

Reactions were performed in glass reactors fitted with septa containing teflon-stoppers. After removing air from the reactor, a solution of Et_2AlCl in hexane (0.15ml of a 1M solution from commercial sources) was added to solution of ester **1** (0.25mmol) as well as the internal standard ($C_{12}H_{26}$, 0.1 mmol) in 2 ml of CH_2Cl_2 (previously distilled over calcium hydride and under argon). The mixture was stirred at –23 °C for 0.5 h. Then, fresh cyclopentadiene (1.25mmol) was added. The resulting solution was stirred at –23 °C for 3 h. Samples for quantitative analysis were taken at definite time intervals. Each aliquot of the reaction mixture was poured into ice water containing 10% sodium bicarbonate. The layers were separated and the aqueous phase was extracted with CH_2Cl_2 and the solvent was evaporated. The residue was injected into a gas chromatograph.

REFERENCES

1. Z. Chen and R. M.Ortuno, *Tetrahedron: Asymmetry*, **5**(3), 371 (1994).
2. Z. Chen and R. M.Ortuno, *Tetrahedron: Asymmetry*, 3(5), 621 (1992).
3. M. Diaz, J. Ibarzo, J. M. Jimenez, R. M. Ortuno, *Tetrahedron: Asymmetry*, **5**(1), 129 (1994).

Enantioselective Hydrogenation of Unsaturated Carboxylic Acids over Palladium Modified by a Substituted Binaphthylene

S.D. Jackson[a], S.R. Watson[a,b], G. Webb[c], P.B. Wells[b] and N.C. Young[a,c]

[a]Synetix,, P.O. Box 1, Belasis Ave., Billingham, Cleveland, TS23 1LB, UK,
[b]Department of Chemistry, University of Hull, Hull, HU6 7RX, UK,
[c]Department of Chemistry, University of Glasgow, Glasgow, G12 8QQ, UK.
e-mail: s_david_jackson@ici.com

Abstract

Enantioselective hydrogenation of tiglic and angelic acids has been achieved using an R(+)- or S(-)-2,2'-diamino-1,1'-binaphthalene modified Pd/carbon catalyst. Enantioselective hydrogenation of methyl tiglate was unsuccessful due to unfavourable steric interactions.

Introduction

A collaborative research programme, aimed at the development of new chiral centres at the surface of heterogeneous catalysts, was established to achieve chiral selectivity with new catalysts, modifiers and reactants. Part of this study involved the hydrogenation of tiglic acid (E-2-methylbut-2-enoic acid, $R_1=CH_3$, $R_2,R_3=H$), angelic acid (Z-2-methylbut-2-enoic acid, $R_2=CH_3$,

R_1, R_3=H), and related ester methyl tiglate (E-methyl-2- methylbut-2-enoate R_1, R_3=CH$_3$, R_2=H) over a 2% Pd/carbon catalyst modified with R(+)- or S(-)-2,2'-diamino-1,1'-binaphthalene.

Experimental

A Pd/carbon catalyst was prepared by the aqueous impregnation of a carbon support with palladium (II) nitrate to yield a nominal 2% Pd loading. The impregnated support was dried at 373K (20 h) and calcined at 523K (20 h). Catalyst activation was achieved by reduction under flowing dihydrogen at 523K (2 h).

Known concentrations of R(+)-2,2'-diamino-1,1'-binaphthalene (diamine) in THF were injected onto the catalyst (typically 2.75 mg modifier per 0.5 g catalyst). The modification proceeded with constant stirring for 24 h. Adsorption isotherms were constructed, 293 K, by periodically removing small amounts of the modifier solution for analysis. After 24 h the modifier solution was removed by centrifuging and decanting. Fresh THF was added and the centrifuging and decanting process repeated.

Hydrogenation was performed in a 250 cm^3 Baskerville autoclave. The reactor was purged with nitrogen and the system heated to reaction temperature (298 K). Once at temperature, the required reactant was dosed into the reactor in solution in THF. The autoclave was then purged and pressurised to the appropriate pressure with hydrogen. Reaction proceeded under a constant pressure of hydrogen with constant stirring for 24 h, unless otherwise stated. Reactants and products were analysed by chiral gas chromatography

Results/Discussion

Isotherms were obtained for the adsorption of (R)- and (S)-diamine onto the Pd/carbon catalyst and onto a carbon support that had been through an identical preparation process to that of the Pd/carbon but without the palladium. Results show that both the active metal and the carbon support adsorbed the modifier (Figure 1). Treating the catalyst with a fresh aliquot of solvent post-modification revealed that no modifier desorbed, indicating that hydrogenation was undertaken with no modifier in solution.

Given that there is no detectable desorption of the modifier it is concluded that the modifier adsorbs strongly on both the metal and the carbon support. Also during hydrogenation no evidence was found for modifier desorption or reaction. Therefore during the reaction the modifier is only present on the

Figure 1. Adsorption of R(+)-2,2'-diamino-1,1'-binaphthalene.

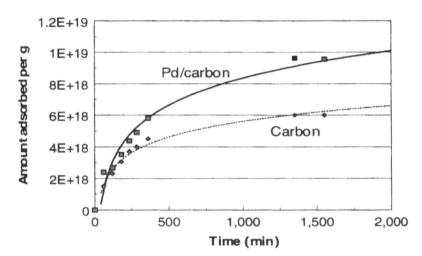

surface of the catalyst. By comparison with adsorption studies of naphthalene and (±)-2,2'-dimethoxy-1,1'-binaphthalene [1], the adsorbed state of the modifier has been determined to be through the diamine groups, with loss of a single hydrogen.

The results obtained from the hydrogenations are summarised in Table 1. Note that where enantiomeric excess is obtained an enhancement of rate is also seen, whereas when there is no enantiomeric excess a decrease in rate is observed. For example, under 10 bar hydrogen pressure, the rate of methyl tiglate hydrogenation over an unmodified catalyst was almost twice that of a modified catalyst (15 mmol.h^{-1}.g^{-1} c.f. 8 mmol.h^{-1}.g^{-1}) [2]. Whereas from Table 1 it can be seen that the conversion obtained after 8 h, with a modified catalyst that gives an enantiomeric excess (entry 5), is greater than that obtained after 20 h with an unmodified catalyst (entry 2). The low optical yields for the hydrogenation of prochiral unsaturated aliphatic alkenoic acids are not uncommon: the highest obtained so far is 53% ee [3], but values of less than 20% ee are typical [4, 5, 6]. There have been few reports of attempts to hydrogenate unsaturated carboxylic esters and those [7, 8], like ourselves, have not been able to obtain any enantioselectivity.

The activation energy was determined to be 7 kJ.mol^{-1} for methyl tiglate hydrogenation and 14 kJ.mol^{-1} for the hydrogenation of tiglic acid. These values are low and would generally indicate a diffusion limited reaction. However tests showed that the rate was proportional to catalyst loading and was independent of stirrer speed. Also as a final check an olefin hydrogenation was performed that gave faster hydrogen up-take than was observed with unsaturated acids/esters and gave the literature value for the activation energy. Hence we are confident that the reaction is under kinetic control.

Table 1. Reaction of unsaturated acids over Pd/carbon.

	Reactant	Modifier	Pressure / bar g	e.e. / %	Conv. / %
1	Methyl tiglate	---	1	0	15 (20 h)
2	Tiglic acid	---	1	0	13 (20 h)
3	Methyl tiglate	(R)-Binaphthyl-diamine	10	0	12 (4 h)
4		(S)-Binaphthyl-diamine	1	0	7 (20 h)
5	Tiglic acid	(S)-Binaphthyl-diamine	1	13 (S)	16 (8 h)
6		(R)-Binaphthyl-diamine	1	6 (R)	98 (20 h)
7		(R)-Binaphthyl-diamine	3	11 (R)	91 (20 h)
8		(R)-Binaphthyl-diamine	5	8 (R)	90 (20 h)
9		(R)-Binaphthyl-diamine	7.5	5 (R)	88 (16 h)
10		(R)-Binaphthyl-diamine	10	<3 (R)	87 (18 h)
11	Angelic acid	(R)-Binaphthyl-diamine	3	8 (R)	99 (20 h)
12		(R)-Binaphthyl-diamine	10	<3 (R)	99 (20 h)

In our previous study of 3-coumaranone hydrogenation [9] we proposed that there was a 1:1 association between the reactant and the adsorbed modifier through a π-π interaction and hydrogen bonding. In this system, knowing the adsorbed state of the modifier, and in keeping with current opinion [10], we have modelled a 1:1 interaction of methyl tiglate and tiglic acid with 1,1'-binaphthyl-2,2'-diamine (Figures 2 and 3). It can be seen that the tiglic acid can achieve a π–π interaction between the C=C bond and the aromatic ring of the diamine. Two bonding interactions can also be formed between the >C=O and an aromatic proton and between the –OH and the -NH. These interactions lock the conformation and ensure hydrogen addition from a single face, thus interpreting the observed sense of the enantioselectivity. Hydrogen addition, after dissociation by the metal, can occur via a hydrogen transfer mechanism similar to that found with alkene hydrogenation by diimide [11]. Such a mechanism would also be expected to exhibit a low activation energy typical of hydrogen transfer reactions.

Figure 2. Interaction of tiglic acid and diamine.

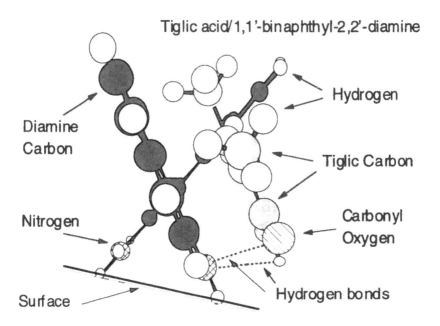

Tiglic acid/1,1'-binaphthyl-2,2'-diamine

With methyl tiglate however, the steric constraints of the methyl group are such that no 1:1 interaction can be forged that allows the formation of the hydrogen bonds or positions the molecule such that selective hydrogen addition can occur. Indeed it can be seen from Figure 3, that a similar orientation as the acid cannot be achieved due to the steric interaction between the methyl group and the surface. Hence no enantiomeric excess is obtained and the modifier only acts to poison the reaction.

Figure 3. Interaction of methyl tiglate and diamine.

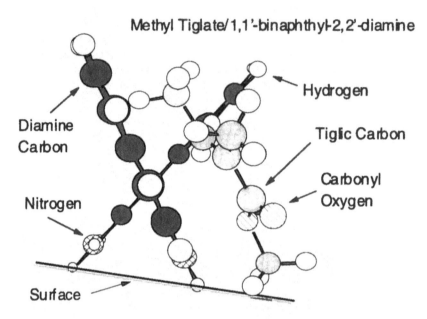

Methyl Tiglate/1,1'-binaphthyl-2,2'-diamine

Hydrogen

Diamine
Carbon

Tiglic Carbon

Carbonyl
Oxygen

Nitrogen

Surface

References

1.   N. C. Young, Ph.D. Thesis, University of Glasgow, (1995).
2.   S. R. Watson, Ph.D. Thesis, University of Hull, (1995).
3.   Y. Nitta and K. Kobiro, *Chem. Lett.*, 165 (1995).
4.   R. L. Beamer, R. H. Belding, and C. S. Ficking, *J. Pharm. Sci.*, **58**, 1142 (1969).
5.   M. Bartok, G. Wittmann, G. B. Bartok, and G. Gondos, *J. Organomet. Chem.*, **384**, 385 (1990).
6.   Y. Nitta, Y. Ueda, and Imanaka, *Chem. Lett.*, 1095 (1994).
7.   A. Tungler, M. Kajtar, T. Mathe, G. Toth, E. Fogassy and J. Petro, Catalysis Today, **5**, 159 (1989).
8.   A. Tungler, T. Tarnai, Y. Mathe, G. Vidra, J. Petro, and R. A. Sheldon, Chem. Ind. (Marcel Dekker), **62**, (*Catal. Org. React.*), 201 (1994).
9.   E. Allan, S. D. Jackson, S. Korn, G. Webb, and N. C. Young, Chem. Ind. (Marcel Dekker), **68**, (*Catal. Org. React.*), 191 (1996).
10.   Y. Nitta and A. Shibata, *Chem. Lett.*, 161 (1998).
11.   S. Siegel, Studies in Surface Science and Catalysis (Elsevier), **59**, (*Heterog. Catal. Fine Chem. II*), 21 (1991).

Enantioselective Catalytic Hydrogenation of (6: 7,8: 9)-Dibenzobicyclo[3, 2, 2]nona-6, 8-dien-2-one on Ru-containing Zeolites

S. Coman[1], E. Angelescu[1], A. Petride[2], M. Banciu[3] and V. I. Pârvulescu[1]

[1] University of Bucharest, Faculty of Chemistry, Department of Chemical Technology and Catalysis, B-dul Regina Elisabeta 4,Bucharest 70346, Romania
[2] Institute of Organic Chemistry, Splaiul Independentei 208, Bucharest, Romania
[3] Polytechnic University of Bucharest, Department of Organic Chemistry, Str. Polizu, Bucharest, Romania

Abstract

Ru/L and Ru/ZSM-5 zeolites were prepared and investigated in the enantioselective hydrogenation of (6: 7, 8: 9)-dibenzobicyclo[3, 2, 2]nona-6, 8-dien-2-one (I). The reaction was carried out in the presence of tartaric acid or cinchona, as modifiers. The results show that no stereospecific interaction was occurring in the hydrogenation of (I) to (6: 7,8: 9)-dibenzobicyclo[3, 2, 2]nona-6, 8-dien-2-ol. But further hydrogenation of (I) to 8:9-benzo-6, 7-cyclohexanbicyclo[3, 2, 2]non-8-en-2-one results in an enatioselective hydrogenation of this to (2S) 8:9-benzo-6, 7-cyclohexanbicyclo[3, 2, 2]non-8-en-2-ol. Tartaric acid was proved to be an effective modifier.

Introduction

During the last decade, the development of heterogeneous catalysts based on supported metals modified by chiral inductors (modifiers) has received increasing attention (1, 2). Because of the practical applications, enantioselective hydrogenation of β-ketoesters and α-ketoesters have been particularly studied (3, 4). However, the mechanism of chiral induction proposed for the catalytic systems are still highly speculative. The e.e. is controlled by many factors which are associated either with the catalyst (e.g., nature of the metal, metal and support morphology) or with the reaction conditions (e.g., temperature, pressure, concentrations of substrate and of chiral additives, nature of the solvent). The transport processes, such as hydrogen availability on the catalyst surface controlled by diffusion, may play a role, as shown recently (5).

483

Chirally modified metals, in which finely divided metals such as Ni, Pt, Pd, have adsorbed on their exterior surfaces a chiral auxiliary. Well-known examples of this kind are the Ni-tartaric acid system for the hydrogenation of β-ketoesters, β-ketones (6) and the Pt- cinchona alkaloid system (7) for the hydrogenation of carbonyl compounds activated by an electron-withdrawing group in an α-position. For both the Ni and Pt treated surfaces, e.e. as high as 95 percent have been reported; but the range of reactants that may be transformed enantioselectively in this way is rather narrow.

Only few studies were reported about Ru catalysts in such kind of enantioselective or diastereoselective reactions. Our previously studies about the hydrogenation of a prostaglandin intermediate on Ru supported molecular sieves, have shown that these catalysts are active and selective, yielding prostaglandin intermediate with natural-like configuration (8). At the same time, the diastereoselective hydrogenation of cyclic β-ketoesters on the above mentionated catalysts, in the presence of tartaric acid or cinchona alkaloid like chiral modifiers, takes place with high diastereoselectivity in (6R,7S)-trans diastereomer product (9).

Our aim was to carry out studies in hydrogenation of large cyclo-ketone (Scheme 1) and to determine the activity and selectivity of Ru-containing zeolites under the same experimental conditions. Furthermore, the effect of modifiers like tartaric acid and cinchona alkaloid was also examined. The present paper discusses the results obtained in enantioselective hydrogenation of (6: 7, 8: 9)-dibenzobicyclo[3, 2, 2]nona-6, 8-dien-2-one.

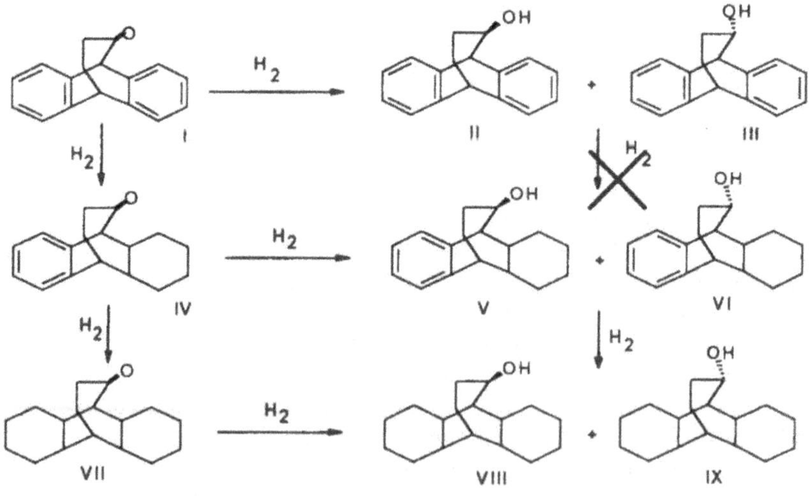

Scheme 1. Pathways in hydrogenation of (6: 7, 8: 9)-dibenzobicyclo[3, 2,2]nona- 6, 8-dien-2-one.

Results and discussion

Catalyst characterisation data are listed in Table 1. As is shown, the reduction degree was higher than 90% only for Ru/L catalyst. In Ru/ZSM-5 catalyst, a substantial fraction of Ru is in an oxidised state, while Ru/L zeolites contain practically only metallic Ru. This behavior should be correlated with the structure and chemical composition of the supports. Structurally, the L zeolite has large and unidimensional pores and low acid strength compared with ZSM-5 zeolite. At the same time, the number of the exchange sites is substantially higher for L zeolite (Si/Al=3.1) than for ZSM-5 (Si/Al=60). The narrow pores of the ZSM-5 do not allow ruthenium to penetrate the structure, therefore the metal is mainly on the external surface where it is more easily reduced. The percentages of the reduced metal, calculated from the H_2-TPR profiles, vary in the order: Ru/ZSM-5 (50%) >12% Ru/L (27%) > 5% Ru/ZSM-5 (14%). After Ru deposition, the zeolites still exhibit high surface areas.

The XRD patterns of Ru-zeolites showed that the structure was not damaged during the thermal treatments. Table 1 gives the particle size of the metallic ruthenium species determined from X-ray line broadening using the Scherrer formula: $d = K\lambda/b\cos\theta$, where d is the mean particle diameter, b is the X-ray broadening, K=0.88 and λ=1,5406 Å. A rather good agreement with the H_2 chemisorption data was obtained. For the same metal content (5%), the dispersion decreased in the order: Ru/L > Ru/ZSM-5. Because of the low dispersion, the metal size determined from the same measurements exceeded 25 nm, and reached about 70 nm in the case of 12% Ru/L. In addition to the accesibility of the metal species which correlates with the pore structure, the strength of the Ru-O bond is also influenced by the support.

Table 2 shows experimental data on the hydrogenation of (6: 7, 8: 9)-dibenzobicyclo[3, 2, 2]nona-6, 8-dien-2-one employing Ru/zeolites (T=353 K, 10 bars, 200 mg catalyst, 10 mg substrate, 4 hours). It is remarkable that from the possibilities indicated in the Scheme 1 only the compounds II, III, IV, V and VI were obtained as major products. As can be seen, in the presence of the cinchona alkaloid, the conversion of cycloketone and the selectivity in alcohol are decreased. At the same time, the resulted alcohol is only in the racemic form. These data indicate that no specific interaction occurs between the substrate and the chiral modifier. Not only the relative position of the receptor sites in the substrate and the chiral modifier, but also the dimension of the two molecules and the rigid conformation of the compound I may contribute to this behavior. The lower conversion can be a consequence of the fact that a non-specific interaction between the two molecules occurs. The hydrogenation of the same molecule in the presence of tartaric acid, and even more in the presence of tartaric acid and pivalic acid leads to an increase of the conversion in comparison with that obtained in the presence of cinchona. However, this

Table 1. Catalysts characterization data

Catalyst	Ru loading wt.%	Langmuir surface area $m^2 g^{-1}$	Reduction degree[1] %	Dispersion[1] %	Particle size Chemisorption nm	XRD nm
Ru/L	5	104	93.8	5.2	25.8	39.1
Ru/L	12	85	91.0	2.5	53.6	46.9
Ru/ZSM-5	5	17	48.2	1.9	70.5	73.0

[1] Data determined from TPO and TPR experiments

Table 2. Experimental data from hydrogenation reaction of (6: 7, 8: 9)-dibenzobicyclo[3, 2, 2]nona-6, 8-dien-2-one

Catalyst	Modifier	Conversion %	Selectivity to II+III, %	E.e. %	Selectivity to IV, %	Selectivity to V+VI, %	E.e. (%)
5% Ru/L	-	85.0	34.20	racemic	40.5	23.3	racemic
	(-)-CD[1]	45.6	14.69	racemic	28.2	56.4	racemic
12% Ru/L	(-)-CD[1]	48.0	11.79	racemic	29.8	56.3	racemic
5% Ru/ZSM-5	(-)-CD[1]	51.8	9.2	racemic	31.8	54.2	racemic
	L(+)-TA[2]	60.5	18.68	racemic	50.3	28.1	18 (S)
	L(+)-TA[2]+PA[3] (TA/PA=0.23)	63.2	25.31	racemic	45.1	27.1	23 (S)
	L(+)-TA[2]+PA[2] (TA/PA=0.13)	70.0	26.71	racemic	46.3	25.7	32 (S)

[1]Cinchonidine; [2]Tartaric acid; [3]Pivalic acid

conversion remains smaller than that in the absence of any modifier, indicating that an interaction exists also in this later case. The resulted alcohol was also the racemic. The increase of the pressure to 40 bars brought no modification in the enantioselectivity to (6: 7, 8: 9)-dibenzobicyclo[3, 2, 2]nona-6, 8-dien-2-ol.

Table 2 also compiles data concerning the selectivity to the compound IV and to its alcohol derivates. Ru is known for its aromatic hydrogenation abilities. The presence of the cinchona improves the selectivity to these compounds, and especially to V and VI, in rapport with experiments made without modifiers, confirming the existence of an interaction. The evolution of the selectivity to these compounds over the three investigated catalysts should be associated with the size of Ru. In the presence of cinchona, only the racemic was obtained irrespective of the catalyst and of the reaction conditions. The tartaric acid or the mixture of the tartaric and pivalic acid lead to a higher selectivity to the compound IV than in the absence of any modifier and also higher than in the presence of cinchona. This can be an indication that in the presence of this modifier a more important part of the catalyst surface is accessible for the interaction with I leading to an advanced hydrogenation of the aromatic ring. Under such conditions the selectivity to V and VI is between those determined in absence of modifiers and that using cinchona. In the presence of the tartaric acid and even more when the pivalic acid is added in the reaction medium the hydrogenation occurs enantioselectively with an e.e. in (2S) 8:9-benzo-6, 7-cyclohexanbicyclo[3, 2, 2]non-8-en-2-ol. We presume that the different conformation of III in relation with I makes more properly a stereoselective interaction with the tartaric and pivalic acid leading to the e.e. given in Table 2.

In conclusion, hydrogenation of dibenzo(6, 7; 8, 9)nonan-2-one over the investigated Ru/zeolites occurs with relatively low selectivities to the dibenzo(6, 7; 8, 9)nonan-2-ol and only the racemic was determined. Further hydrogenation to 8:9-benzo-6, 7-cyclohexanbicyclo[3, 2, 2]non-8-en-2-one results in an enatioselective hydrogenation to (2S) 8:9-benzo-6, 7-cyclohexanbicyclo[3, 2, 2]non-8-en-2-ol., but only in the presence of tartaric acid. The addition of pivalic acid improves the e.e.

Experimental

The reaction conditions were as follows: 10 mg ketone, 353 K reaction temperature and 10 ml of ethanol (for the case cinchona was used as modifier) or 10 ml THF (for the case L(+)-tartaric acid was used as modifier). In order to improve the selectivity, some experiments were carried out using L(+)-tartaric acid and pivalic acid modified catalyst. The molar ratio L(+)-tartaric acid/pivalic acid was in the range 0.13-0.23. The catalyst (200 mg), the substrate and the solvent were placed in the reaction vessel and the mixture was stirred for 4 hours under hydrogen pressure between 5 and 40 bars. Both the temperature of the reaction mixture and the hydrogen pressure were kept constant throughout the reaction. Chromatography was conducted with an Carlo-Erba HRGC 5300 chromatograph equipped with two capillary columns, with dimethyl-penthyl-β-cyclo-dextrin and with permethyl-β-cyclo-dextrin, respectively. The products formed were identified on the basis of the retention times of authentic compounds. The reaction rate was calculated from the GC-MS data processed with a Spectra Physics Chrom. Jet. integrator. In the case of the addition of the pivalic acid, prior to analysis, the reaction product was neutralized with K_2CO_3, extracted in 100 ml anhydrous ethylic ether and dried on Na_2SO_4.

Ru supported catalysts have been obtained by deposition of Ru on zeolites (L, ZSM-5) in the K form, from a solution containing 0.4 M ruthenium (III) chloride following a procedure reported elsewhere (10). Before the reaction, the catalysts were reduced in flowing hydrogen (30 ml min^{-1}) at 573 K for 6 hours. The heating rate was of 1 K min^{-1}. Samples were characterized by elemental analysis, adsorption-desorption of N_2 at 77 K, H_2 chemisorption, H_2-TPR, XRD, and XPS. Details on these experimental procedures were reported previously (11).

References

1. R. A. Sheldon, in "Chirotechnology: Industrial Synthesis of Optically Active Compounds", Dekker, New York, 1993.
2. K. T. Wan and M. E. Davis, *J. Catal.*, **152**, 25 (1995).
3. T. Osawa, T. Harada, and A. Tai, *J. Mol. Catal.*, **87**, 333 (1994).
4. B. Minder, M. Schurch, T. Mallat, and A. Baiker, *Catal. Lett.*, **31**, 143 (1995).
5. U. K. Singh, R. N. Landau, Y. Sun, C. LeBlond, D. G. Blackmond, S. K. Tanielyan, R. L. Augustine, *J. Catal.*, **154**, 91 (1995).
6. A. Baiker and H. U. Blaser, in : G. Ertl, H. Knozinger, J. Weitkamp (Eds), *"Handbook of Heterogeneous Catalysts"*, VCH, Weinheim, Vol. **51**, (1996) p. 2422.
7. J. M. Thomas, *Angew. Chem., Adv. Materials*, **101**, 1105 (1989).
8. F. Cocu, S. Coman, C. Tanase, D. Macovei, and V. I. Parvulescu, *Stud. Surf. Sci. Catal.*, **108**, 207 (1997).
9. S. Coman, C. Bendic, M. Hillebrand, Em. Angelescu, V. I. Parvulescu, A. Petride, and M. Banciu, in: F. Herkes (Ed), *"Catalysis in Organic Reactions"*, Dekker, New York, (1998) p. 145.
10. V. I. Parvulescu, V. Parvulescu, S. Coman, D. Macovei, and Em. Angelescu, *Stud. Surf. Sci. Catal.*, **91**, 561 (1995)
11. V. I. Parvulescu, S. Coman, P. Palade, D. Macovei, C. M. Teodorescu, G. Filoti, R. Molina, G. Poncelet, and F. E. Wagner, *Appl. Surf. Sci.*, **141**, 164 (1999).

Enantioselective Hydrogenation of α-Ketoesters over a Pt/Al₂O₃ Catalyst: Effect of Steric Constraints on the Enantioselection

Kornél Szőri[2], Béla Török[1]*, Károly Felföldi[2] and Mihály Bartók[1,2]

[1]Organic Catalysis Research Group of the Hungarian Academy of Sciences and
[2]Department of Organic Chemistry, University of Szeged, H-6720 Szeged, Dóm tér 8, Hungary, e-mail: torok@chem.u-szeged.hu, fax:36-62-425-768

Abstract

The enantioselective hydrogenation of several α-ketoesters with varying alkyl substituents has been studied over cinchona alkaloid-modified Pt/Al₂O₃ catalyst. The steric constraints caused by the bulky alkyl groups strongly affected the outcome of the catalytic reaction. The increasing substituent size, in the case of the ester group, however, resulted in only slight changes in the reaction rates and enantioselectivity. On the basis of the effect of the substituents an important contribution is proposed to refine the existing mechanistic picture.

1. Introduction

The asymmetric heterogeneous catalytic hydrogenation is one of the most promising processes in the synthesis of chiral pharmaceuticals and agrochemicals [1]. The cinchona alkaloid modified platinum catalyst system developed by Orito [2] is a successful example in this field with wide interest. The method was found to be of excellent performance in the hydrogenation of α-ketoesters (Scheme 1) [3] and very recently α-ketoacetals [4].

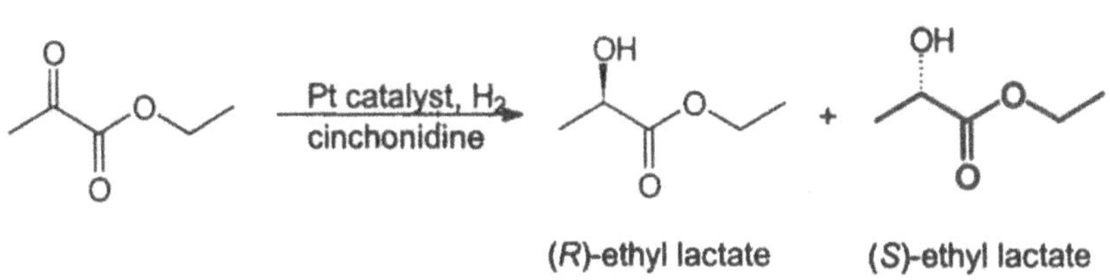

(R)-ethyl lactate   (S)-ethyl lactate

Scheme 1

489

This unique catalytic system is still in the focus of interest. Recently, mainly two major routes are followed, first new application possibilities opened up widening the classes of substrates [4, 5]. On the other hand, extensive efforts were carried out to get more insight into of the mechanism [1, 6].

Numerous papers were concerned in revealing the mechanism of the enantiodifferentiation, however, according to recent reviews [7] there still are untouched items. One of these is the effect of steric hindrance of the substrate including both the alkyl and the ester groups.

As a result, the principal reactant of the Pt-cinchona system, namely the ethyl pyruvate was systematically modified on both its alkyl and esters groups. Bulkier substituents were introduced step by step such as methyl, ethyl, i-propyl, t-butyl, phenyl and phenyl-ethyl. Our major goal was to learn more about the effect of steric constraints at both side of the reactant in order to get more insight into the molecular movements during the enantiodifferentiation.

2. Experimental

2.1. Materials
Ethyl and methyl pyruvate were of analytical grade and purchased from Fluka, while acetic acid with minimum purity of 99.5% was also a Fluka products. The cinchonidine (CD) and cinchonine (CN) used as modifiers (minimum purity >98%) were purchased from Fluka. The further α-ketoesters were synthesized according to the literature [8]. The catalyst used in the hydrogenations was a 5% Pt/Al$_2$O$_3$, (Engelhard 4759, D=0.37 after 400 °C reductive pretreatment D=0.28).

2.2. Methods

2.2.1. Hydrogenations under atmospheric H$_2$ pressure   The hydrogenation reactions were carried out at room temperature (25 °C) in a conventional atmospheric batch reactor system equipped with a rubber septum. High purity, oxygen-free hydrogen was prepared by a Whatman Model 75-34 hydrogen generator directly connected to the system. In a typical run 50 mg of the catalyst and 5 mg of cinchonidine was suspended in 5 ml of acetic acid and the reactor was flushed with hydrogen for 15 min. Then the catalyst was activated in a closed system by stirring for 1 h. After the activation period 2.25 mmole substrate was injected into the reactor during vigorous stirring and the hydrogen uptake was followed by a gas burette.

2.2.2. Product analysis   The substrates and products were identified on the basis of their mass spectra using a HP 5890 GC/HP 5970 MS system equipped with a 50 m long HP-1 capillary column. The quantitative analysis including enantiomeric separation was performed with a HP 5890 GC-FID gas chromatograph using 30m long Cyclodex-B (J&W Scientific) and 30m long Lipodex A (Macherey-Nagel) capillary columns. The optical yield is expressed as the enantiomeric excess (ee%) of the (R)-(+)- or (S)-(-)-ethyl lactate (indicated in Tables) and calculated according to the literature (ee%= (| [R]-[S] |) x 100 / ([R]+[S])).

3. Results and Discussion

As mentioned in the introduction although sporadical results [9] can be found with different substituted α-ketoesters, no systematic studies are available in the literature concerning the steric effect of the substrate. As a result, special, modified "pyruvate type" α-ketoesters were synthesized. The hydrogenation of these compounds was then studied. Although, a wide variety of solvents were tested in the literature acetic acid was found to be the best solvent. As a result, it was selected for this study. According to a recent study by Sun *et al.* in this solvent the modifier can easily be hydrogenated and so in part lose its ability to induce enantiodifferentiation [10]. The relatively high CD concentration used prevents this effect. The reaction scheme and the schematic representation of the substrates are shown in Scheme 2.

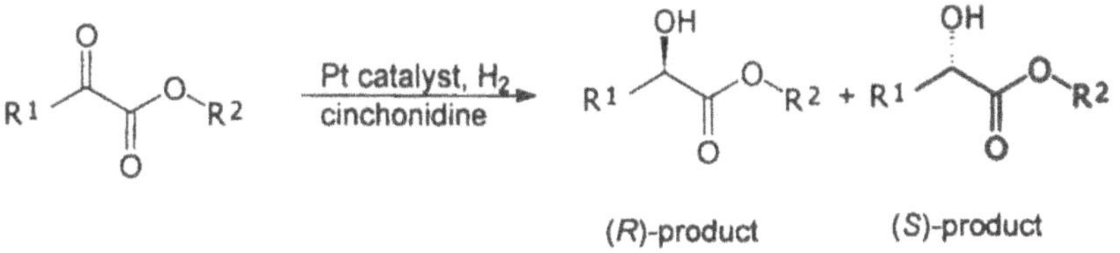

	R^1	R^2		R^1	R^2
1	Me	Me	4	iPr	Et
2	Me	Et	5	tBu	Et
3	Me	Pr	6	Ph	Et
			7	$PhCH_2CH_2$	Et

Scheme 2

3.1. Effect of the variation of the ester group

First, the effect of the size of the ester group was studied. The substituents varied on the carboxyl group were Me, Et, and iPr. The results of the hydrogenations are summarized in Table 1.

As the data show, the increase in the size of the ester group does not significantly decrease the catalytic performance. Both the hydrogenation rates and the enantiomeric excesses are comparable to those obtained with methyl pyruvate (1) having the smallest ester group. In addition, the two modifiers used, produced nearly the same results except for the configuration of the hydroxyester products.

Table 1 Enantioselective hydrogenation of α-ketoesters over a 5% Pt/Al$_2$O$_3$ catalyst modified by cinchonidine and cinchonine in acetic acid (25 °C, 1 bar hydrogen pressure)

Substrate	Modifier	r_0 (mmolmin^{-1}g$_{cat}^{-1}$)	Conversion (%)	Product configuration	ee (%)
1	CD	1.25	100	(R)	81
1	CN	1.46	100	(S)	83
1	–	0.41	100	racem.	0
2	CD	1.04	100	(R)	86
2	CN	1.25	100	(S)	81
2	–	0.39	98	racem.	0
3	CD	0.83	100	(R)	80
3	CN	0.88	100	(S)	74
3	–	0.12	100	racem.	0

3.2. Effect of the variation of the alkyl group

After testing the size of the ester group, the effect of the alkyl group was studied. In this case the substituents changed from methyl to *tert.*-butyl, and aromatic substituents such as Ph- and PhCH$_2$CH$_2$- were also tested. The results of the hydrogenations are tabulated in Table 2.

Table 2 Enantioselective hydrogenation of α-ketoesters over a 5% Pt/Al$_2$O$_3$ catalyst modified by cinchonidine and cinchonine in acetic acid (25 °C, 1 bar hydrogen pressure)

Substrate	Modifier	r_0 (mmolmin^{-1}g$_{cat}^{-1}$)	Conversion (%)	Product configuration	ee (%)
2	CD	1.04	100	(R)	86
2	CN	1.25	100	(S)	81
2	–	0.39	100	racem.	0
4	CD	1.04	100	(R)	62
4	CN	0.72	100	(S)	63
4	–	0.41	97	racem.	0
5	CD	0.33	95	(R)	47
5	CN	0.25	95	(S)	15
5	–	0.08	95	racem.	0
6	CD	0.50	100	(R)	75
6	CN	0.49	100	(S)	42
6	–	0.33	96	racem.	0
7	CD	0.51	100	(R)	75

7	CN	0.66	98	(S)	76
7	–	0.16	100	racem.	0

As the reaction rate and enantioselectivity data express, the increase in the size of the alkyl part of the substrate resulted in much more characteristic changes. As a function of size of the substituent both r_0 and ee values are dramatically decreased. In the case of methyl group, for example, the ee was as high as 81% using cinchonine. However, testing a substrate with bulky *tert.*-butyl group the enantioselectivity was decreased to 15%.

This decrease could be observed in the case of phenyl substituent as well, although only in a small extent due to its flexible nature. Increasing the distance between the carbonyl and the phenyl groups with the insertion of a -CH_2CH_2- moiety the enantioselectivity is also close to that obtained with ethyl pyruvate (**2**) (76% to 81%).

3.3. Discussion

Summarizing the results described in sections 3.1. and 3.2. we must make it absolutely clear that the role of the ester (denoted as "head") and the alkyl groups (denoted as "tail") are quite different.

First, it should be mentioned that our present proposal are based on the former *modified catalyst model* [1, 7] and refine it concerning the position of the substrate. As shown, the increase in the size of the "head" affects the selectivity only in a negligible extent, while introducing bulky groups into the tail position dramatically decreases the optical yields. Although, the numerical values are different using cinchonidine or cinchonine, the trends are unambiguously clear and same. In our point of view these data allow us to propose a geometrical arrangement during the molecular movements leading to the formation of the transition complex. It is suggested that the complex formation and thus the enantioselection takes place when the substrate comes to the adsorbed modifier with its "tail". This proposal are illustrated in Figure 1.

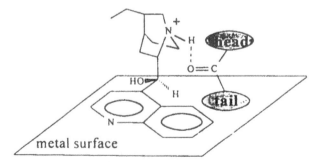

Figure 1 The proposed geometrical arrangement of the substrate and modifier in the enantioselective hydrogenation of α-ketoesters over cinchona modified Pt catalysts

As consequence, the "tail" should be of small size or flexible nature to get high enantioselectivities. Otherwise (e.g. *t*Bu group in the tail, Table 2, 5) the transition complex formation is not favorable and the product selectivity will be determined by the non-modified so called *racemic* hydrogenation.

4. Conclusions

The first experimental evidence is provided to determine the geometrical arrangement of the substrate during the formation of the transition complex in the cinchona-modified Platinum-catalyzed enantioselective hydrogenation of α-ketoesters. The results obtained with several substituted α-ketoesters unequivocally showed that the substrate comes with its alkyl group close to the modifier.

ACKNOWLEDGMENT
Financial support (OTKA T031707) is highly appreciated.

REFERENCES
1. A. Baiker, H.-U. Blaser, in Handbook of Heterogeneous Catalysis (Eds.: G. Ertl, H. Knözinger, J. Weitkamp), Wiley-VCH, 1997, Vol. 5, p. 2422 and references cited therein; G. V. Smith, F. Notheisz, Heterogeneous Catalysis in Organic Chemistry, Academic Press, San Diego, 1999.
2. Y. Orito, S. Imai, S. Niwa, J. Chem. Soc. Jpn., (1979) 1118, C.A., 91 (1979), 192483h.
3. B. Török, K. Felföldi, G. Szakonyi, K. Balázsik and M. Bartók, *Catal. Lett.* 52, 81(1998);
4. B. Török, K. Felföldi, K. Balázsik, M. Bartók, *Chem. Commun.*, 1725 (1999), M. Studer, S. Bukhardt, H.-U. Blaser, *Chem. Commun.*, 1727 (1999).
5. M. Studer, V. Okafor, H.-U. Blaser, *Chem. Commun.*, 1053 (1998); N. Kunzle, A. Szabo, M. Schurch, G. Z. Wang, T. Mallat, A. Baiker, *Chem. Commun.*, 1377 (1998).
6. R. L. Augustine, S. K. Tanielyan, L. K. Doyle, *Tetrahedron: Asymmetry* **4**, 1803 (1993); Y. Sun, J. Wang, C. LeBlond, R. N. Landau and D. G. Blackmond, *J. Catal.* **161**, 759 (1996); M. Bartók, K. Felföldi, B. Török, T. Bartók, *Chem. Commun.*, 2605 (1998); T. Bürgi, A. Baiker, *J. Am. Chem. Soc.*, **120**, 12920 (1998). M. Bartók, K. Felföldi, Gy. Szöllösi, T. Bartók, *Catal. Lett.*, **61**, 1 (1999).
7. A. Baiker, *J. Mol. Catal. A*, **115**, 473 (1997); H. U. Blaser, H. P. Jalett, M. Müller, M. Studer, *Catal. Today*, **37**, 441 (1997).
8. K. Szőri, B. Török, K. Felföldi, M. Bartók, *J. Catal.* (in preparation).
9. H.-U. Blaser, H.-P. Jalett, J. Wiehl, *J. Mol. Catal.*, **68**, 215 (1991); M. Schürch, N. Künzle, T. Mallat, A. Baiker, *J. Catal.*, **176**, 569 (1998); X. Zuo, H. Liu, G. Guo, X. Yang, *Tetrahedron*, **55**, 7787 (1999).

10.   C. Leblond, J. Wang, J. Liu, T. Andrews, Y.-K. Sun, *J. Am. Chem. Soc.*, **121**, 4920 (1999).

Anchored Homogeneous Catalysts

R. L. Augustine, Setrak K. Tanielyan, Stephen Anderson, Hong Yang and Yujing Gao

Center for Applied Catalysis, Seton Hall University, South Orange, NJ 07079, USA

Abstract

Almost all previous attempts to design a "heterogenized homogeneous catalyst" involved first attaching the ligand to a solid support material and then preparing the heterogeneous catalyst using this solid ligand. In contrast, we have succeeded in attaching preformed homogeneous complexes to a solid support using a heteropoly acid as the anchoring agent. The heteropoly acid is attached to the support material and the homogeneous catalyst is anchored by an interaction between the heteropoly acid and the metal atom of the complex.

We have now used such anchored rhodium and ruthenium complexes for a number of chiral and achiral hydrogenations. We have established that the amount of metal lost in these reactions is less than 1 ppm.

Introduction

Enantioselective syntheses using chiral catalysts have been the object of extensive research for a number of years (1). Most of this work has used soluble chiral organometallic complexes to promote a variety of synthetically useful reactions. Since these catalysts are molecular species they have a single 'active site' surrounded by an appropriate chiral pocket defined by the ligands present in the complex. With the appropriate chiral ligand such species are usually very selective, frequently giving products having enantiomeric excesses (ee's) in the 95%-99% range. These catalysts, being in solution with the reaction products, can be difficult to separate from the reaction products and once separated are not easily reconstituted for potential re-use. From an economic standpoint, this loss of the catalyst can be appreciable, not so much because of the loss of the metal but, more importantly, the chiral ligand which can be many times more expensive than the metal. One method for minimizing this factor has been to use these soluble catalysts with a very high substrate/catalyst ratio with the catalyst frequently being present at only a few hundred ppm. This low catalyst load, however, can lead to prolonged reaction times.

Heterogeneous catalysts, on the other hand, are insoluble in the reaction medium and are, therefore, easily removed from the product. There are, however, only two viable enantioselective heterogeneous catalyst systems. The first, nickel metal modified with tartaric acid and sodium bromide, is useful only for the hydrogenation of β ketoesters and β diketones to the corresponding chiral β hydroxy esters, β hydroxy ketones or β dihydroxy compounds (2). With the addition of pivalic acid the enantioselective hydrogenation of methyl ketones can also be accomplished (2). The second system, platinum modified with a cinchona alkaloid, is useful only for the enantioselective hydrogenation of α keto esters and acids (3).

This dearth of viable enantioselective heterogeneous catalysts has prompted the search for a method to convert the more generally selective homogeneous catalysts into species which are insoluble in the reaction mixture in order to facilitate the separation of the catalyst. While the first attempt at doing this was made over twenty-five years ago with the quest still going on, there has been a general lack of success in this area (4-9).

The most common method used to 'heterogenize' a homogeneous catalyst has been to attach a ligand to a solid support material and then react these ligands with a metallic species to prepare the supported complex. There are, however, several problems associated with this approach, the most important of which is that the metal ion is attached to the support by bonding through the ligand and is, therefore, prone to leaching, frequently to a rather large extent. Further, these catalysts are usually less active than the corresponding homogeneous species. They also frequently lose their activity and selectivity on separation and attempted re-use. This attachment of the ligand to a solid support is not applicable to all types of ligands and can be particularly difficult when one is attempting to produce an analog of an effective chiral ligand.

For general use in enantioselective synthesis, then, what is needed is a method by which a homogeneous catalyst can be anchored to a support material without the necessity of any ligand modification. Ideally, one should be able to anchor a pre-formed homogeneous catalyst to a support while retaining the activity and selectivity of the corresponding homogeneous catalyst even on extensive re-use.

The catalysts described here not only meet these criteria but are frequently more active and selective than the homogeneous species on re-use (10,11). In our preliminary report on the use of these catalysts, we described some of the reactions for which they were used without disclosing the nature of the anchoring agent (12).

Experimental

These anchored homogeneous catalysts are prepared by adding a solution of the heteropoly acid (20 μmoles in 2.5 mL of alcohol) with vigorous stirring to a suspension of the support material (300mg in 10 mL of alcohol) with stirring continued for about three hours followed by the removal of the liquid and thorough washing of the solid. This solid is then suspended in another 10mL of degassed alcohol and a solution of the homogeneous catalyst (20 μmoles in 1 mL of alcohol) is added under an inert atmosphere, with stirring, over a 30 min period. Stirring is continued for 8 to 12 hours under an inert atmosphere, the liquid removed and the solid washed thoroughly until no color is observed in the wash liquid. This material can be used directly or dried for future use.

In some of our initial work with alumina supports, a slight color was sometimes imparted to the reaction mixture which, at first, was thought to indicate some leaching of the complex. However, analysis of these reaction mixtures showed that they contained not only rhodium but also tungsten and aluminum which indicated that this loss occurred by attrition of the alumina and not by any leaching of the complex. Using ethanol in the preparation and reaction procedures has successfully removed this problem.

The hydrogenations were run at 25°C and 1 atm H2 using 20 umoles of supported complex to saturate 0.35 mmoles of substrate in 15 mL of ethanol. The data shown in Table 1 and Fig. 5 were obtained for the hydrogenation of 0.8 mmole of substrate in 15 mL of ethanol.

Nature of the Catalyst

Our anchored homogeneous catalysts are composed of three components; a catalytically active organometallic complex, a support material and a heteropoly acid which is used to anchor the complex to the support as depicted in Fig. 1.

Heteropoly acids are polyprotic mixed oxides that are composed of a central metal ion or ions which are bonded to an appropriate number of oxygens and surrounded by a near spherical shell of octahedral oxometallic species joined

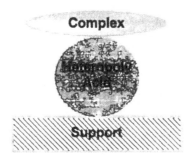

Fig. 1. A cartoon showing a heteropoly acid as the anchoring agent
for a homogeneous complex.

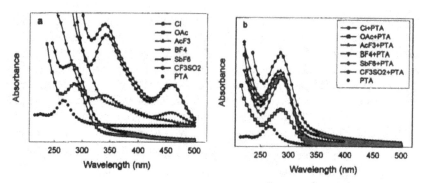

Fig. 2. a) UV spectra of several Rh(DiPAMP)$^+$ complexes having
          different counterions. b) UV spectra of Rh(DiPAMP)/PTA
          complexes prepared from the complexes shown in a.

together by shared oxygens (13). The central atom is commonly a cation having
a +3 to +5 oxidation state such as P^{+5}, As^{+5}, Si^{+4} or Mn^{+4}. The metallic
species associated with the octahedra are usually Mo, W or V. The most
common of these heteropoly acids are phosphotungstic acid (PTA),
phosphomolybdic acid (PMA), silicotungstic acid (STA) and silicomolybdic
acid (SMA), which are referred to as Keggin acids.

Attachment of the heteropoly acid to a support such as alumina takes
place by interaction of the acidic protons of the acid with basic sites on the
support (14). Further, it has been reported that Rh and Ir complexes react with a
heteropoly acid to give a material in which the metal atom is attached to the
heteropoly acid through the surface oxygen atoms on the acid (15-17). Further
support of this type of interaction between a complex and a heteropoly acid
comes from the UV data shown in Fig. 2. In Fig 2a are depicted the UV spectra
of several soluble Rh(DiPAMP)$^+$ complexes with varying counterions. In those
species having the BF_4^-, SbF_6^- and triflate counterions the rhodium is present as
a cationic species having a distinct doublet in the UV spectra. Those complexes
with the Cl$^-$ and OAc$^-$ have these counterions bound directly to the Rh atom so
there is no formal charge on the metal. These species show a shorter wavelength
singlet in the UV. In Fig. 2b are shown the UV spectra of these complexes after
they have been treated with phosphotungstic acid. In every instance, the UV
spectra are almost identical to those in Fig. 2a for the complexes in which there
is no Rh cation.

However, EXAFS data obtained after heating a mixture of a heteropoly
acid and a rhodium complex indicated that an ion pair was present and not a
direct bond between the heteropoly acid and the complex (18). Obviously, the

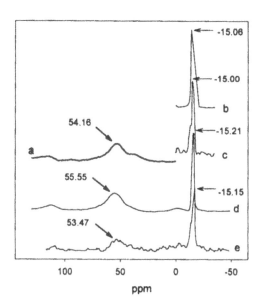

Fig 3. ^{31}P MAS-nmr spectra of: a) Rh(DiPamp)$^+$, b) PTA, c) PTA
on alumina d) Rh(DiPamp)/PTA and e)
Rh(DiPamp)/PTA/alumina.

exact nature of the interaction between a homogeneous complex and a supported
heteropoly acid needs to be clarified.

Fig. 3 shows the ^{31}P MAS-nmr data for the unsupported PTA, PTA on
alumina and the anchored Rh(DiPAMP) complex on the PTA/alumina. Also
given are the ^{31}P nmr's for the solid Rh(DiPAMP)$^+$ and the PTA/Rh(DiPAMP)
complex (19). These data show that there does not seem to be any significant
change in the PTA during any of these steps. Some shift for the ligand
phosphines may be present but the broad nature of these peaks precludes any
definitive statement in this regard.

Reaction Data

$$H_2C\overset{CO_2Me}{\underset{NAc}{\diagdown}} \xrightarrow[\text{1 Atm. R.T.}]{H_2 \quad \text{Catalyst}} H_3C\overset{H}{\underset{NAc}{\diagdown}}CO_2Me + H_3C\overset{H}{\underset{NAc}{\diagdown}}CO_2Me \qquad (1)$$

$$\mathbf{1} \qquad\qquad\qquad (R) \qquad\qquad (S)$$

The data presented in Table 1 show the reaction rate and product ee for
successive hydrogenations of methyl 2-acetamidoacrylate (**1**) (Eqn. 1) run over a

Table 1. Reaction rate and product ee data from the multiple hydrogenations of **1** over a Rh(DiPamp)/ PTA/Montmorillonite K catalyst.

Use Number	Rate [a]	Product ee (%)
Homogeneous	0.25	76
1	0.18	67%
2	1.20	92
3	1.26	94
6	1.49	96
9	1.29	97
15	b	97

[a] moles of H_2 / mole Rh / min.
[b] Rate data for this run were lost due to a computer malfunction.

Rh(DiPAMP)$^+$ complex supported on PTA treated Montmorillonite K as well as the corresponding data obtained using the homogeneous catalyst. The first use of the heterogeneous catalyst was slower than that observed with the homogeneous catalyst and the product ee was also lower. However, when the first product mixture was removed from the reactor and a fresh solution of the starting material added to the heterogeneous, anchored catalyst, subsequent re-use proceeded significantly faster than the homogeneously catalyzed reaction with higher product ee's observed as well. This catalyst was re-used fifteen times with no loss of activity or selectivity. Analysis of the combined reaction mixtures from the last ten cycles showed that the rhodium loss in these reactions amounted to less than 1 ppm.

While we are not certain of the reason for the increase in activity after the first use of the heterogeneous catalyst, the fact that in some reactions the catalyst changed from yellow to a light gray-violet on use and back to yellow on exposure to air suggested that this activation may be the result of some partial reduction of the tungsten in the PTA. Some support for this hypothesis was found in the ^{31}P MAS-nmr taken of a catalyst sample isolated after the first use, dried and kept in an inert atmosphere. This spectrum showed a small splitting of the PTA peak which may be the result of a change in oxidation state of some of the tungsten ions surrounding the central phosphorous in the PTA. To further test this hypothesis, a sample of the catalyst was stirred under hydrogen at room

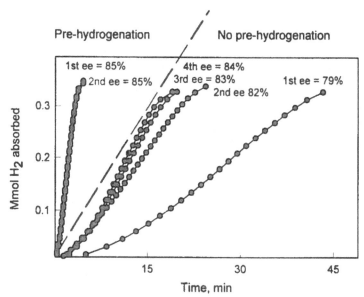

Fig. 4. Hydrogen uptake curves for the hydrogenation of **1** over
Rh(DiPAMP)/PTA/Al$_2$O$_3$ catalysts which had been
either pre-hydrogenated or not treated with hydrogen
before the introduction of the substrate.

temperature and atmospheric pressure for several hours before the introduction
of the acrylate substrate, **1**. Fig. 4 shows the hydrogen uptake curves for acrylate
hydrogenation over Rh(DiPAMP)/PTA/Al$_2$O$_3$ catalysts, one which was pre-
hydrogenated and one which was not. These data show that pre-hydrogenation
does activate the catalyst and, further, if it is done before the substrate is added,
the effect appears to be more pronounced.

A similar effect was observed when other heteropoly acids were used as
anchoring agents with pre-hydrogenation resulting in the activation of the
catalyst. Table 2 shows the reaction data for the hydrogenation of **1** over
Rh(DiPAMP)$^+$ anchored on alumina using the more common of the Keggin type
heteropoly acids. In every instance, pre-hydrogenation resulted in catalyst
activation. In most cases, pre-hydrogenation produced a more active catalyst
than that obtained after multiple use.

One of the advantages of this approach to anchoring homogeneous
catalysts is its apparent generality in that this procedure can be used to anchor a
variety of pre-formed active homogeneous catalysts onto a number of different
supports. In Table 3 are listed the reaction rate and product ee data observed for

Table 2. Heteropoly acid effect on the hydrogenation of **1** over Rh(DiPAMP)$^+$ anchored to alumina using different heteropoly acids.

Heteropoly Acid	Non-prehydrogenated			Pre-hydrogenated		
	Use #	Rate [a]	ee %	Use #	Rate [a]	ee %
Phosphotungstic	1	0.36	79	1	3.73	85
	4	0.80	84	4	3.73	85
Silicotungstic	1	0.32	79	1	1.33	81
	4	1.37	80	3	1.33	81
Phosphomolybdic	1	0.38	86	1	1.37	87
	5	1.00	86	4	1.37	87
Silicomolybdic	1	0.21	84	1	0.79	85
	5	0.93	85	4	0.79	85

[a] mmole H$_2$/mmole Rh/min.

the hydrogenation of **1** over several of Rh complexes anchored to alumina along with the corresponding homogeneous catalyst data. In Table 4 are given the data for the hydrogenation of **1** using Rh(DiPAMP)$^+$ anchored on different supports.

Table 3. Reaction rate and product ee data from the multiple hydrogenations of **1** over Rh(Ligand)/ PTA/alumina catalysts along with the data from the homogeneously catalyzed hydrogenations.

Ligand	Use #	Anchored		Homogeneous	
		Rate[a]	ee %	Rate[a]	ee %
DiPamp	1	0.32	90	0.25	76
	3	1.67	93		
Prophos	1	2.0	68	0.26	66
	3	2.6	63		
Me-Duphos	1	1.8	83	3.3	96
	3	4.4	95		
BPPM	1	3.75	21	7.4	84
	3	8.15	87		

[a] mmole H$_2$/mmole Rh/min.

Table 4. Reaction rate and product ee data from the multiple hydrogenations of **1** over a Rh(DiPamp)$^+$ catalyst on different PTA modified supports.

Support	Use #	Rate[a]	ee %
Montmorillonite K	1	0.18	67
	3	1.26	94
Carbon	1	0.07	83
	3	0.41	90
Alumina	1	0.32	90
	3	1.67	93
Lanthana	1	0.38	91
	3	0.44	92

[a] mmole H$_2$/mmole Rh/min.

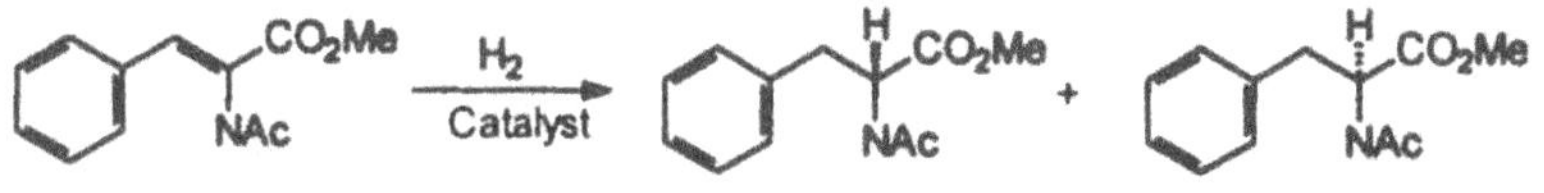

(2)

We have also used these anchored homogeneous catalysts for the hydrogenation of a number of other prochiral substrates such as methyl 2-acetamidocinnamate, 2-actamidocinnamic acid, dimethyl itaconate and 2-methyl-2-hexenoic acid. Fig. 5 shows the hydrogen uptake curves and product ee's for the hydrogenation of methyl 2-acetamidocinnamate (**2**) (Eqn. 2) run over Rh(Me-DuPHOS) anchored to PTA modified alumina. An anchored ruthenium catalyst containing a proprietary ligand has been used for the repeated hydrogenation of dimethyl itaconate (**3**) (Eqn. 3) with the reaction data shown in Fig. 6. No ruthenium was detected in these reaction mixtures.

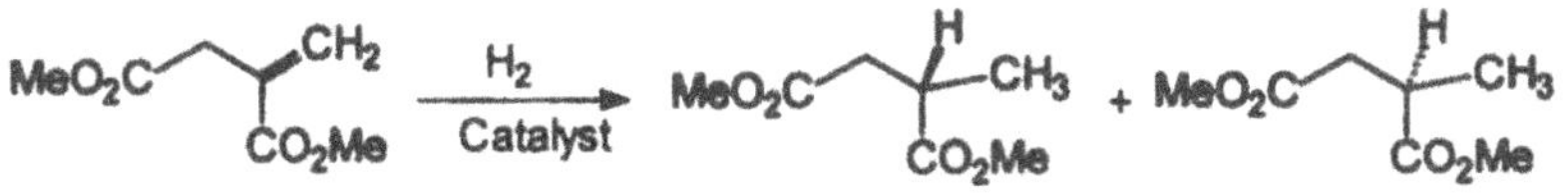

(3)

In addition to these chiral hydrogenations, a number of achiral complexes were also anchored to alumina and used for the hydrogenation of 1-hexene. A PTA anchored Wilkinson's catalyst has been successfully reused several times for this reaction with no detectable leaching of the catalyst. This catalyst was also used in a hexene hydrogenation with a substrate/catalyst ratio

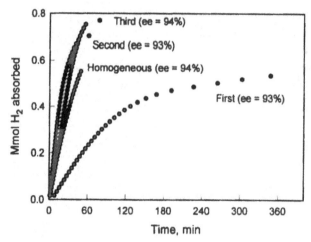

Fig. 5. Hydrogen uptake curves for the hydrogenation of 2 over
Rh(Duphos)/PTA/Al$_2$O$_3$.

of 12,000. The hydrogenations of four 5 ml portions of 1-hexene were run over
PTA anchored Rh(bppb), again, with the amount of rhodium in the reaction
mixtures below the level of detection which corresponds to less than 1 ppm.

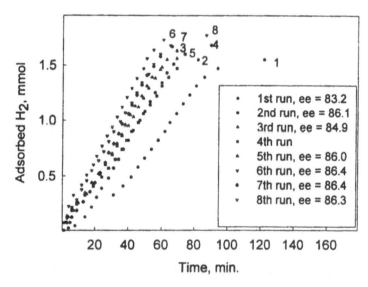

Fig. 6. Hydrogen uptake data for the hydrogenation of 3 over an anchored
Ru complex. Hydrogenations were run at 50°C and 55 psig.

While the general procedure for anchoring a complex is to stir a solution of the complex in a slurry of a modified support, we recognized that there would be times when it would be more desirable to add a ligand to an anchored complex precursor. We have successfully anchored Rh(COD)$_2$ and Ru(p-cymene) precursors to PTA modified alumina and have treated these anchored precursors with a number of different ligands to give anchored catalysts essentially indistinguishable from those prepared by anchoring the preformed complex directly. A more complete discussion of this approach is described in another paper in this volume (20).

Conclusions

It appears, then, that these heteropoly acid anchored homogeneous catalysts do meet the requirements listed in the introduction for a viable supported homogeneous catalyst. The preparation procedure used enables one to anchor a variety of preformed complexes so ligand modification is not involved. On re-use these catalysts are at least as active and selective as the homogeneous analogs frequently becoming more active and selective. While the lifetime of these catalysts in a flow system is only now being investigated, we have shown that they are stable to extensive re-use in a batch process as long as the activated catalyst is not exposed to air. Metal leaching was not observed.

Acknowledgements

The development of these catalysts was funded by grants CTS-9312533 and CTS-9708227 from the US National Science Foundation and financial assistance from the Engelhard Corporation.

References

1. I. Ojima, *Catalytic Asymmetric Synthesis*, VCH, New York, 1993.
2. T. Osawa, T. Harada and A. Tai, *Catal. Today*, 37, 465 (1997).
3. H.-U. Blaser, H.-P Jalett, M. Muller and M. Studer, *Catal. Today*, 37, 441 (1997).
4. D. C. Bailey and S. H. Langer, *Chem. Rev.*, 81, 109 (1981).
5. J. P. Collman, L. S. Hegedus, M. P. Cooke, J. R. Norton, G. Dolcetti, D. N. Marquardt, *J. Am. Chem. Soc.*, 94, 1789 (1972).
6. W. Dumont, J.-C. Poulin, T. P. Daud and H. B. Kagan, *J. Am. Chem. Soc.*, 95, 8295 (1973).
7. L. L. Murrell, in *Advanced Materials in Catalysis*, Academic Press, New York, 1977, Chapter 8.

8. V. Isaeva, A. Derouault and J. Barrault, *Bull. Soc. Chim., Fr.*, **133**, 351 (1996).

9. U. Nagel and J. Leipold, *Chem. Ber.*, **129**, 815 (1996).

10. S.K. Tanielyan and R.L. Augustine, U.S. Pat. 6,005,148 (to Seton Hall University); PCT Int. Appl., WO-9828074; *Chem. Abstr.*, **129**, 109217 (1998).

11. R. Augustine and S. Tanielyan, *Chem. Commun.*, 1257 (1999).

12. S. K. Tanielyan and R. L. Augustine, *Chem. Ind. (Dekker)*, **75** (Catal. Org. React.) 101 (1998).

13. Y. Izumi, K. Urabe and M. Onaka, *Zeolite, Clay and Heteropoly Acid in Organic Reactions*, VCH, New York, 1992, Chapter 3.

14. Y. Izumi, R. Hasere, K. Urabi, *J. Catal.*, **84**, 402 (1983).

15. Y. Lin, K. Nomiya and R.G. Finke, *Inorg. Chem.*, **32**, 6040 (1993).

16. M. Pohl, D. K. Lyon, N. Mizuno, K. Nomiya, R. G. Finke, *Inorg. Chem.*, **34**, 1413 (1995).

17. T. Nagata, M. Pohl, H. Weiner and R.G. Finke, *Inorg. Chem.*, **36**, 1366 (1997).

18. A.R. Seidle, W.B. Gleason, R.A. Mewmark, R.P. Skarjune, P.A. Lyon, C.G. Markell, K. O. Hodgson and A.L. Roe, *Inorg. Chem.*, **29**, 1667 (1990).

19. The MAS-nmr spectra were obtained by Xiaolin Yang, Engelhard Corp., Iselin, NJ, USA.

20. S. Anderson, H. Yang, S. K. Tanielyan and R. L. Augustine, This Volume.

Immobilization of Palladium Catalysts for Trost-Tsuji C-C and C-N Bond Formation: Which Method?

S. dos Santos,[b] J. Moineau,[a] G. Pozzi,[a] F. Quignard,[b] D. Sinou[a] and A. Choplin[b]

[a]Laboratoire de Synthèse Asymétrique, CNRS-UCBL,43 bd du 11 Novembre 1918, 69 622 Villeurbanne Cedex , France
[b]Institut de Recherches sur la Catalyse-CNRS, 2 Avenue A. Einstein, 69 626 Villeurbanne Cedex, France

Abstract

Three methods of immobilization of Pd(0) catalysts were tested for the Tsuji-Trost type reaction of C-C and C-N bond formation by allylic substitution of cinnamyl carbonates by active methylene compounds (ethyl acetoacetate and dimethyl malonate) and morpholine respectively. A comparison is made in terms of activity, selectivity and capacity of the catalytic systems to be recycled (stability, Pd leaching).

Introduction

Many efforts have been devoted to the immobilization of molecular catalysts during the last twenty years. It is still nowadays a major challenge, particularly with catalysts having poor activity, but excellent selectivity. Their recycling must then be made possible with conservation of their properties; this is most important where the metal and/or ligands are very expensive.

Two major fields have been rapidly growing (1): - heterogeneization of the molecular catalysts (the so-called "Supported Homogeneous Catalysts") by formation of a bond between an atom of a solid surface and either the metal center or an atom of one of the ligands, - immobilization in a two-phase liquid medium ("Biphasic Catalysis") (2-4). Both fields are now mature; their respective advantages and drawbacks are known, at least for a number of model reactions such as the hydroformylation of olefins, the hydrogenation of α,β unsaturated aldehydes... It is clear that the method of immobilization must be chosen in accordance with the mechanism of the target reaction and the properties of the reactants and products under the reaction conditions so as to minimize the loss of activity and/or selectivity, the metal leaching and the irreversible destruction of the active species to metal particles.

509

During the last five years, we focused on one of the most powerful reactions of carbon-carbon and carbon-heteroatom bond formations, the so-called Tsuji-Trost reaction (5). It is best catalyzed by palladium(0) stabilized by phosphine ligands in THF as the solvent. The mechanism implies two major steps (scheme) (6-9):

Scheme: simplified mechanism of allylic substitution

The first step involves the simultaneous dissociation of a phosphine and coordination of the olefin, which then adds oxidatively to Pd to give a π-allylic complex. Attack of the nucleophile on the π-allyl complex followed by dissociation of the alkylated product gives the desired compound and regenerates the active species. Because of this mechanism, the direct anchoring of the complex by a covalent bond between a surface atom of a solid support and the metal center cannot be used as a mean of immobilization of the catalyst. In accordance, all published heterogeneisation procedures are based on the covalent anchoring of the phosphine ligands. Two reports are typical of this approach: both concern the grafting of Pd(PPh$_3$)$_4$ on modified polystyrene (10-11). Bergbreiter et al., (11) who use a temperature dependent soluble polymer, claim an easy recycling of the catalyst as long as the system is maintained oxygen free.

Biphasic catalysis seems a priori well adapted to this mechanism. This method should introduce only changes in the catalytic properties directly related to the ligand modifications which are necessary to allow the solubilization of the complex in the medium of interest and to the diffusion of the reactants to the catalyst phase. Numerous variations of this concept can be envisaged, including heterogeneization of one phase as with the Supported Aqueous Phase Catalysts (12) or with the thermomorphic polymer supported catalysts (13).

Very few examples can be found in the literature which allow for a rigorous comparison of the precise advantages of one or the other method of immobilization, in terms of activity, selectivity and reuse capacity. We wish to report here the results of our research in this field, using as a test reaction the allylic substitution of cinnamyl carbonates by various nucleophiles i.e. ethyl acetoacetate, dimethyl malonate and morpholine. Three methods of immobilization of the palladium(0) active complex were tested, which are all based on the principles of biphasic catalysis. Thus, the active entity is solubilised either in water or in a perfluorinated solvent, non miscible to the organic solvent containing the reactants and the reaction products. The first approach uses a water-nitrile medium, the second a perfluorinated solvent-tetrahydrofuran mixture. The third method can be considered as the heterogeneization of the first medium: the aqueous phase containing the catalyst is supported as a thin film on a high surface area hydrophilic silica. Part of these results were recently published (14-16); here we wish to analyze all these data together.

Experimental Section

All experiments were carried out under nitrogen using Schlenk techniques. The water soluble phosphine TPPTS [(P(C$_6$H$_4$-m-SO$_3$Na)$_3$)], the silica 200 MP (185 m^{2}g^{-1}, pore diameter: 24 nm) and the perflurophosphine P[C$_6$H$_4$-p-(CH$_2$)$_2$C$_6$F$_{13}$]$_3$ were generous gifts from Rhône Poulenc, Grace and Pr. Dr. W. Leitner (Max Planck Institute, Muhlheim, Germany) respectively. The catalytic tests and the conditions of analysis of the products by chromatography were reported previously (14-16).

Results

Synthesis and characterization of the catalytic systems

The water soluble palladium(0) species, Pd(TPPTS)$_3$ was synthesized *in situ* by reacting Pd(OAc)$_2$ with TPPTS (5 equivalents). The reaction was monitored by ^{31}P NMR spectroscopy. The coordination of two phosphines to Pd(II) is immediate and leads to the formation of Pd(OAc)$_2$(TPPTS)$_2$, characterized by a ^{31}P NMR peak at 29 ppm. The reduction of Pd(II) to Pd(0) and the simultaneous oxidation of the phosphine is a slow process, which can be followed by the intensity increase of the peak at 34.6 ppm characteristic of OTPPTS and the intensity decrease of the peak at 29 ppm. The formation of Pd(TPPTS)$_3$ is evidenced by the appearance of a peak of growing intensity and decreasing width, shifting from –2 ppm to 19 ppm with time. This time dependent behavior can be attributed to a fast exchange between TPPTS coordinated to Pd(0) (in Pd(TPPTS)₃) and free TPPTS, on the basis of similar experiments performed at lower temperature or in ethylene glycol, both conditions slowing down the rate of the exchange process (15) (eq. 1):

$$Pd(OAc)_2 \xrightarrow[H_2O]{+5\ TPPTS} Pd(OAc)_2(TPPTS)_2$$
$$\rightarrow OTPPTS + Pd(TPPTS)_3 + TPPTS + 2AcO^- \quad (1)$$

This reaction is completed within 12 h at room temperature and within 1 h at 50°C.

The Supported Aqueous Phase Catalyst (SAPC) was prepared by the incipient wetness impregnation of a mesoporous silica (pre-evacuated at 200°C overnight) by an aqueous solution of Pd(OAc)₂/5 TPPTS. After evacuation to dryness, the SAPC contains typically 0.76% wt. Pd and ca 3% wt. H_2O. The ^{31}P NMR MAS spectrum of this solid confirms the presence of Pd(TPPTS)₃ as the main surface phosphine palladium species ($\delta= 22.2$ ppm). The mobility of this complex on the surface of silica increases with the amount of water; this is suggested by the observation of a pseudo-liquid phase spectrum as soon as the water amount is close to ca. 20% wt. The exchange process between non coordinated and palladium coordinated TPPTS, which was evidenced in solution, is detected on the solid at the higher water contents and as long as free TPPTS is present. A semi-quantitative analysis of the spectra reveals that approximately 80% of the palladium is reduced to Pd(0) in the SAP catalysts.

The solubilization of phosphinated complexes in perfluorinated solvents can be achieved by partial fluorination of the ligands (17-19). We have used a triarylphosphine, bearing a -(CH₂)₂(CF₂)₆F substituent. It has been demonstrated that this specific substitution pattern provides efficient protection of the catalytic center from steric and /or electronic influence of the perfluoroalkyl solubilizers (20). The catalytic Pd(0)(PAr*₃)₃ complex was prepared in situ by treating Pd₂(dba)₃ (dba = dibenzylideneacetone) by the perfluorinated phosphane (3 equiv./Pd) in perfluoromethylcyclohexane. Although we did not further characterize this system, one may reasonably assume that the substitution of the dba ligands by the phosphine is an easy process.

Catalytic properties of the Pd(OAc)$_2$/nPAr$_3$* systems for the allylic substitution reaction

The catalytic properties of the different systems were evaluated for the reaction of allylic substitution of cinnamyl carbonates by various nucleophiles (eq. 2), which leads to the formation of C-C or C-N bonds.

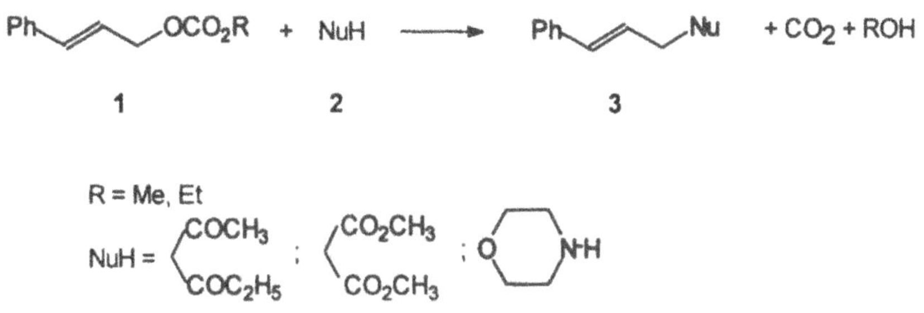

$$Ph\diagup\diagdown\diagup OCO_2R \ + \ NuH \ \longrightarrow \ Ph\diagup\diagdown\diagup Nu \ + CO_2 + ROH$$

1   2   3

R = Me, Et

$$NuH = \begin{cases} COCH_3 \\ COC_2H_5 \end{cases} ; \begin{cases} CO_2CH_3 \\ CO_2CH_3 \end{cases} ; \quad O\diagup\diagdown NH$$

(2)

We will first consider separately the behavior of the three nucleophiles, which differ significantly by their intrinsic reactivity towards the allylic carbonates and by their solubility in the different media tested.

with ethyl acetoacetate

Although the experimental conditions are not rigorously identical and although all the reaction parameters have not yet been fully optimized, some important remarks can be made when considering the data given in table 1. As expected, the activity of the catalytic system generated from Pd(OAc)$_2$ and 5 (or 6) equivalents of TPPTS drops considerably when the homogeneous reaction medium CH$_3$CN/H$_2$O is replaced by the biphasic medium PhCN/H$_2$O. A temperature increase from 25°C to 50°C is necessary in the latter case to achieve full carbonate conversion within 24 h. But a further temperature increase up to 80°C is detrimental for the selectivity of the reaction and for the stability of the catalyst, in both media. Then, large amounts of cinnamyl alcohol are formed and black metal particles precipitate, a phenomenon which hampers definitely the reuse of the catalytic system.

These drawbacks can be overcome when the catalytic system is immobilized on silica using the SAP methodology. The water content of the SAP catalyst is a crucial parameter since its increase is beneficial to the activity (Fig. 1). Nevertheless, this content must be lower than 50% wt., a value above which major problems of diffusion are sharply suspected given the poor dispersion of the solid in the reaction medium. The simple scheme proposed by Davis *et al.* (12) for the working mode of these catalysts can explain the following observations: in CH_3CN, the activity of SAPC is lower than that of the equivalent homogeneous catalytic system, but in PhCN, its activity is higher as soon as the water content of the solid exceeds 20% wt. In the latter case, the desired monoalkylated product is obtained within 1 hour with a high selectivity (95 %). These data may be simply interpreted on the basis of a decrease of the interphase surface area through the presence of silica in CH_3CN and an increase in PhCN up to a value close to the specific surface area of silica.

Table 1 Influence of the immobilization procedure on the catalytic properties of $Pd(OAc)_2/$ 5 TPPTS for reaction (2), R = Et, NuH = ethyl acetoacetate.

Medium	T (°C)	t	conv. (%)	yield (%)[c]	sel. (%)	[1][c]
CH_3CN/H_2O[a]	25	24 h	100	80[b]	100	1
"[a]	50	12 h	100	88[b]	88	1
"	80	8 h	57	28[c]	50	0.06
$PhCN/H_2O$[a]	25	24 h	30	30	100	1
"[a]	50	24 h	100	94	100	1
"	80	4 h	38	3[e]	8	0.06
CH_3CN/SAP[f]	80	8 h	38	38	100	0.05
$PhCN/ SAP$[f]	80	50 min	100	95	95	0.05
$C_7F_{14}/$[THFb, g]	50	15min	100	100	100	0.11

Exp. cond. : [carbonate]/[nucleophile]/[Pd]/[P] = 30/25/1/5 except for ^{a} = 30/25/1/6 and ^{b} = 30/20/1/3. ^{c}: expressed in mol.L^{-1} ; ^{d}: determined by GC ; ^{e}: secondary product is cinnamyl alcohol ; ^{f}: water amount of SAP catalyst : 32 % wt. ^{g}: catalyst is Pd$_2$dba$_3$/P(C$_6$H$_4$-p-(CH$_2$)$_2$C$_6$F$_{13}$)$_3$.

The activity and selectivity for the monoalkylated product of the perfluorinated catalyst are high. This may be explained by the fact that the mutual miscibility of perfluorocarbon and THF becomes high under the reaction conditions and perhaps by the differences in electronic and steric effects induced by the perfluorinated ligands. Interestingly, the catalytic system could be

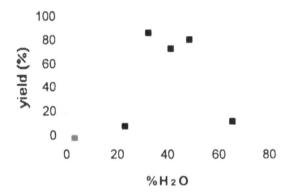

Figure 1 Influence of the water amount on the catalytic properties of the SAP catalyst for the reaction of cinnamyl ethyl carbonate with ethyl acetoacetate in PhCN (for the experimental conditions, see table 1). t_R = 40 min.

recycled 8 times, without significant loss of activity, by simple decantation at 0°C. After that, a sharp decrease of the activity was observed which, although not yet fully understood, may be simply correlated to accidental introduction of O$_2$, known to dissolve well in the perfluorinated solvents.

with morpholine

The reactivity of morpholine towards ethyl cinnamyl carbonate is very high under homogeneous conditions (Table 2). Thus, in CH$_3$CN/H$_2$O and THF/C$_7$F$_{14}$, full conversion is achieved within 5 to 15 minutes at 50°C; the small variations in activity may be attributed to small differences of the [substrate]/[Pd] ratios. Under biphasic conditions, i.e. in PhCN/H$_2$O, the reaction is slower despite the fact that morpholine is water soluble. Yet, the selectivities for the alkylated product are in all cases close to 100% and no palladium particles are detected *de visu*. This may be explained by the lower temperature used in these experiments when compared to that necessary in the case of ethyl acetoacetate.

The silica supported catalyst is most efficient when used in PhCN as the organic solvent. Its water content is again an important parameter: it is optimal for 30-40% wt. H_2O. Then, the SAP catalyst is as efficient as the related homogeneous catalyst used in CH_3CN/H_2O. But it presents a major advantage which is its capacity to be used in a continuous flow process. To this end, the problem of water leaching, which is the main source of activity loss (15), had to be solved. Since Pd leaching is negligibly small (< limit of detection: 1ppm, which corresponds to 0.4% of the initial amount Pd) we found that, in the present case, the utilization of water saturated benzonitrile was sufficient to maintain constant the activity of the SAPC during 12 h; a productivity of 2200 moles of carbonate transformed per mole of palladium was obtained (21).

Table 2 Influence of the immobilization procedure on the catalytic properties of $Pd(OAc)_2$/5 TPPTS for reaction (2). R = Et, NuH = morpholine.

Medium	t (min)	conv. (%)	yield %	sel. (%)	[1]
CH_3CN/H_2O	5	95	92	97	0.036
$PhCN/H_2O$	360	70	67	97	0.036
CH_3CN/SAP^c	20	100	95	95	0.03
$PhCN/ SAP^c$	5	100	100	100	0.03
PhCN/SAPa	15	100	100	100	0.03
$THF/C_7F_{14}^{b,d}$	15	100	100	100	0.11

Exp. cond. : T = 50°C ; [nucleophile]/[carbonate]/[Pd]/[P] = 30/25/1/5 except for [a] = 240/200/1/5 and [b] = 120/80/1/3; [c]: % H_2O of SAP catalyst : 49 % wt. ; [d]: catalyst is Pd_2dba_3/3P$(C_6H_4$-p-$(CH_2)_2C_6F_{13})_3$, R = Me.

The fluorinated system could also be recycled, after simple decantation at 0°C. Thus up to five quantitative recyclings were readily performed using palladium loadings as low as 1 mol %. As in the case of ethyl acetoacetate, a significant drop of activity was observed afterwards, most probably for the reasons given previously. We are currently investigating the degree of leaching of palladium and/or of the fluorous solvent into THF.

with dimethyl malonate

The case of dimethyl malonate is the most interesting. Indeed, this substrate is alkylated easily in acetonitrile with a high reaction rate and a high selectivity to the monoalkylated product (Table 3). However, when a water based biphasic medium is used, the performance of the related catalytic system is poor: the desired reaction does not occur at all. This was attributed to the higher pKa of dimethyl malonate as compared to that of ethyl acetoacetate; the presence of water is thus disadvantageous for the formation of the enolate. But, addition of a strong base such as DBU allows for the formation of the monoalkylated product, but also of cinnamyl alcohol (22).

When supported on silica, the same catalytic system works well without the addition of any base; a high selectivity to the monoalkylated product is achieved. This may be related to the dramatic shifts of the following equilibria, when the water amounts involved in typical experiments vary from 7 mL (biphasic medium) to 250 μL (SAP conditions):

$$RO^- + H_2O \rightleftharpoons ROH + OH^- \qquad (3)$$

$$OH^- + NuH \rightleftharpoons Nu^- + H_2O \qquad (4)$$

$$RO^- + NuH \rightleftharpoons Nu^- + ROH \qquad (5)$$

According to the pK_a scale, equilibria (4) and (5) should lie well over to the right (23). In the absence of water, the alcoholate RO^- resulting from the formation of the allylic complex (scheme) deprotonates dimethyl malonate; this is in agreement with the data obtained in CH_3CN. With large amounts of water, equilibrium (3) takes place; it is shifted to the right while equilibrium (4) should be shifted to the left. As nucleophilic substitution by OH^- at palladium has so far never been reported, it is reasonable to assume that the hydroxide anion catalyzes only the hydrolysis of cinnamyl ethyl carbonate to cinnamic alcohol, a reaction favoured by a high temperature.

Table 3 Influence of the immobilization procedure on the catalytic properties of $Pd(OAc)_2/$ 5 TPPTS for reaction (2), R = Et, NuH = dimethyl malonate.

Medium	T (°C)	t	conv. (%)	yield (%)	sel. (%)	[1]
CH_3CN^c	80	5 min	100	95	95	0.5
CH_3CN/H_2O^a	50	24 h	0	0	0	0.5
$CH_3CN/^{H2Oa, d}$	50	24 h	100	67	67	0.5
$PhCN/H_2O$	80	24 h	44	0	0^g	0.05
$PhCN/ SAP^e$	80	8 h	80	70	87	0.05
$THF / C_7F_{14}^{b, f}$	50	15 min	100	100	100	0.11

Exp. cond.: [nucleophile]/[carbonate]/[Pd]/[P] = 30/25/1/5 except for: [a] =30/25/1/9 and [b] = 60/40/1/3.; [c]: $Pd(OAc)_2/5$ PPh_3; [d]: with added DBU (1 equiv.); [e]: % H_2O: 48% wt. ; [f]: $Pd_2dba_3/3P(C_6H_4$-p-$(CH_2)_2C_6F_{13})_3$; [g]: cinnamyl alcohol is the only detected product.

The perfluorinated medium is in the case of dimethyl malonate the most efficient, already at moderate temperatures: the related catalyst is highly active and selective. It could be recycled five times without significant loss of all its properties.

Discussion and Conclusions

The immobilization of homogeneous catalytic systems becomes necessary when the recycling of the catalysts and/or their use in a continuous process are the priorities. We have demonstrated for an important reaction such as the Trost-Tsuji reaction that this can be successfully achieved by taking advantage of either biphasic or supported aqueous phase strategies. The Trost –Tsuji reaction, widely used by organic chemists, is efficiently catalysed by Pd(PPh₃)₄ or palladium (0) complexes stabilized by ligands adapted to the target selectivity (enantioselectivity for example). Unfortunately, optimization of the reaction conditions has often not been performed and most of the reported data concern systems tested for low ratios of [substrate]/[Pd] (typically comprised between 10 and 50) and with long reaction time (typically 24h) to ensure completion of the reaction.

Our approach allows one to compare for the first time the behavior of catalysts, differing only by the substituents of the phosphine ligands; these modifications are necessary to achieve solubilization of the active entity in the chosen medium (water of fluorocarbon). The generalization of these methods to other reactants or reactions relies thus on the availability of such modified phosphines. Their panel is increasing very fast in both cases (4, 17-20), even if the synthesis are not always straightforward, particularly in the case of chiral ligands.

Although both solvents are given as non-toxic, water is the most environmentally and human friendly; unfortunately, it is not neutral towards all reactions. The case of the allylic alkylation of dimethyl malonate illustrates perfectly this statement; water clearly inhibits the formation of the target monoalkylated product under biphasic conditions. On the contrary, the FBS (Fluorous Biphase System) is very efficient and selective for this reaction; the catalytic phase can be reused for several times, after decantation at room temperature. Studies are currently under investigation to replace THF by a solvent non miscible to the perfluorinated solvent even at reaction temperature (T>50°C) to allow for continuous separation of the products from the catalyst. At present, the SAP (Supported Aqueous Phase) immobilization is the most appealing method: the stability of these catalysts towards palladium leaching allowed its use in a continuous flow process and therefore a important increase of the productivity of the catalyst.

Yet, it is still very difficult to choose *a priori* the method of immobilization of a given catalytic system; this is exemplified by the data recently published concerning the immobilization of similar palladium catalysts by the SAP methodology and their properties for the closely related Heck reaction; their capacity to being recycled is reported as poor (24). This latter result may be correlated to the complexity of the catalytic system which implies *inter alia* the presence of a base. It is evident that future work in the field of catalysis and fine chemicals implies first the optimization of the catalytic homogeneous systems and a better understanding of the role of each actor of the reaction before heterogeneisation can be safely envisaged.

References

1.   see for example : A. Choplin and F. Quignard, *Coord. Chem. Rev.* **178-180**, 1679 (1998) and references therein.
2.   W.A. Herrmann and C. Kohlpainter, *Angew. Chem. Intern. Ed. Engl.* **32**, 1524 (1993).
3.   F. Joo and A. Katho, *J. Mol. Catal. A : Chemical* **116**, 3 (1997).

4.   B. Cornils, *Aqueous Phase Organometallic Catalysis. Concepts and Applications*, VCH, Weinheim, 1996.

5.   J. Tsuji, H. Takahashi and M. Morikawa, *Tetrahedron Lett.* 4387 (1965)

6.   J. Tsuji, *Palladium Reagents and Catalysts*, Wiley and Sons, New York, (1995).

7.   G.C. Frost, J. Horwarth and J.J.M. Williams, *Tetrahedron : Asymmetry* 3, 1089 (1992).

8.   B.M. Trost, *Acc. Chem Res.* 29, 355 (1996).

9.   B.M. Trost and D.L. van Vranken, *Chem. Rev.* 96, 395 (1996).

10.   B.M. Trost and E. Keinan, *J. Amer. Chem. Soc.* 100, 7779 (1978).

11.   D.E. Bergbreiter and D.A. Weatherford, *J. Org. Chem.* 54, 2726 (1989).

12.   J.P. Arhancet, M.E. Davis, J.S. Merola and B.E. Hanson, *Nature*, 339 (1989)

13.   D.E. Bergbreiter, Y.S. Liu and P.L. Osburn, *J. Amer. Chem. Soc.* 120, 4250 (1998).

14.   E. Blart, J.P. Genêt, M. Safi, M. Savignac and D. Sinou, *Tetrahedron* 50, 505 (1994).

15.   S. dos Santos, Y. Tong, F. Quignard, A. Choplin, D. Sinou and J.P. Dutasta, *Organometallics* 17, 78 (1998).

16.   R. Kling, D. Sinou, G. Pozzi, A. Choplin, F. Quignard, S. Busch, S. Kainz, D. Koch and W. Leitner, *Tetrahedron Lett.* 39, 9439 (1998).

17.   I.T. Horvath and J. Rabai, *Science*, 266, 72 (1994).

18.   I T. Horvath, *Acc. Chem. Res.* 31, 641 (1998).

19.   M. Cavazzini, F. Montanari, G. Pozzi and S. Quici, *J. Fluorine Chem.* 94, 183 (1999).

20.   S. Kainz, D. Koch, W. Beumann and W. Leitner, *Angew. Chem. Int. Ed. Engl.* 36, 1628 (1997).

21.   S. dos Santos, F. Quignard, A. Choplin and D. Sinou, (*to be published*).

22.   S. Sigismondi and D. Sinou, *J. Mol. Catal.* 116, 289 (1997).

23.   J March, *Advanced Organic Chemistry, Third Edition* Wiley, New York, 1985, pp. 220.

24.   L. Tonks, M.C. Anson, K. Hellgardt, A.R. Mirza, D.F. Thompson and J.M.J. Williams, *Tetrahedron Lett.* 38, 4319 (1997).

Two-phase Hydroaminomethylation of Olefins with Ammonia

B. Zimmermann[a], J. Herwig[b], M. Beller[c]*

[a] *Technische Universität München, Lichtenbergstr. 4, D-85747 Garching*
[b] *Celanese GmbH – Werk Ruhrchemie, D-46128 Oberhausen*
[c] *Institut für Organische Katalyseforschung an der Universität Rostock, Buchbinderstr. 5-6, D-18055 Rostock*

Abstract

We describe the selective hydroaminomethylation of aliphatic olefins with ammonia to form linear primary and secondary aliphatic amines. A dual metal catalytic system based on rhodium and iridium is used in combination with a two-phase reaction system which leads to easily adjustable selectivities. In addition, for the first time the hydroaminomethylation of styrene with ammonia giving pharmaceutically interesting primary amines is shown. It is possible to recycle the aqueous catalyst phase with no significant loss of activity and selectivity, respectively, which is important for the industrial application of this reaction.

Introduction

Linear aliphatic amines are important bulk and fine chemicals for the chemical and pharmaceutical industry (1). From both economic and scientific points of view there exist considerable interest in developing versatile and direct preparation routes from economically attractive feedstock. In this context hydroamination (2-5) and hydroaminomethylation (6-19) of olefins to amines represent atom-economic and efficient syntheses of this class of compounds. While the direct hydroamination of olefins still remains one of the important challenges in catalysis, the hydroaminomethylation seems to be closer to industrial realization. As shown in Scheme 1 the hydroaminomethylation has to be regarded as a domino reaction consisting of an initial hydroformylation of an olefin to an aldehyde and a subsequent reductive amination.

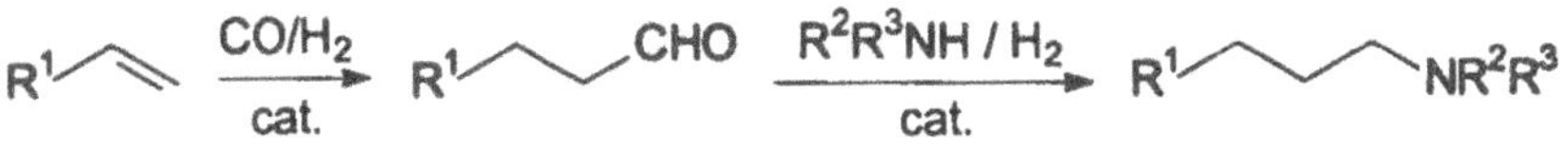

$$R^1 \diagup\!\!\diagdown \xrightarrow[\text{cat.}]{CO/H_2} R^1 \diagup\!\!\diagdown\!\!\diagup CHO \xrightarrow[\text{cat.}]{R^2R^3NH\,/\,H_2} R^1 \diagup\!\!\diagdown\!\!\diagup\!\!\diagdown NR^2R^3$$

Scheme 1: Hydroaminomethylation of olefins (R^1-R^3=H, alkyl).

521

The hydroaminomethylation reaction was originally discovered by Reppe at BASF and described first in 1949 (6). Here, the reaction of an olefin, ammonia, water and carbon monoxide in the presence of $Fe(CO)_5$ yielded mixtures of primary, secondary and tertiary amines. Since its discovery other catalyst systems have been shown to catalyze hydroaminomethylation reactions (Table 1).

As shown in Table 1 the industrial interest in this reaction is obvious. Hence, considerable amount of work in this field is found in patents (11, 13, 18). Apart

Table 1. Hydroaminomethylation of olefins with different catalysts.

year	Author (ref.)	company	catalyst	conditions	products			NH-compound
					prim	sec	tert	
1949	Reppe (6)	BASF	$Fe(CO)_5$	150°C - 200 bar CO/H_2O	(+)	+	+	NH_3, RNH_2, R_2NH
1971	Iqubal (7)	Monsanto	Rh_2O_3	170°C - 140 bar CO/H_2O	-	-	+	R_2NH
1980	Laine (8)	SRI Calif.	Rh/Fe Fe/Ru	150°C - 60 bar CO/H_2O	-	-	+	R_2NH
1982	Jachimo-wicz (10)	Grace	Rh(I) / Rh_2O_3	140°C - 70 bar CO/H_2O	(+)	-	+	NH_3, R_2NH
1985	Knifton (11)	Texaco	Ru	170°C - 100 bar CO/H_2	-	+	(+)	RNH_2
1987	Knifton (26,27)	Texaco	$Co_2(CO)_8$/ PR_3	200°C - 140 bar CO/H_2	+	(+)	-	NH_3
1989	Jones (17)	BP	Rh(I)	115°C - 60 bar CO/H_2	-	+	-	RNH_2
1992	Kalck (15,16)	ENS Toulouse	$Rh(I)/PPh_3$	80°C - 10 bar CO/H_2	-	+	+	RNH_2, R_2NH
1992	Drent (13)	Shell	Ru/NR_3 $Rh/Ru/NR_3$	200°C - 100 bar CO/H_2; CO/H_2O	-	+	(+)	NH_3, RNH_2, R_2NH
1995	Diekhaus (18)	Hoechst	Rh/Ru	CO/H_2	-	+	+	RNH_2, R_2NH
since 1997	Eilbracht (20-23)	University Dortmund	Rh(I)	135°C - 60 bar CO/H_2	-	+	+	NH_3, RNH_2, R_2NH
1999	Keim (19)	University Aachen	Ru	120°C - 45 bar CO/H_2	(+)	+	+	NH_3, RNH_2, R_2NH

from the typical hydroformylation catalysts such as cobalt and rhodium phosphine systems, also ruthenium catalysts and mixed metal systems (Fe/Rh (7, 8); Fe/Ru (8)) were applied in the hydroaminomethylation. The early catalyst systems used water as hydrogen source (watergas shift reaction) whereas syn gas (H_2/CO) was throughout used for the reaction. Except for some rhodium catalysts (15-17) the reaction temperature usually was fairly high (> 140°C) and activities in general were quite low for most of the homogeneous catalysts shown in Table 1. Since only Knifton investigated ammonia as nitrogen source to yield primary amines (26, 27), secondary and tertiary amines are the preferred products of the hydroaminomethylation obtained with the listed catalysts. Whereas typical selectivities for tertiary amines reach 100% at high yields (20), secondary amines were obtained with 98% selectivity and 60-99% yield (e.g. 17,23). In general activities of the catalysts are low and usually do not exceed a turnover frequency of 100 h^{-1} for the homogeneous reactions.

In spite of the advantages of hydroaminomethylation, for example availability of starting materials and atom efficiency, comparatively few applications in organic synthesis are known. Only very recently the synthetic potential of this domino reaction was demonstrated elegantly by Eilbracht et al. (20-23) and others (24, 25).

Due to reasons of selectivity primary and secondary amines were almost exclusively used as amine components in hydroaminomethylations. The synthesis of industrially more important primary amines from ammonia and olefins has only been investigated by Knifton et al. (26, 27) and very recently by us (28).

Knifton and co-workers used $Co_2(CO)_8$ in the presence of basic phosphines at 200°C and 140 bar pressure (26, 27). While selectivities to the primary amine were good for low conversions, at higher conversion (85%) only small amounts (32%) of various primary amines were obtained. However, most problematic for an industrial realization were the extremely low activities of the parent catalyst system (turnover frequency TOF = 9 h^{-1}).

In order to use the hydroaminomethylation reaction of olefins with ammonia on an industrial scale obviously there is a need a) to increase the unsatisfactory n:iso selectivity of the hydroformylation step, b) to increase the selectivity towards the primary amine at high conversions, c) to increase the catalyst activity, and d) to develop an efficient catalyst recycling with the known homogeneous catalyst systems.

Our approach (28) to solve these problems was twofold: 1) In order to increase the rate of product formation especially the hydrogenation of the corresponding imine to the amine must be faster. Here, addition of an active imine hydrogenation catalyst (Ir) should be favorable. 2) In order to simplify catalyst separation and to control the selectivity with respect to the formation of primary, secondary, and tertiary amines (29) the use of the principle of two-phase catalysis should lead to better results (30-33).

The use of dual Rh/Ir catalyst systems for the first time for hydroaminomethylation stemmed from the idea that the current catalysts based on rhodium and cobalt are in the presence of an excess of ligand no longer sufficiently active for the hydrogenation of the C-N double bond. On the other hand iridium is known to be a very active catalyst for the hydrogenation of C-N double bonds (34-37).

Experimental Section

Two-phase hydroaminomethylation of olefins is typically carried out in the following manner: 35.7 mmol olefin, 2.3 mg [Rh(cod)Cl]$_2$ (4.6 μmol, 0.026 mol-%), 25.0 mg [Ir(cod)Cl]$_2$ (0.037 mmol, 0.21 mol-%), 7.2 g aqueous TPPTS solution (0.55 mol/kg, 3.96 mmol, P:Rh = 425, P:Ir = 52), 21.5 ml aqueous NH$_3$ (25%, 288 mmol, olefin:NH$_3$ = 1:8), degassed methyl-tert.-butyl ether (MTBE, 25 ml), and degassed isooctane (internal GC standard, 2.5 ml) were placed under an inert gas atmosphere into a 200-ml stainless steel autoclave with a glass liner, baffle unit, and magnetic stirrer. The autoclave was sealed, flushed once with synthesis gas, pressurized to 13 bar CO and 65 bar H$_2$ (total pressure (cold): 78 bar, CO:H$_2$ = 1:5), and a reaction temperature of 130°C was set (total pressure (warm): 120 bar). After a reaction time of 10 h the autoclave was cooled to 0 °C and slowly depressurized. The organic phase was analyzed by gas chromatography (HP5860 Series, column: Optima-1 Macherey-Nagel, 50 m, 0.23 mm, 5.0μm) (olefin quantification), and the aqueous phase was separated, treated with 0.5 g NaOH, and extracted with MTBE (2x 2mL). The combined organic phases were analyzed by gas chromatography (amine quantification).

For lower ammonia concentrations the proportionate amounts of standard solutions were used and the volume of the aqueous phase increased to 20 mL with degassed water.

The liquefied gases (propene and butene) were first condensed into a steel cylinder and then introduced into the autoclave under pressure.

Styrene was used with no further treatment and purification of the reaction product was achieved by column chromatography (silica, ethylacetate:triethylamine = 96:4, R$_f$ = 0.25). Reaction conditions for styrene: p(cold) = 78bar, CO:H$_2$ = 1:5, NH$_3$:olefin = 8:1, t = 10h, 0.052 mol-% Rh, 0.21 mol-% Ir, ligand = TPPTS (1), P/Rh = 106, P/Ir = 26.

All products were unambiguously identified by GC/MS. After purification of the products by distillation or column chromatography NMR spectra consistent with literature were obtained.

BINAS was made from NAPHOS by sulfonation according to Herrmann et al.(38).

Results and Discussion

Initial studies (28) to verify our concept were performed using 1-pentene as olefin. Indeed the hydroaminomethylation of 1-pentene with synthesis gas ($CO:H_2$ = 1:5) and ammonia in the presence of an Rh/Ir/TPPTS catalyst system under standard hydroformylation reaction conditions (130°C, 120 bar) in an aqueous two-phase system gives amines in 79% yield (Table 2). While only small amounts of the secondary hexylamines di-n-hexylamine, di(2-methylpentyl)amine, and N-n-hexyl-N-(2-methylpentyl)amine were formed, the major reaction products are primary hexylamines. Only trace amounts of tertiary amines, imines, enamines, and higher boiling aldol condensation products were observed. Reduction of the reaction temperature to 110 °C led to somewhat lower overall yields and to moderately increased selectivities with respect to the primary amine (Table 2, entry 2). At temperatures below 100°C the hydroformylation reaction becomes very slow and a dramatic decrease of conversion is observed. At higher temperatures above 130°C (Table 2, entry 5) ligand degradation and metal leaching occurred. Furthermore, the extent of isomerization of the terminal olefin to the less reactive internal olefin increases with temperature.

Table 2. Hydroaminomethylation of 1-pentene:[a] variation of temperature.

entry	T [°C]	Conv. [%]	yield [%] 2-pentene	amines	n/i prim.	prim./sec. amine
1	90	<5	3	-	-	-
2	110	87	12	72	86:14	86:14
3	120	91	14	77	86:14	83:17
4	130	95	16	79	84:16	81:19
5[b]	140	98	27	71	85:15	81:19
6[c]	130	90	12	<5%	-	-

[a]Conditions: p(cold) = 78bar, $CO:H_2$ = 1:5, NH_3:olefin = 8:1, t = 10h, 0.026 mol-% Rh, 0.21 mol-% Ir, ligand = TPPTS (1), P/Rh = 425, P/Ir = 52.
[b]Significant catalyst/ligand degradation and leaching.
[c]Experiment performed without iridium.

Applying the new dual metal catalyst system significantly better selectivities were obtained for the primary amines (91 versus 32 %) in comparison to the results of Knifton (26, 27). If the experiment was performed with rhodium only and without iridium as selective imine hydrogenation catalyst very low yields were obtained and mainly imines as well as other high boiling side products were observed (Table 2 entry 6). This result shows the correctness of our assumptions: the fast and selective hydroformylation but slow hydrogenation catalyst rhodium is combined with a fast and selective hydrogenation catalyst (iridium) in order to establish an effective reaction from olefin to amine.

Table 3. Hydroaminomethylation of 1-pentene:[a] variation of CO/H_2-ratio.

entry	CO/H_2	conv. [%]	yield [%] 2-pentene	amines	pentane	n/i prim.	prim./sec. Amine
1	1:1	95	10	39	4	78:22	76:24
2	1:2	95	10	71	4	80:20	89:11
3	1:5	94	11	75	7	84:16	90:10

[a]Conditions: p(cold) = 78bar, T=125°C, NH_3:olefin = 8:1, t = 10h, 0.026 mol-% Rh, 0.21 mol-% Ir, ligand = TPPTS (1), P/Rh = 425, P/Ir = 52.

For the reaction a stoichiometric amount of 2 moles hydrogen and 1 mole carbon monoxide is needed for one equivalent of olefin. As shown in Table 3 best yields of amines are obtained if an excess of hydrogen is used (entry 3). However hydrogenation of the olefin rises slightly. If understoichiometric amounts of hydrogen are applied the yield in amines drops significantly from 70 to 39 % (Table 3, entry 1).

Although the primary amine is more nucleophilic compared to ammonia and thus reacts faster with the aldehyde, both primary and secondary amines can be formed with greater than 90% selectivity by variation of the olefin/ammonia ratio (Table 4, entry 1 and 4). Already in the presence of 2 equivalents of ammonia the primary amine is formed preferentially (prim./sec. = 72:28).

Table 4. Hydroaminomethylation of 1-pentene:[a] selective synthesis of primary and secondary amines.

entry	NH_3: olefin	conv. [%]	yield [%] amines	n/iso prim.	prim./sec.
1	8	84	69	84:16	91:9
2	4	90	56	87:13	86:14
3	2	90	51	87:13	72:28
4	0,5	95	60	84:16	6:94

[a]Conditions: T = 130°C, p(cold) = 78bar, CO:H_2 = 1:5, t = 10h, 0.026 mol-% Rh, 0.21 mol-% Ir, ligand = TPPTS (1), P/Rh = 425, P/Ir = 52.

In the presence of 0.5 equivalents ammonia per equivalent of olefin the aldehyde reacts with ammonia and subsequently with the newly formed primary amine, leading to a preferential formation of the secondary amine (Scheme 2).

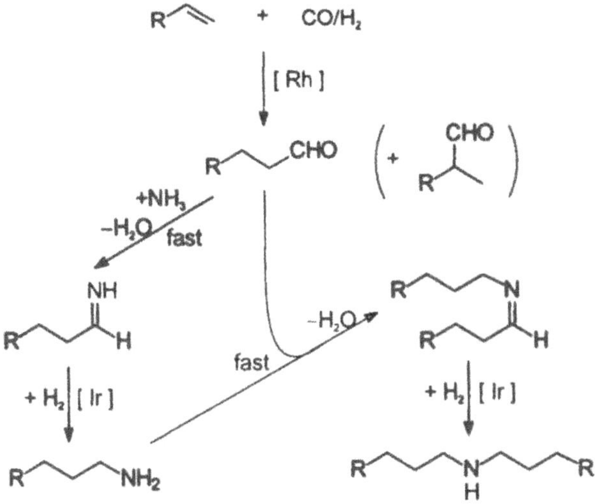

$R\diagdown\diagup$ + CO/H_2

[Rh]

$R\diagdown\diagup CHO$ (+ $R\diagdown CHO$)

+NH_3
$-H_2O$ fast

$-H_2O$ fast

NH
$R\diagdown\diagup\diagup H$

$R\diagdown\diagup\diagup N\diagdown$
$R\diagdown\diagup\diagup H$

$+ H_2$ [Ir]

$+ H_2$ [Ir]

$R\diagdown\diagup\diagup NH_2$

$R\diagdown\diagup\diagup N\diagdown\diagup R$
H

Scheme 2: Formation of primary and secondary amines by hydroamino-methylation of olefins.

We thought that the competitive reactions of ammonia and the primary amine with the aldehyde are controlled by the actual concentration of both NH compounds in the hydrophilic phase because the hydroformylation step takes place there. Depending on the two-phase system especially the concentration of the primary amine in the hydrophilic phase will vary. Clearly this should offer an additional selectivity control and thus allow higher selectivities for the primary amine by choosing the right combination of a hydrophilic and hydrophobic solvent. It emerged, that by lowering the polarity of the organic phase the selectivity with respect to the primary amine increased (Table 5, entry 1-3). Due to an extraction of the primary amine from the aqueous phase it does no longer compete with ammonia in the aqueous catalyst phase during imine formation. The lower the polarity of the organic solvent, the better the hydrophobic primary amine is extracted in comparison to ammonia; this leads to an improvement in selectivity! A similar behavior is observed when the polarity difference between the aqueous and organic phase is increased by addition of salt to the water phase (Table 5, entry 4). *These results make clear that a two-phase system does not only permit an easier catalyst recycling (see below), but even more interesting allows a selectivity control of the formation of primary amines versus secondary amines.* We believe that this inherent advantage of two-phase catalysis compared to homogenous catalysis will be much more exploited in the future for other reactions, too (29).

Table 5. Hydroaminomethylation of 1-pentene:[a] variation of organic solvent.

entry	organic solvent	conv. [%]	yield [%] 2-pentene	amines	n/i prim.	prim./sec.
1	MTBE	98	21	78	87:13	69:31
2	Anisole	97	20	76	88:12	78:22
3	Toluene	96	26	69	88:12	82:18
4	MTBE/ salt[b]	96	23	68	96:4	76:24

[a]Conditions: T = 130°C, p(cold) = 78bar, CO:H$_2$ = 1:5, NH$_3$:olefin = 8:1, t = 10h, 0.026 mol-% Rh, 0.21 mol-% Ir, ligand = TPPTS (1), P/Rh = 425, P/Ir = 52.
[b]Aqueous phase: 15% Na$_2$SO$_4$ solution.

Apart from 1-pentene also industrially important olefins such as propene, 1-butene (Table 6) and styrene react similarly with ammonia in the two-phase system. In all cases the corresponding amines were obtained in high yields and selectivities. Nevertheless, a significant effect of the olefin chain length on the selectivity of primary amine formation is observed. The amount of primary amine decreases with decreasing olefin chain length. This correlates with the increased water solubility of the primary amine formed. Due to a higher concentration of the primary amine in the water phase and a lower extraction effect into the organic solvent higher yields of secondary amines are observed.

Ar = C$_6$H$_4$-m-SO$_3$Na

TPPTS (1)   **BINAS (2)**

So far we focussed our attention mainly to the important selectivity question of primary versus secondary amine formation. Besides, the regioselectivity of the amine formation is also of importance. It is interesting to note that the n:iso selectivities (76:24 to 87:13) of the primary amines obtained in the presence of the ligand TPPTS (1) are slightly lower than those obtained for similar two-phase hydroformylations. Hence, the presence of ammonia or amines led to an increase of branched aldehydes, which might be explained by a replacement of phosphine ligands on the central rhodium atom by sterically less hindered ammonia.

Table 6. Hydroaminomethylation: variation of olefins and ligands.

olefin	NH$_3$: olefin	ligand: TPPTS			ligand: BINAS		
		yield (amine)	n/iso	prim.: sec.	yield (amine)	n/iso	prim.: sec.
Propene	8	80 %	76:24	72:28	90 %	99:1	77:23
1-butene	8	80 %	85:15	78:22	85 %	99:1	78:22
1-pentene	8	75 %	84:16	90:10	75 %	99:1	87:13
Propene	0,5	90 %	82:18	20:80	95 %	99:1	1:99
1-butene	0,5	90 %	87:13	38:62	95 %	99:1	1:99
1-pentene	0,5	85 %	82:18	48:52	90 %	99:1	10:90

^{a}Conditions: T = 130°C, p(cold) = 78 bar (pentene/butene), 60 bar (propene), CO:H$_2$ = 1:5, t = 10h, 0.026 mol-% Rh, 0.21 mol-% Ir, TPPTS (1): P/Rh = 425, P/Ir = 52, BINAS (2): P/Rh = 140, P/Ir = 18.

In general high regioselectivities are not absolutely necessary for industrial applications of the hydroaminomethylation since branched primary products are also of commercial importance. Nevertheless, it is of fundamental interest to control also the regioselectivity of this reaction. It is well known that the structure of the phosphine ligand used is the main influencing factor on the n:iso selectivity of hydroformylations. Therefore, we used the ligand BINAS (2) developed by Herrmann et al. (38) in the hydroaminomethylation of propene, 1-butene, and 1-pentene (Table 6). Excellent n:iso selectivities of 99:1 were achieved in all cases.

To determine the possibility of catalyst recycling the aqueous catalyst phase was reused for five subsequent runs under identical conditions only exchanging the organic phase (solvent, olefin, products) and adding the consumed ammonia (Figure 1).

The conversion rate (45 – 50%) remains nearly constant showing no significant loss of activity of the catalyst. In addition, the n/iso-selectivity remains constant (89 – 91%) which indicates that no significant degradation of the phosphine ligand occurs. The slight increase of side reactions (hydrogenation and isomerization of the olefin) is not clear so far. The drop in selectivity for primary/secondary amine is attributed to a certain loss of ammonia during manipulation with the autoclave (depressurizing, purging, etc.) since only one equivalent of ammonia was added after each run.

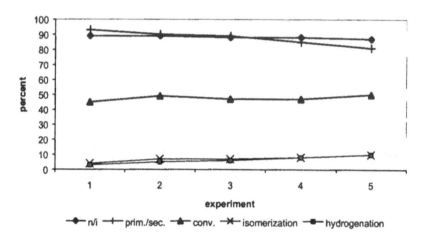

Figure 1. Two-phase hydroaminomethylation of 1-penten: recycling of the aqueous catalyst phase.

It is clear that two-phase catalysis with water as hydrophilic phase is applicable only to such olefins which are sufficiently soluble in the water phase, e.g. propene, butene and pentene. Due to a decreasing solubility in water higher olefins like 1-octene normally do not react under standard two-phase reaction conditions (39). This fact poses a severe limitation to the scope of the starting material for the two-phase hydroaminomethylation. However, by adding tensides or co-solvents also hydrophobic olefins might be hydroaminomethylated using our catalyst system. Here, in a first set of experiments we were interested in the reaction of aromatic olefins such as styrene derivatives because the reaction products represent pharmaceutically interesting intermediates. As an example the hydroaminomethylation of styrene yielded 2-phenylpropylamine in 48% isolated yield (purified by column chromatography) with a typical n:iso selectivity of 5:95 (Scheme 3).

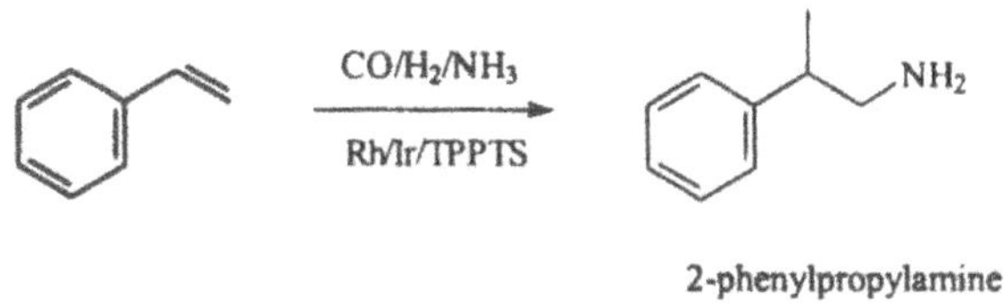

2-phenylpropylamine

Scheme 3: Hydroaminomethylation of styrene.

This reaction shows that it is indeed possible to form primary amines in preparative useful yields from relatively hydrophobic olefins, syn gas, and ammonia. Further studies towards the synthesis of other functionalized and unfunctionalized olefins are in progress.

Acknowledgments

We thank Celanese AG for supplies of TPPTS and NAPHOS; Dr. R. Fischer (Celanese AG), Dr. H. Geissler (Clariant AG), and Prof. Dr. Kühlein (Hoechst AG) for initial discussions; and the Bundesministerium für Bildung und Forschung (BMBF) for financial support. We also thank Prof. Dr. O. Nuyken (TU München) for the availability of laboratory space and additional infrastructure.

References

1. G. Heilen, H. J. Mercker, D. Frank, A. Reck and R. Jäckh, in *Ullmanns Encyclopedia of Industrial Chemistry, Vol. A2,* 5th ed., VCH, Weinheim, 1985, p.1.
2. T. Müller and M. Beller, *Chem. Rev.* **98**, 675 (1998).
3. J.-J. Brunet, D. Neubecker and F. Niedercorn, *J. Mol. Catal.* **49**, 253 (1989).
4. D. M. Roundhill, *Chem. Rev.* **92**, 1 (1992).
5. D. Steinborn and R. Taube, *Z. Chem.* **26**, 349 (1986).
6. a) W. Reppe, *Experientia* **5**, 93, (1949); b) W. Reppe, H. Vetter, *Liebigs Ann. Chem.* **582**, 133 (1953).
7. A. F. M. Iqbal, *Helv. Chim. Acta* **45**, 1440 (1971).
8. R. M. Laine, *J. Org. Chem.* **45**, 3370 (1980).
9. K. Murata, A. Matsuda and T. Matsuda, *J. Mol. Catal.* **23**, 121 (1984).
10. F. Jachimowicz and J. W. Raksis, *J. Org. Chem.* **47**, 44 (1982).
11. E. MacEntire and J. F. Knifton (Texaco Inc.), *EP* 240193, (1987) [Chem. Abstr. **110**, 134785h (1989)].
12. J. J. Brunet, D. Neibecker, F. Agbossou and R. S. Srivastava, *J. Mol. Catal.* **87**, 233 (1994).

13. E. Drent and A. J. M. Breed (Shell Int. Res. M.), *EP* 457386, (1992) [Chem. Abstr. 116, 83212h (1992)].
14. S. Törös, I. Gemes-Pesci, B. Heil, S. Maho and Z. Tuber, *J. Chem. Soc., Chem. Commun.* **1992**, 585.
15. T. Baig and P. Kalck, *J. Chem. Soc., Chem. Commun.* **1992**, 1373.
16. T. Baig, J. Molinier and P. Kalck, *J. Organomet. Chem.* **455**, 219 (1993).
17. M. D. Jones, *J. Organomet. Chem.* **366**, 403 (1989).
18. G. Diekhaus, D. Kampmann, C. Kniep, T. Müller, J. Walter and J. Weber (Hoechst AG), *DE* 4334809 (1993) [Chem. Abstr. 122, 314160g (1995)].
19. H. Schaffrath and W. Keim, *J. Mol. Catal.* **140**, 107 (1999).
20. a) T. Rische, L. Bärfacker and P. Eilbracht, *Eur. J. Org. Chem.* 653 (1999); b) P. Eilbracht, C. L. Kranemann and L. Bärfacker, *Eur. J. Org. Chem.* 1907 (1999).
21. T. Rische, B. Kitsos-Rzychon and P. Eilbracht, *Tetrahedron* **54**, 2723 (1998).
22. C. L. Kranemann and P. Eilbracht, *Synthesis* **1998**, 71.
23. T. Rische and P. Eilbracht, *Synthesis* **1997**, 1331.
24. B. Breit, *Tetrahedron Lett.* **39**, 5163 (1998).
25. E. Nagy, B. Heil and S. Törös, *J. Organomet. Chem.* **586**, 101 (1999).
26. J. F. Knifton, *Catalysis Today* **36**, 305 (1997).
27. J.-J. Lin, R. Rock and J. F. Knifton (Texaco Inc.), US 4794199, (1988) [Chem. Abstr. 110, 215203u (1989)].
28. B. Zimmermann, J. Herwig and M. Beller, *Angew. Chem. Int. Ed.* **38**, 2372 (1999).
29. T. Prinz, W. Keim and B. Drießen-Hölscher, *Angew. Chem.* **108**, 1835 (1996); *Angew. Chem. Int. Ed. Engl.* **35**, 1708 (1996).
30. B. Cornils and W. A. Herrmann (eds.): *Aqueous phase organometallic catalysis: concepts and applications*, Wiley-VCH, Weinheim, 1998.
31. W. A. Herrmann and C. W. Kohlpaintner, *Angew. Chem.* **105**, 1588 (1993); *Angew. Chem. Int. Ed. Engl.* **32**, 1524 (1993).
32. I.T. Horvath (ed.) *J. Mol. Catal.* **116** (1-2) (1993).
33. P. Kalck and F. Monteil, *Adv. Organomet. Chem.* **34**, 219 (1992).
34. Y. N. C. Chan, D. Meyer and J. A. Osborn, *J. Chem. Soc., Chem. Commun.* **1990**, 869.
35. A. Benyei, F. Joo, *J. Mol. Catal.* **58**, 151 (1990).
36. R. V. Chaudhari, B. M. Bhanage, S. S. Divekar and R. M. Deshpande (Council Sci. Ind. Res.), *US* 5650546 (1997) [Chem. Abstr. 127, 161588d (1997)].
37. F. Spindler, B. Pugin and H.-U. Blaser, *Angew. Chem.* **102**, 561 (1990); *Angew. Chem. Int. Ed. Engl.* **29**, 558 (1990).
38. W. A. Herrmann, G. P. Albanese, R. B. Manetsberger, P. Lappe and H. Bahrmann, *Angew. Chem.* **107**, 893 (1995); *Angew. Chem. Int. Ed. Engl.* **29**, 558 (1995).
39. H. Bahrmann and S. Bogdanovic, in *Aqueous phase organometallic catalysis: concepts and applications*, B. Cornils, W. A. Herrmann (eds.) Wiley-VCH, Weinheim, 1998.

Novel Noble-Metal-Based Catalysts for Liquid-Phase Acetone Condensation

A. A. Nikolopoulos,[a] G. B. Howe,[a] B. W.-L. Jang,[a] R. Subramanian,[a,d]
J. J. Spivey,[a] D. J. Olsen,[b] T. J. Devon,[c] and R. D. Culp[c]

[a] Research Triangle Institute, P.O. Box 12194, RTP, NC 27709-2194
[b] Eastman Chemical Company, P.O. Box 1972, Kingsport, TN 37662-5150
[c] Eastman Chemical Company, P.O. Box 7444, Longview, TX 75607-7444

Abstract

The effects of reaction temperature and reduction procedure on the activity and selectivity of the single-stage liquid-phase condensation of acetone to methyl isobutyl ketone (MIBK) were examined on a 0.1wt.% Pd-supported hydrotalcite catalyst. A reaction temperature increase enhances the conversion of acetone but strongly suppresses the MIBK yield, due to enhanced over-condensation to undesirable (mainly C_9 cyclic) by-products. The MIBK yield exhibits a maximum with reaction temperature and is strongly correlated with the relative contribution of the competing condensation vs. hydrogenation functions of a particular catalyst. A reduction (*in-situ* or *ex-situ*) treatment is required for catalyst activation, leading to enhanced activity for MIBK formation.

Introduction

The condensation of alcohols, aldehydes and ketones, commonly known as aldol condensation, is an extensively used reaction process for the synthesis of higher molecular weight compounds of these organic species. In particular, the liquid-phase condensation of acetone is a well-known synthesis route for the industrial production of various commodity chemicals such as diacetone alcohol (DAA), mesityl oxide (MO), mesitylene, isophorone, 3,5-xylenol, and methyl isobutyl ketone (MIBK) (1,2).

Aliphatic ketones such as MIBK are industrially important solvents commonly used in paints, resins, and coatings. In addition, MIBK is used as an extracting agent in the production of lubricating oils and antibiotics (2-4). The current annual US production of MIBK is ca. 210 million lb and more than 60% of its production comes from acetone condensation (4). The condensation / hydrogenation of acetone to MIBK is shown schematically in Figure 1.

---

[d] Current address: Precious Metals Division, Johnson Matthey, 2001 Nolte Drive, West Deptford, NJ 08066

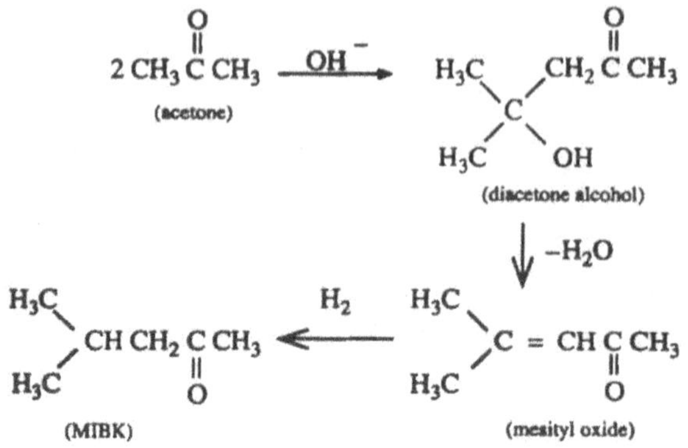

Figure 1   Schematic of the acetone condensation/hydrogenation to MIBK.

Typical industrial practice for aldol condensation reactions involves the use of homogeneous liquid-based catalysts, like sodium and calcium hydroxide (NaOH, Ca(OH)$_2$). Such basic catalysts are used in the first step of the 3-step acetone-to-MIBK process, i.e., the dimerization of acetone to DAA. The subsequent dehydration of DAA to MO requires acidic catalysts. Finally, specific metal sites are needed for the selective hydrogenation of the C=C double bond in MO to MIBK (2,3).

The conventional 3-step homogeneous acetone condensation process has some serious disadvantages (2,3). The condensation of acetone to MO is limited by thermodynamic equilibrium to about 20% acetone conversion at 120°C (2,5). Additionally, the use of alkali hydroxide catalysts generates a wastewater stream containing substantial amounts of dissolved salts per lb of product. This waste stream must be neutralized and then properly disposed (2). This requires the addition of product separation and waste disposal stages to the overall process, adding to the cost.

A single-step heterogeneously catalyzed condensation/hydrogenation of acetone to MIBK can successfully overcome these disadvantages (2,3). The kinetically fast hydrogenation of MO to MIBK shifts the equilibrium of the acetone condensation towards MO, increasing the overall conversion of acetone (2). Furthermore, the presence of a non-homogeneous (solid) catalyst facilitates product separation and eliminates the generation of a wastewater stream, therefore making this single-step process more economically favorable and environmentally benign (3).

Various solid catalysts containing acidic and/or basic sites have been examined for acetone condensation, such as alkali (1), alkaline earth (6-10), and transition metal (11,12) oxides and phosphates (13-16), zeolites (17,18), ion-exchange resins (13,19) and hydrotalcites (20-24). Hydrogenation of various unsaturated ketones (for example, MO to MIBK) has been reported using Ni, Co, Cu, Zn, Sn, and noble metals (mainly Pd) dispersed these acid/base supports. Pd-supported hydrotalcites (Pd/HTC) have been shown to be promising catalysts for acetone and phenol condensation reactions (23-26). Recent results on acetone condensation (26) indicate that low-loading (ca. 0.1wt.%) Pd/HTC catalysts show high acetone conversion (up to 38%) and high selectivity to MIBK (ca. 82%) in a liquid-phase batch micro-reactor at 118°C and 400 psig (28 atm). The selectivity to MIBK *versus* various condensation/hydrogenation byproducts is a function of the reaction conditions and the balance between the acid/base and hydrogenation properties of the catalyst (26).

The objective of this work was to investigate the effects of reaction temperature and catalyst reduction procedure (i.e., the state of the metal sites) on the activity/selectivity of Pd/HTC catalysts for the acetone-to-MIBK reaction. A better understanding of the effect of the physical and chemical properties of the catalyst on the condensation and hydrogenation reactions could enable the design of an optimized catalytic process for maximizing the formation of MIBK.

Experimental

Catalyst Synthesis and Characterization

A magnesium-aluminum (Mg-Al) hydroxyl-carbonate of hydrotalcite structure (HTC) was obtained in powder form from LaRoche Industries Inc. (ID#95-194-HT-001, sample 570) and was used as catalyst support. The Pd was dispersed on the HTC support by wet impregnation of aqueous solutions of palladium chloride ($PdCl_2$) of various concentrations. The catalysts were dried at 120°C for 2 hours, calcined at 350°C for 2 hours, and finally crushed and sieved to a particle size of 20-40 mesh (370-840 μm).

The Pd/HTC catalysts were characterized by various methods. Their surface area was measured by N_2 physisorption (BET method) using a Quanta Chrome – NOVA 1000 instrument. The actual metal loading was quantified by ICP/OES. The reducibility of the dispersed metal was determined by TPR (temperature-programmed reduction) under 10% H_2/Ar up to 600°C, using an AMI-100 characterization system (Zeton-Altamira, Pittsburgh, PA). The TPR results were used to determine the appropriate reduction temperature for catalyst activation. Catalyst acidity and basicity were estimated by NH_3 and CO_2 TPD (temperature-programmed desorption) up to 400°C with a heating rate of 10°C/min, on samples reduced in 10% H_2/Ar at 450°C for 8 hours and treated in 10% NH_3/He and 10% CO_2/He, respectively (26).

The typical catalyst treatment procedure prior to reaction involved the *ex-situ* reduction under pure H_2, by heating to 450°C at 5°C/min and holding for 8 hours (26). An *in-situ* reduction procedure was performed on selected samples by applying the same heat treatment inside the reactor, before adding the reactant acetone. *In-situ* reduction avoids any possible re-oxidation of the catalyst by exposure to the air during its transfer from the reduction unit to the reactor. In addition, an *ex-situ* heat treatment under N_2 instead of H_2 (no reduction) was also performed on selected catalyst samples.

Reaction Studies

The reaction experiments were performed in a high-pressure 50-ml batch reactor (Autoclave Engineers Inc., Erie, PA) equipped with a 900-W electric heater. The liquid reactants and products were stirred using a magnedrive impeller; its speed was controlled by a motor (Saftronics non-regenerative DC Drive) and indicated by a digital tachometer. The reactor pressure was monitored using an analog pressure gauge and a pressure transducer. The reactor temperature was continuously monitored and digitally displayed.

The catalyst was loaded into a high-temperature, high-pressure micro-Robinson catalyst basket, with a 50-mesh opening and 8.4-ml nominal internal volume. The flow rates of the feed gases (H_2 and N_2) were controlled by metering valves and measured by electronic mass flow meters. All the flow, temperature, pressure, and rotating speed data were collected in a PC through a data acquisition interface. A more detailed description of the reaction system is presented elsewhere (26).

Experiments were typically performed by loading ca. 2.7 g of catalyst (reduced *ex-situ*) into the catalyst basket which was then placed into the reactor. Acetone (20 ml, 270 mmol) was then added. The reactor was purged with N_2 to remove air and pressurized (by closing the reactor vent valve) in a 25% H_2 - 75% N_2 blend to 350 psig while heating to the required reaction temperature and stirring continuously at 200 rpm. This continuous stirring was used to minimize diffusion limitations between the bulk liquid phase and the catalyst particles.

After reaching the desired reaction conditions (typically 350 psig and 118°C), the N_2 flow was stopped while the H_2 flow was kept open and the reaction pressure was increased to 400 psig (28 atm). The pressure was maintained by adding H_2 as makeup for H_2 consumed in the reaction, thus maintaining constant reactant composition (26). The experiment was typically carried out for 5 hours under these conditions. Analysis of the liquid was performed in a GC with FID and a capillary column (HP-1 methyl siloxane) for separation of compounds. Methanol was used as an internal standard.

Liquid-Phase Acetone Condensation

Results and Discussion

Catalyst Characterization

BET surface area measurements indicated that all three Pd/HTC catalysts (0.1, 0.7, and 1.5wt.% loading) had significantly lower surface area than the parent HTC material. This could be related to removal or substitution of the stabilizing ions of the HTC structure by the impregnation process (in the form of water or CO_2). Also, the NH_3 and CO_2 adsorption measurements showed that the Pd/HTC samples possess a moderately higher acidity (by ca. 33%) but a substantially lower basicity (by more than 50%) with respect to the parent HTC (26). These results are apparently related to the presence of residual amounts of chloride ions in the catalyst, particularly for the higher loading samples. A more detailed analysis of these characterization studies is given elsewhere (26).

Preliminary Reaction Studies

Preliminary reaction experiments were conducted at a reaction temperature of 118°C on the parent HTC and all Pd/HTC catalysts. The acetone conversion on HTC (with no added Pd) was 20%, corresponding to the thermodynamic equilibrium of acetone condensation to MO under the conditions examined (2,5). The products were DAA (15% selectivity), indicating acetone dimerization on the basic sites, and MO (85% selectivity), implying strong DAA dehydration on acidic sites. Only minimal amounts of MIBK (< 1%) were detected, as expected since HTC lacks hydrogenation sites (26).

The presence of hydrogenation sites on the Pd/HTC samples resulted in the complete hydrogenation of MO to MIBK (82% selectivity on 0.1% Pd/HTC) and a significantly higher acetone conversion (38% on the same catalyst), due to fast hydrogenation of MO. The higher-loading Pd/HTC samples gave lower acetone conversion and MIBK selectivity, along with increasing isopropanol (IPA) formation (26). Experiments conducted on 0.1% Pd/HTC under the same reaction conditions but with double the stirring speed (440 rpm) gave essentially identical conversion and selectivity values, thus indicating the absence of diffusion limitations between the liquid phase and the catalyst particles.

Effect of Reaction Temperature

Acetone condensation experiments were performed on the 0.1wt.% Pd/HTC catalyst in the temperature range of 99-153°C and using the typical reaction procedure discussed above. The objective of this series of experiments was to evaluate the effect of reaction temperature on the selectivity to MIBK with respect to other condensation/hydrogenation byproducts. The results of this study are presented in Table 1.

Table 1   Effect of reaction temperature on the acetone condensation activity and selectivity of 0.1% Pd/HTC at 400 psig (28 atm)

Temp. (°C)	Convers. (%)	Selectivity (mol%)						
		DAA	MO	MIBK	IPA	DMH	DIBK	Other
99	26.0	22.6	0	72.8	3.3	0.5	0.8	0
110	29.4	19.3	0	75.3	3.9	0.6	1.0	0
118	37.5	14.7	0	82.4	0	1.2	1.7	0
128	41.3	14.3	0	82.7	0	1.2	1.9	0
137	41.6	14.2	0	82.9	0	0.3	2.7	0
153	42.6	12.3	10.3	21.2	0	0.6	0.5	55.1[a]

DAA: diacetone alcohol (4-hydroxy-4-methyl-2-pentanone)
MO: mesityl oxide (4-methyl-3-penten-2-one)
MIBK: methyl isobutyl ketone (4-methyl-2-pentanone)
IPA: isopropanol (2-propanol)
DMH: dimethyl heptanone (4,6-dimethyl-2-heptanone)
DIBK: diisobutyl ketone (2,6-dimethyl-4-heptanone)
[a] Other: isophorone (3,5,5-trimethyl 2-cyclohexenone): 45.1%; other C_9: 10%

The conversion of acetone was increased monotonically with temperature (from 26% at 99°C to 43% at 153°C). However, the selectivity to MIBK decreased drastically at the highest temperature examined (only 21% at 153°C, compared to 83% at 137°C). These results clearly indicate that the formation of MIBK is not favored at elevated temperatures due to enhanced undesirable formation of byproducts. Product analysis at 153°C indicated the formation of only minimal amounts of DIBK, DMH, and DIBC (typical byproducts at lower temperatures). Other by-products were isophorone (3,5,5-trimethyl 2-cyclohexenone) and its hydrogenated counterpart (3,5,5-trimethyl cyclohexanone), with isophorone being the major by-product (selectivity of ca. 45%). These C_9 cyclic species were identified by off-line gas chromatography - mass spectroscopy (GC-MS). Therefore, elevated temperatures favor the over-condensation of acetone to its trimers (mainly cyclic C_9 species) rather than its hydrogenation to IPA or that of its dimer (MO) to MIBK.

The yield (conversion x selectivity) of MIBK is plotted as a function of reaction temperature in Figure 2 (filled squares). [Some of the data points of this MIBK yield curve have already been presented (26) and are included here for the sake of completeness.] In addition, MIBK yields of various Pd-supported catalysts, i.e., ion-exchange resins (IER), zirconium phosphate (ZrP), niobium oxide (Nb_2O_5), and alumina (Al_2O_3) as reported in the literature (12-14,19) is also presented for comparison. [A number of studies performed at atmospheric

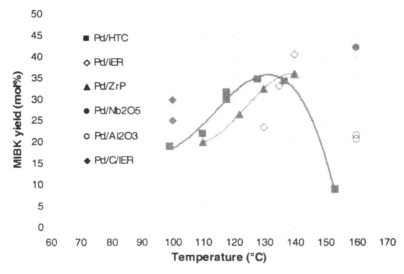

Figure 2 MIBK yield vs. reaction temperature on Pd-supported catalysts.
Pd/HTC: Present study (0.1% Pd, batch reactor, 28 atm)
Pd/IER: Ref. 13 and refs. within (0.05-3% Pd, flow reactor, 30-100 atm)
Pd/ZrP: Ref. 14 (0.5% Pd, flow reactor, 20 atm)
Pd/Nb$_2$O$_5$:   Ref. 12 (0.1% Pd, batch reactor, 20 atm)
Pd/Al$_2$O$_3$:   Ref. 12 (0.1-0.3% Pd, batch reactor, 20 atm)
Pd/C/IER: Ref. 19 (Pd/H$^+$ = 0.1, batch reactor, 3.5 atm)

pressure (e.g. as in Ref. 24) are not included in Fig. 2 because they are also typically carried out at higher temperatures and are not directly related to this (high pressure) study]. The comparison of the MIBK yield of the 0.1% Pd/HTC and the other Pd-supported catalysts is not direct due to variations in the reaction conditions (reactor type and pressure). However, Figure 2 does provide a comparison of the effect of temperature in the present study to data reported by others for supported Pd catalysts.

The MIBK yield shows a maximum with reaction temperature (36% at 133°C) due to preferential formation of over-condensation byproducts at high temperatures, as previously shown (Table 1). It should be stressed that the position of the MIBK yield curve appears to be correlated with the relative importance of two catalyst properties that are necessary for this reaction: condensation and hydrogenation. An increase in the hydrogenation activity (e.g., a higher Pd loading for the present study) would tend to shift the MIBK yield curve to higher temperatures, so as to limit the enhanced acetone hydrogenation to IPA and enhance its condensation to MO. Indeed, the MIBK yield for the catalysts containing Pd higher loadings (0.7 and 1.5%) at 118°C was found to be lower than that of the 0.1% Pd/HTC due to enhanced IPA formation and minimal over-condensation activity (26). On the other hand, a decrease in hydrogenation activity or an increase in the (acid-base) condensation activity of a catalyst would be expected to shift the MIBK yield maximum to lower reaction temperatures, all other reaction parameters (i.e., pressure) being kept constant. Although the MIBK yield of 0.5% Pd/ZrP increases monotonically up to 140°C (14), a decrease in MIBK yield due to enhanced over-condensation would be expected at higher temperatures. Taking these variations in catalyst characteristics and reaction conditions into account, it is evident that the 0.1% Pd/HTC is a relatively active and selective catalyst for the acetone-to-MIBK reaction, though the MIBK yield is not as high as for the 0.1% Pd/Nb$_2$O$_5$.

Effect of Reduction Procedure

The acetone condensation activity/selectivity of 0.1% Pd/HTC were evaluated on samples from the same batch that were reduced *in-situ* prior to reaction (in order to avoid any possible re-oxidation of the catalyst before the reaction), rather than reduced *ex-situ* (450°C under H$_2$ for 8 h) as in all other runs. For the sample reduced *in-situ*, the same reduction was applied inside the batch reactor before adding the reactant acetone. To test the effect of using a reduced catalyst compared to a catalyst that was not reduced, another 0.1% Pd/HTC sample was treated *ex-situ* with the same heat treatment procedure but under an inert (N$_2$) flow instead of the standard reduction (H$_2$) flow. The results of this study are presented in Figure 3, expressed in terms of the obtained MIBK yield at typical reaction conditions (118°C, 400 psig).

The 0.1% Pd/HTC treated under N$_2$ (no reduction) exhibited moderately higher acetone conversion (25.5%) compared to the equilibrium-determined 20% of the parent HTC, and a relatively low MIBK yield (16%). Obviously this unreduced sample is substantially inferior compared to the reduced ones (Fig. 3). However, this limited formation of MIBK indicates that the unreduced sample possesses some limited hydrogenation activity. Temperature-programmed reduction (TPR) experiments on the 0.1% Pd/HTC catalyst indicated that the reduction of dispersed Pd (measured as consumption of H$_2$) starts at temperatures as low as 100°C. Thus it is quite possible that in the progress of the reaction at 118°C some limited reduction of Pd sites actually takes place, enabling hydrogenation of the already formed MO to MIBK.

The 0.1% Pd/HTC sample reduced *in-situ* showed only slightly higher acetone conversion (43%) and MIBK yield (33%) compared to the sample reduced *ex-situ*. Therefore, the state of Pd appears to be only minimally affected by exposure to air at ambient conditions after reduction at 450°C and before the start of the acetone condensation reaction.

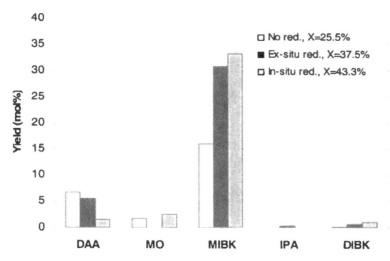

Figure 3 Effect of reduction procedure on the product yield for acetone condensation on 0.1% Pd/HTC at 118°C and 400 psig (28 atm).

Conclusions

The results of this study indicate the complexity of maximizing the formation of MIBK by the single-step acetone condensation/hydrogenation on multi-functional catalytic systems. The Pd/HTC catalysts examined appear to possess the required acidity/basicity for the condensation of acetone to MO, as well as the metal-site activity for the hydrogenation of MO to MIBK. Reduction of these catalysts (*in-situ* or *ex-situ*) prior to reaction is required for their activation, in order to exhibit enhanced activity for MIBK formation.

The yield and selectivity of MIBK show a maximum with reaction temperature on a 0.1% Pd/HTC catalyst. An increase in reaction temperature from 99°C to 153°C results in enhanced acetone conversion, but also tends to shift product selectivity towards undesirable over-condensation products (mainly C_9 cyclic species) above ca. 130°C , thus severely limiting the yield of MIBK.

The MIBK yield appears to be strongly correlated with the relative contribution of the condensation *versus* the hydrogenation activity of a given catalyst. This relative contribution is also affected by reaction temperature, since a temperature increase strongly favors condensation over hydrogenation. Consequently, the yield of MIBK is expected to show a maximum with reaction temperature on any multi-functional acetone condensation catalyst, but the position of this maximum is determined by the acidic/basic and hydrogenation properties of that particular catalyst.

References

1.   G.S. Salvapati, K.V. Ramanamurty and M. Janardanarao, *J. Mol. Catal.* **54**, 9 (1989).
2.   J. Braithwaite, in *Kirk Othmer Encyclopedia of Chemical Technology*, 4th ed., Vol. 14, 1995, pp. 978-1021.
3.   S. Wilde and D. Sommer, *Petroleum Technology Quarterly* **Spring 1998**, 127 (1998).
4.   *Chemical Market Reporter*, **255(10)**, 37 (1999).
5.   F.G. Klein and J.T. Banchero, *Ind. Eng. Chem.* **48(8)**, 1278 (1956).
6.   H. Dabbagh and B.H. Davis, *J. Molec. Catal.* **48**, 117 (1988).
7.   L.M. Gandia and M. Montes, *Appl. Catal. A* **101**, L1 (1993).
8.   J.I. Di Cosimo, V.K. Diez and C.R. Apesteguia,. *Appl. Catal. A* **137**, 149 (1996).
9.   K.-H. Lin and A.-N. Ko, *Appl. Catal. A* **147**, L259 (1996).
10.   B.-Y. Coh, J.M. Hur and H.-I. Lee, *Korean J. Chem. Eng.* **14(6)**, 464 (1997).
11.   L.M. Gandia and M. Montes, *React. Kinet. Catal. Lett.* **53(2)**, 261 (1994).
12.   Y. Higashio and T. Nakayama, *Catal. Today* **28**, 127 (1996).
13.   Y. Onoue, Y. Mizutani, S. Akiyama, Y. Izumi, Y. Watanabe and J. Maekawa, *Bull. Japan Petr. Inst.* **16(1)**, 55 (1974).
14.   Y. Watanabe, Y. Matsumura, Y. Izumi and Y. Mizutani, *Bull. Chem. Soc. Japan* **47**, 2922 (1974).
15.   Y. Watanabe, Y. Matsumura, Y. Izumi and Y. Mizutani, *J. Catal.* **40**, 76 (1975).
16.   Y. Watanabe, Y. Matsumura, Y. Izumi and Y. Mizutani, *Bull. Chem. Soc. Japan* **50**, 1539 (1977).
17.   L. Novakova and. L. Kubelkova, *J. Catal.* **126**, 689 (1990).
18.   H. Hattori, *Chem. Rev.* **95**, 537 (1995).
19.   C.U. Pittman, Jr. and Y.F. Liang, *J. Org. Chem.* **45**, 5048 (1980).
20.   W.T. Reichle, *J. Catal.* 1985, **94**,547 (1985).
21.   A. Guida, M.H. Lhouty, D. Tichit, F. Figueras and P. Geneste, *Appl. Catal. A* **164**, 251 (1997).
22.   K.K. Rao, M. Gravelle, J.S. Valente and F. Figueras, *J. Catal.* **173**, 115 (1998).
23.   S. Narayanan and K. Krishna, *Appl. Catal. A* **174**, 221 (1998).
24.   Y.Z. Chen, C.M. Hwang and C.W. Liaw, *Appl. Catal. A* **169**, 207 (1998).
25.   Y.Z. Chen, C.W. Liaw and L.I. Lee, *Appl. Catal. A* 1999, **177**, 1 (1999).

26. A.A. Nikolopoulos, B.W.-L. Jang, R. Subramanian, J.J. Spivey, D.J. Olsen, T.J. Devon and R.D. Culp, in *Green Chemistry: Recent Advances in Chemical Syntheses and Processes*, accepted for publication.

Rhodium Catalyzed Methanol Carbonylation: New Low Water Technology

Noel Hallinan and James Hinnenkamp

Millennium Petrochemicals, Inc., 11530 Northlake Drive, Cincinnati, OH 45249

Abstract

Rhodium catalyzed carbonylation of methanol to acetic acid is typically carried out at high reactor water concentration due to catalyst activity and stability limitations. High reaction rates and low catalyst usage can be achieved at low reactor water concentration by the introduction of tertiary phosphine oxide additives. The success of these additives results from their ability to replace water as a means of maintaining the rhodium in its catalytically active form.

Introduction

Rhodium catalyzed carbonylation of methanol, first developed by Monsanto in the late '60s (1), is still the dominant method of acetic acid manufacture today. While the general features of this continuous homogeneous process have remained the same, the efficiency in terms of catalyst stability and activity has been greatly improved in recent years by various acetic acid manufacturers (2). Historically, the barrier to higher production rates has been the requirement of a high reactor water concentration (> 7M) to minimize rhodium precipitation and maximize rhodium activity, and thus, the subsequent necessity of costly water removal in the purification stage. This paper reports on new low water technology which allows the levels of catalyst stability and activity observed at 7M reactor water concentration to be maintained or even exceeded at 3M water.

Experimental

Results discussed herein were obtained from experiments conducted in a Hastelloy C-276 300 ml reactor operated in batch mode. Using a custom built gear pump, also constructed of Hastelloy C-276, reactor solution was monitored by continuous circulation through a high pressure cell enclosed in a process hardened infrared spectrometer. Rhodium acetate (Engelhard) and iridium acetate and ruthenium acetylacetonate (Alfa Aesar) were used as soluble catalyst sources.

In a typical reaction study, the reaction solution components, minus catalyst, were added to the reactor. After leak testing with N_2 and purging with CO, the reactor was heated to the desired temperature at a CO pressure of $7 - 14$ atm.

The reaction was started by injecting the catalyst into the reactor and raising the pressure to $25 - 30$ atm. The reaction proceeded at constant reactor pressure maintained by feeding CO from a high pressure reservoir *via* a regulator. At appropriate time intervals, infrared spectra were recorded on-line and liquid samples were removed for gas chromatographic (GC) analysis. All rate measurements quoted are for the initial period of reaction only, where component concentrations have not deviated significantly from the starting values. Unless otherwise noted, iodide was charged to the reactor as hydriodic acid (HI). Methyl iodide (MeI) concentrations, discussed in the text and given in tabular form, arise from an equilibrium reaction of HI and methyl acetate (MeOAc) upon heating reactor solutions.

For purposes of material balance, acetic acid (HOAc) formation is reported only in terms of CO uptake as in eq. 4. HOAc formed before catalyst injection *via* the equilibrium in eq. 3 is not included.

Reaction Chemistry

In its simplest form, the net reaction producing (HOAc) is as shown below:

$$\text{CH}_3\text{OH} + \text{CO} \xrightarrow[\text{CH}_3\text{I Promoter}]{\text{Rh Catalyst}} \text{CH}_3\text{COOH} \qquad (1)$$

In reality the reaction has been shown to involve a number of steps as outlined in equations 2 – 4.

$$\text{CH}_3\text{OH} + \text{CH}_3\text{COOH} \rightleftharpoons \text{CH}_3\text{COOCH}_3 + \text{H}_2\text{O} \qquad (2)$$

$$\text{CH}_3\text{COOCH}_3 + \text{HI} \underset{\text{(2 steps)}}{\overset{\text{H}_2\text{O}}{\rightleftharpoons}} \text{CH}_3\text{I} + \text{CH}_3\text{COOH} \qquad (3)$$

$$\text{CH}_3\text{I} + \text{CO} \overset{\text{"Rh"}}{\rightleftharpoons} \{\text{CH}_3\text{COI}\} \overset{\text{H}_2\text{O}}{\rightarrow} \text{CH}_3\text{COOH} + \text{HI} \qquad (4)$$

The only side reaction of note is the water gas shift (WGS) reaction, in which CO and H_2O are consumed as shown in equation 5.

$$\text{CO} + \text{H}_2\text{O} \overset{\text{"Rh"}}{\rightleftharpoons} \text{CO}_2 + \text{H}_2 \qquad (5)$$

It is believed that $[\text{Rh(CO)}_2\text{I}_2]^-$, a square planar Rh^I species, is the active catalytic species for the reactions in eqs. 1 and 5. A simplified representation of the reaction cycles involved is given in Figure 1, where the rhodium species typically observed in reactor solutions are identified. The rate determining step in HOAc formation is the oxidative addition of MeI to Rh^I.

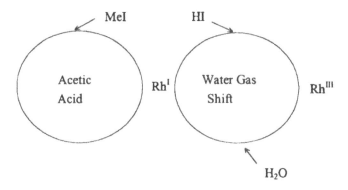

$$Rh^I = [Rh(CO)_2I_2]^-, \quad Rh^{III} = [Rh(CO)_2I_4]^-$$

Figure 1 Simplified scheme of the rhodium catalyzed acetic acid reaction and the competing water gas shift reaction.

The original rate equation for the Monsanto process is of the form:

$$\frac{d[HOAc]}{dt} = k\,[Rh_T]\,[I_T]$$

Where: Rh_T = Total rhodium present
I_T = Total iodide present

A limitation of this rate equation is that it does not contain a term for water dependence, yet it has been extensively documented (3) that lowering reactor water concentration adversely impacts reaction rate.

Results and Discussion

The first step in development of the low water technology described herein was an investigation of the effect of water upon the reaction rate. This was greatly facilitated by the availability of close to real time infrared analysis of the reactor solution under reaction conditions. A kinetic study performed in the range 0.49 – 9.1M H_2O showed rate behavior similar to that seen by other investigators (3). A close to exponential dependence of rate on water concentration is observed, as shown in Figure 2.

Typical catalyst behavior for a batch reactor run is as shown in the time profile in Figure 3. At reaction outset the Rh^I concentration is at its highest relative to Rh^{III}. As the reaction progresses and as HI builds up in solution there is a shift in equilibrium favoring Rh^{III}.

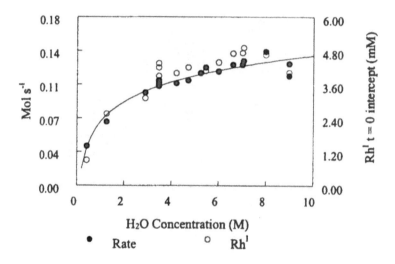

Figure 2 Effect of H_2O concentration on initial rate and on initial Rh^I concentration. (mM = millimolar).

The initial Rh^I concentration in the various reactor runs at different starting water exhibits the same exponential dependence on water concentration as initial rate (see Figure 2). This strongly suggests that the reason for loss of catalyst activity and stability is due to an increasingly unfavorable Rh^{III}/Rh^I ratio as water concentration is lowered. As shown in Figure 1, oxidative addition of HI to Rh^I in the WGS cycle leads to formation of a Rh^{III} species which is subsequently reduced back to Rh^I. This reduction is facilitated by H_2O. As the H_2O concentration in the reactor is lowered the concentration of Rh^{III} relative to Rh^I increases. This leads to a drop in rate and an increase in rhodium precipitation as Rh^{III} species can easily lose CO leading to insoluble neutral complexes.

The inactive Rh^{III} species forms as a result of oxidative addition of HI. While this reaction is undesirable, the presence of HI is important because it

serves to convert MeOAc to MeI, required for rate promotion in the formation
of HOAc. Thus, it is desirable to find a suitable HI complexing agent, which
will prevent oxidative addition of HI to Rh^I while leaving the iodide in a form
available to react with MeOAc so that rate promotion is not compromised. A
further criterion for such a scavenger is that it not decompose under the
aggressive conditions of the acetic acid process.

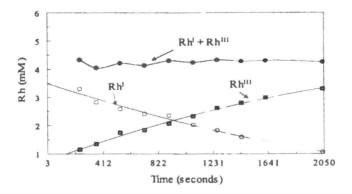

Figure 3 Typical catalyst behavior in a batch reactor run.

Previous results (4) involving non-batch reactor studies had indicated that
certain inorganic salts of weak acids are capable of imparting stability to Rh^I.
Several of these salts were tested in the batch reactor to examine the effect on
rate and Rh^I concentration. These included LiOAc, Na_2HPO_4 and
$(CH_2COONa)_2$. Similar effects were observed in all cases: an increase in initial
Rh^I concentration and Rh^I longevity coupled with a decrease in initial rate
compared to runs without additives. The magnitude of these effects was a
function of salt concentration. Catalyst time profiles for runs with variable
concentrations of $(CH_2COONa)_2$ are shown in Figure 4. GC data indicated that
the initial MeI concentration showed an inverse dependence on $(CH_2COONa)_2$
concentration (see Table 1), thus explaining the decrease in rates observed even
at high Rh^I concentration. The inorganic iodides (MI), formed from reaction of
these salts with HI, show less tendency to react with MeOAc than does HI. We
and others (5) have shown that the MI + MeOAc reaction to generate MeI has
an equilibrium constant, K of only approx. 0.03, whereas the HI + MeOAc
reaction has a K of approx. 90. The unfavorable equilibrium is ultimately a
consequence of the basicity of the above inorganic salts, with pK_b's ranging
from 5 - 15.

Table 1 Rate and MeI concentration data for batch reactor runs with variable $(CH_2COONa)_2$ concentration. Initial conditions were: 175C, 6.6 mM Rh, 1M MeI, 0.7M MeOAc, 2.5M H_2O.

$(CH_3COONa)_2$	Initial Rate (Mol/l/h)	Initial MeI, M
0.0	3.0	0.80
0.1	2.8	0.74
0.3	1.4	0.36
0.5	0.6	0.18

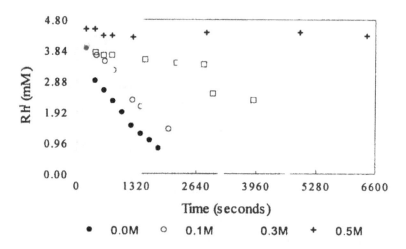

Figure 4 Effect of $(CH_2COONa)_2$ concentration on initial Rh^I and on Rh^I longevity.

Thus, our search for a suitable HI scavenger shifted to more weakly basic species. Tertiary phosphine oxides ($R_3P=O$) are very weak bases with pK_b's ranging from –2 to –4. These compounds were considered to be particularly attractive candidates for the acetic acid process because of high thermal stability with a phosphoryl group (P=O) bond energy in the 120 – 150 kcal/mole range (6). Because of this high bond energy, the chemistry of the phosphoryl group does not typically include hydrolysis and reactivity is more or less limited to the phosphoryl group acting as a nucleophile.

It has been reported in the literature (7) and shown by us (4) that phosphine oxides form 2:1 adducts with HI of the type shown below:

$$(R_3P=O-H—O=PR_3)$$
$$I$$

The degree of interaction is expected to be determined by the basicity of the phosphine oxide. The greater the double bond character of the phosphorus to oxygen bond the lower will be the basicity. A requirement of these phosphine oxides as additives is that the R_3PO/HI complexes – while preventing HI from reacting with Rh^I – must be capable of reacting with MeOAc at a rate at least equal to the reaction of HI with MeOAc to form MeI, as shown in equation 6.

$$MeOAc + [(R_3PO)_2H]I \rightarrow 2 R_3PO + MeI + HOAc \qquad (6)$$

Much of our initial batch reactor work with phosphines (8) focussed on tributylphosphine oxide (Bu_3PO) and triphenylphosphine oxide (Ph_3PO). These compounds were chosen because of commercial availability and because these represent opposite ends of the phosphine oxide basicity scale, the triphenyl analog being less basic. These initial studies involved charging iodide to the reactor as the preformed $[(R_3PO)_2H]I$ complex. Thus, in order for HOAc formation to take place, this complex must be capable of reacting with MeOAc to form MeI. Invariably it was found that CO uptake did indeed take place, with the uptake rate being dependent on the basicity and concentration of phosphine oxide.

In a series of experiments carried out at 5M H_2O, it was found that the presence of R_3PO led to a significantly enhanced initial Rh^I concentration and Rh^I longevity. Differing effects on initial rate were observed however. A 25% drop in rate was observed for Bu_3PO compared to a 25% increase for Ph_3PO as illustrated in Figure 5. A third compound, triphenylphosphate ($(PhO)_3PO$) which is a considerably weaker base than Ph_3PO, showed identical rate and catalyst behavior to the run without additive.

Chromatographic data showed a 45% drop in initial MeI concentration for the Bu_3PO run compared to the other runs. Thus, even in the phosphine oxide series (which are all weak bases) there is a window of basicity which allows optimal catalyst stability and activity. $(PhO)_3PO$ is too weakly basic to provide any catalyst protection and behaved merely as a spectator. Bu_3PO provided the best catalyst protection but behaved more like the inorganic salts described earlier. Namely, the ionizable iodide concentration was not sufficient to maintain MeI concentration. On the other hand, Ph_3PO is capable of imparting enhanced Rh^I

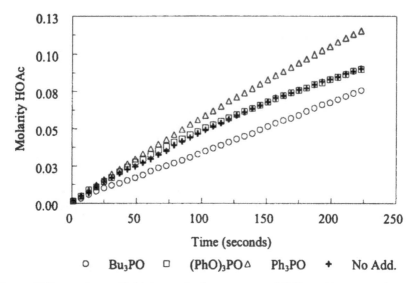

Figure 5 Comparison of initial rates in the presence of different R_3PO species. Conditions: 1M R_3PO, 0.5M HI, 0.7M MeOAc, 5M H_2O, 175C, 4.4 mM Rh

stability without impacting MeI concentration.

A further study was carried out with Ph_3PO in which runs with 3M and 7M H_2O with variable phosphine oxide concentration were completed. The data in Table 2 show that the more than 50% decrease in initial rate observed at 3M H_2O and no additive compared to 7M H_2O can be eliminated by addition of sufficient Ph_3PO. This table also shows that the increased rates of reaction at lower water concentration with increasing Ph_3PO concentration are due to the protection of Rh^I.

Measurement of solution CO_2 concentration by on-line infrared showed the expected trend in rates of water gas shift (see table). Increasing Ph_3PO concentration leads to decreased rates of CO_2 formation, consistent with an implied lower HI concentration. Control experiments with calibration gas mixtures showed that the solubility of CO_2 in reactor solution was unaffected by Ph_3PO in the 0 – 1.5M range.

It should be noted that there is no evidence for direct interaction of these phosphine oxides with catalyst, as evidenced by infrared measurements. The nucleophilicity of the phosphoryl oxygen is apparently not great enough for complexation with $[Rh(CO)_2I_2]^-$, though there is evidence in the literature for complexation of tertiary phosphine oxides with other Rh^I species (9, 10).

Table 2 Effect of Ph_3PO on initial rate, MeI and Rh^I concentration.
Conditions: 1.3M MeI, 0.7M MeOAc, 175C, 4.4mM Rh.
Rate units are Mol/l/h. "Methyl" = MeOAc + MeI.

H_2O, M	Ph_3PO, M	Rate	% Rh as Rh^I	% "Methyl" as MeI	Initial Rel. Rate CO_2
7	0.0	5.27	96	65	1.5
3	0.0	2.38	57	62	1.0
3	0.5	3.19	66	60	0.6
3	1.0	3.49	76	64	0.4
3	1.5	5.04	94	63	0.3

An experimental design was performed to determine whether a synergism exists between Ph_3PO and H_2O in stabilizing Rh^I. The relative stability of Rh^I at various Ph_3PO and H_2O concentrations is presented in Figure 6.

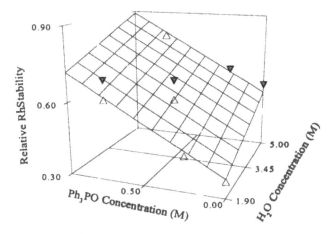

Figure 6 Relative Rh^I stability vs. H_2O and Ph_3PO concentration.
Conditions: 0.25M HI, 0.6M MeOAc, 175C, 4.4 mM Rh

These relative Rh^I stabilities were obtained by integration of the measured Rh^I concentration over the course of each experiment and comparison of the areas to a hypothetical run where the Rh^I concentration remains at its initial value throughout. There is no apparent curvature in the plane nor is there any evidence of an interaction between Ph_3PO and H_2O, i.e. neither requires the other to be effective. Thus Ph_3PO can be used as a straightforward substitute for H_2O. For example, in the set of reaction conditions used in Figure 6, Ph_3PO is approximately 6 times more effective at stabilizing Rh^I than is H_2O. In practice, an effectiveness ratio of $> 10:1$ has been demonstrated in our pilot and manufacturing plants.

In a broad study, the effect of trivalent tertiary phosphites, $(RO)_3P$, phosphines, R_3P and phosphates, $(RO)_3PO$, on rate was examined. On-line infrared measurements showed that phosphites and phosphines undergo coordination to Rh^I. Furthermore, quaternization of phosphines with MeI and hydrolysis of the alkoxy groups of tertiary alkyl phosphites occurred. In all cases, measured rates were substantially lower than runs without additives.

In the absence of other effects, one might expect a correlation between P=O stretching frequency and rate. Clearly this is not the case as shown in Table 3, where initial rates for phosphates, phosphine oxides, phosphinic acids and phosphoramides are reported. Basicity alone does not dictate rate in the case of phosphine oxides with functionalized side groups. The propensity of alkoxy groups to hydrolyze and of amino groups to quaternize masks any correlation that may exist.

Table 3 Infrared P=O stretches and initial rates for tertiary phosphine oxides, phosphates and phosphoramides.
Conditions: 0.5M HI, 0.7M MeOAc, 5M H_2O, 175C, 4.4 mM Rh.

Species	P=O Infrared Freq. (cm^{-1})	Rate, Mol/l/h
$(PhO)_3PO$	1299	1.98
$(EtO)_3PO$	1270	0.60
$(BuO)_3PO$	1235	0.11
$(BuO)_2(OH)PO$	1231	0.43
$((CH_3)_2N)_3PO$	1205	0.42
$Ph(OH)HPO$	1195	0.12
Ph_3PO	1190	2.48
No Additive	-------	2.02
Bu_3PO	1155	1.53

In order to benchmark the phosphine oxide technology described herein against alternative low water technologies (2), a series of batch reactor experiments were performed under similar conditions. In particular, as shown in Table 4, this involved using iridium or iridium/ruthenium combinations as the metal catalyst or using lithium iodide as the additive. The relative rates in Table 4 show that under the experimental conditions used, the Rh/Ph₃PO combination is superior to all others examined.

Table 4 Relative rate data for batch reactor experiments with various catalysts and additives. Initial conditions were: 175C, 0.7M MeOAc, 1M MeI, 4.4 mM catalyst, 3M H_2O.

Catalyst	Additive (1M)	Rel. Rate
Rh	------	1
Rh	Ph₃PO	1.28
Rh	LiI	0.90
Ir	------	0.30
Ir/Ru[a]	------	0.76
Ir	Ph₃PO	0.15
Ir/Ru[a]	Ph₃PO	0.23
Ir	Bu₃PO	0.10

[a] Ru in 5 fold molar excess over Ir

Summary

Rate and catalyst stability observed at 7M reactor H_2O concentration can be maintained at 3M H_2O by addition of tertiary phosphine oxides. The success of phosphine oxides results from in situ generation of an alternate form of iodide from HI, thereby protecting the active catalyst, Rh[I]. The favorable chemical changes introduced by these compounds makes these ideal additives to further low water technology in the acetic acid process

Acknowledgements

We are grateful to Dr. Steve Augustine for help in setting up the experimental design and in interpretation of kinetic results and to Dr. Ronnie Hanes for helpful discussions.

References:

1.   F.E. Paulik and J.R. Roth, *J. Chem. Soc., Chem. Comm.*, 1578 (1968).
2(a) B.L. Smith, P. Torrence, A. Aguilo and J. Alder, Hoechst Celanese Corporation, *US Pat.* 5,026,908 (1991).
2(b) K.E. Clode and D.J. Watson, The British Petroleum Company, *Eur. Pat.* 616,997 (1994).
2(c) P. M. Maitlis, A. Haynes, G. Sunley and M.J. Howard, *J. Chem. Soc., Dalton Trans.*, 2187 (1996).
3(a) J. Hjortkjaer and O.R. Jensen, *Ind. Eng. Chem. Prod. Res. Dev.* **16**, 281 (1977).
3(b) B.L. Smith, G.P. Torrence, M.A. Murphy and A. Aguilo, *J. Organomet. Chem.*, **303**, 257 (1986).
3(c) J.S. Kim, K.S. Ro and S.I. Woo, *J. Mol. Catal.*, **69**, 15 (1991).
4.   Research performed at Millennium Petrochemicals Inc.
5.   B.L. Smith, P. Torrence, M.A. Murphy and A. Aguilo, *J. Mol. Catal.* **39**, 115 (1986).
6.   S. Hartley, W. Holmes and J. McCombrey, *Quart. Rev.*, **17**, 204 (1963).
7.   A. Bertoluzza, S. Bonora and M. Morelli, *Can. J. Spectrosc.*, **32**, 107 (1987).
8.   J. Hinnenkamp and N. Hallinan, Millenium Petrochemicals Inc., *US. Pat.*, 5,817,869 (1998).
9.   G. Bandoli and D. Clemente, *J. Organomet. Chem.*, **71**, 125 (1974).
10. R.W Wegman, A.G. Abatjoglou and A.M. Harrison, *J. Chem. Soc., Chem. Comm.*, 1891 (1987).

Anchored Homogeneous Catalyst Precursors

Stephen Anderson, Hong Yang, Setrak K. Tanielyan and
Robert L. Augustine

*Center for Applied Catalysis, Seton Hall University, South Orange, NJ 07079,
USA*

Abstract

Virtually all previously reported attempts at attaching homogeneous catalysts to
a solid support have been concerned with making a solid ligand and then using
this 'heterogeneous ligand' to prepare the active catalyst. Such an approach can
require considerable synthetic efforts to produce the 'heterogeneous ligand'
particularly if one wants to incorporate onto a solid one of the more complex
chiral ligands in use today. Our approach to anchoring homogeneous catalysts
involves the attachment of the homogeneous catalyst to the support using a
heteropoly acid which interacts with both the support material and the metal
atom of the complex.

We have used this procedure to anchor catalyst precursors, such as
$Rh(COD)_2$, and Ru(p-cymene) to an alumina support and then treated these
materials with a number of different chiral and achiral ligands to prepare
anchored homogeneous catalysts.

Introduction

For the past twenty-five years or so many attempts were made to 'heterogenize'
homogeneous catalysts in order to combine the activity and selectivity of the
homogeneous species with the ease of separation and capacity for potential re-
use qualities of heterogeneous catalysts (1-6). The most common method used
to accomplish this goal has been to incorporate a ligand into or onto a solid
support and then to react the resulting solid ligand with an appropriate metallic
species to prepare the tethered complex. There are, however, several problems
associated with this approach, the most important of which is that the metal ion
is attached to the support by bonding through the ligand and is, therefore, prone
to leaching, frequently to a rather large extent. Further, these catalysts are
usually less active than the corresponding homogeneous species. Also they
frequently lose their activity and selectivity on separation and attempted re-use.
This attachment of the ligand to a solid support is not generally applicable to all
types of ligands and can be particularly difficult when one is attempting to
produce an analog of an effective chiral ligand which could be attached to a
support material.

We have recently described a new method for anchoring homogeneous catalysts to solid support materials using a heteropoly acid as the anchoring agent (7-10). In these catalysts the heteropoly acid is first attached to the support material by interaction of the acidic protons on the acid with the basic sites on the support material. The homogeneous complex is then anchored by an interaction between the heteropoly acid and the metal atom of the complex.(10)

We have used this approach to anchor a number of rhodium and ruthenium complexes onto several different support materials. They have been found to retain this activity and selectivity on extended re-use. Frequently, these anchored catalysts are even more active and selective than the homogeneous analogs. Further, there is no evidence of any metal leaching from these catalysts on re-use.(10)

Since the anchoring is considered to be through the metal atom of the complex, the nature of the ligands on the complex is not critical. Since, in many instances, commercial use of homogeneous catalysts involves complexes having proprietary ligands, it was thought that instead of trying to devise procedures for anchoring each of these different types of complexes, it would be beneficial if one could produce an anchored homogeneous catalyst precursor to which appropriate ligands could be attached prior to reaction. We describe here the attachment of catalyst precursors such as $Rh(COD)_2$ and $Ru(p\text{-cymene})$ to an alumina support and the use of these precursors to prepare chiral and achiral anchored homogeneous catalysts.

Experimental

These anchored homogeneous precursors are prepared by adding a solution of the heteropoly acid (20 µmoles in 2.5 mL of alcohol) with vigorous stirring to a suspension of the support material (300mg in 10 mL of alcohol) with stirring continued for about three hours followed by the removal of the liquid and thorough washing of the solid. This solid is then suspended in another 10mL of degassed alcohol and a solution of the homogeneous precursor. The $Rh(COD)_2^+BF_4^-$ is prepared by adding a THF solution of $(Rh(COD)Cl)_2$ containing an excess of COD to a THF solution of $AgBF_4$ and removing the AgCl formed by filtration. The $Rh(COD)_2^+BF_4^-$ or $RuCl_2(p\text{-cymene})_2$ (20 µmoles in 1 mL of alcohol) is added to the heteropoly acid modified alumina under an inert atmosphere, with stirring, over a 30 min period. Stirring is continued for 8 to 12 hours under an inert atmosphere, the liquid removed and the solid washed thoroughly until no color is observed in the wash liquid. This material can be used directly or dried for future use. Stirring an alcoholic suspension of this anchored precursor with 1.1 equivalents of the ligand in an inert atmosphere for 8 to 12 hours completes the ligand exchange and gives the

Scheme 1

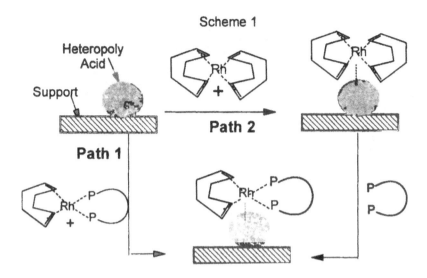

anchored homogeneous catalyst. Ligand exchange on the anchored Ru(p-cymene) is best accomplished by heating the suspension at 50°C for two hours.

Table 1. Reaction rates and product ee's obtained on hydrogenation of **1** over alumina anchored Rh(DiPAMP) catalysts prepared from a supported precursor or from the complex.[a]

	Prepared through Path 2[b]		Prepared through Path 1[b]	
Use Number	Rate[c]	ee%	Rate[c]	ee%
1	0.28	79	0.39	84
2	0.48	85	0.49	87
3	0.49	87	0.54	87
4	0.58	87		
5	0.55	85		

[a] Hydrogenations run at 25°C under 1 atm H_2 using 20 µmoles of supported Rh(DiPAMP) to saturate 0.35 mmole of **1** for each run.

[b] See Scheme 1.

[c] moles H_2/mole Rh/min.

Table 2. Reaction rates and product ee's obtained on hydrogenation of **1** over alumina anchored Rh(BPPM) catalysts prepared from a supported precursor or from the complex.[a]

Use Number	Prepared through Path 2[b]		Prepared through Path 1[b]	
	Rate[c]	ee%	Rate[c]	ee%
1[d]	2.3	32	3.5	21
2[e]	11.6	80	7.2	82
3[e]	8.9	79	7.9	83
4[e]	7.8	80	6.1	84
5[e]	7.5	80		

[a] Hydrogenations run at 25°C under 1 atm H_2 using 20 μmoles of supported Rh(BPPM) to saturate 0.35 mmole of **1** for each run.
[b] See Scheme 1.
[c] moles H_2/mole Rh/min.
[d] Run at 25°C.
[e] Run at 50°C.

Results

Scheme 1 depicts the general procedures which can be used to prepare an anchored homogeneous catalyst using our technology. One path involves the preparation of the catalytically active complex which is then attached to the heteropoly acid treated support (Path 1). The other depicts the reaction of the precursor with the heteropoly acid treated support followed by ligand exchange to give the anchored homogeneous catalyst (Path 2). Table 1 lists the comparative reaction data for the hydrogenation of methyl 2-acetamidoacrylate (**1**) over tethered Rh(DiPAMP) catalysts prepared either by the anchoring of the preformed catalyst or by reaction the tethered Rh(COD)$_2$ precursor with the DiPAMP ligand.

Table 2 shows the hydrogenation rate and ee data for the hydrogenation of **1** over anchored Rh(BPPM) catalysts prepared by direct anchoring of the Rh(BPPM) complex (Path 1) and a ligand reaction between BPPM and the anchored Rh(COD)$_2$ precursor (Path 2).

Fig. 1 shows the hydrogen uptake data for the hydrogenation of methyl 2-acetamidocinnamate (**2**) over an anchored Rh(DiPAMP) catalyst prepared from the anchored precursor.

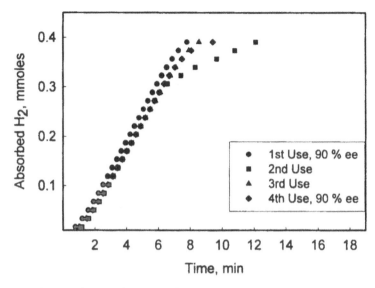

Fig.1. Hydrogenation uptake curves for the hydrogenation of **2** over
Rh(DiPAMP)/PTA/Al$_2$O$_3$ prepared by reaction of DiPAMP with
anchored Rh(COD)$_2$. Hydrogenations run at 25°C under 1 atm H$_2$ using
20 μmoles of Rh(DiPAMP) to saturate 0.4 mmole of **2** for each run.

Achiral ligands such as dppe and dppb were also added to the anchored
Rh(COD)$_2$ with the resulting catalysts used for the hydrogenation of 1-hexene.
The hydrogenation of four 5 mL portions of 1-hexene over the anchored
Rh(dppb) prepared from the precursor proceeded readily. The total TON in this
reaction sequence is about 8,000. Analysis of the reaction mixtures showed that
if any Rh were present, it was there in an amount below the detection limit of the
analytical procedure, which in this instance is less than 1 ppm.

We have also recently anchored a Ru(p-cymene) precursor to a
phosphotungstic acid treated alumina and used the resulting species to prepare
an anchored Ru(Binap) catalyst. When used for the hydrogenation of dimethyl
itaconate there was no difference in either activity or reaction selectivity between
the anchored Ru(Binap) catalyst prepared from the anchored Ru(p-cymene) and
one made by the reaction of a preformed Ru(Binap) complex with the
phosphotungstic acid treated alumina. More work is presently underway in using
this anchored Ru(p-cymene) to prepare other supported Ru complex catalysts.

Conclusions

Our technique for anchoring homogeneous catalysts which involves the interaction of a supported heteropoly acid with the metal atom of a homogeneous complex makes it possible to anchor precursor materials which can then be used to prepare the heterogenized homogeneous catalyst in the catalytic reactor before it is used. While the data presented only concerns anchored $Rh(COD)_2$ and, to a lesser extent, anchored Ru(p-cymene), it does show that the concept is a valid one and that this approach has the potential to be applied to a variety of different catalyst precursors in the preparation of a large number of anchored homogeneous catalysts.

Acknowledgements

The development of these catalysts was funded by grants CTS-9312533 and CTS-9708227 from the US National Science Foundation.

References

1. 1. D. C. Bailey and S. H. Langer, *Chem. Rev.*, **81**, 109 (1981).
2. J. P. Collman, L. S. Hegedus, M. P. Cooke, J. R. Norton, G. Dolcetti, D. N. Marquardt, *J. Am. Chem. Soc.*, **94**, 1789 (1972).
3. W. Dumont, J.-C. Poulin, T. P. Daud and H. B. Kagan, *J. Am. Chem. Soc.*, **95**, 8295 (1973).
4. L. L. Murrell, in *Advanced Materials in Catalysis*, Academic Press, New York, 1977, Chapter 8.
5. V. Isaeva, A. Derouault and J. Barrault, *Bull. Soc. Chim., Fr.*, **133**, 351 (1996).
6. U. Nagel and J. Leipold, *Chem. Ber.*, **129**, 815 (1996).
7. S.K. Tanielyan and R.L. Augustine, U.S. Pat. Appl., 08/994,025; PCT Int. Appl., WO-9828074; *Chem. Abstr.*, **129**, 109217 (1998).
8. R. Augustine and S. Tanielyan, *Chem. Commun.*, 1257 (1999).
9. S. K. Tanielyan and R. L. Augustine, *Chem. Ind. (Dekker)*, 75 (Catal. Org. React.) 101 (1998).
10. R.L. Augustine, S.K. Tanielyan, S. Anderson, H. Yang and Y. Gao, This volume.

Fibre-based catalysts: applications as heterogeneous and supported homogeneous catalysts for the fine chemicals industry

C. F. J. Barnard[a], K. G. Griffin[a], J. Froelich[a], K. Ekman[b], M. Sundell[b] and R.Peltonen[b]

[a] Johnson Matthey Technology Centre, Blounts Court, Reading, RG4 9NH, UK
[b] Smoptech Ltd , Virusmaentie 65, FIN-20300 Turku, Finland

Abstract

Insoluble polymer supports in the form of beads have found wide application in synthesis as a means of readily separating reagents from reaction products. However, these materials are not physically robust and cannot be used with conventional agitation. An alternative fibrous grafted copolymer has been explored as a support material for catalysts, either for particulate metal catalysts or for metal complexes that are active homogeneous catalysts. The use of these materials in a number of applications is described, highlighting their ease of recovery and re-cycle.

Introduction

Insoluble polymer supports have been employed for some time to provide a means of easily separating reagents from reaction products (1,2). The most commonly used material is styrene-divinylbenzene copolymer in the form of beads. This can be modified with a wide variety of functional groups to provide anchoring for either the reactant molecule for use with soluble reagents, or for supporting the reagents for use with soluble reactants. However, these beads generally suffer from a lack of physical strength, such that they are degraded to powder by agitation. Also, diffusion of the reactants into and out of the pores of the polymer beads can limit the reaction rate (3). Clearly, alternative forms of polymer support that overcome these limitations are required. Johnson Matthey and Smoptech are developing supported reagents using a fibrous grafted co-polymer. Using polyethylene as the base polymer means that that the fibres are insoluble in all solvents. The fibres can be cut to suitable lengths to allow different reactor and recycling options to be used.

Results and Discussion

A number of studies have been carried out to exemplify the properties of the fibrous polymer materials, with particular emphasis on their application as catalyst supports (4). Attention has been paid to reaction rates using these catalysts in comparison with bead-supported materials and homogeneous catalysts. The influence of the support on the selectivity of particulate metal catalysts and the potential for recycling is also addressed.

Kinetics

Using graft co-polymers allows a high density of active functional sites to be generated on the polymer. In addition these sites are all located on the surface of the polymer eliminating the potentially rate limiting effects of diffusion into and out of pores associated with polymer beads. This can be illustrated by ion exchange using a polycarboxylic acid polymer.

In a batch reactor 1 litre of copper nitrate solution was stirred with stoichiometric amounts of wet ion exchangers (in sodium form). An acrylic acid grafted polyethylene fibre was compared with beads, Dowex MWC-1. Samples were withdrawn at intervals and the solution was analysed by atomic absorption. The ready accessibility of the ion exchange sites on the fibre is evident from the rapid reduction in copper concentration in solution.

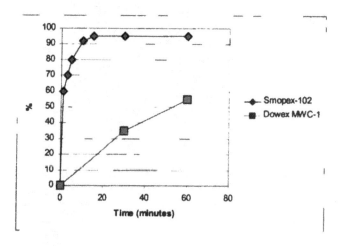

Figure 1: Ion Exchange of Cu^{2+} from $Cu(NO_3)_2$ solution

The accessibility of the acid sites on this fibrous polymer was also evident by its effectiveness as an acid catalyst for esterification (5).

In an isothermal batch reactor methanol and the acetic acid were heated at 55°C and the esterification was commenced by adding strong acid ion exchangers (-SO₃H). The catalysts used were Smopex-101 (3.2mmol/g H⁺) and Amberlyst-15 (4.9mmol/g H⁺). Samples were taken from the reactor and unreacted acetic acid was determined by titration with sodium hydroxide. The fibre catalyst was as effective as the bead catalyst but at a much lower loading.

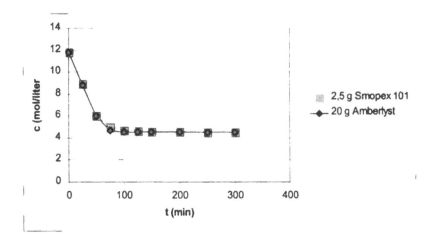

Figure 2: Esterification of acetic acid with methanol

Selectivity

The selectivity of heterogeneous catalysts can be strongly influenced by the characteristics of the support material, e.g. by the availability of acid or basic sites (6). The preparation of platinum group metal heterogeneous catalysts on grafted copolymer supports allows the incorporation of a range of functionality into the catalyst which could normally only be achieved by the addition of soluble reagents into the process in addition to the heterogeneous catalyst.

This is illustrated by a study of the hydrogenation of acetophenone. When prepared on a basic support such as grafted polyvinylpyridine copolymer (Smopex-105) the catalyst was inactive for this conversion. Using a weakly acidic support (polyacrylic acid copolymer Smopex-102) the alcohol was obtained as the major product. When the acidity of the polymer was increased by the incorporation of some sulphonic acid groups (i.e. a mixed carboxylic acid, sulphonic acid polymer Smopex-107) a mixture of alcohol and alkane products was obtained. Using a strongly acid support having only sulphonic acid groups (Smopex-101) gave the alkane.

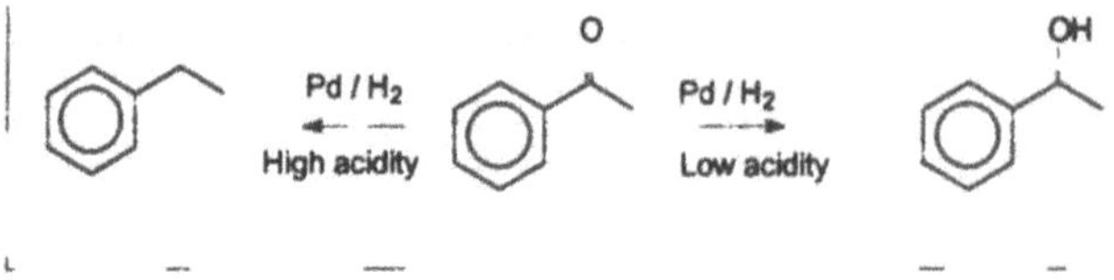

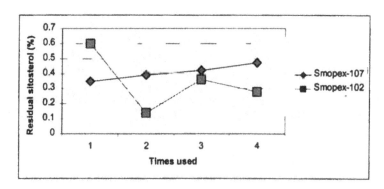

Figure 3: Influence of acidity on the hydrogenation of acetophenone

Recycle

Platinum group metal catalysts prepared on polymer supports can be handled in a similar way to conventional carbon-supported catalysts. The durability of the fibrous support also allows for easy recovery and washing of the catalyst for recycle.

This is shown below for the hydrogenation of sitosterol. The sitosterol:Pd weight ratio was 1000:1 in each case. Two types of support polymer were used with carboxylic acid functional groups (Smopex-102) and mixed carboxylic and sulphonic acid groups (Smopex-107). Palladium loading on the polymer was *ca.* 5% in each case.

Figure 4: Catalyst recycle in the hydrogenation of Sitosterol

After hydrogenation the catalyst was recovered and re-used in the next hydrogenation. The catalyst was used in total four times. The reaction time in each case was 90 minutes and the graph shows percent residual starting material for each reaction as determined by gas chromatography.

Supported Homogeneous Catalysts

The ready accessibility of the functional groups of these grafted copolymer supports makes them ideal for the anchoring of homogeneous catalysts. Also, the wide range of functional groups that can be incorporated by choice of the appropriate vinyl monomer means that there are many possibilities for the linker chemistry, making it adaptable to different catalyst complexes. The activity and selectivity of the homogeneous catalyst may then be obtained with the ease of handling and recovery of a heterogeneous catalyst.

As a first example of this type of catalyst, we have prepared supported palladium catalysts and evaluated them for carbon-carbon coupling reactions, for example the coupling of aryl and alkenyl halides with alkenes (the Heck reaction)[7]. A variety of palladium catalysts or precursors have been reported for these reactions but the most commonly used are [Pd(PPh₃)₄] or the *in situ* combination of [Pd(OAc)₂]₃ and PPh₃. We have compared the activity of this latter combination with our supported catalysts.

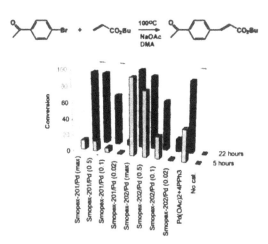

Figure 5: Heck coupling using supported palladium catalysts

Catalysts were prepared using benzyl (Smopex-201) or phenyl (Smopex-202) linkers to diphenylphosphine groups (*ca.* 1mmol/g). These samples were reacted with palladium to bind all available phosphines (Smopex-201,202/max) or at lower Pd levels (0.5, 0.1 and 0.02 versus P). An equal

conversion achievable if longer times were allowed. After each reaction the recovered catalysts were combined with fresh reactants for the next cycle. The homogeneous control sample (the *in situ* combination of $[Pd(OAc)_2]_3$ and PPh_3) was prepared with fresh catalyst each time. A bead-supported palladium catalyst (PPh_2 binding with benzyl linker, *ca.* 1mmol/g polymer) was also prepared and tested in the same way. However, it was found that the beads were degraded during the reactions so that they could not be recycled effectively and hence activity was rapidly lost. The fibrous catalysts were not physically degraded and were recycled efficiently each time. There was only a slight loss of activity over the series of experiments (except for Smopex-203, {proprietary phosphite, *ca.* 0.7mmol P/g polymer} where activity remained constant) which is attributed to chemical degradation of the palladium catalyst during the reaction and recycles.

Covalent linking of the palladium catalyst to the polymer support ensures that there is only very low leaching of the palladium into the solution during the course of the reaction. Samples of the reaction medium were analysed by ICP-MS and the palladium concentration was found to be less than 2ppm. When compared with the palladium content of each reaction this corresponded to losses of only a few percent of the total palladium over the course of the experiment.

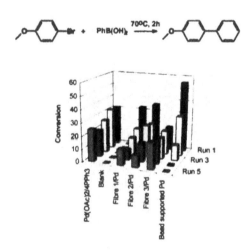

Figure 6: Suzuki reaction using recycled supported palladium catalysts

The operating temperature range for these polymer materials is limited by the nature of the supporting polymer. For polyethylene-based materials, the

weight of fibre catalyst was used in each reaction so that the substrate to palladium ratio varied from *ca.* 200:1 to approximately 10000:1. The coupling of 4-bromoacetophenone with butyl acrylate was studied. The reaction was monitored at 5 and 22 hours. The *in situ* combination of $[Pd(OAc)_2]_3$ and PPh_3 (substrate:Pd = 500:1) was used as reference. The results show that the rate increases with the amount of palladium used and that the supported catalysts have similar activity to the homogeneous catalyst. Smopex-201 shows an initially slow reaction compared with Smopex-202 but both achieved a high level of coupling after the full reaction time.

Leaching and Recycle

These supported palladium catalysts can be easily separated from the reaction mixture by filtration. After washing and drying they may be re-used in further reactions.

This was studied for a Suzuki coupling reaction in which a boronic acid is reacted with an aryl halide to form a biaryl compound (7). The coupling of 4-bromoanisole with phenylboronic acid was chosen as a suitable model reaction.

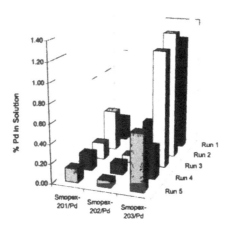

Figure 7: Leaching of Palladium into solution during Suzuki coupling

The substrate to palladium ratio was approximately 100:1 in each reaction. The reactions were heated at 70°C for 2 hours and then analysed. The results obtained therefore reflect the relative reaction rates and not the

upper limit is *ca.* 125°C in common organic solvents, but some variation can be achieved by including cross-linking in the fibre.

Conclusions

Grafted copolymer supports are robust and versatile materials that can be used as supports for both particulate metals and metal complexes to create a new generation of catalysts. These catalysts can be recovered and recycled allowing them to be used in a cost-effective manner. The catalysts are not physically degraded during use and low levels of metal leaching are observed. The properties of the polymer support can be used to influence the performance of particulate metal catalysts, reducing the need for the addition of further reagents to the process.

The polymer supports provide opportunities for many different types of linkers for anchoring the wide variety of metal complexes that function as homogeneous catalysts. Johnson Matthey and Smoptech will be developing a series of such catalysts.

Experimental

Ion exchange: Copper nitrate solution (0.15M, 1 litre) was stirred at 500rpm in a closed glass vessel. Stoichiometric amounts of wet ion exchangers were added to the reactor. Samples withdrawn from the reactor were analysed for copper by atomic absorption spectroscopy.

Esterification: Methanol (180ml) and the ion exchange resin catalyst (Amberlyst-15, 0.35-1.2mm beads, 4.9mmol/g, 20g or Smopex-101, 3.2mmol/g, 2.5g) were heated to 55°C in an isothermal batch reactor fitted with a reflux condenser. A stoichiometric amount of pre-heated acetic acid was added. Samples were withdrawn from the reactor and acetic acid content was determined by titration with NaOH.

Hydrogenation of Acetophenone: Catalyst (5% Pd on support; 5mg) was added to acetophenone solution (0.5M in ethanol containing mesitylene as internal standard; 1ml). Samples were hydrogenated at 30 psig, ambient temperature for 3 hours. Product analysis was carried out by GC.

Hydrogenation of Sitosterol: Sitosterol solution (20 wt% in n-propanol) was hydrogenated using 5% Pd on support at 75°C and 1.1 bar H_2 for 1.5 hours. Residual sitosterol was determined by GC. The catalyst was recovered by filtration and re-used 3 times under the same conditions.

Heck Coupling: Each catalyst (20mg supported Pd materials or 0.71ml 10mmol solution of $[Pd(OAc)_2]_3/4PPh_3$) was mixed with 4-bromoacetophenone (2M, 1.79ml) n-butyl acrylate (2.8M, 1.79ml) and sodium acetate (0.32g). All solutions were prepared in N,N-dimethylacetamide. A reaction blank was prepared by omitting the catalyst. All samples were stirred and heated at 100°C for 22 hours under nitrogen. Samples were taken by syringe after 5 hours and at the end of the reaction, and analysed by GC.

Suzuki coupling: Each catalyst (20mg supported Pd materials or 1.65ml 10mmol solution of $[Pd(OAc)_2]_3/4PPh_3$) was mixed with 4-bromoanisole (1M in toluene, 1.65ml) phenylboronic acid (0.305g) and potassium carbonate (0.449g) in toluene (3.35ml). The mixtures were stirred and heated at 70°C for 2 hours under nitrogen. Samples were centrifuged and the supernatant analysed by GC. Samples of the supernatant liquor were also treated by acid digestion and analysed for palladium content by ICP-MS. The supported catalysts were recovered by filtration, washed and dried *in vacuo*. They were then re-used under the same reaction conditions. This recycle was repeated four times, with a fresh homogeneous control reaction being prepared each time.

Samples of Smopex catalysts were provided by Smoptech Ltd and are available from Johnson Matthey, Orchard Road, Royston SG8 5HE, UK.

References

1. Polymer-supported reactions in organic synthesis, P. Hodge and D. C. Sherrington, eds., Wiley-Interscience, New York, 1980.

2. N. K. Mathur, C. K. Narang and R. E. Williams, Polymers as Aids in Organic Chemistry, Academic press, New York, 1980.

3. J. Jagur-Grodzinski, Heterogeneous Modification of Polymers, John Wiley and Sons, Chichester, England, 1997.

4. A. D. Pomogailo, Catalysis by Polymer-Immobilized Metal Complexes, Gordon and Breach Science Publishers, Australia, 1998.

5. G. A. Olah, T. Keumi and D. Meidar, Synthesis, 929-930, 1978.

6. Handbook of Heterogeneous Catalysis, vol. 5, G. Ertl, H.Knozinger and J. Weitkamp, eds., VCH, Weinheim, Germany, 1997.

7. Diederich and P. J. Stang (eds.), Metal-catalyzed Cross-coupling Reactions, Wiley-VCH, Weinheim, Germany, 1998.

Tethering Technology: Anchored Homogeneous Catalysts

J. A. M. Brandts[a], J. G. Donkervoort[a], C. Ansems[a], P. H. Berben[a], A. Gerlach[b] and M. J. Burk[b]

[a]Engelhard de Meern B.V., Strijkviertel 67, P.O. Box 19, 3454 ZG De Meern, The Netherlands
[b]Chirotech Technology Limited, Cambridge Science Park, Milton Road, Cambridge CB44WE, United Kingdom

Abstract

A method is reported, in which preformed homogeneous catalysts are anchored on a support material that can successfully be applied in certain hydrogenation reactions. These so-called 'tethered' catalysts can be re-used several times with constant high activity and show high chemo- and, if chiral catalysts are applied, enantioselectivity.

Introduction

There is a growing need to perform reactions more selectively, i.e. reduce the formation of unwanted side-products. As a result less purification steps are necessary to obtain the pure and wanted products. This is especially true for the synthesis of chiral building blocks, which are, for example, used in the production of pharmaceuticals and agro-chemicals (1). Homogeneous catalysts could potentially be used as they can be more selective and operate at milder conditions than heterogeneous catalysts. The ligand systems present in homogeneous catalysts provide the opportunity to 'fine-tune' the formation of wanted products. When the optimal ligand-metal combination is used, regio-selective reactions but also enantioselective hydrogenations, isomerizations, epoxidations and hydroformylations can be performed with high control.

However, a major drawback in using homogeneous catalysts is the difficulty in separating product from catalyst. Moreover, quite often the price of the homogeneous catalyst, which is used to make a reaction product, contributes significantly to the total product costs. Therefore, it would be attractive if the homogeneous catalyst could be used several times. The obvious advantages would be (a) easy separation of the catalyst from the product is possible, (b) the catalyst can be re-used, and as a result (c) the costs for using this catalyst are reduced.

573

A number of attempts have been made in which enantioselective reactions were performed using anchored homogeneous catalysts (2,3,4). All of the systems reported have several drawbacks such as leaching, decreasing activity and selectivity upon re-using the catalyst multiple times (5,6).

Recently, a new technique was reported that circumvents the aforementioned problems (7). It allows preformed homogeneous catalysts that consist of a cationic metal ion, a neutral mono- or bidentate ligand and an anionic counter ion, i.e. $M^+(L)_nX^-$ (M^+= Rh, Ru, Ir, Pd etc.; L = P-P, P-N, N-N, O-N etc.; X^-= Cl, BF_4, ClO_4, SbF_6 etc.) to be anchored on a support. Heteropoly acids (HPA's) are used as the anchoring species between the support material and the homogeneous catalyst. Interestingly, these anchored homogeneous catalysts maintain a high degree of activity and selectivity and show no leaching. The homogeneous catalyst precursor, the type of HPA or the support material used, e.g. γ-Al_2O_3, carbon, SiO_2 or clays (8), are parameters, which can be used to fine-tune the characteristics of the anchored homogeneous catalysts.

The mechanism of the binding between the homogeneous catalyst and the HPA is still unclear. It could, for example, involve an acid-base reaction between the homogeneous catalyst, $M^+(L)_nX^-$, and the acidic protons of the HPA. This reaction would result in the formation of the anchored homogeneous catalyst and HX (X = Cl, BF_4, ClO_4, SbF_6, etc.). Investigations are in progress to verify this hypothesis, which will be reported in due course.

A   B

Figure 1 The bidentate diphosphine ligands **A** (R,R-Me-DuPHOS) and **B** (DiPFc) used to prepare the Rh-complexes.

In this paper, we present the tethering and application of two different homogeneous catalysts, which exemplify the great advantages of the tethering technique. The first tethered homogeneous catalyst that is reported is [Rh(R,R-Me-DuPHOS)(COD)BF_4] (Fig. 1, **A**). This chiral complex was specifically chosen for its wide applicability in chiral hydrogenation reactions (9). In addition, the use of a chiral homogeneous catalyst allows making the distinction between the hydrogenation over the anchored chiral homogeneous catalyst and a non-chiral rhodium complex, e.g. formed through metal reduction and/or ligand dissociation. For the anchored chiral homogeneous catalyst, enantioselectivity should be observed similar to the homogeneous catalyzed reaction.

As a second example, the non-chiral [Rh(DiPFc)(COD)BF$_4$] (Fig. 1, **B**) is used for its interesting chemoselective behavior in certain hydrogenation reactions (*vide infra*) (10).

Both homogeneous catalysts have been successfully tethered and applied in hydrogenation reactions. In the following examples, γ-Al$_2$O$_3$ was used as the support material and phosphotungstic acid (PTA), phosphomolybdic acid (PMA) or silicotungstic acid (STA) were used as the anchoring agent.

Results and Discussion

Rh(*R,R*-Me-DuPHOS)(COD)BF$_4$

Complexes γ-Al$_2$O$_3$/PTA/[Rh(*R,R*-Me-DuPHOS)(COD)]$^+$ (1), γ-Al$_2$O$_3$/STA/[Rh(*R,R*-Me-DuPHOS)(COD)]$^+$ (2) and γ-Al$_2$O$_3$/PMA/[Rh(*R,R*-Me-DuPHOS)(COD)]$^+$ (3) were prepared in order to be able to compare the use of different HPA's. The hydrogenation of methyl-2-acetamidoacrylate (MAA) was used as the model reaction. After full conversion, the products were separated from the catalyst and analyzed for Rh and W content. The catalyst was re-used. Results are given in Table 1 and graphically presented in Figures 2, 3 and 4.

Table 1 Data generated by the use of **1,2** and **3** showing the differences caused by the use of different HPA's.

Catalyst	Run nr.	e.e. (%)	Rh leach (μg) [% tot. amount Rh]	W or Mo leach (μg) [% tot. amount HPA]
1[a]	1	91.7	n.d.[b] [<1]	222 [0.05]
	2	94.2	n.d. [<1]	88 [0.02]
	3	95.1	n.d. [<1]	16 [0.003]
	4	94.5	n.d. [<1]	n.d. [<0.002]
2[a]	1	94.1	11 [1]	183 [0.04]
	2	93.5	n.d. [<1]	89 [0.02]
	3	94.8	n.d. [<1]	25 [0.005]
	4	93.6	n.d. [<1]	16 [0.003]
3[a]	1	96.1	87 [8.5]	880 [0.4]
	2	94.0	59 [6.3]	567 [0.3]
	3	85.3	32 [3.6]	251[0.1]
	4	66.2	27 [2.9]	299 [0.2]

[a] MAA, 10 psi H$_2$-pressure, r.t., 1000 rpm., S/C =100, 25 mL EtOH, [MAA] = 0.025 M.
[b] n.d. = not detected.

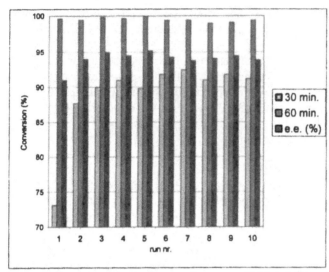

Figure 2 Ten successive hydrogenation runs of MAA followed in time using **1** as the tethered catalyst.

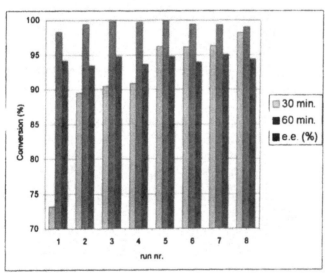

Figure 3 Eight successive hydrogenation runs of MAA using **2** as the tethered catalyst.

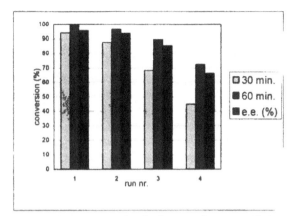

Figure 4 Four successive hydrogenation runs of MAA using **3** as the tethered catalyst.

The data in Table 1 clearly show that the differences between the use of PTA and STA as the anchoring agent are small. Both complexes **1** and **2** show comparable high enantioselectivities, which are as high as obtained with the equivalent homogeneous catalyst. In going from run 1 to 2 to 3 a significant increase in activity is observed. This can be explained by assuming that all Rh centers are fully activated by COD hydrogenation after the third run. The stability of the activated tethered catalysts is also maintained as can be concluded from the results obtained in run 5-8 in Figure 3. These runs were performed after overnight storage under a N_2-atmosphere and show an even improved activity and constant high selectivity. For the specific hydrogenation of MAA the addition of H_2O (4%) further improves the catalytic activity (Figure 3, run 8).

The data in Table 1 also clearly show the differences caused by the use of various HPA's. Upon using PTA or STA, no rhodium leaching is observed whereas for PMA significant amounts of rhodium were detected in the filtrate. As a consequence, complex **1** and **2** maintain their activity and selectivity in each successive run, whereas for **3** a decrease in activity is observed. Rhodium leaching can take place either by breaking the bond between the Rh-complex and the HPA or by breaking the bond between the Rh/HPA structure and the support material. As the Rh:PMA ratio in the filtrate is higher than 20 (see Table 1), this indicates that most probably the interaction between the Rh-complex and the HPA is broken during the reaction.

Rh(DiPFc)(COD)BF₄

The same tethering technique can also be applied for achiral homogeneous catalysts. As an example, Rh(DiPFc)(COD)BF₄ was anchored on γ-Al₂O₃/PTA. This homogeneous catalyst is specifically of interest due to its chemoselective and sulfur tolerant behavior (11). As a test reaction the tethered complex γ-Al₂O₃/PTA/[Rh(DiPFc)(COD)]⁺ (4) was used in the reduction of thiophene-2-carboxaldehyde (TCHO). All reactions were carried out in an *i*-PrOH/H₂O (11,12) solvent mixture to avoid acetal formation catalyzed by some remaining HPA acidity. After full conversion the products were separated from the catalyst and analyzed for Rh and W content. Results are given in table 2 and graphically presented in Figure 5.

Table 2 Data generated by four successive runs catalyzed by **4**.

Catalyst	Run nr.	Rh leach (μg) [% tot. amount Rh]	W leach (μg) [% tot. amount PTA]
4ᵃ	1	17 [1.7]	68 [0.01]
	2	15 [1.5]	110 [0.02]
	3	n.d.ᵇ[<1]	36 [0.005]
	4	n.d.[<1]	22 [0.003]

ᵃ TCHO, 100 psi H₂-pressure, r.t., 1000 rpm, S/C =330, 25 mL *i*-PrOH/H₂O (1:1), [TCHO] = 0.0625 M.
ᵇ n.d. = not detected

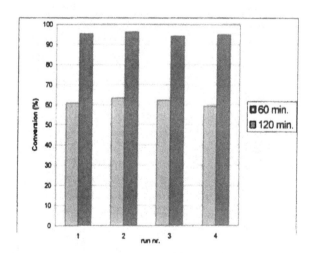

Figure 5 Four successive runs of TCHO catalyzed by **4**.

From the results presented in Figure 5, it can be seen that this tethered homogeneous catalyst maintains its activity through 4 consecutive runs. The only product observed is the corresponding thiophenic alcohol in quantitative yield. Conventional heterogeneous catalysts gave only limited activity due to the poisoning effect of the sulfur.

Other substrates were tested and gave results comparable to TCHO (see Table 3). Of particular interest is the observation that the chemoselectivity of this catalyst allows aldehyde reduction in the presence of aromatic bromides (entry 2) and aromatic nitro groups (entry 3). Again, conventional heterogeneous catalysts do not give these types of chemoselectivity.

Although many advantages can be given for this tethering technology, it also has limitations: (a) a maximum catalyst loading of ~0.5 % w/w of Rh, (b) not all solvents can be used because some solvents (e.g. MeOH) cause leaching, and (c) certain functional groups cause leaching, e.g. quaternary ammonium moieties.

In conclusion, tethering is a technique that has several advantages over the use of the corresponding homogeneous catalysts. Multiple re-uses are possible, the product is less contaminated, sometimes higher activity and selectivity are observed and the system can be fine-tuned using several options (solvent, support material, HPA).

Experimental

All reactions were carried out in an inert atmosphere (N_2) and all solvents were deoxygenated before use. The γ-Al_2O_3/HPA complexes were prepared according to literature procedures (8). Tethered catalysts were prepared by the slow addition of a solution of the homogeneous pre-catalyst (200 μmol) to stirred slurry of γ-Al_2O_3/HPA (10 g, contains ~10 % w/w elemental tungsten or ~5.5 % w/w elemental molybdenum; determined by inductively coupled plasma measurements) in 50 mL of MeOH. After reaction overnight the upper layer was removed and the colored precipitate extracted with MeOH four times or until the washings were colorless. γ-Al_2O_3/PTA/[Rh(R,R-Me-DuPHOS)(COD)]$^+$ (1); 0.34 % w/w Rh, 9.8 % w/w W, γ-Al_2O_3/STA/[Rh(R,R-Me-DuPHOS)(COD)]$^+$ (2); 0.39 % w/w Rh, 10.0 % w/w W, γ-Al_2O_3/PMA/[Rh(R,R-Me-DuPHOS)(COD)]$^+$ (3); 0.26 % w/w Rh, 5.5 % w/w Mo, γ-Al_2O_3/PTA/[Rh(DiPFc)(COD)]$^+$ (4); 0.17 % w/w Rh, 9.7 % w/w W. Hydrogenation reactions were carried out in a 50 mL stirred stainless steel autoclave. The reactor was loaded with the appropriate amount of catalyst (~ 10 μmol Rh), purged with nitrogen, and a 25 mL solution containing the appropriate amount of substrate (100-1000 equivalents) was added via a cannula. The mixture was purged with H_2 (5 pressurization (100 psi)/release cycles with short time stirring), the reactor was put under H_2 pressure and the reaction was started by stirring. The reaction progress was followed in time by taking samples at predetermined time intervals. Products were analyzed by (chiral) GC, HPLC and Inductively Coupled Plasma (ICP).

Table 3 Catalytic reduction of 4-nitrobenzaldehyde and several S-containing aldehydes by 4.[a] The product distribution (%) is shown between brackets.[b]

Substrate	Products		
thiophene-CHO	thiophene-CH$_2$OH (100 %)		
Br-thiophene-CHO (c)	Br-thiophene-CH$_2$OH (60 %)	thiophene-CHO (d)	
benzaldehyde with NO$_2$	CH$_2$OH with NO$_2$ (100 %)	CHO with NH$_2$ (n.d.)	CH$_2$OH with NH$_2$ (n.d.)
CHO with SMe	CH$_2$OH with SMe (100 %)		
S-CH$_2$CH$_2$-CHO	S-CH$_2$CH$_2$-CH$_2$OH (100 %)		

[a] 100 psi H_2, S/C = 300-1000, 0.1 M in 2-propanol-water (1:1), r.t., 18 h.
[b] n.d. = not detected by GC after 18 hours of reaction time.
[c] After 18 hours 40 % of starting material was found.
[d] < 2%

References

1.   I. Agranat and H. Caner, *Drug Discovery Today*, **4**, 313 (1999).
2.   D. J. Bayston, J. L. Fraser, M. R. Ashton, A. D. Baker, M. E. C. Polwka and E. Moses, *J. Org. Chem.*, **9**, 7313 (1998).
3.   C. Langham, D. Bethell, D. F. Lee, P. McMorn, P. C. Bulman Page, D. J. Willock, C. Sly, F. E. Hancock, F. King and G. J. Hutchings, *Appl. Cat. A.*, **182**, 85 (1999).
4.   R. Margalef-Català, P. Salagre, E. Fernández and C. Claver, *Cat. Letters*, **60**, 121 (1999).
5.   F. Gelman, D. Avnir, H. Schumann and J. Blum, *J. Mol. Cat. A.*, **146**, 123 (1999).
6.   H. Gao and R. J. Angelici, *Organometallics*, **18**, 989 (1999).
7.   S. K. Tanielyan and R. L. Augustine, US Pat. 6,025,295 to Seton Hall University.
8.   F. Cavani, *Catalysis Today*, **41**, 73 (1998).
9.   M. J. Burk in Handbook of Chiral Chemicals (Marcel Dekker), 339-359 (1999).
10.   M. J. Burk, T. P. G. Harper, J. R. Lee, C. Kalberg, *Tetrahedron Lett.*, **35**, 4963 (1994).
11.   M. J. Burk, A. Gerlach and D. Semmeril, *J. Org. Chem.*, submitted.
12.   M. J. Burk, A. Gerlach, Patent pending.

The Effect of Rare-Earth Elements on the Catalytic and Adsorptive Activity of Supported Transition Metals

A. J. Dyakonov[1], D. A. Grider[1], B. J. McCormick[2], P. K. Kahol[3]

[1]Lorillard Tobacco Company, 420 English Street, Greensboro, NC 27405
WS University, Department of [2]Chemistry and [3] Physics, Wichita, KS 67208

Abstract

Supported transition metal (TM) catalysts modified with rare-earth elements (REE) were synthesized. Oxygen chemisorption, high temperature oxidation and static magnetic measurements were applied to characterize the catalysts. Catalytic activities of these systems were studied in CO_x hydrogenation, CO oxidation, and NH_3 synthesis from H_2 and N_2. A physical model describing the promotion activity of REE on TM via the dipole magnetic interaction and the resulting increase in the probability of a catalytic act is suggested.

Introduction

The incorporation of REE into TM catalysts is known to promote their catalytic activity in hydro-dehydrogenation (1-5). The effect of REE on catalysis on a supported nickel was studied in (2). The magnetic interaction between TM metallic clusters and REE ions and the resulting increase in the probability of chemisorntion was discussed. This approach to the activation of a TM catalyst was verified in catalytic processes involving CO_x, NH_3, H_2 and O_2. The approach was found to be promising for the design of active adsorbents of CO in the combustion gases where the low temperature of gas-solid contact is critical.

Experimental

A series of TM catalysts were prepared; the incorporated metals and their ratios are shown in Table 1. Silica (BET-area 180 m^2/g), and a Y zeolite, both in the form of 0.1-0.4 mm particle fraction, were used as substrates; Nd_2O_3 was also used as a substrate in a powder form. TM and REE were supported by precipitating the corresponding chlorides in a 10% excess aqueous $(NH_4)_2CO_3$. A pulse technique was used to study the oxidation of Ni. Oxidation at room temperature was done for the calculation of the Ni^0 surface area. A similar procedure, performed at 450°C, provided an estimation of the total Ni^0 as referenced in (2). The magnetic characteristics of the samples were determined by a Faraday method in the field up to 15 kOe from 77 to 700K. Catalytic activity was GC- measured in a 1 ml flow reactor, for the following reactions:

hydrogenation of CO_x at 100-400°C, $H_2:CO_x$ = 1.5 to 50, gas flow rate of 6000 ml of the reaction gas per 1ml of catalyst per 1hour (hrs^{-1}); oxidation of CO at 100-450°C, He:CO:O_2 = 93:2:5, gas flow rate 9000 hrs^{-1}; synthesis of NH_3 at 200-700°C, $H_2:N_2$ = 3 to 50, gas flow rate 3000 hrs^{-1}. The conditions in these experiments were verified to provide a kinetic regime in a near-differential type reactor. Testing pH paper was applied to the reaction gas of NH_3 synthesis to determine the onset of nitrogen hydrogenation.

Table 1   Ni- catalysts. Annealed 6 hrs at 300°C, these catalysts were reduced in 2000 hrs^{-1} H_2 flow at a specified temperature.

No	Catalyst Samples	Ni_{total} wt%	wt%Ni^0 at $T_{red.}$		Ni area, m^2/g	
			400°C	500°C	400°C	500°C
1	Ni/SiO$_2$	1.8	0.3	1.0	1.4	5.9
2	Ni/SiO$_2$+Pr	2.8	-	0.5	-	2.2
3	Ni/SiO$_2$+Nd	2.7	-	0.4	-	1.9
4	Ni/SiO$_2$+Sm	2.5	-	0.5	-	2.5
5	Ni/SiO$_2$+Dy	2.2	0.063	0.5	0.55	2.0
6	Ni/SiO$_2$+Ho	1.9	-	0.5	-	1.5
7	Ni/SiO$_2$+Er	3.1	-	0.6	-	2.0
8	Ni/SiO$_2$+Tm	2.6	-	0.7	-	2.6
9	Ni/SiO$_2$+Yb	2.4	-	0.6	-	2.9
10	Ni-CaY	1.4	-	-	-	-
11	Ni/Nd$_2$O$_3$	1.0	0.4	0.4	1.0	1.2

Results and Discussion

Metal-support interactions between two strong magnetic phases, such as Ni and Nd_2O_3, were studied by thermo-magnetic measurement of individual oxides and sample 11. The slope of the magnetization curves provided a measure of the magnetic and/or structural isolation of Ni^{2+} ions. For example, NiO-CaY (sample 10) showed a homogeneous distribution of Ni^{2+} ions within zeolite cavities. NiO/SiO$_2$ (sample 1), in the contrary, showed an antiferromagnetic behavior, complicated by the structural defects. REE promoted, Ni, Fe and Co oxide catalysts showed about 30% decrease in the activation energy of CO oxidation at 400-700°C because of disorder, introduced by a strong paramagnetic REE. Magnetic measurements provided a determination of the crystal size of reduced nickel in these catalysts. This method works accurately for superparamagnetic systems with small nickel particles (5-8). Reduced samples 1 through 10 in Table 1 were found to be superparamagnetic under the applied conditions, and may be described in terms of the Langevin equation. This yielded an average diameter of the f.c.c. crystalline Ni^0 particles, which were found to be as small as 1 and 3 nm for the reduction temperatures of 400 and 500°C respectively. The results of a study of all REE catalysts listed in Table 1 suggested, however, the existence of dipol interactions between Ni^0 particles and promoters. This interaction occurred more significant for the 3 nm particle Ni^0 clusters formed at 500°C, causing a low-temperature deviation from the Curie law. In the case of smaller size of crystals in Ni-Dy catalyst reduced at 400°C, the numbers of parallel ordered atoms were estimated as 260 and 14 respectively.

CO hydrogenation on Ni/SiO_2 and Ni/SiO_2+Dy was characterized by a negative order of -0.5 and $E_{act.}$= of 100 kJ/mol for samples 1 and 5. Unlike the CO_2+H_2 system, the CO+H_2 reaction was found to be a size-sensitive. Table 2 shows that the reaction rate of non-promoted Ni/SiO_2 (sample 1) has increased by a factor of approximately 2 with the Ni^0 particle enlarged from 1 to 3 nm. The behavior of Ni/SiO_2 + Dy catalysts, however, contrasted with these results. Despite a significant increase in the catalyst activity, the size sensitivity of the promoted catalyst was reversed.

Table 2   A-factor of CO+H_2 reaction at C_{in} initial concentration of CO.

Sample	$T_{red.}$,	\multicolumn{4}{c|}{$A*10^{-10}$, molecule/Ni-atom*s}			
Number	°C	C_{in} = 2.0	6.7	20.1	36.4%
1	400	1.7	-	-	0.4
1	500	3.1	1.7	1.2	0.6
5	400	17.3	-	-	5.0
5	500	7.2	-	3.1	1.6

Unlike the unpromoted Ni/SiO_2 catalyst, increasing Ni^0 particle size has decreased the activity of surface Ni^0 in sample 5. This shows a higher promotion activity of REE on small Ni crystals, and is consistent with the CO_2 hydrogenation data, shown in Table 3. Measurement of the rates of CO+H_2 catalysis with and without incorporated REE provided the same $E_{act.}$ and the order; suggesting unaltered properties of a single Ni^0 catalytic site. Increase in the A-factor due to the REE is evident of a greater number of active sites.

CO_2 hydrogenation was another reaction, used to study TM-REE catalysts. The kinetic order of approximately 1 was determined for CO_2 from 17 to 240 torr pressure, shown in Table 3 as C_{in}, mol % at total atmospheric pressure. Activation energy of 100±10 kJ/mol was found regardless of the history of the catalyst, which suggests that the incorporation of REE did not affect the properties of active catalytic sites. However, A-factors indicated a significant change in a surface concentration of these sites. CO_2 hydrogenation reaction rate was found to be invariant to the size of Ni^0 particle. The differences in the activity, shown in Table 3, however, were attributed to the magnetic distortion of the Ni^0 particle by the surrounding paramagnetic REE. The comparison of the activities of catalysts with 1 and 3 nm of Ni crystals, reduced at 400 and 500°C respectively, showed the increase in activity by a factor of 2 for larger metal particles, whereas the small particles ($T_{red.}$=400°C) were 24 times more active than sample 11. This suggests that REE - induced magnetic distortion within a superparamagnetic Ni^0 crystal, which must be maximal for the small crystals, caused the observed activation. In the case of Ni^0/Nd_2O_3, for example, this distortion could be described in terms of a strong metal-support interaction. It is noteworthy that the dipole magnetic interaction is usually too weak to influence strong

Table 3   A-factor for $CO_2 +H_2$ reaction

Sample Number	$T_{red.}$, °C	$A*10^{-10}$, molecule/Ni-atom*s	
		$C_{in} = 2.3$	31.2%
1	400	-	0.64
1	500	0.058	0.64
5	400	-	15.6
5	500	0.14	1.5
11	400	-	7.4
11	500	-	7.4

interactions within a working catalysis system. These interactions may although cause a change in the electronic structure of catalyst and/or in the rate of energy exchange between collided molecule and catalyst crystal *via* the magnetic interaction between elements.

Measurements of a retention time of nitrogen pulse by the columns filled with the catalyst samples at high temperatures were undertaken to verify the increase of probability of chemisorption on Ni^0 in contact with Dy. N_2 pulses were passed through a 1x0.03 m chromatographic heated column filled with catalyst sample 1 or 5. The retention times were recorded, being a characteristic of probability of adsorption. The retention was found greater for the Ni+Dy catalyst sample 5 than for unpromoted sample 1. This difference increased at lower temperatures, which suggests that a relatively weak magnetic interaction between Ni and Dy at high temperature has increased with the decreasing temperature. NH_3 synthesis from N_2 and H_2 was studied up to 530°C. The rate of this reaction depends on the probability of N_2 adsorption. The sensitivity of the measurements of ammonia formed corresponds to approximately 1% of conversion. The temperature of the onset of nitrogen hydrogenation process was then recorded. The experiment showed that Ni/SiO_2 catalyst did not form ammonia until ~500°C, whereas Ni/SiO_2+Dy initiated NH_3 formation at around 300°C. Incorporation of Dy into the Ni catalyst resulted in an increased probability of nitrogen adsorption and, hence, the increased activity in ammonia synthesis, which is consistent with the above model.

Nickel-based catalysts are commonly active for dehydrogenation of hydrocarbons, and particularly, cyclohexane. The catalyst activity in the side exothermic reaction of hydrogenolysis, however, often controls the system, leading to the formation of C_1 products on a locally self-overheated surface. The kinetics of the system above the temperature of such an overheating is controlled by diffusion only. This situation was observed in the case of 57%Ni/SiO$_2$ catalyst (2). The benzene selectivity dependence was found broken due to catalyst surface excessive local heating at about 280°C. Control, or suppression, of the local heating was achieved by doping with small amounts of Pr compound, which was incorporated on the surface of reduced Ni crystals. A Ni-REE catalyst described above, modified with praseodymium chloride, showed an increased selectivity of benzene formation. The corresponding selectivity curve did not experience a breakage, indicating that excessive heating did not occur. This result may be explained by the increased rate of dissipation of the exothermic reaction heat in the catalyst lattice, caused by the presence of the rare earth element. This prevents overheating of the catalyst active site and the initiation of the hydrogenolysis. It is noteworthy that the lower the temperature, the greater the promotion effect of REE on the selectivity of benzene since the dipole interaction of Ni-to-REE becomes more significant at lower temperatures.

Summarizing the discussion we highlight that the incorporation of REE into the reduced Ni and other transition metal containing supported catalysts resulted in an increase of the effective number of active sites on them. This is indicated by the increase in preexponential factor in the temperature dependence of CO, CO$_2$ hydrogenation rate, and by the decrease in the temperature of onset of NH$_3$ synthesis when REE were added. This was interpreted as an elevation of the probability of chemisorption of gas molecules on Ni^0 surface, which is controlled by the rate of chemisorption heat dissipation from a collided gas molecule to the catalyst body. This heat transfer process can be facilitated by and via magnetic interaction between superparamagnetic crystal (e.g., f.c.c. Ni^0 crystal) and strong paramagnetic additive REE. The interaction was found to be more influential in hydrogenation in the case of smaller Ni^0 particles, or when the magnetic interaction energy has increased compare to the anisotropic energy of ferromagnetic nickel particle.

Magnetic interaction itself is too weak to be able to influence any of catalytic acts. It, however, may increase the overall catalysis rate by accelerating the heat exchange within a tiny moment of gas molecule collision to the metal surface, providing more molecules remaining on surface to participate in the following catalysis. This phenomenon must prevail at lower temperatures, when magnetic interaction is less disturbed by heat, which was indicated for cyclohexane dehydrogenation. The catalytic property of a single active site, which is determined by the nature of TM, is not affected by the presence of REE, which is indicated by unchanged activation energy of CO/CO$_2$

hydrogenation. Oxide-state catalysts of the same content, however, are deemed to be different also in their crystal structure. This was indicated by a decrease in activation energy of CO oxidation on their surfaces. Strong paramagnetic REE compounds may efficiently oppose the formation of, for example, cubic NiO lattice on a stage of a catalyst preparation. This would facilitate the formation of more amorphous TM oxide.

Conclusions

Incorporation of REE into the reduced Ni and other transition metal containing supported catalysts result in an increase of the effective number of active sites on them. Magnetic interaction may increase the overall catalysis rate by accelerating the heat exchange within a tiny time moment of gas molecule collision to a metal surface, providing more molecules remaining on surface to participate in the following catalysis. This phenomenon must prevail at a lower temperature, when magnetic interaction is less disturbed by heat. The catalytic property of a single active site, which is determined by the nature of TM, is not affected by the presence of REE, which is indicated by unchanged activation energy of CO and CO_2 hydrogenation. Oxide-state catalysts of the same content, however, are deemed to be different also in their crystal structure. This was indicated by a decrease in activation energy of CO oxidation on their surfaces. Strong paramagnetic REE compounds may efficiently oppose the formation of, for example, cubic NiO lattice on a stage of a catalyst preparation. This would facilitate the formation of more amorphous TM oxide.

References

1.   X. Zhang, A.B. Walters, M.A. Vannice, *J. Catalysis*, **155**, 290 (1995).
2.   A.J. Dyakonov, *Sov. J. Chem. Physics*, **9**, 2543 (1992).
3.   H. Immamura, Y. Miura, K. Fujita, Y. Sakata, S. Tsuchia, *J.Molecular Catalysis A: Chemical*, **140**, 81 (1999).
4.   K.O. Xavier, R. Sreekala, K.K.A. Rashid, K.K.M. Yusuff, B. Sen, *Catalysis Today*, **49**, 17 (1999).
5.   W. Teunissen, A.A. Bol, J.W. Geus, *Catal. Today*, **48**, 329 (1999).
6.   K. Kili, F. Le Normand, *J. Molecular Catalysis A: Chemical*, **140**, 267 (1999).
7.   J. Barroult., *Appl.Catal.*, **21**, 307 (1986).
8.   A.L. Borer, R. Prins, *J. Catalysis*, **144**, 439 (1993).

Synthesis and Study of Amino Acid-Based Ligands for Late Transition Metal Complexation

Patrycja V. Galka and Heinz-Bernhard Kraatz

Department of Chemistry, University of Saskatchewan, 110 Science Place, Saskatoon, SK S7N 5C9 CANADA

Abstract

Phosphinito amino acids are readily prepared from serine and tyrosine derivatives and chlorodiphenylphosphine. These new ligands were fully characterized and readily form metal complexes with Pt(II) precursors. The X-ray structure of cis-{Boc-Ser(OPPh$_2$)-OMe}PtCl$_2$ is reported, exhibiting H-bonding between the phosphino amino acid ligands. The complex exhibits a low reactivity towards the hydrosilation of styrene.

Introduction

A major research direction pursued in our laboratories is the development of phosphino- and phosphinito-peptides as chiral ligands for catalytic asymmetric hydrogenation and hydrosilation of olefins. Although used as chiral auxiliaries (1), peptides have been completely neglected as chiral ligand in homogeneous catalytic systems and offer the unique opportunity to develop a flexible approach to the development of chiral and catalytically active transition metal phosphine and phosphinito complexes. Being made up of amino acids building blocks, peptides can be rationally designed to give molecules with particular shapes, sizes and solubility properties. There are numerous examples of aminophosphine- and amidophosphine-phosphinito ligands based on hydroxyproline, its derivatives (2, 3), other amino alcohols and of sugar-based (3) ligands that are being used for a variety of asymmetric catalytic transformations (2b). As of yet, these systems have not been incorporated into larger peptidic frameworks, nor have small phosphinito-peptides been prepared. Only recently, Gilbertson reported an example of a helical peptide carrying a phosphino amino acid (3, 5). Only preliminary coordination and reactivity studies with were reported (3, 5). Chiral phosphine and phosphinito ligands based on hydroxyproline, such as those described by Mortreux and others have been used for catalytic transformations, such as asymmetric hydrogenation and hydroformylation of olefins and ketones (2). To the best of our knowledge, there are no reports in the literature investigating phosphino- or phosphinito-peptides as chiral reaction vessels for transition metal catalyzed asymmetric transformations.

In this paper, we will describe the first step in our studies, which is the preparation of simple amino acid-based phosphinito ligands, their characterization, coordination chemistry with Pt(II) and their relevance to our overall goal of designing larger peptidic cavitants from small phosphinito (and phosphino) amino acid building blocks. Using these building blocks, we will assemble more elaborate systems using peptide-coupling strategies currently used in our laboratories. Some preliminary results of the hydrosilation of styrene using one of the Pt(II)- complexes will also be discussed.

Results and Discussion

Starting from the readily available fully protected amino acid Boc-Ser(OH)-OMe, we were able to obtain the corresponding phosphinito serine derivative, which readily form platinum complexes when reacted with (cod)PtCl$_2$ in THF (Scheme 1).

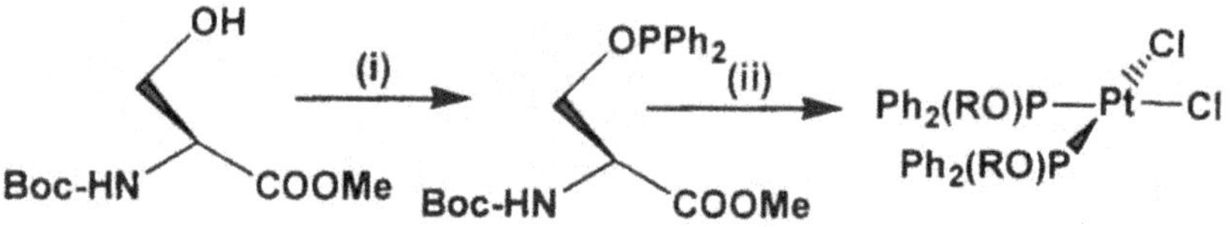

Scheme 1: Synthesis of Boc-Ser(OPPh$_2$)-OMe and cis-{Boc-Ser(OPPh$_2$)-OMe}PtCl$_2$. (i) ClPPh$_2$, CH$_2$Cl$_2$, Et$_3$N, DMAP; (ii) (cod)PtCl$_2$, THF

Other N-protecting groups such as the Z group and N-linked peptides do not interfere with the reaction, allowing the syntheses of Z-Ser(OPPh$_2$)-OMe (2) and of the phosphinito-peptide Z-β-Ala-Tyr(OPPh$_2$)-OMe (3) in quantitative yields. These new ligands were fully characterized by NMR and MS. All ligands form colorless complexes with Pt(II). For cis-{Boc-Ser(OPPh$_2$)-OMe}PtCl$_2$ (4), single crystals were obtained from a CH$_2$Cl$_2$/hexanes mixture and subjected to X-ray crystallographic analysis. The ^{31}P NMR spectrum of solutions of 4 and cis-{Z-β-Ala-Tyr(OPPh$_2$)-OMe}PtCl$_2$ (5) exhibit similar spectra showing singlets around δ 84 with Pt satellites having a J$_{Pt-P}$ typical of cis arrangement of the ligands in the complex.

A view of 4 is shown in Figure 1. Rather than discussing the structure in detail, we would like to point out the general features. As expected for a Pt(II) complex, the coordination geometry is close to square planar having the two phosphinito amino acid ligands *cis* and compares well with other known Pd(II) and Pt(II) phosphinite structures (2, 6). The Pt-P distances are slightly shorter than those reported by Mortreux and coworkers for [PtCl₂((S)-Ph,Ph-ProNOP)] (2), indicating stronger binding and decreasing the ability of the phosphinite ligand to dissociate. In addition, the two phosphinito ligands are linked at the C-termini by 2 H-bonds locking the ligand into position. At present we have no evidence to suggest that H-bonding is present in solution.

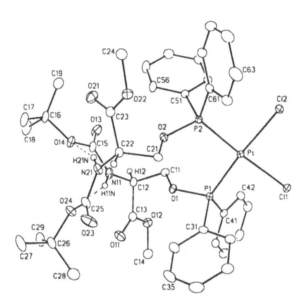

Figure 1   ORTEP view of **4**. Selected bond distances and angles: Pt-Cl(1) 2.368(5) Å, Pt-Cl(2) 2.360(4) Å, Pt-P(1) 2.226(4) Å, Pt-P(2) 2.223(3)°, Cl(1)-Pt-Cl(2) 86.2 (2)°, P(1)-Pt-P(2) 95.8(2)°, Cl(1)-Pt-P(2) 174.2(2)°, Cl(2)-Pt-P(1) 173.1(2)°, H11N-O24 2.41 Å, H21N-O14 2.28 Å

However, the structural features help to explain the observed reactivity of **4** towards the hydrosilation (7). Preliminary results of the hydrosilation of styrene in THF solution are presented in Table 1. The Pt-P bond distances indicate a strong Pt-P bond, which decreases the ligand's ability to dissociate from the metal center. This ability is decreased even further by the formation of H-bonds, locking the ligands in place. Since ligand dissociation to form a vacant coordination site on the metal center is key for olefin coordination and hence for efficient catalysis, decreasing the ability of the ligand to dissociate will result in a net decrease of the catalyst's efficiency which is reflected in the low overall yield of the hydrosilation reaction using **4** as the catalyst, ranging from 1-4% (in THF). In general, the reaction in THF is slow. Expectedly, longer reaction times and heating will increase the yields. Furthermore, both the regioisomers of the saturated (**B + C**) and unsaturated (**A**) hydrosilation product are obtained. With **5** as catalyst, we observe only **B** and **C**. We find no evidence for the formation of the unsaturated hydrosilation product

A. At present we cannot rationalize the apparent change in reactivity of **1** versus **5**. We will carry out more detailed studies in order to explore the differences in reactivity further. In particular, we wish to address the question of stereoselectivity of the hydrosilation reaction using **1** and related phosphinito amino acids as building blocks for larger phosphinito peptide frameworks.

Table 1 Hydrosilation of styrene in THF.

Run	ligand	T (in °C)	t (in h)	A	B+C	Overall Yield
1	1	20	12	12	88	1
2	1	20	48	5	95	2
3	1	60	48	4	96	4
4	1	20	12	12	88	15[b]
5	3	20	12	100	-	4

[a] Yields of A, B, and C are based on GC yields.
[b] Addition of ca. 1 equivalent of **1** to Cl$_2$Pt(cod). The overall yield is comparable to that obtained by the reaction of of Cl$_2$Pt(cod) with styrene and Et$_3$SiH. However, under these conditions, the regioselectivity is lost and the ration of A to B+C is ca 1.

Experimental

General Synthesis of Phosphinito Amino Acids and Peptides: Boc-Ser(OPPh$_2$)-OMe (1): Via Hamilton syringe, Ph$_2$PCl (4.46 mmol; 0.82 mL) is added dropwise to a solution of Boc-Ser-OMe (4.56 mmol; 1 g), triethylamine (13.68 mmol; 1.9 ml) and DMAP (0.05 g) in dry dichloromethane (20 ml). The mixture was stirred overnight. After 12 h, the reaction mixture was filtered and all volatiles were evaporated in vacuo, giving 1.52 g of an off-white solid. The solvent was removed *in vacuo* giving rise to a white-yellowish solid. Yield 83% ^{31}P{^{1}H} NMR (δ, C$_6$D$_6$): 117.6. ^{1}H NMR (δ, C$_6$D$_6$): 7.48-7.40 (4H, m, Ph of PPh$_2$), 7.16-7.00 (6H, m, Ph of PPh$_2$), 5.69 (1H, d, J$_{HH}$ = 9 Hz, NH), 4.61 (1H, m, CH of Ser), 4.10 (2H, m, CH$_2$ of Ser), 3.20 (3H, s, OCH$_3$), 1.38 (9H, s, CH$_3$ of Boc). ^{13}C{^{1}H} NMR (δ, CDCl$_3$): 170.8 (C=O), 154.9 (C=O), 143.2, 130.7, 130.4, 128.6, 129.3 (all C$_{ar}$), 80.2 (CH$_2$), 70.0 (C), 56.1 (CH), 52.6 (OCH$_3$), 28.4 (CH$_3$).

Z-Ser(OPPh$_2$)-OMe (2): ^{31}P{^{1}H} NMR (δ, C$_6$D$_6$): 118.1 (s). ^{1}H NMR (δ, C$_6$D$_6$): 7.53-7.17 (15H, m, Ph of PPh$_2$ and Ph of Z), 6.06 (d, (1H, d, J$_{HH}$ = 9 Hz, NH), 5.10 (2H, d, J$_{HH}$ = 6 Hz, CH$_2$), 4.69 (1H, m, CH), 4.18 (2H, m, CH$_2$), 3.28 (3H, s, OCH$_3$). ^{13}C{^{1}H} NMR (δ, CDCl$_3$): 176.8 (C=O), 174.9 (C=O), 153.2 (C$_{ar}$), 130.3, 130.1, 129.6, 128.4, 128.1, 127.9 (C$_{ar}$), 64.3 (CH$_2$), 57.9 (CH), 54.9 (OCH$_3$).

Z-β-Ala-Tyr(OPPh$_2$)-OMe (3): ^{31}P{^{1}H} NMR (δ, C$_6$D$_6$): 111.5 (s). ^{1}H NMR (δ, C$_6$D$_6$): 7.51-7.34 (10H, m, CH of Ph), 7.27 (5H, m, Ch of Ph and Tyr), 6.73 (4H, m, CH of Ph and Tyr), 6.59 (1H, d, J$_{HH}$ = 9 Hz, NH of Tyr), 5.58 (1H, m, NH of Ala), 5.00 (2H, s, CH$_2$ of Z), 4.69 (m, 1, CH of Tyr), 3.61 (s, 3,OCH$_3$), 3.35 (m, 2,CH$_2$ of Ala), 2.94 (m, 2, CH$_2$ of Tyr), 2.35 (m, 2,CH$_2$ of Ala). ^{13}C{^{1}H} NMR (δ, CDCl$_3$): 171.7(C$_{ar}$), 171.15 (C=O$_2$), 156.4 (C=O$_{Ala}$), 155.8 (C$_{ar}$), 140.5, 140.2 (C), 136.4 (C=O$_{Tyr}$), 130.6, 130.3, 128.5, 128.4 (C$_{ar}$), 126.2 (C), 118.5 (CH), 66.0 (CH$_2$ of Z), 53.5 (CH), 53.3 (CH$_2$ of Tyr), 51.8 (OCH$_3$), 36.4, 36.9 (CH$_2$ of Ala).

General Synthesis of cis-Cl$_2$Pt(ligand): of cis-Cl$_2$Pt(1) (4): To a stirring solution of 1 (0.26 mmol; 0.105 g) in THF (10 mL) a suspension of Cl$_2$Pt(cod) (0.26 mmol; 0.05 g) was added dropwise. The solvent was evaporated in vacuo, resulting in a white solid. Recrystallization from CH$_2$Cl$_2$/hexane resulted in the deposition of colorless plates. ^{31}P NMR (δ, CDCl$_3$): 84.8 (J$_{Pt-P}$ = 3471 Hz). cis-Cl$_2$Pt(3) (5): ^{31}P NMR (δ ppm, CDCl$_3$): 84.1 (J$_{Pt-P}$ = 3510).

Hydrosilation of styrene: To a Schlenk flask containing (ligand)PtCl$_2$ (0.053 mmol) in THF (20 mL), 5.3 mmol, 0.6 mL of the styrene and 5.3 mmol, 0.85 mL of triethylsilane were added. The temperature of the reaction vessel was kept at 20°C or 60°C (water bath) and the progress of the reaction was checked by GC chromatography (HP, 30 m crosslinked 5% PH ME Siloxane capillary column, FID) and GC-MS (THE70250S/SEQ Double Focussing E/B Sector, VG Analytical 70/20 VSE chromatograph, Fision 8060, 8000 series)

Acknowledgements

We thank the Natural Sciences and Engineering Research Council of Canada for financial support in the form of an operating grant (#218857-99) and the Department of Chemistry for additional financial support. We also wish the express our gratitude to Bob McDonald (University of Alberta) for the data collection of **4**.

References

1. H. Nitta, D. Yu, M. Kudo, A. Mori and S. Inoue, *J. Am. Chem. Soc.*, **114**, 7969 (1992), and references therein.
2. a) S. Naili, J.-F. Carpentier, F. Agbossou, A. Mortreux, G. Nowogrocki and J.-P. Wignacourt, *Organometallics*, **14**, 401 (1995), and references therein.
   b) A. Roucoux, L. Thieffry, J.-F. Carpentier, M. Devocelle, C. Méliet, F. Agbossou and A. Mortreux, *Organometallics*, **15**, 2440 (1996), and references therein.
   c) F. Agbossou, J.-F. Carpentier and A. Mortreux, *Chem. Rev.* **95**, 2485 (1995).
   d) G. Parrinello and J. K. Stille, *J. Am. Chem. Soc. 109*, 7122 (1987).
   e) K. Tani, K. Suwa, E. Tanigawa, T. Ise, T. Yamagata, Y. Tatsuno, S. Otsuka, *J. Organomet. Chem.* **370**, 203 (1989).
3. S. R. Gilbertson and D. Xie, *Angew. Chem. Int. Ed. Engl.* **38**, 2750 (1999) and references therein.
4. T. V. RajaBabu, T. A. Ayers, G. A. Halliday, K. K. You and J. C. Calabrese, *J. Org. Chem.*, **62**, 6012 (1997).
5. a) S. R. Gilbertson and R. V. Pawlick, *Angew. Chem. Int. Ed. Engl.*, **35**, 902 (1996).
   b) S. R. Gilbertson and X. Wang, *J. Org. Chem.*, **61**, 434 (1996).
6. E. Cesarotti, M. Grassi, L. Prati and F. Demartin, *J. Chem. Soc. Dalton Trans.*, 2073 (1991).
7. Ojima, I. Catalytic Aymmetric Synthesis, VCH, New York, 1993.

DBU as Nucleophilic Catalyst in the Baylis Hillman Reaction

Varinder K. Aggarwal and Andrea Mereu

Department of Chemistry, University of Sheffield, Brook Hill, Sheffield, UK, S3 7HF

Abstract

Nucleophilic and unhindered catalysts are normally required for the Baylis Hillman reaction. We have discovered that DBU, which is normally regarded as a hindered and non-nucleophilic base, is in fact the optimum catalyst for this reaction providing adducts at much faster rates than using DABCO or 3HQD. The scope of the Baylis Hillman reaction is enhanced using this catalyst and implications of this finding are discussed.

Introduction

The Baylis Hillman reaction (1) has great synthetic utility as it converts simple starting materials into densely functionalised products (2).

The RDS of the Baylis Hillman reaction is the reaction of the aldehyde **4** with the ammonium enolate **3** (3). To obtain faster rates, higher concentrations of the ammonium enolate are required. Amines which can shift the equilibrium towards the ammonium enolate **3** by stabilising this species should achieve this.

595

Results and Discussion

In this paper we describe our study on the nature of the catalyst and the discovery that much higher rates can be achieved with alternative structures.

We started to screen a range of amines which all had the potential for the positive charge on nitrogen to be stabilised through conjugation with another heteroatom (Table 1). These catalysts were screened in our standard test reaction.

Of the aromatic heterocyclic catalysts (runs 1, 2) only Dimethylaminopyridine (DMAP) gave any Baylis Hillman adduct, but at a rate only slightly higher than DABCO (Table 2, runs 1, 3). None of the amidine catalysts were stable under the reaction conditions except for DBU 7 which not only gave a clean reaction but also the fastest rate (Table 1, run 7; Table 2, run 6).

Table 1 Amine Catalysts in the Baylis-Hillman reaction[a]

Run	Nucleophilic compound	Time [h]	Yield [%]
1	DMAP	96	87
2	1-Methylimidazole	120	0
3	2-Methyl-2-oxazoline	120	0
4	N-Methyl-4,5-dihydroimidazole	1	10
5	DBN[b]	2 min	13[c]
6	DBU[d]	6	89
7	Substituted guanidine[e]	48	30

[a] Reaction of methyl acrylate and benzaldehyde in presence of amine (1 eq. each); no solvent used.
[b] 1,5-Diazabicyclo[4,3,0]non-5-ene.
[c] A fast reaction occurred but the catalyst decomposed rapidly.
[d] 1,8-Diazabicyclo[5,4,0]undec-7-ene.
[e] 1,3,4,6,7,8-Hexahydro-1-methyl-2H-pyrimido [1,2-a]pyrimidine.

In comparison with other commonly used catalysts DBU was over an order of magnitude faster than the current best catalyst (run 4, 3-hydroxyquinuclidine 3-HQD) and superior to the DABCO-La(OTf)$_3$-triethanolamine system that we have developed (run 2) (4). Even at 10 mol% loading, DBU was superior to both stoichiometric DABCO or 3-HQD (Table 2, run 5).

Table 2 Comparison of DBU with the other Catalytic Systems[a]

Run	Catalytic system	Reaction rate[b]	k_{rel}	Time [h]	Yield [%]
1	DABCO	0.016	1	120	91
2	DABCO-La(OTf)$_3$-N(CH$_2$CH$_2$OH)$_3$[c]	0.511	31.9	12	83
3	DMAP	0.038	2.4	96	87
4	3-HQD	0.076	4.8	30	91
5	DBU (0.1 eq.)	0.163	10.2	24	75
6	DBU	0.762	49.5	6	89

[a] Reaction of methyl acrylate and benzaldehyde in presence of amine (1 eq. each); no solvent used).
[b] % Product per minute.
[c] DABCO 1 eq.,La(OTf)$_3$ 0.05 eq. and N(CH$_2$CH$_2$OH)$_3$ 0.5 eq. .

We investigated the scope of the DBU-catalysed Baylis Hillman reaction (5) by reacting benzaldehyde with a range of Michael acceptors (Table 3), and methyl acrylate (Table 4) and 2-cyclohexen-1-one (Table 5) with a range of electrophiles.

Notable examples from Table 3 include a fast reaction with tert-butyl acrylate (run 3) (6), and a very rapid reaction with 2-cyclohexen-1-one (run 5).

Table 3 DBU Catalysed Reactions of Activated Alkenes with Benzaldehyde

Run	Alkenes	Time [h]	Yield [%]
1	Methyl acrylate	6	89
2	Ethyl acrylate	24	80
3	tert-Butyl acrylate	72	74[a]
4	Acrylonitrile	3	92
5	2-Cyclohexen-1-one	0.5	60[b]

[a] No side products; 20-25% of benzaldehyde and acrylate recovered.
[b] No benzaldehyde left, 30% of 2-cyclohexen-1-one recovered.

Table 4 DBU Catalysed Reactions of Electrophiles with Methyl Acrylate

Run	Aldehyde	Time [h]	Yield [%]
1	Benzaldehyde	6	89
2	2-Nitrobenzaldehyde	1.5	95
3	4-Nitrobenzaldehyde	1	95
4	4-Anisaldehyde	48	62
5	Propionaldehyde	24	17[a]
6	Trimethylacetaldehyde[b]	70	20[a]
7	2,2,2-Trifluoroacetophenone	2	60[c]
8	2,2,2-Trifluoroacetophenone[d]	48	78

[a] Decomposition of aldehyde (aldol reactions) (7).
[b] Reaction performed in presence of 0.05 eq. La(OTf)$_3$.
[c] Product unstable in presence of high concentration of catalyst.
[d] 0.1 eq DBU used.

Notable examples from Table 4 include reaction with 4-anisaldehyde which gave a good yield after just 2 days (run 4) (8) and for the first time reactions with pivaldehyde (run 6) (9) and trifluoroacetophenone (runs 7, 8). Pivaldehyde required La(OTf)$_3$ to promote the reaction as no adduct was obtained without it. La(OTf)$_3$ often enhances the rates and gives higher yields of adducts (Table 4, run 6; Table 5, run 4).

Table 5 DBU Catalysed Reactions of Aldehydes with 2-Cyclohexen-1-one

Entry	Aldehyde	Time	Yield [%]
1	Benzaldehyde	1 h	65
2	2-Anisaldehyde	50 min	70
3	Cyclohexanecarboxaldehyde	7 h	73
4	Trimethylacetaldehyde[a]	21 h	75

[a] 1.2 eq. 2-cyclohexen-1-one and 0.05% La(OTf)$_3$ used.

Notable examples from Table 5 include high yielding reactions with all the difficult aldehydes (2-anisaldehyde, run 2; pivaldehyde, run 4; and even the aliphatic enolisable aldehyde cyclohexanecarboxaldehyde, run 3), demonstrating the effectiveness and superiority of this catalyst. Evidently with faster reacting enones compared to acrylates, enolisable aldehydes are better tolerated.

DBU is considered to be a non-nucleophilic hindered base (10,11); features that are diametrically opposite to what is normally required of amine catalysts for the Baylis Hillman reaction. DABCO, for example is one of the best amine catalysts and is an unhindered, nucleophilic base. In contrast, DBU stabilises the intermediate β-ammonium enolate through conjugation which increases its equilibrium concentration and this results in significantly enhanced rates.

These studies reveal that to achieve high rates in the Baylis Hillman reaction the nucleophilicity of the amine is much less important than factors which stabilise the intermediate β-ammonium enolate.

Experimental

All reactions performed on a 2 mmol scale using a 1:1:1 ratio of carbonyl compound:alkene:catalyst (except were specified otherwise); no solvent were used. Kinitic data (Table 2) obtained by GC-MS analysis (Perkin Elmer instrument)

To a stirred mixture of methyl acrylate (0.71 ml, 7.87 mmol) and benzaldehyde (0.8 ml, 7.87 mmol) at room temperature under nitrogen was added DBU (1.18 ml, 7.87 mmol). After 6h the reaction was stopped by diluting with ether (30 ml) and washed with HCl (2 M, 20 ml) followed by water (20 ml). After drying over Na$_2$SO$_4$, filtration and evaporation, the crude mixture was purified by column chromatography eluting with petroleum ether/diethyl ether (2:1) to give the adduct (1.3 g, 89%) as an oil.

Acknowledgements

We thank EPSRC for financial support for this work.

References

1   M. E. D. Hillman, and A. B. Baylis, *Chem. Abstr.*, 77, 34174q (1972).
2   For reviews see: a) S. E. Drewes, and G. H. P. Roos, *Tetrahedron*, 44,
         4653-4670 (1988); b) D. Basavaiah, P. D. Rao, and R. S. Hyma,
         Tetrahedron, 52, 8001-8062 (1996); c) E. Ciganek, *Organic Reactions*,
         51, John Wiley, 201-350 (1997).
3   J. S. Hill, and N. S. Isaacs, *J. Phys. Org. Chem.*, 3, 285 (1990).
4   V. K. Aggarwal, A. Mereu, G. J. Tarver, and R. McCague, *J. Org.
         Chem.*, 63, 7183-7189 (1998).
5   There is one isolated example on the use of DBU as a catalyst to obtain
         a the Baylis-Hillman product (10). However, they exclude the
         possibility of a Baylis Hillman mechanism putting foward a mechanism
         (in which DBU acts as a base) that is not consistent with our own
         observations (also we have not been able to reproduce some of their
         results): J. R. Hwu, G. H. Hakimelahi, and C. T. Chou, *Tetrahedron
         Lett.*, 33, 6469 (1992).
6   DABCO gives 65% yield after 28d: Y. Fort, M. C. Berthe, and P.
         Caubere, *Tetrahedron*, 48, 6371 (1992).
7   Low yields of adducts with enolisable aldehydes using DBU as a
         catalyst has been reported.   See: a) D. Basavaiah, and V. V. L.
         Gowriswari, *Synth. Commun.*, 17, 587 (1987); b) P. Auvray, P.
         Knochel, and J. F. Normant, *Tetrahedron*, 44, 6095 (1988).
8   DABCO gives 90% after 20d: A. Foucaud, and F. El Guemmout, *Bull.
         Chim. Soc. Fr*, 3, 403 (1989).
9   Reaction using DABCO reportedly failed.   See: M. C. Berthe, P.
         Caubere, Y. Fort, *Eur. Patent Appl.* , EP 465,293 (1992); [*Chem.
         Abstr.*, 116, 152605c (1992)]; U. S. Patent, 5,332,836 (1994).
10   For a review on the use of DBU as a base see: H. Oediger, F. Moller,
         K. Eiter, *Synthesis*, 591 (1972).
11   For contrary examples, see: a) M. Chakrabarty, A. Batabyal, and A.
         Patra, *J. Chem. Research (S)*, 190 (1996); b) L. Ma, and D. Dolphin,
         Tetrahedron, 52, 849 (1996); c) R. Reed, R. Rean, F. Dahan, and G.
         Bertrand, *Angew. Chem. Int. Ed.*, 32, 399 (1993); *Angew. Chem.* 1993,
         105, 464; d) M. G. Johnson, and R. J. Foglesong, *Tetrahedron Lett.*,
         38, 7003 (1997); e) R. D. Chambers, A. J. Roche, A. S. Batsanov, and
         J. A. K. Howard, *J. Chem. Soc. Chem. Comm.*, 2055 (1994).

Bifunctional Copper Catalysts, Part III: Carvenone Synthesis Starting from Limonene Oxide

F. Zaccheria[a], R. Psaro[a], N. Ravasio[a], L. De Gioia[b]

[a]C.N.R. Centro CSSCMTBSO, c/o Dipartimento di Chimica I.M.A.,
Via Venezian 21, 20133 Milano, Italy
[b]Dipartimento di Biotecnologie e Bioscienze, P.za della Scienza, 2
Università degli Studi di Milano-Bicocca, 20126 Milano, Italy

E-mail: labcat@csmtbo.mi.cnr.it

Abstract

Different silica-aluminas were shown to be effective catalysts for the opening of limonene oxide. Moreover, the two geometric isomers react at different rates to give different products; this allowed a kinetic resolution. Some silica-alumina cogels can also be used as supports for copper based catalysts allowing the direct transformation of limonene oxide into carvenone.

Introduction

We recently focused our research activity on the design of bifunctional catalysts allowing the reduction of the number of steps in organic synthesis or chemical transformation (1, 2). Moreover, we are continuously looking for heterogeneous alternatives to the use of homogeneous Lewis acids in order to minimize inorganic salts and/or toxic waste production. We already reported that dihydrocarvone 1 could be converted into carvenone 2 in the presence of copper catalysts supported on different silica aluminas (3) (Scheme 1).

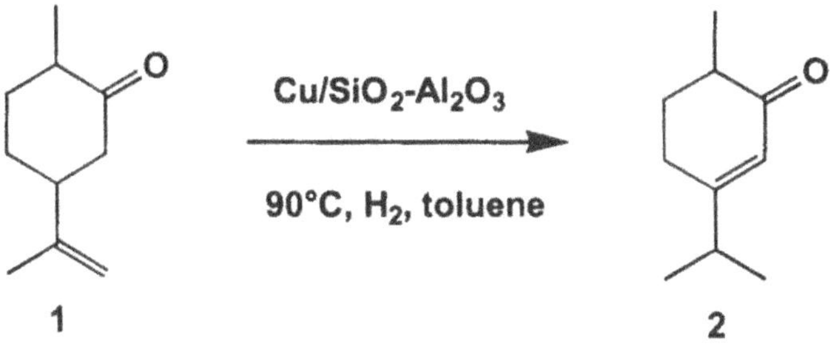

Scheme 1

It should be stressed that this reaction requires the presence of both reduced copper and molecular hydrogen, thus suggesting that isomerization proceeds via hydride intermediate.

We also investigated the activity of silica aluminas as solid Lewis acids in the opening of α-pinene oxide and we found that they are effective and selective towards the production of α-campholenic aldehyde, a valuable intermediate in the synthesis of α and β santalol (4) (Scheme 2). Due to the interest in carvenone 2 in the fragrance industry, we planned to obtain 2 from readily available limonene oxide 3 by opening the epoxide in the presence of silica alumina and promoting the conjugation process by means of a copper catalyst. A better way should be to carry out the two steps with the same catalyst, given that the presence of the metal doesn't inhibit the epoxide opening (Scheme 2).

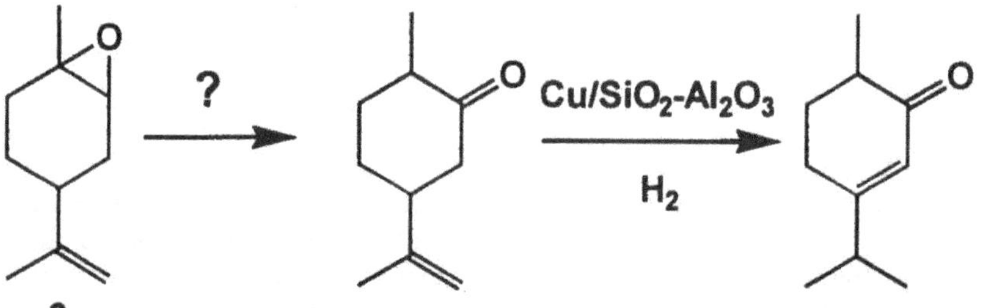

Scheme 2

Experimental

Silica-alumina cogels, (A, 13% Al_2O_3, N_2-BET=400 m^2/g, PV=1.1 ml/g; B, 13% Al_2O_3, N_2-BET=475 m^2/g, PV=0.8 ml/g; C, 25% Al_2O_3, N_2-BET=400 m^2/g, PV=1.0 ml/g; D, 1.5% Al_2O_3, N_2-BET=485 m^2/g, PV=0.79 ml/g), hereafter called respectively A, B, C, D, were obtained from Grace Davison (Worms, D). The (+)-cis and (+)-trans limonene 1,2-epoxide (>99%) were purchased from Fluka and the mixture of cis and trans (+)-limonene oxide (>97%) from Aldrich. The solid acids were treated at 270°C for 20 minutes in air and for 20 minutes under reduced pressure at the same temperature. The 8% copper catalysts, prepared as already reported (5), underwent a further treatment of reduction at 270°C with H_2 at atmospheric pressure. For the epoxide opening, the substrates (0.05 g) were dissolved in toluene dehydrated over molecular sieves (4 ml) and the solution transferred under N_2 into a glass reaction vessel where the catalyst (0.05 g) had been previously dehydrated. Reactions were carried out at room temperature with magnetic stirring under N_2. After epoxide opening, carried out as written above, H_2 was charged in the glass vessel and the reaction carried out at 90°C under magnetic stirring.

Reaction mixtures were analyzed by GC using a crosslinked 5% phenyl methyl silicone (HP 5 M.S., 30 m) or a nonbonded, poly(80% biscyanopropyl/20% cyanopropylphenyl siloxane; SP2330, 60 m) capillary column. Reaction products were identified through their IR, MS (HP 5971 series) and ^{1}H NMR spectra (Bruker 300 MHz). In Table 2 only significative products are reported, the others being rearrangement products.

Molecular mechanics and dynamics calculations were carried out on a Silicon Graphics Indigo R10000 workstation, using the Cerius2 software package (Molecular Simulations Inc., San Diego, CA). Molecular mechanics geometry optimizations were carried out using the conjugate gradient algorithm requiring that the residual gradient of energy did not exceed 0.01 kcal mol^{-1} Å^{-1}. Molecular dynamics calculations were carried out to sample the potential hypersurface at the costant temperature of 300, 500 and 700 K, respectively. The Universal Force Field (6) was used in all calculations.

Results

1-Epoxide opening: the study of the reaction of a 1:1 *cis:trans* mixture of limonene oxide **3** in the presence of silica aluminas, revealed a different reactivity of the two isomers. To have a deeper insight into the behaviour of these geometric isomers, we studied them separately. Results are reported in Table 1. It appears that not only the *cis* isomer reacts much faster, but also that the reaction products are different. The *cis* gave mainly aldehyde **4**, whereas **1** was obtained with fairly good selectivity from the *trans*. As far as the catalyst influence is concerned, only minor differences were observed with *trans* **3**, while in the reaction of *cis* **3**, catalyst B produced aldehyde **4** with lower selectivity. On the other hand, significant amounts of exo-carveol **6** were formed in this reaction.

Only few studies deal with the isomerization of **3** (7-9); in particular ZnBr$_2$, which is the most widely used Lewis acid in terpene chemistry, was found to give **4** and **1** together with the cyclopentylacetone derivative. Tanabe (10) explored the reactivity of different solid acids and found that silica alumina gives mainly **4** and **1**, while alumina gives **5** and **6** and sulfuric acid treated silica para-cymene (4-isopropyl toluene) and carvenone **2**. Therefore, results obtained in the isomerization of *cis* **3** with B seem to be in agreement with a less acidic character of this support.

The different reactivity of the two geometric isomers has never been reported. Only in the presence of 5 M LiClO$_4$·Et$_2$O (11) was *trans* **3** found to react specifically, giving *trans* **1** and leaving the *cis* unreacted. To justify their results, the authors invoke the rule of diaxial ring opening. In the presence of our solid catalyst, the opposite behaviour is observed the *cis* **3** reacting much faster. In our opinion, this is due to its easier access to the catalytic site, thus comparison between the two more stable conformers shows the oxygen atom as more hindered in the *trans* isomer (Figure 1). The sharp difference in the reaction rate suggested that we look for the conditions that will allow the kinetic resolution of the 3 mixture. This can be conveniently realized by using catalyst B (entry 10); almost all the *trans* 3 can be recovered unaffected. On the other hand, people interested in aldehyde 4 can obtain it with 42% yield in the presence of catalyst A through reaction of the commercial mixture of isomers, much cheaper than the pure *cis* isomer.

Table 1- Isomerization of limonene oxide in presence of silica-alumina catalysts under N_2, at room temperature in toluene.

Entry	Substrate	Catalyst	t (h)	% trans 3	% 4	%1	% 5	% 6
1	cis 3	A	1	-	77	7	15	
2	cis 3	B	2	-	48	13	16	21
3	cis 3	C	1	-	74	9	12	
4	cis 3	D	1.5	-	72	9	9	
5	trans 3	A	>4	-	-	69	13	
6	trans 3	B	>4	-	-	67	15	
7	trans 3	C	>4	20	-	52	11	
8	trans 3	D	>4	-	-	70	10	
9	mix 1:1	A	0.5	27	42	26	2	
10	mix 1:1	B	0.5	50	18	8	7	11
11	mix 1:1	C	0.5	44	29	12	5	
12	mix 1:1	D	0.5	40	31	16	6	
13	mix 1:1	D[a]	0.5	48	27			

[a] Substrate:Catalyst = 2:1, 3% *cis* 3 unreacted

trans 3   *cis* 3

Figure 1
The two more stable conformers of *cis*-3 and *trans*-3 calculated with Cerius2.

2- Bifunctional catalysis: in order to set up the bifunctional catalytic process, we first investigated the influence of supported copper on the catalyst activity and selectivity, that is if high amount of 1 can be obtained in the presence of copper catalyst. Thus, in the case of α-pinene oxide, a lower selectivity in the desired aldehyde was obtained when working with copper catalyst (12). Results are reported in Table 2.

Table 2 – Isomerization of limonene epoxide in the presence of copper/silica-alumina catalysts

Entry	Catalyst	Conditions	t (h)	Conv.	%4	%1	%2	Arom.
1	Cu/A	N_2, room T	4	27	15	4	-	-
		H_2, 90 °C	24	100	29	37	-	14
2	Cu/B[a]	N_2, room T	4	2				
		H_2, 90 °C	7	100	-	-	65	22
3	Cu/B	N_2, room T	4	50[b]	19	22	-	-
		H_2, 90 °C	4	100	-	-	42	47
4	Cu/B	H_2, 90 °C	4	100	-	-	14	80
5	Cu/C	N_2, room T	4	25	11	4	-	4
		H_2, 90 °C	24	100	24	42	-	9
6	Cu/D[a]	N_2, room T	4	64	-	46	-	-
		H_2, 90 °C	5	100	-	6	60	17
7	Cu/D	N_2, room T	4	100	20	57	-	-
		H_2, 90 °C	7		-	-	43	30
8	Cu/D	H_2, 90 °C	6	100	-	-	47	30

[a] Substrate: *trans* 3
[b] *trans* 3 unreacted

In the present work, copper deposition seems to suppress the acidity of both A and C. Thus, after 4 hours under inert conditions only very low conversions were observed (entries 1 and 5), that is also reaction of *cis* 3 is inhibited. Switching to H_2 atmosphere, promoted the epoxide opening, but carvenone 2 was never formed even after very long reaction times. Reaction of *trans* 3 is very much slowed down in the presence of Cu/B and, to a lower extent, of Cu/D. However, by introducing H_2 and raising the temperature to 90°C in the reaction mixture even at low conversion, good yields in 2 were obtained (entries 2 and 6, Scheme 3). It should be stressed that *trans* 3 can be easily obtained by resolution of the mixture as shown in Table 1.

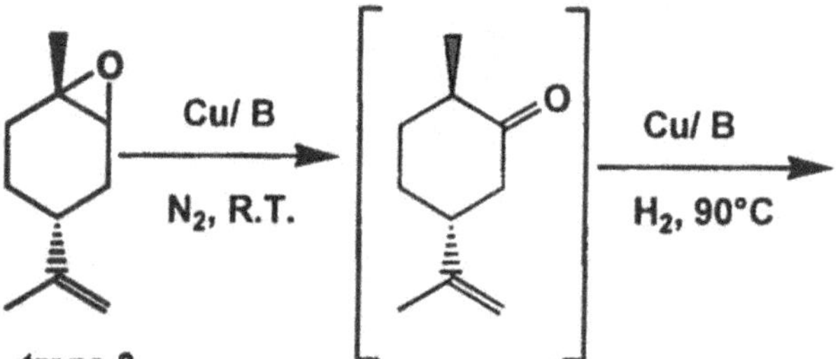

Scheme 3

It is also possible to start from the commercial mixture of isomers. With these two catalysts (entries 3 and 7) we obtained ≈40% of 2 together with a mixture of aromatic hydrocarbons mainly composed of para-cymene. The later compound is used in the fragrance industry as starting material for the synthesis of musks and as intermediate in the production of para-cresol (13). The formation of para-cymene and carvenone, according to Tanabe (10), suggests that this kind of copper catalyst, under these conditions, exhibits Brønsted acid character.

We also tried to realize the transformation in one step, by carrying out the reaction directly under H_2 (entries 4 and 8), Under these conditions Cu/B becomes too acidic and the main products are aromatics hydrocarbons. On the contrary, the use of Cu/D allows to obtain 2 in good yield and in a shorter time with respect to the two-step reaction.

Only a direct transformation of 3 into 2 as been reported. This was carried out in the presence of synthetic zeolites, but under much more demanding conditions (14), while 2 can be obtained with a 28% yield through reaction of 3,4-epoxycarane with $BF_3 \cdot Et_2O$ (15).

The experiments carried out with Cu/A and Cu/C under H_2 at 90°C to obtain 2, as already mentioned above, were always unsuccessful. TPR profiles of our copper containing catalysts show that only Cu/B and Cu/D are completely reduced at 270°C. Therefore Cu/A and Cu/C aren't able to give **2** as full reduction of Cu is a prerequisite for isomerization reaction.

Conclusions

We have shown that silica-alumina cogel catalysts can be effectively used as solid acid catalysts in the opening of epoxides and can be proposed as a *clean* alternative to the use of $ZnBr_2$ or $BF_3 \cdot Et_2O$. Moreover, in the particular case of **3** they can be used for a kinetic resolution of the geometric isomers. A bifunctional catalyst can be obtained by supporting copper on some of these solids, thus allowing the formation of carvenone in one step (Scheme 4).

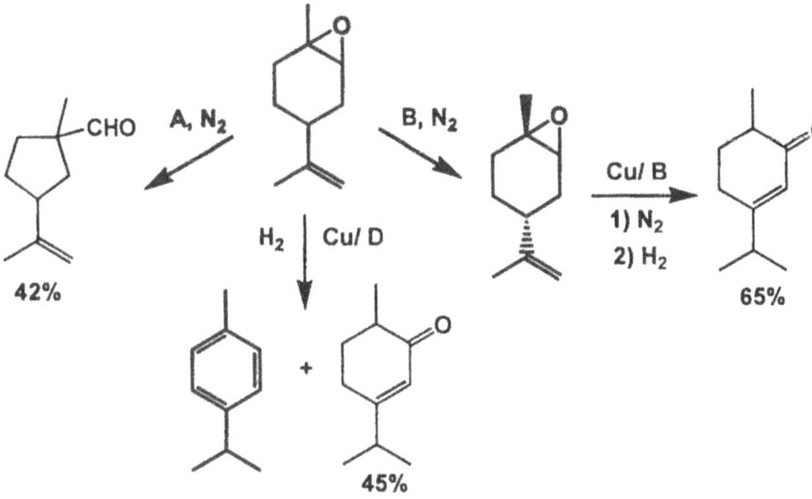

Scheme 4

Acknowledgments

The authors gratefully acknowledge the CNR-MURST for financial support through the "Program Chemistry LAW 95/95-I year".

References

(1) Part I: N. Ravasio, V. Leo, F. Babudri, M. Gargano, *Tetrahedron Lett.*, **38**, 7103 (1997); Part II: N. Ravasio, N. Poli, R. Psaro, M. Saba, F. Zaccheria, *Topics in Catalysis* accepted.

(2) N. Ravasio, *Recent Res. Devel. Organic Chem.*, **3**, 79 (1999).

(3) N. Ravasio, M. Antenori, M. Gargano, Chem. Ind. (Marcel Dekker), **68**, (*Catal. Org. React.*), 413 (1996).

(4) N. Ravasio, M. Finiguerra, M. Gargano, Chem. Ind. (Marcel Dekker), **75**, (*Catal. Org. React.*), 513 (1998).

(5) F. Boccuzzi, G. Martra, C. Partipilo Papalia, N. Ravasio, *J. Catal.*, **184**, 327 (1999).

(6) A. K. Rappe', C. J. Casewit, K. S. Colwell, W. A. Goddard, W. M. Skiff, *J. Am. Chem. Soc.*, **114**, 10024, (1992)

(7) R. L. Settine, G. L. Parks, G. L. K. Hunter, *J. Org. Chem.*, **29**, 616 (1964).

(8) T. Takanami, R. Hirabe, M. Ueno, F. Hino, K. Suda, *Chem. Lett.*, 1031 (1996).

(9) A. Kergomard, M. T. Geneix, *Mem. Soc. Chim. Fr.*, 390 (1957).

(10) K. Arata, S. Akutagawa, K. Tanabe, *J. Catal.*, **41**, 173 (1976).

(11) R. Sudha, K. Malola Narasimhan, V. Geetha Saraswathy, S. Sankararaman, *J. Org. Chem.*, **61**, 1877 (1996).

(12) N. Ravasio, M. Finiguerra, M. Gargano, *Supported Reagents and Catalysts in Chemistry*, B. K. Hodnett, A. P. Kybett, J. H. Clark and K. Smith, ed., Royal Society of Chemistry, 231 (1998).

(13) D. Buhl, P. A. Werich, W. M. H. Sachtler, W. F. Holderich, *Appl. Catal. A: Gen.*, **1141**, 1 (1998).

(14) Fujiwara, Y., Nomura, M., Igawa, K., JP Pat: 62, 114, 926 [87, 114, 926], *CAS*, **109**, 728 (1988).

(15) P. J. Kropp, *J. Am. Chem. Soc.*, **88**, 4926 (1966)

Active Sites on Well-Characterized Pt/SiO$_2$ Determined by (-)-Apopinene Deuteriumation, Cyclohexene Hydrogenation, and CS$_2$ Titration

Gerard V. Smith and Daniel Ostgard[*]

Department of Chemistry and Biochemistry, Southern Illinois University, Carbondale IL 62901-4409, United States of America

Abstract

Cyclohexene and (-)-apopinene were hydrogenated over various Pt/SiO$_2$ catalysts, and their turnover frequencies (TOFs) and numbers of active sites, from CS$_2$ titration (CS$_2$-sites), are reported.

Introduction

Historically, stereochemistry has contributed much to understanding organic reaction mechanisms in solution; similarly, during the past forty-five years, stereochemistry has made contributions to understanding organic reaction mechanisms on surfaces (1). An important goal is to understand mechanisms of selectivity on surfaces, especially on metal surfaces, which catalyze much of the fine chemicals and specialty chemicals processes. In this study we examine the differences in hydrogenation and deuteriumation between two different molecular probes, cyclohexene and (-)-apopinene. Cyclohexene is a relatively flat and flexible molecule with both sides of its double bond able to adsorb on the surface, while (-)-apopinene is a rigid, inflexible molecule with one side of its double bond blocked from adsorption by a *gem*-dimethyl group.

Experimental

The data were obtained on a series of carefully prepared Pt/SiO$_2$ catalysts with dispersions ranging from 21.5%D to 80.3%D and characterized by both hydrogen chemisorption and electron microscopy (2, 3, 4). Meticulous

---

[*] Current Address: Degussa-Huels AG, P.O. 1345, D63457 Hanau, Germany

purifications of reactants were performed prior to hydrogenations and deuteriumations. Experiments showed the rates of hydrogenation and deuteriumation to be identical and careful precautions assured the absence of diffusion control (5). Carbon disulfide titration revealed the relative number of active sites. A series of ion exchanged and impregnated Pt/SiO_2 catalysts were provided by the laboratories of Professors Burwell and Butt of Northwestern University. Details about these catalysts can be found in their publication (6). Additionally, each catalyst was characterized in our laboratory by hydrogen chemisorption and electron microscopy (7) as shown in Table 1.

Table 1   Dispersions of Pt/SiO_2 Catalysts.

%Pt	%D_{chem}	%D_{EM}[c]	Pt particle size range (nm)	avg. Pt particle size (nm)	avg. Pt particle size stand. dev. (nm)
3.7[a]	14[a]				
1.48	21.5[b]	28.1	1.20-10.0	4.24	2.13
1.48	27.0[b]	23.5	1.18-19.1	5.07	1.97
0.8[a]	34[a]				
1.17	40.7[b]	31.8	2.08-9.37	3.75	1.07
0.53[a]	56[a]				
2.3[a]	62[a]				
0.48	62.1[b]	74.6	1.10-2.80	1.6	0.44
1.46[a]	80[a]				
0.83	80.3[b]	82.9	0.60-2.0	1.44	0.45
0.38[a]	100[a]				
1.5[a]	100[a]				

[a] Catalyst data from reference (8); %D determined by hydrogen-oxygen titration.
[b] Determined by hydrogen chemisorption.
[c] Determined by electron microscopy.

Cyclohexene, A.C.S. Reagent grade (99%), Fisher Scientific Co., was distilled in an argon atmosphere. The (-)-apopinene was synthesized by 5% $Pd/BaSO_4$ (Aldrich) catalyzed decarbonylation of (-)-myrtenal (Aldrich) and distilled from the reactor as formed. The crude (-)-apopinene was purified three times by gas chromatography, the last such purification was performed within 30 minutes of running a hydrogenation. As a final precaution to remove traces of oxygen and peroxides (9), the reactants were filtered through activated alumina under argon just prior to hydrogenation. The alumina (80-200 mesh) was activated by heating at $400^{\circ}C$ for 3 hours in flowing argon. Zero grade argon and helium were obtained from MG Industries Gas Products and purified immediately before entering the reaction system by passing first through a purifier containing molecular sieves and activated carbon and then through an Altech Oxytrap, which lowered the oxygen concentration to less than 0.1 ppm. Ultra high purity hydrogen (MG Industries, minimum purity 99.999%) was used for chemisorption characterizations and for liquid phase hydrogenations of cyclohexene. This hydrogen was purified by passage through the two purifiers described above. Additionally, all the above-mentioned gases were further purified by passage through an MnO/SiO_2 trap (remove traces of oxygen) before passing into the reactor. Ultra high purity hydrogen and CP grade deuterium were used in the hydrogenation and deuteriumation of (-)-apopinene. They were purified through a 12 cm column of 0.5%Pd on Linde SK300 Zeolite followed by a 23 cm column of Linde 3A molecular sieve. The apparatuses for 1 atm. liquid phase hydrogenations and the hydrogenation procedures have been previously described (3, 10, 11). Preliminary experiments showed that small amounts of catalysts agitated at 1900 rpm in a vortex manner produced reaction rates uninfluenced by diffusion (5). Additionally, the reaction system met the Koros Novak criteria for absence of diffusion (12) as well as exhibiting activation energies averaging 29.03±0.34 KJ/mole, which is close to 28.28 KJ/mole reported for different catalyst dispersions (8). Using hydrogenation reactions as indicators, titrations of Pt surfaces were conducted by measuring the hydrogenation rate followed by repeated injections of small amounts of CS_2 dissolved in cyclohexane and remeasuring the rate. This addition of CS_2 continued until approximately 20% of the activity was remaining and these points were extrapolated to zero activity to determine the amount of CS_2-active sites (5, 7, 11, 13, 14, 15).

Results

Results of these hydrogenations and deuteriumations are shown in Figures 1 and 2. CS_2-active sites are CS_2 molecules divided by total Pt surface atoms.

Discussion

These results reveal subtle steric interactions between the substrate and the metal surfaces. Based on the assumption that the smaller metal crystallites contain more edges, [2]M Siegel sites (16), and corners, [3]M, while the larger crystallites contain more planes, [1]M, one might expect these molecular probes to reveal which of these various sites best catalyze hydrogenation. Indeed, this was our reasoning in our earlier attempts to characterize some highly dispersed Pd/SiO₂ and Pt/SiO₂ catalysts with (+)-apopinene (2). These present results are basically the same for apopinene as the earlier report in which both the CS_2 active sites and the TOFs increased to 60%D and then decreased. Therefore, hydrogenation of apopinene is mildly structure sensitive. However, we see that cyclohexene does not follow this pattern. CS_2 active sites for cyclohexene do indeed increase and go through a broad maximum from approximately 40%D to 60%D, after which they decrease, but TOFs for cyclohexene are constant over the range of

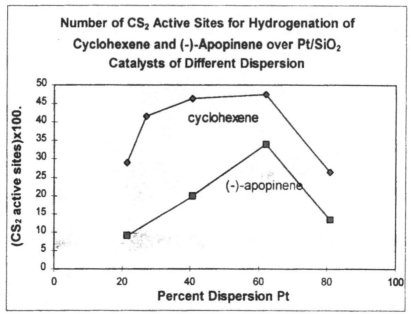

Fig. 1 CS$_2$-sites for Hydrogenation of Cyclohexene (♦) and (-)-Apopinene (■) determined by dividing number of CS$_2$ molecules by total Pt surface atoms.

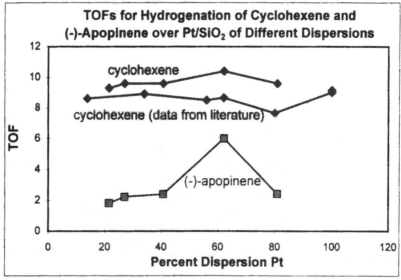

Fig. 2   TOFs for Hydrogenation of Cyclohexene (♦) and (-)-Apopinene (■).

dispersion. Hydrogenation of cyclohexene is structure insensitive as previously reported (8). Thus we see that these CS$_2$ titrations reveal different results depending on the indicator molecule used.

At every dispersion, CS$_2$ titration reveals more active sites for cyclohexene than for apopinene, Figure 1, yet only apopinene TOFs correlate with these sites and go through the same maximum as the CS$_2$ sites, Figure 2. Clearly, cyclohexene and apopinene interact with the same Pt surface in different ways.

Assuming the same mechanism for hydrogenation of both molecules, at least part of this difference can be explained by stereochemical differences between them. For example cyclohexene, with little steric congestion near the double bond, should adsorb more readily than apopinene on the ^{1}M sites of catalysts with low dispersion. In contrast, apopinene, with steric hindrances above and below the plane of its double bond, should adsorb with difficulty on ^{1}M sites but with relative ease on the ^{2}M sites of the highly dispersed catalysts. Consequently, apopinene exhibits maximum activity on catalysts with dispersions near 60%D, and as the number of CS$_2$-active sites decreases, so does the apopinene TOF. But the question remains: why does cyclohexene maintain a constant TOF at all dispersions even though the number of active sites varies?

Apparently all Pt surface sites are capable of catalyzing hydrogenation at the same rate (i.e., cyclohexene); however, steric factors in the structures of substrates alters their surface concentrations and therefore modify their rates on the various sites. In the case of cyclohexene, all Pt sites are accessible, so the rate is the same on all sites providing hydrogen is adequately supplied to all sites. But apopinene is more discriminating. Its structure inhibits adsorption on the ^{1}M sites of the low dispersed catalysts and allows adsorption on ^{2}M sites. Although its structure should also allow it to adsorb readily on the ^{3}M sites of the highly dispersed catalysts, CS$_2$ titration reveals them to be less active. So why does cyclohexene TOF not decrease also when titrations reveals ^{3}M sites to be less active?

Purely steric factors must not be the only factors affecting rates. Electronic factors associated with the catalyst and the catalyst-substrate interactions must also be important. For example, Goodman and colleagues recently found a quantum effect (electronic effect) in small (<30Å) Pd particles (17) that greatly influence their ability to catalyze reactions, and Saltsburg reported a size effect (~30Å) in Layered Synthetic Microstructures of Ni and silica (18), which exhibited significantly higher catalytic activities. Both of these effects are seen in a slightly larger size range than in our studies (~20Å) of the hydrogenations of the apopinenes but not in the hydrogenation of cyclohexene. In this respect, the more complicated structure of the apopinenes makes them more sensitive molecular probes than the less complex structure of cyclohexene.

Likely, adsorbing CS_2 on Pt not only destroys active sites, but also causes changes in electronic distributions at the surface (19). Therefore it seems reasonable that adsorption of CS_2 changes the "activity" of the remaining active sites. These changes are more noticeable when cyclohexene hydrogenation is used to monitor CS_2 adsorption (poisoning) than when apopinene hydrogenation is used to monitor. For example, Figure 2 shows cyclohexene undergoes hydrogenation (deuteriumation) at the same rate regardless of the kind of surface site available; in other words, it is structure insensitive. In contrast, cyclohexene hydrogenation becomes mildly structure sensitive on the same catalysts when CS_2 is adsorbed (Figure 1). Yet apopinene still exhibits similar mild structure sensitivity even when CS_2 is adsorbed. It is less sensitive to the electronic effects caused by CS_2 adsorption. Why do these electronic changes affect cyclohexene hydrogenation and not apopinene?

The structures of cyclohexene and apopinene impart previously unrecognized abilities to discriminate between electronic and structural effects. Cyclohexene is a relatively flat and flexible molecule with both sides of its double bond able to adsorb on the surface. It can adsorb on most any surface site. On the other hand, apopinene is a rigid, inflexible molecule with one side of its double bond completely blocked from adsorption by a *gem*-dimethyl group and the other side of its double bond partially blocked by a bridging methylene group(Figure 3). It can adsorb with difficulty on plane sites (^1M) but more readily of edge (^2M) and corner (^3M) sites.

Fig. 3   Three depictions of (-)-Apopinene

Adsorption of CS_2 renders ^3M sites on the highest dispersed catalysts much less active than before adsorption because on these small crystallites each adsorbed CS_2 molecules is close to each Pt atom, so its electronic effect is most pronounced. Since apopinene, because of its unique structure, finds these sites less active anyway, it is not greatly influenced by this electronic perturbation. However, cyclohexene finds these sites much less favorable for hydrogenation than before CS_2 adsorption. On the large crystallites, (lowest dispersions) containing high proportions of ^1M sites, cyclohexene is apparently again more affected than apopinene. However, apopinene is also affected, but not as much. This effect may be the physical influence of CS_2 blocking sites on the planes in addition to modifying the electronic characteristics of the surface. Such a blocking effect should impact both cyclohexene and apopinene approximately the same. So the unique structure of apopinene masks the effect which is more pronounced in cyclohexene. Clearly, the more complicated unique structure of apopinene renders it a different probe than cyclohexene, and caution must be exercised in deducing surface information from only one molecular probe. Rather, a variety of molecular probes should be chosen to characterize surfaces.

References

1. G.V. Smith and F. Notheisz, Heterogeneous Catalysis in Organic Chemistry, Academic Press, San Diego, 1999; M. Bartók, J. Czombos, K. Felföldi, L. Gera, Gy. Göndös, Á. Molnár, F. Notheisz, I. Pálinkó, Gy. Wittmann, Á.G. Zsigmond, Stereochemistry of Heterogeneous Metal Catalysis, John Wiley & Sons, New York (1985); G.V. Smith, Catalysis of Organic Reactions (R.E. Malz, Jr., Ed.), Marcel Dekker, Inc., New York (1996) p 1.
2. G.V. Smith, F. Notheisz, Á.G. Zsigmond, D. Ostgard, T. Nishizawa, and M. Bartók, Proceedings of the Ninth International Congress on Catalysis, 3, Calgary, June 26-July1, The Chemical Institute of Canada (1988), p. 1066.
3. G.V. Smith, Á. Molnár, M.M. Khan, D. Ostgard, and N. Yoshida, J. Catal., 98 (1986) 502.
4. F. Notheisz, M. Bartók, D. Ostgard, and G.V. Smith, J. Catal., 101 (1986) 212.
5. F. Notheisz, Á. Zsigmond, M. Bartók, Zs. Szegletes, and G.V. Smith, Applied Catalysis A: General, 120 (1994) 105.
6. T. Uchijima, J.M. Herrmann, Y. Inoue, R.L. Burwell, J.B. Butt, and J.B. Cohen, J. Catal. 50 (1997) 464.
7. G.V. Smith, F. Notheisz, Á.G. Zsigmond, D. Ostgard, T. Nishizawa, and M. Bartók, Proceedings of Ninth International Congress on Catalysis, Calgary, The Chemical Institute of Canada, (1988), pp. 1066-1073.
8. R.J. Madon, J.P. O'Connell, M. Boudart, AIChE Journal, 24 (1978) 104.
9. E. Segal, R. J. Madon, and M. Boudart, J. Catal., 52 (1978) 462.
10. G.V. Smith, J. Stoch, S. Tjandra, and T. Wiltowski, Catalysis of Organic Reactions, (M.G. Scaros and M.L. Prunier, Eds.), Marcel Dekker, Inc., New York, 1995, p. 403
11. G.V. Smith, O. Zahara, A. Molnar, M.M., Khan, B. Richter, and W. E. Brower, J.Catal., 83, (1983) 238.
12. R.M. Koros, and E.J. Nowak, Chem. Eng. Sci., 22 (1967) 470.
13. G.V. Smith, M. Bartók, F. Notheisz, Á.G. Zsigmond, and I. Pálinkó, J. Catal., 110 (1988) 203.
14. G.V. Smith, D.J. Ostgard, F. Notheisz, A.G. Zsigmond, I. Palinko, and M. Bartok, Catalysis in Organic Reactions (D.W. Blackburn, Ed.), Marcel Dekker, Inc. New York, 1990, p. 157.
15. T. Wiltowski, G.V. Smith, and D. Ostgard, Catalysis of Organic Reactions, (W.E. Pascoe, Ed.), Marcel Dekker, Inc., New York, 1992, p. 143.

16. G.V. Smith and F. Notheisz, Heterogeneous Catalysis in Organic Chemistry, Academic Press, San Diego, 1999, p. 23.
17. C. Xu, X. Lai, G.W. Zajac, D.W. Goodman, Phys. Rev. B, 56 (1997) 13464.
18. I. Zuburtikudis and H. Saltsburg, Science, 258 (1992) 1337.
19. J. Oudar, in, Deactivation and Poisoning of Catalysts, J. Oudar and H. Wise eds., Marcel Dekker, Inc., New York (1985) Chapter 2.

Application of Simmons-Smith Reagents to Epoxidation and Aziridination

V. K. Aggarwal,[a] M. P. Coogan,[a] R. A. Stenson,[a] R. V. H. Jones,[b] and R. Fieldhouse[b]

[a]The University of Sheffield, Brook Hill, Sheffield, UK, S3 7HF.
[b]Zeneca Agrochemicals, Process Technology Department, Earls Road, Grangemouth, Stirlingshire, UK, FK3 8XG.

Abstract

Treatment of aldehydes or imines with sulfur ylides, generated by the reaction of Simmons-Smith carbenoids with sulfides, provides terminal epoxides and terminal aziridines respectively, in high yields. Using chiral sulfides epoxides are obtained with enantioselectivities of up to 54%.

Introduction

Epoxides and aziridines are present in many natural products and are themselves extremely important synthetic intermediates. A novel procedure to synthesise such heterocycles, which employs Simmons-Smith carbenoids (1) generated by the Furukawa protocol (2), has been developed (Scheme 1).

Scheme 1 Reaction cycle

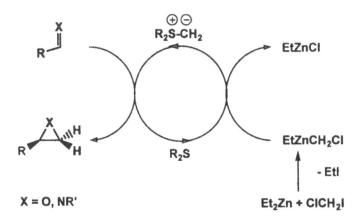

X = O, NR'

Results and Discussion

We report herein that Simmons-Smith reagents react with sulfides to generate ylides, which in turn react with aldehydes. Mild reaction conditions provide terminal epoxides in good yield for both aromatic and aliphatic substrates (Figure 1). When unsaturated aldehydes are employed no products resulting from cyclopropanation are detected. α-Amino aldehydes react smoothly to give a mixture of diastereoisomers. No racemisation at the α-centre is observed.

Figure 1 Application of Simmons-Smith reagents to epoxidation

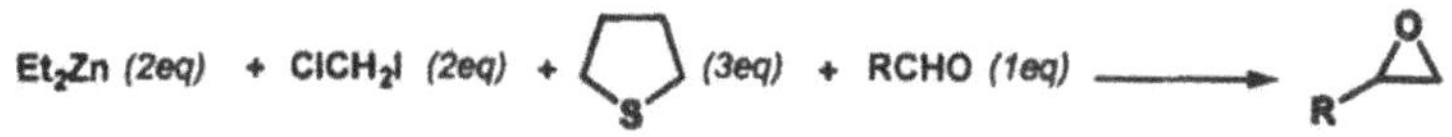

Entry	Substrate	Product[a]	% Yield[b]
1	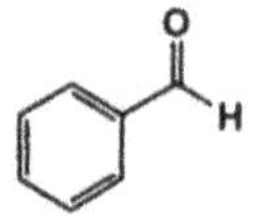	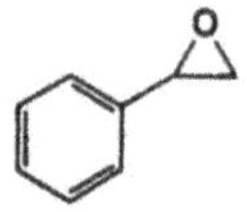	74
2	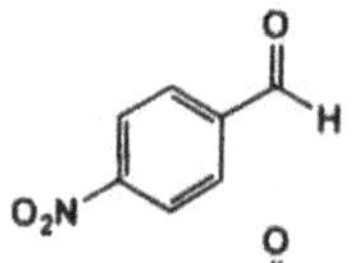	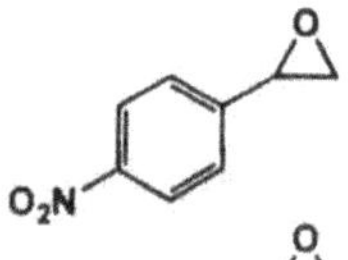	95
3	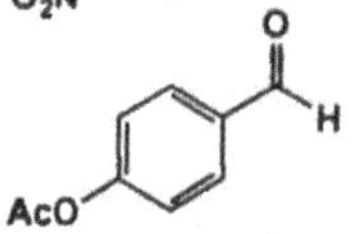	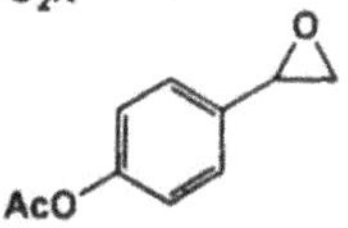	58
4	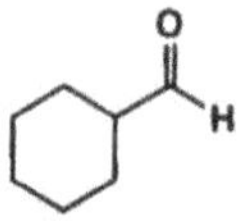	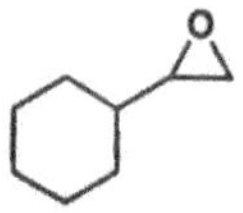	60
5	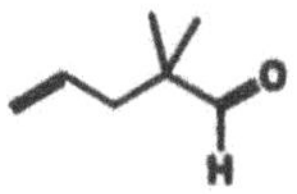	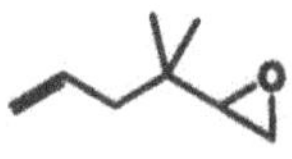	70
6	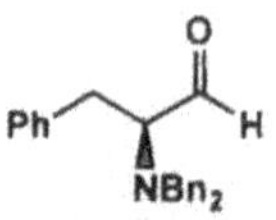	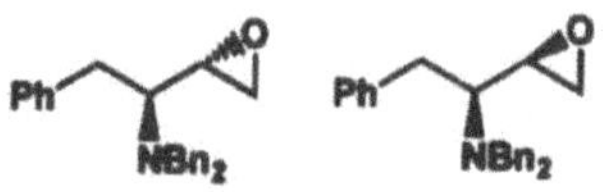 5 : 1	84

[a]Product ratios determined by gas chromatography/^{1}H NMR.
[b]Isolated yields.

The use of readily available chiral ligands (Figure 2, Entries 1 & 2) as additives provided non-racemic styrene oxide, albeit with low enantiomeric excess. Selectivity is thought to arise *via* association of the zinc with the aldehyde and/or sulfur ylide during the enantiodifferentiating step.

Figure 2 Addition of chiral ligands

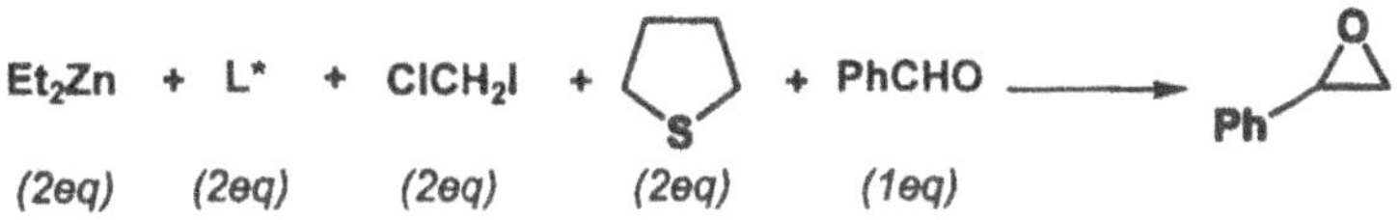

Entry	L*	% ee	% Yield
1	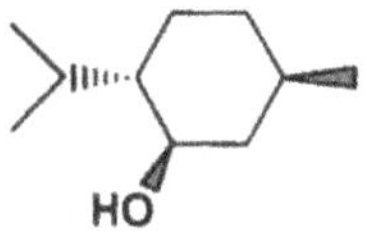	11	95
2	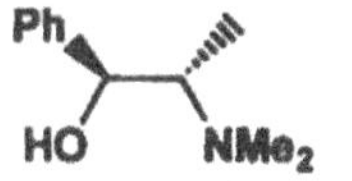	11	98

In an attempt to increase enantioselectivity chiral sulfides (1-3) were employed (Figure 3). However, sulfide 1 provided styrene oxide in low yield and with poor enantiomeric excess. Camphor-based sulfides 2 and 3 failed to epoxidise benzaldehyde. It is possible that sulfide 2 decomposed under reaction conditions and sulfide 3 bound too strongly to zinc, thus inhibiting ylide formation.

Figure 3 Addition of chiral sulfides

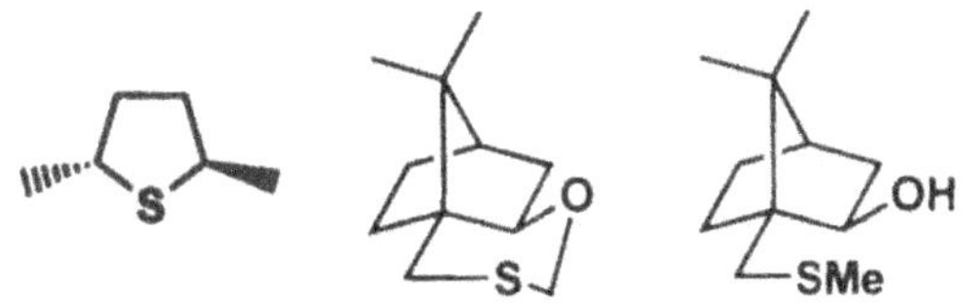

1	2	3
23%, 12% ee	0%	0%

A novel class of bidentate chelating sulfides (4) containing the bis-oxazoline skeleton were considered. Sulfide **4** binds to zinc which ensures that the zinc ion is held in close proximity to the sulfur ylide. Lewis acid co-ordination to the aldehyde should promote a *pseudo*-intramolecular pathway, in which the conformations at all centres of the chelated species were controlled. It was hoped that this strategy would lead to an increase in asymmetric induction.

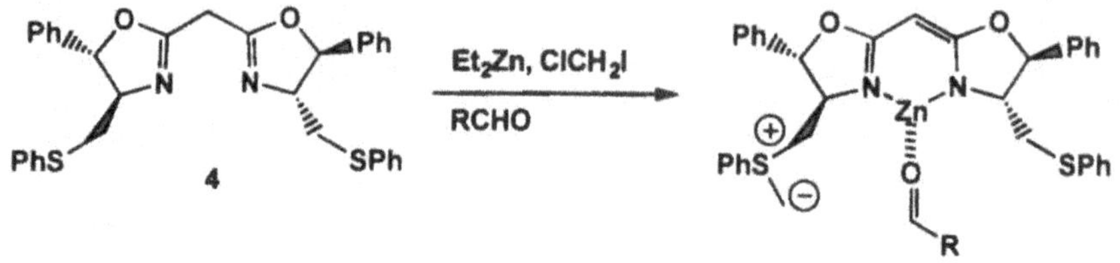

Gratifyingly, an epoxidation system employing sulfide **4** was shown to epoxidise both aromatic and aliphatic aldehydes (Figure 4) with moderate enantioselectivities. To our knowledge, these enantiomeric excesses are the highest reported for the preparation of terminal epoxides *via* sulfur ylide mediated chemistry.

Figure 4 Asymmetric induction using chiral bis-oxazoline-based sulfides

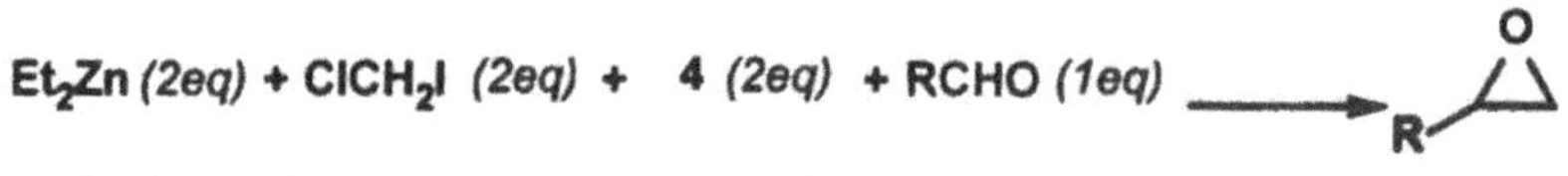

Entry	Substrate	Product	% ee[a]	% Yield[b]
1	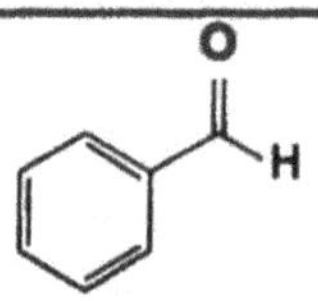	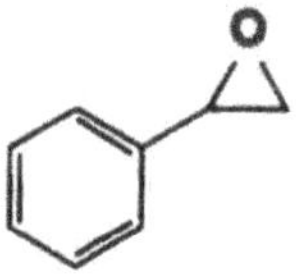	47	54
2	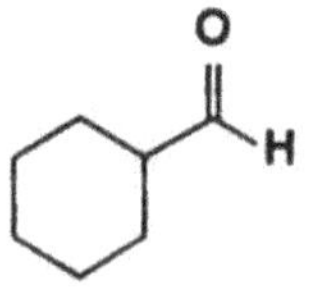	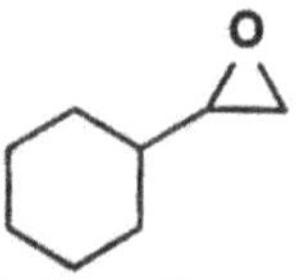	54	30

[a]Enantiomeric excesses determined by chiral gas chromatography (α-cyclodextrin column).
[b]Isolated yields.

The Simmons-Smith based methodology may also be extended to the synthesis of terminal aziridines. A range of *N*-tosyl substituted imines were aziridinated in good yields (Figure 5).

Figure 5 Application of Simmons-Smith reagents to aziridination

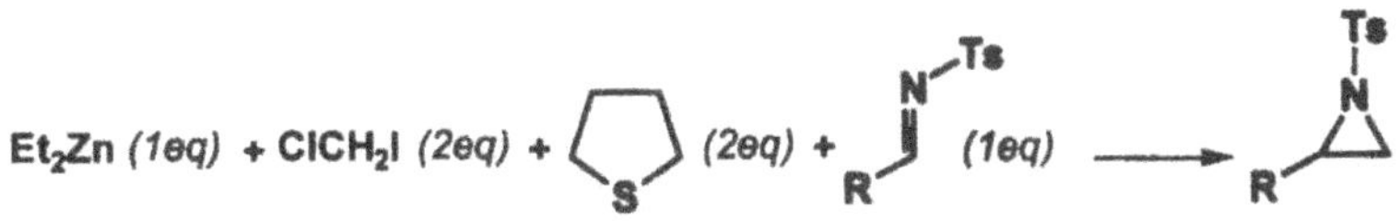

Entry	Substrate	Product	% Yield[a]
1	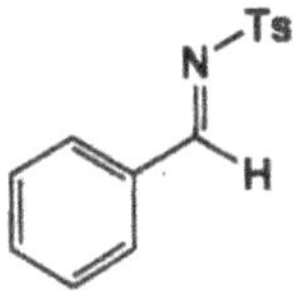	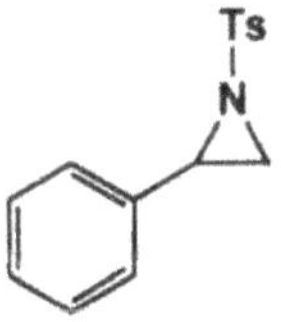	68
2	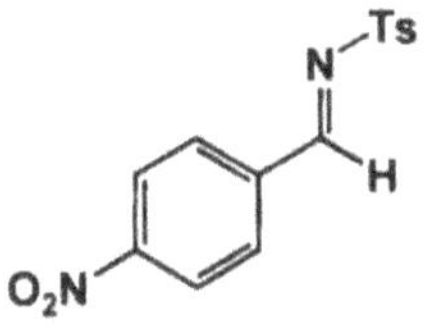	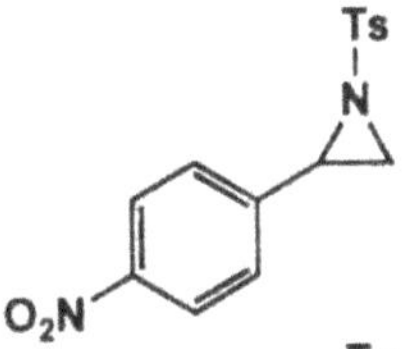	66[b]
3	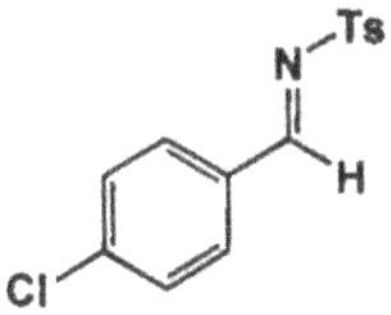	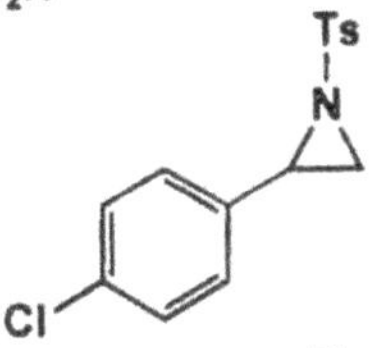	71
4	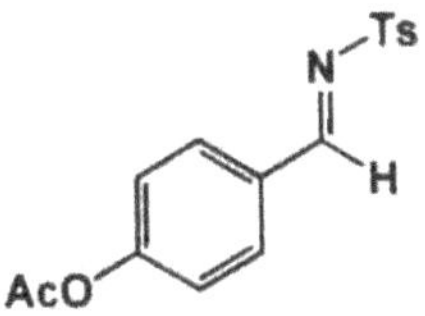	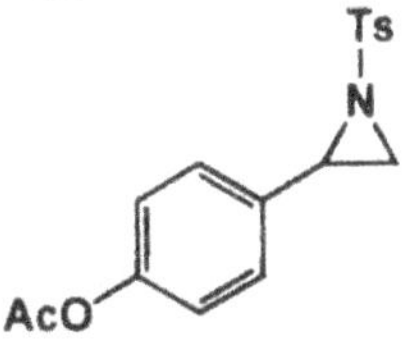	68
5	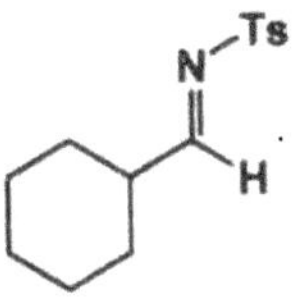	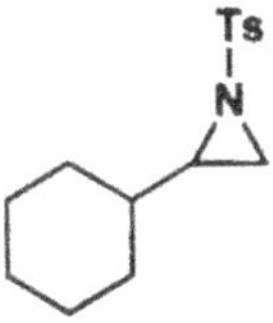	72

[a]Isolated yields.
[b]2.5 equivalents $ClCH_2I$.

In conclusion, a general and mild method for the preparation of both terminal epoxides and terminal aziridines is presented. Good to excellent chemical yields are obtained. Zinc participation in the crucial stereodifferentiating step has enabled asymmetric variants of the process to be developed. The use of simple chiral sulfides gave epoxides with low enantiomeric excess. However, sulfides tethered to bis-oxazolines gave significant improvements in enantioselectivity. A reagent system employing sulfide **4** epoxidised both aromatic and aliphatic aldehydes with enantioselectivities of up to 54%.

Experimental

Chloroiodomethane (0.14 mL, 2.0 mmol) was added to a stirred solution of diethyl zinc (1.8 mL, 1.1M solution in toluene, 2.0 mmol) in 1,2-dichloroethane (6 mL) at -15 °C, under nitrogen. After 15 min tetrahydrothiophene (0.26 mL, 3.0 mmol) and benzaldehyde (0.10 mL, 1.0 mmol) were added. The reaction mixture was allowed to warm to ambient temperature, stirred for 48 h then diluted with CH_2Cl_2 (10 mL) and washed with saturated aqueous NH_4Cl (5 mL). The aqueous phase was extracted with CH_2Cl_2 (3 x 20 mL) and the organic extracts were combined, dried ($MgSO_4$) and concentrated *in vacuo*. The residue was purified by flash chromatography on silica gel (eluent: 10% EtOAc/Petroleum ether 40-60 fraction) to afford *styrene oxide* as a colourless oil (78 mg, 74%): δ_H (250 MHz; $CDCl_3$) 7.25 (5H, m), 3.85 (1H, dd, J = 4.5, 3.0 Hz, CHOCH$_2$), 3.30 (1H, dd, J = 5.0, 4.5 Hz, CH$_2$O), 2.85 (1H, dd, J = 5.0, 3.0 Hz, CH$_2$O).

Acknowledgements

We thank Dr. A. Ali for carrying out preliminary studies, the Engineering and Physical Sciences Research Council and Zeneca Agrochemicals P.T.D. for their support.

References

1.   H. E. Simmons and R. D. Smith, *J. Am. Chem. Soc.*, **81**, 4256 (1959).
2.   J. Furukawa, N. Kawabata, and J. Nishimura, *Tetrahedron*, **24**, 53 (1968).
3.   V. K. Aggarwal, A. Ali, and M. P. Coogan, *J. Org. Chem.*, **62**, 8628 (1997).
4.   V. K. Aggarwal, L. Bell, M. P. Coogan, and P. J. Jubault, *J. Chem. Soc., Perkin Trans 1.*, 2037 (1998).

Index

Milton Keynes UK
Ingram Content Group UK Ltd.
UKHW020003071024
449327UK00031B/2638